Undergraduate Texts in Mathematics

Undergraduate Texts in Mathematics

Undergraduate Texts in Mathematics are generally aimed at third- and fourth-year undergraduate mathematics students at North American universities. These texts strive to provide students and teachers with new perspectives and novel approaches. The books include motivation that guides the reader to an appreciation of interrelations among different aspects of the subject. They feature examples that illustrate key concepts as well as exercises that strengthen understanding.

More information about this series at http://www.springer.com/series/666

Peter D. Lax · Maria Shea Terrell

Multivariable Calculus
with Applications

 Springer

Peter D. Lax
Courant Institute of Mathematical Sciences
New York University
New York, NY
USA

Maria Shea Terrell
Department of Mathematics
Cornell University
Ithaca, NY
USA

ISSN 0172-6056   ISSN 2197-5604   (electronic)
Undergraduate Texts in Mathematics
ISBN 978-3-319-74072-0   ISBN 978-3-319-74073-7   (eBook)
https://doi.org/10.1007/978-3-319-74073-7

Library of Congress Control Number: 2017963518

Mathematics Subject Classification (2010): 93C35, 0001, 97xx

Printed on acid-free paper

This Springer imprint is published by Springer Nature
The registered company is Springer International Publishing AG
The registered company address is: Gewerbestrasse 11, 6330 Cham, Switzerland

Preface

Our purpose in writing a multivariable calculus text has been to help students learn that mathematics is the language in which scientific ideas can be precisely formulated and that science is a source of mathematical ideas that profoundly shape the development of mathematics.

In calculus, students are expected to acquire a number of problem-solving techniques and to practice using them. Our goal is to prepare students to solve problems in multivariable calculus and to encourage them to ask, Why does calculus work? As a result throughout the text we offer explanations of all the important theorems to help students understand their meaning. Our aim is to foster understanding.

The text is intended for a first course in multivariable calculus. Only knowledge of single variable calculus is expected. In some explanations we refer to the following theorems of calculus as discussed for example in *Calculus With Applications*, Peter D. Lax and Maria Shea Terrell, Springer 2014.

- **Monotone Convergence Theorem** A bounded monotone sequence has a limit.
- **Greatest Lower Bound and Least Upper Bound Theorem** A set of numbers that is bounded below has a greatest lower bound. A set of numbers that is bounded above has a least upper bound.

Chapters 1 and 2 introduce the concept of vectors in $\mathbb{R}^n$ and functions from $\mathbb{R}^n$ to $\mathbb{R}^m$. Chapters 3 through 8 show how the concepts of derivative and integral, and the important theorems of single variable calculus are extended to partial derivatives and multiple integrals, and to Stokes' and the Divergence Theorems.

To do partial derivatives without showing how they are used is futile. Therefore in Chapter 8 we use vector calculus to derive and discuss several conservation laws. In Chapter 9 we present and discuss a number of physical theories using partial differential equations. We quote a final passage from the book:

We observe, with some astonishment, that except for the symbols used, the equations for membranes in which the elastic forces are so balanced that they do not vibrate, and heat-conducting bodies in which the temperature is so balanced that it does not change, are *identical*.

There is no physical reason why the equilibrium of an elastic membrane and the equilibrium of heat distribution should be governed by the same equation, but they are, and so

Their mathematical theory is the same.

This is what makes mathematics a universal tool in dealing with problems of science.

We thank friends and colleagues who have given us encouragement, helpful feedback, and comments on early drafts of the book, especially Louise Raphael of Howard University and Laurent Saloff-Coste and Robert Strichartz of Cornell University. We also thank Cornell students in Math 2220 who suggested ways to improve the text. We especially thank Prabudhya Bhattacharyya for his careful reading and comments on the text while he was an undergraduate Mathematics and Physics major at Cornell University.

The book would not have been possible without the support and help of Bob Terrell. We owe Bob more than we can say.

New York, USA   Peter D. Lax
Ithaca, USA   Maria Shea Terrell

Contents

Chapter 1
Vectors and matrices

Abstract The mathematical description of aspects of the natural world requires a collection of numbers. For example, a position on the surface of the earth is described by two numbers, latitude and longitude. To specify a position above the earth requires a third number, the altitude. To describe the state of a gas we have to specify its density and temperature; if it is a mixture of gases like oxygen and nitrogen, we have to specify their proportion. Such situations are abstracted in the concept of a vector.

1.1 Two-dimensional vectors

Definition 1.1. An ordered pair of numbers is called a *two-dimensional vector*. We denote a vector by a capital letter

$$\mathbf{U} = (u_1, u_2).$$

The numbers u_1 and u_2 are called the *components* of the vector $\mathbf{U}$. The set of all two-dimensional vectors, denoted $\mathbb{R}^2$, is called two-dimensional space.

We introduce the following algebraic operations for two-dimensional vectors

(a) The *multiple* of a vector $\mathbf{U} = (u_1, u_2)$ by a number c, $c\mathbf{U}$, is defined as the vector obtained by multiplying each component of $\mathbf{U}$ by c:

$$c\mathbf{U} = (cu_1, cu_2). \tag{1.1}$$

(b) The *sum* of vectors $\mathbf{U} = (u_1, u_2)$ and $\mathbf{V} = (v_1, v_2)$, $\mathbf{U} + \mathbf{V}$, is defined by adding the corresponding components of $\mathbf{U}$ and $\mathbf{V}$:

$$\mathbf{U} + \mathbf{V} = (u_1 + v_1, u_2 + v_2). \tag{1.2}$$

© Springer International Publishing AG 2017
P. D. Lax and M. S. Terrell, *Multivariable Calculus with Applications*,
Undergraduate Texts in Mathematics, https://doi.org/10.1007/978-3-319-74073-7_1

We denote $(0,0)$ as $\mathbf{0}$ and call it the *zero vector*. Note that $\mathbf{U} + \mathbf{0} = \mathbf{U}$ for every vector $\mathbf{U}$. The symbol $-\mathbf{U}$ denotes the vector $(-u_1, -u_2)$. The vector $\mathbf{V} - \mathbf{U}$ defined as $\mathbf{V} + (-\mathbf{U})$ is called the *difference* of $\mathbf{V}$ and $\mathbf{U}$.

Multiplication by a number (or *scalar*) and addition of vectors have the usual algebraic properties:

$$
\begin{aligned}
\mathbf{U} + \mathbf{V} &= \mathbf{V} + \mathbf{U} & \text{commutative}\\
(\mathbf{U} + \mathbf{V}) + \mathbf{W} &= \mathbf{U} + (\mathbf{V} + \mathbf{W}) & \text{associative}\\
c(\mathbf{U} + \mathbf{V}) &= c\mathbf{U} + c\mathbf{V} & \text{distributive}\\
(a + b)\mathbf{U} &= a\mathbf{U} + b\mathbf{U} & \text{distributive}\\
\mathbf{U} + (-\mathbf{U}) &= \mathbf{0} & \text{additive inverse}
\end{aligned}
$$

In Problem 1.6 we ask you to verify these properties. Vectors $\mathbf{U} = (x, y)$ can be pictured as points in the Cartesian x, y plane. See Figure 1.1 for an example of two vectors $(3, 5)$ and $(7, 2)$ and their sum.

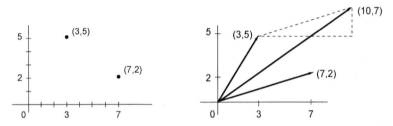

Fig. 1.1 *Left:* Points in the plane. *Right:* Addition of vectors $(3, 5)$ and $(7, 2)$.

By visualizing vectors as points in the plane, multiplication of a vector $\mathbf{U}$ by a number c and the addition of two vectors $\mathbf{U}$ and $\mathbf{V}$ have the following geometric interpretation.

(a) For a nonzero vector $\mathbf{U}$ and a number c, the point $c\mathbf{U}$ lies on the line through the origin and the point $\mathbf{U}$. Its distance from the origin is $|c|$ times the distance of $\mathbf{U}$ from the origin. The origin divides this line into two rays; when c is positive, $\mathbf{U}$ and $c\mathbf{U}$ lie on the same ray; when c is negative, $\mathbf{U}$ and $c\mathbf{U}$ lie on opposite rays. See Figure 1.2.
(b) If the points $\mathbf{0}$, $\mathbf{U}$, and $\mathbf{V}$ do not lie on a line, the four points $\mathbf{0}$, $\mathbf{U}$, $\mathbf{V}$, and $\mathbf{U} + \mathbf{V}$ form the vertices of parallelogram. (We ask you to prove this in Problem 1.7.) See Figure 1.3.
(c) For c between 0 and 1 the points $\mathbf{V} + c\mathbf{U}$ lie on the line segment from $\mathbf{V}$ to $\mathbf{V} + \mathbf{U}$. That side of the parallelogram is parallel to the segment from $\mathbf{0}$ to $\mathbf{U}$ and has the same length. The directed line segment from $\mathbf{V}$ to $\mathbf{V} + \mathbf{U}$ is another way to visualize the vector $\mathbf{U}$. See Figure 1.4.

We can visualize vector addition and multiplication by a number in two dimensions. But we will see in Section 1.4 that in dimensions higher than three it is the

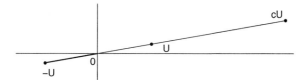

Fig. 1.2 Points **0**, **U**, and c**U** are on a line, $c > 0$.

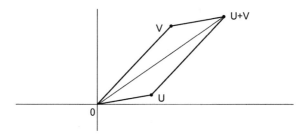

Fig. 1.3 **0**, **U**, **U** + **V**, and **V** form a parallelogram.

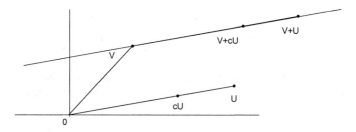

Fig. 1.4 For $0 \leq c \leq 1$ the points **V** + c**U** lie on a line segment from **V** to **V** + **U**.

algebraic properties of vectors that are most useful. Two basic concepts we will use are *linear combination* and *linear independence*.

Definition 1.2. A *linear combination* of two vectors **U** and **V** is a vector of the form

$$a\mathbf{U} + b\mathbf{V} \tag{1.3}$$

where a and b are numbers.

Example 1.1. The vector

$$\mathbf{U} = (5, 3)$$

is a linear combination of $(1, 1)$ and $(-1, 1)$ since

$$\mathbf{U} = 4(1, 1) - (-1, 1).$$

□

Example 1.2. Every vector (x,y) is a linear combination of $(1,0)$ and $(0,1)$ since

$$(x,y) = x(1,0) + y(0,1).$$

□

You might wonder if every vector in $\mathbb{R}^2$ can be obtained as some combination of two given vectors **U** and **V**. As we will see in Theorem 1.1, the answer depends on whether or not **U** and **V** are linearly independent.

Definition 1.3. Two vectors **U** and **V** are called *linearly independent* if the only linear combination $a\mathbf{U} + b\mathbf{V}$ of them that is the zero vector is the trivial linear combination with $a = 0$ and $b = 0$.

Example 1.3. Are the vectors $(1,0)$ and $(0,1)$ are linearly independent? Suppose

$$a(1,0) + b(0,1) = (0,0).$$

Then $(a,0) + (0,b) = (a,b) = (0,0)$. This implies

$$a = 0, \qquad b = 0.$$

Therefore the only linear combination of $(1,0)$ and $(0,1)$ that is $(0,0)$ is the trivial one $0(1,0) + 0(0,1)$. The vectors $(1,0)$ and $(0,1)$ are linearly independent.   □

Two vectors are called linearly *dependent* if they are not independent.

Example 1.4. Are $\mathbf{U} = (1,2)$ and $\mathbf{V} = (2,4)$ linearly independent? Suppose

$$a(1,2) + b(2,4) = (0,0).$$

Then $(a + 2b, 2a + 4b) = (0,0)$. This is true whenever $a = -2b$. For example take $a = 2$ and $b = -1$. Then

$$2\mathbf{U} + (-1)\mathbf{V} = 2(1,2) + (-1)(2,4) = (0,0).$$

The vectors $(1,2)$ and $(2,4)$ are linearly dependent.   □

The next theorem tells us that if **C** and **D** are linearly independent then we can express every vector **U** in $\mathbb{R}^2$ as a linear combination of **C** and **D**.

Theorem 1.1. *Given two linearly independent vectors* **C** *and* **D** *in* $\mathbb{R}^2$, *every vector* **U** *in* $\mathbb{R}^2$ *can be expressed uniquely as a linear combination of them:*

$$\mathbf{U} = a\mathbf{C} + b\mathbf{D}.$$

Proof. Neither of the vectors is the zero vector, for if one of them, say **C** were, then

$$1\mathbf{C} + 0\mathbf{D} = \mathbf{0}$$

would be a nontrivial linear combination of $\mathbf{C}$ and $\mathbf{D}$ that is the zero vector. Next we show that at least one of the vectors $\mathbf{C}$ or $\mathbf{D}$ has nonzero first component. For if not, $\mathbf{C}$ and $\mathbf{D}$ would be of the form

$$\mathbf{C} = (0, c_2), \quad \mathbf{D} = (0, d_2), \quad c_2 \neq 0, \quad d_2 \neq 0.$$

But then $d_2\mathbf{C} - c_2\mathbf{D} = (0, d_2c_2) - (0, c_2d_2) = (0,0) = \mathbf{0}$, a nontrivial linear relation between $\mathbf{C}$ and $\mathbf{D}$.

Suppose the first component c_1 of $\mathbf{C}$ is nonzero. Then we can subtract a multiple $a = \dfrac{d_1}{c_1}$ of $\mathbf{C}$ from $\mathbf{D}$ and obtain a vector $\mathbf{D}'$ whose first component is zero:

$$\mathbf{D}' = \mathbf{D} - a\mathbf{C},$$

say $\mathbf{D}' = (0, d)$. Since $\mathbf{D}'$ is a nontrivial linear combination of $\mathbf{C}$ and $\mathbf{D}$, $\mathbf{D}'$ is not the zero vector, and $d \neq 0$. We then subtract a multiple of $\mathbf{D}'$ from $\mathbf{C}$ to obtain a vector $\mathbf{C}'$ whose second component is zero and whose first component is unchanged:

$$\mathbf{C}' = \mathbf{C} - b\mathbf{D}' = (c_1, 0).$$

Since c_1 and d are not zero, every vector $\mathbf{U}$ can be expressed as a linear combination of $\mathbf{C}'$ and $\mathbf{D}'$. Since $\mathbf{C}'$ and $\mathbf{D}'$ are linear combinations of $\mathbf{C}$ and $\mathbf{D}$, so is every linear combination of them.

To check uniqueness suppose there were two linear combinations of $\mathbf{C}$ and $\mathbf{D}$ for a vector $\mathbf{U}$,

$$\mathbf{U} = a\mathbf{C} + b\mathbf{D} = a'\mathbf{C} + b'\mathbf{D}.$$

Subtract to get

$$(a - a')\mathbf{C} + (b - b')\mathbf{D} = \mathbf{0}.$$

Since $\mathbf{C}$ and $\mathbf{D}$ are linearly independent, this linear combination must be the trivial one with

$$a - a' = b - b' = 0.$$

That proves $a' = a$ and $b' = b$. This completes the proof of Theorem 1.1. □

A basic tool for studying vectors and functions of vectors is the notion of a *linear function*.

Definition 1.4. A function ℓ from $\mathbb{R}^2$ to the set of real numbers $\mathbb{R}$ whose input $\mathbf{U}$ is a vector and whose value $\ell(\mathbf{U})$ is a number is called *linear* if

(a) $\ell(c\mathbf{U}) = c\ell(\mathbf{U})$ and
(b) $\ell(\mathbf{U} + \mathbf{V}) = \ell(\mathbf{U}) + \ell(\mathbf{V})$

for all numbers c and vectors $\mathbf{U}$ and $\mathbf{V}$.

Combining these two properties of a linear function ℓ we deduce that for all numbers a, b and all vectors $\mathbf{U}$, $\mathbf{V}$

$$\ell(a\mathbf{U} + b\mathbf{V}) = \ell(a\mathbf{U}) + \ell(b\mathbf{V}) = a\ell(\mathbf{U}) + b\ell(\mathbf{V}). \tag{1.4}$$

Theorem 1.2. *A function ℓ from $\mathbb{R}^2$ to $\mathbb{R}$ is linear if and only if it is of the form*

$$\ell(x,y) = px + qy. \tag{1.5}$$

for some numbers p and q.

Proof. Suppose ℓ is linear. Take $\mathbf{E}_1$ and $\mathbf{E}_2$ to be the vectors $(1,0)$ and $(0,1)$. We can express the vector (x,y) as $x\mathbf{E}_1 + y\mathbf{E}_2$. By linearity

$$\ell(x,y) = \ell(x\mathbf{E}_1 + y\mathbf{E}_2) = x\ell(\mathbf{E}_1) + y\ell(\mathbf{E}_2).$$

Let $p = \ell(\mathbf{E}_1)$ and $q = \ell(\mathbf{E}_2)$, then $\ell(x,y) = px + qy$ for all (x,y) in $\mathbb{R}^2$.

Conversely, we ask you in Problem 1.12 to show that every function of the form $\ell(x,y) = px + qy$ is linear. $\qquad\square$

Problems

1.1. Use a ruler to estimate the value c shown in Figure 1.2.

1.2. Make a sketch of two linearly dependent nonzero vectors $\mathbf{U}$ and $\mathbf{V}$ in $\mathbb{R}^2$.

1.3. Let $\mathbf{U} = (1,-1)$ and $\mathbf{V} = (1,1)$.

(a) Find all numbers a and b that satisfy the equation

$$a\mathbf{U} + b\mathbf{V} = \mathbf{0}.$$

Prove that $\mathbf{U}$ and $\mathbf{V}$ are linearly independent.
(b) Express $(2,4)$ as a linear combination of $\mathbf{U}$ and $\mathbf{V}$.
(c) Express an arbitrary vector (x,y) as a linear combination of $\mathbf{U}$ and $\mathbf{V}$.

1.4. Find a number k so that the vectors $(k,-1)$ and $(1,3)$ are linearly dependent.

1.5. Find a linear function ℓ from $\mathbb{R}^2$ to $\mathbb{R}$ that satisfies $\ell(1,2) = 3$ and $\ell(2,3) = 5$.

1.6. Let $\mathbf{U} = (u_1, u_2)$, $\mathbf{V} = (v_1, v_2)$, and $\mathbf{W} = (w_1, w_2)$ be vectors in $\mathbb{R}^2$ and let a, b, and c be numbers. Use the definitions $\mathbf{U} + \mathbf{V} = (u_1 + v_1, u_2 + v_2)$, $c\mathbf{U} = (cu_1, cu_2)$, and $-\mathbf{U} = (-u_1, -u_2)$ to show the following properties.

(a) $\mathbf{U} + \mathbf{V} = \mathbf{V} + \mathbf{U}$

(b) $\mathbf{U} + (\mathbf{V} + \mathbf{W}) = (\mathbf{U} + \mathbf{V}) + \mathbf{W}$
(c) $c(\mathbf{U} + \mathbf{V}) = c\mathbf{U} + c\mathbf{V}$
(d) $(a + b)\mathbf{U} = a\mathbf{U} + b\mathbf{U}$
(e) $\mathbf{U} + (-\mathbf{U}) = \mathbf{0}$

1.7. Suppose that the points $\mathbf{0} = (0,0)$, $\mathbf{U} = (u_1, u_2)$, and $\mathbf{V} = (v_1, v_2)$ do not all lie on a line. Show that the quadrilateral with vertices $\mathbf{0}$, $\mathbf{U}$, $\mathbf{U} + \mathbf{V}$, and $\mathbf{V}$ is a parallelogram by proving the following properties.

(a) the line through $\mathbf{0}$ and $\mathbf{U}$ is parallel to the line through $\mathbf{V}$ and $\mathbf{U} + \mathbf{V}$,
(b) the line through $\mathbf{0}$ and $\mathbf{V}$ is parallel to the line through $\mathbf{U}$ and $\mathbf{U} + \mathbf{V}$.

1.8. (a) Make a sketch of two nonzero vectors $\mathbf{U}$ and $\mathbf{V}$ in $\mathbb{R}^2$ such that $\mathbf{U}$ is not a multiple of $\mathbf{V}$.
(b) Using $\mathbf{U}$ and $\mathbf{V}$ from part (a) make a sketch of the vectors $\mathbf{U} + \mathbf{V}$, $-\mathbf{V}$, and $\mathbf{U} - \mathbf{V}$.

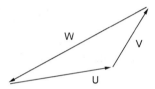

Fig. 1.5 Vectors in Problem 1.9.

1.9. Three vectors $\mathbf{U}$, $\mathbf{V}$, and $\mathbf{W}$ are drawn as directed segments between points in the plane in Figure 1.5. Express $\mathbf{W}$ in terms of $\mathbf{U}$ and $\mathbf{V}$, and show that $\mathbf{U} + \mathbf{V} + \mathbf{W} = \mathbf{0}$.

1.10. Several vectors are drawn in Figure 1.6 as directed segments between points in the plane.

(a) Express $\mathbf{Y}$ as a linear combination of $\mathbf{U}$ and $\mathbf{V}$ and verify that $\mathbf{U} + \mathbf{V} + \mathbf{Y} = \mathbf{0}$.
(b) Express $\mathbf{Y}$ as a linear combination of $\mathbf{W}$ and $\mathbf{X}$ and verify that $\mathbf{W} + \mathbf{X} - \mathbf{Y} = \mathbf{0}$.
(c) Show that $\mathbf{U} + \mathbf{V} + \mathbf{W} + \mathbf{X} = \mathbf{0}$.

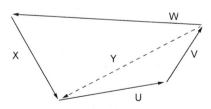

Fig. 1.6 A polygon of vectors in Problem 1.10.

1.11. Let $\mathbf{U} = (u_1, u_2)$. Show that the function $\ell(\mathbf{U}) = u_1 - 8u_2$ is linear.

1.12. Let ℓ be a function from $\mathbb{R}^2$ to $\mathbb{R}$ of the form $\ell(x,y) = px + qy$ where p and q are numbers. Show that ℓ is linear by showing that for all vectors $\mathbf{U}$ and $\mathbf{V}$ in $\mathbb{R}^2$ and all numbers c, the following properties hold.

(a) $\ell(c\mathbf{U}) = c\ell(\mathbf{U})$
(b) $\ell(\mathbf{U} + \mathbf{V}) = \ell(\mathbf{U}) + \ell(\mathbf{V})$

1.13. Write the vector equation

$$(4,5) = a(1,3) + b(3,1)$$

as a system of two equations for the two unknowns a and b.

1.14. Consider the system of two equations for the two unknowns x and y,

$$3x + y = 0$$
$$5x + 12y = 2.$$

The word "system" means that we are interested in numbers x,y that satisfy *both* equations.

(a) Write this system as a vector equation $x\mathbf{U} + y\mathbf{V} = \mathbf{W}$.
(b) Solve for x and y.

1.15. Let $\mathbf{U} = (1,2)$ and $\mathbf{V} = (2,4)$. Find two ways to express the vector $(4,8)$ as a linear combination

$$(4,8) = a\mathbf{U} + b\mathbf{V}.$$

Are $\mathbf{U}$ and $\mathbf{V}$ linearly independent?

1.16. Consider the vectors $\mathbf{U} = (1,3)$ and $\mathbf{V} = (3,1)$.

(a) Are $\mathbf{U}$ and $\mathbf{V}$ linearly independent?
(b) Express the vector $(4,4)$ as a linear combination of $\mathbf{U}$ and $\mathbf{V}$.
(c) Express the vector $(4,5)$ as a linear combination of $\mathbf{U}$ and $\mathbf{V}$.

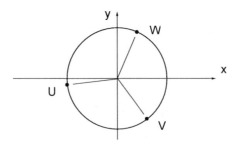

Fig. 1.7 Three points as in Problem 1.17.

1.17. Let **U**, **V**, and **W** be three points on the unit circle centered at the origin of $\mathbb{R}^2$, that divide the circumference into three arcs of equal length. See Figure 1.7.

(a) Show that rotation by 120 degrees around the origin carries $\mathbf{U} + \mathbf{V} + \mathbf{W}$ into itself. Conclude that the sum of the vectors **U**, **V**, and **W** is **0**.

(b) Conclude that

$$\sin(\theta) + \sin\left(\theta + \tfrac{2\pi}{3}\right) + \sin\left(\theta + \tfrac{4\pi}{3}\right) = 0$$

for all θ.

(c) Show that $\displaystyle\sum_{k=1}^{n} \cos\left(\theta + \tfrac{2k\pi}{n}\right) = 0$ for all θ and all $n = 2, 3, \ldots$.

1.18. Let $f(\mathbf{U})$ be the distance between the points **U** and **0** in $\mathbb{R}^2$.

(a) For what numbers c is $f(c\mathbf{U}) = cf(\mathbf{U})$ true?
(b) Is f a linear function?

1.19. Suppose f is a linear function and $f(-.5, 0) = 100$. Find $f(.5, 0)$.

1.20. Suppose f is a linear function and $f(0, 1) = -2$, $f(1, 0) = 6$.

(a) Find $f(1, 1)$.
(b) Find $f(x, y)$.

1.2 The norm and dot product of vectors

Definition 1.5. The *norm* of $\mathbf{U} = (x, y)$, denoted as $\|\mathbf{U}\|$, is defined as

$$\|\mathbf{U}\| = \sqrt{x^2 + y^2}.$$

A *unit vector* is a vector of norm 1.

Applying the Pythagorean theorem to the right triangle whose vertices are $(0, 0)$, $(x, 0)$, and (x, y), (see Fig. 1.8), we see that the norm of (x, y) is the distance between (x, y) and the origin. The norm of **U** is also sometimes called the *length* of **U**.

Example 1.5. The norm of $\mathbf{U} = (1, 2)$ is

$$\|\mathbf{U}\| = \sqrt{1^2 + 2^2} = \sqrt{5}.$$

The norm of $\mathbf{V} = \left(\tfrac{\sqrt{2}}{2}, \tfrac{\sqrt{2}}{2}\right)$ is

$$\|\mathbf{V}\| = \sqrt{\left(\tfrac{\sqrt{2}}{2}\right)^2 + \left(\tfrac{\sqrt{2}}{2}\right)^2} = 1,$$

so **V** is a unit vector. □

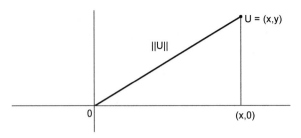

Fig. 1.8 $\|\mathbf{U}\|$ is the distance between $\mathbf{U}$ and the origin.

A concept related to norm is the *dot product* of two vectors $\mathbf{U}$ and $\mathbf{V}$.

Definition 1.6. The dot product of $\mathbf{U} = (u_1, u_2)$ and $\mathbf{V} = (v_1, v_2)$ is

$$\mathbf{U} \cdot \mathbf{V} = u_1 v_1 + u_2 v_2. \tag{1.6}$$

The dot product has some of the usual properties of a product.

(a) Distributive: for vectors $\mathbf{U}$, $\mathbf{V}$, and $\mathbf{W}$

$$\mathbf{U} \cdot (\mathbf{V} + \mathbf{W}) = \mathbf{U} \cdot \mathbf{V} + \mathbf{U} \cdot \mathbf{W},$$

(b) Commutative: for vectors $\mathbf{U}$ and $\mathbf{V}$

$$\mathbf{U} \cdot \mathbf{V} = \mathbf{V} \cdot \mathbf{U}.$$

In Problem 1.21 we ask you to verify the distributive and commutative properties.

It follows from Definitions 1.5 and 1.6 of the norm and dot product that the dot product of a vector with itself is its norm squared:

$$\mathbf{U} \cdot \mathbf{U} = \|\mathbf{U}\|^2. \tag{1.7}$$

We have shown in Theorem 1.2 that every linear function from $\mathbb{R}^2$ to $\mathbb{R}$ is of the form $\ell(\mathbf{U}) = \ell(x, y) = px + qy$. This result can be restated in terms of the dot product:

Theorem 1.3. *A function ℓ from $\mathbb{R}^2$ to $\mathbb{R}$ is linear if and only if it is of the form*

$$\ell(\mathbf{U}) = \mathbf{C} \cdot \mathbf{U},$$

where $\mathbf{C} = (p, q)$ is some vector in $\mathbb{R}^2$.

Example 1.6. Let ℓ be a linear function for which

$$\ell(1, 1) = 5 \quad \text{and} \quad \ell(-1, 1) = -1.$$

Let's find the vector $\mathbf{C} = (p,q)$ so that $\ell(\mathbf{U}) = \mathbf{C} \cdot \mathbf{U}$. By Theorem 1.3 we have

$$5 = \ell(1,1) = (p,q) \cdot (1,1) = p+q, \qquad -1 = \ell(-1,1) = (p,q) \cdot (-1,1) = -p+q.$$

Solving for p and q we get $2q = 4$, $q = 2$, and $p = 3$, so

$$\ell(x,y) = (3,2) \cdot (x,y) = 3x+2y.$$

$\square$

An interesting relation between norm and dot product follows from the distributive and commutative laws applied to the dot product. Using the distributive law we see

$$(\mathbf{U}-\mathbf{V}) \cdot (\mathbf{U}-\mathbf{V}) = \mathbf{U} \cdot (\mathbf{U}-\mathbf{V}) - \mathbf{V} \cdot (\mathbf{U}-\mathbf{V}) = \mathbf{U} \cdot \mathbf{U} - \mathbf{U} \cdot \mathbf{V} - \mathbf{V} \cdot \mathbf{U} + \mathbf{V} \cdot \mathbf{V}.$$

Using the notation of norm and commutativity of the dot product, $\mathbf{U} \cdot \mathbf{V} = \mathbf{V} \cdot \mathbf{U}$, we can rewrite the equation above as

$$\|\mathbf{U}-\mathbf{V}\|^2 = \|\mathbf{U}\|^2 + \|\mathbf{V}\|^2 - 2\mathbf{U} \cdot \mathbf{V}. \tag{1.8}$$

Since $\|\mathbf{U}-\mathbf{V}\|^2$ is nonnegative, it follows from (1.8) that

$$\mathbf{U} \cdot \mathbf{V} \le \tfrac{1}{2}\|\mathbf{U}\|^2 + \tfrac{1}{2}\|\mathbf{V}\|^2. \tag{1.9}$$

We show next that an even sharper inequality holds:

Theorem 1.4. *For all vectors $\mathbf{U}$ and $\mathbf{V}$ in $\mathbb{R}^2$ the following inequality holds.*

$$\mathbf{U} \cdot \mathbf{V} \le \|\mathbf{U}\|\|\mathbf{V}\|. \tag{1.10}$$

Proof. If either $\mathbf{U}$ or $\mathbf{V}$ is the zero vector, inequality (1.10) holds, because both sides are zero. If both $\mathbf{U}$ and $\mathbf{V}$ are unit vectors, inequality (1.10) follows from (1.9). For all nonzero vectors $\mathbf{U}$ and $\mathbf{V}$,

$$\frac{1}{\|\mathbf{U}\|}\mathbf{U}, \quad \frac{1}{\|\mathbf{V}\|}\mathbf{V}$$

are unit vectors and therefore by inequality (1.9)

$$\frac{\mathbf{U} \cdot \mathbf{V}}{\|\mathbf{U}\|\|\mathbf{V}\|} \le 1,$$

from which inequality (1.10) follows. $\square$

For all vectors $\mathbf{U}$ and $\mathbf{V}$,

$$0 \le (\|\mathbf{U}\| - \|\mathbf{V}\|)^2 = \|\mathbf{U}\|^2 + \|\mathbf{V}\|^2 - 2\|\mathbf{U}\|\|\mathbf{V}\|.$$

Therefore

$$\|\mathbf{U}\|\,\|\mathbf{V}\| \le \tfrac{1}{2}\|\mathbf{U}\|^2 + \tfrac{1}{2}\|\mathbf{V}\|^2.$$

By (1.10) we see that

$$\mathbf{U}\cdot\mathbf{V} \le \|\mathbf{U}\|\,\|\mathbf{V}\| \le \tfrac{1}{2}\|\mathbf{U}\|^2 + \tfrac{1}{2}\|\mathbf{V}\|^2$$

Thus (1.10) is a "sharper" inequality than (1.9).

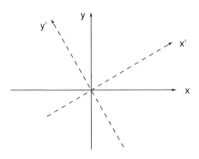

Fig. 1.9 Coordinate axes x, y, and x', y'.

Suppose we replace the coordinate axes by another pair of perpendicular lines through the origin. See Figure 1.9. Let x' and y' be the coordinates in the new system of a vector $\mathbf{U}$ whose coordinates in the original system were x and y. Then

$$x'^2 + y'^2 = x^2 + y^2$$

because both sides express the square of the distance between the point $\mathbf{U}$ and the origin.

The dot product of two vectors calculated in the new coordinates is equal to their dot product calculated in the old coordinates. To see that we note that formula (1.8) holds in both coordinate systems. The term on the left and the first two terms on the right are norms and therefore have the same value in both coordinate systems; but then the remaining term on the right, two times the dot product of $\mathbf{U}$ and $\mathbf{V}$, must also have the same value in both coordinate systems.

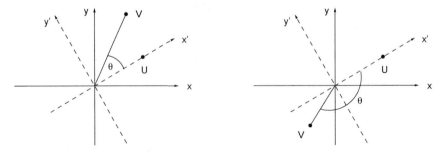

Fig. 1.10 The angle θ between vectors $\mathbf{U}$ and $\mathbf{V}$.

The independence of the dot product of the coordinate system suggests that the dot product of two vectors $\mathbf{U}$ and $\mathbf{V}$, introduced algebraically, has a geometric significance. To figure out what it is we use a coordinate system where the new positive x' axis is the ray through the origin and the point $\mathbf{U}$; see Figure 1.10. In this new coordinate system the coordinates of $\mathbf{U}$ and $\mathbf{V}$ are:

$$\mathbf{U} = (\|\mathbf{U}\|, 0), \qquad \mathbf{V} = (\|\mathbf{V}\|\cos\theta, \|\mathbf{V}\|\sin\theta),$$

where θ is the *angle* between $\mathbf{U}$ and $\mathbf{V}$. That is, the angle θ between the positive x' axis and the line through $\mathbf{0}$ and $\mathbf{V}$, where $0 \le \theta \le \pi$. So in this coordinate system the dot product of $\mathbf{U}$ and $\mathbf{V}$ is

$$\mathbf{U} \cdot \mathbf{V} = \|\mathbf{U}\|\|\mathbf{V}\|\cos\theta.$$

Since the dot product in our two coordinate systems is the same, this proves the following theorem.

Theorem 1.5. *The dot product of two nonzero vectors $\mathbf{U}$ and $\mathbf{V}$ is the product of the norms of the two vectors times the cosine of the angle between the two vectors,*
$$\mathbf{U} \cdot \mathbf{V} = \|\mathbf{U}\|\|\mathbf{V}\|\cos\theta.$$

In particular if two nonzero vectors $\mathbf{U}$ and $\mathbf{V}$ are perpendicular, $\theta = \frac{\pi}{2}$, their dot product is zero, and conversely. When the dot product of vectors $\mathbf{U}$ and $\mathbf{V}$ is zero we say that $\mathbf{U}$ and $\mathbf{V}$ are *orthogonal*.

Problems

1.21. Let $\mathbf{U} = (u_1, u_2)$, $\mathbf{V} = (v_1, v_2)$, and $\mathbf{W} = (w_1, w_2)$. Prove

(a) the distributive property $\mathbf{U} \cdot (\mathbf{V} + \mathbf{W}) = \mathbf{U} \cdot \mathbf{V} + \mathbf{U} \cdot \mathbf{W}$.
(b) the commutative property $\mathbf{U} \cdot \mathbf{V} = \mathbf{V} \cdot \mathbf{U}$.

1.22. Which vectors are orthogonal?

(a) (a, b), $(-b, a)$
(b) $(1, -1)$, $(1, 1)$
(c) $(0, 0)$, $(1, 1)$
(d) $(1, 1)$, $(1, 1)$

1.23. Which of these vectors are unit vectors?

(a) $(\frac{3}{5}, \frac{4}{5})$
(b) $(\cos\theta, \sin\theta)$
(c) $(\sqrt{.8}, \sqrt{.2})$
(d) $(.8, .2)$

1.24. Use equation (1.8) and Theorem 1.5 to prove the Law of Cosines: for every triangle in the plane with sides a, b, c and angle θ opposite side c, (see Figure 1.11),

$$c^2 = a^2 + b^2 - 2ab\cos\theta.$$

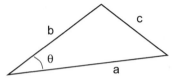

Fig. 1.11 A triangle in Problem 1.24.

1.25. Let ℓ be a linear function from $\mathbb{R}^2$ to $\mathbb{R}$ for which $\ell(2,1) = 3$ and $\ell(1,1) = 2$. Find the vector $\mathbf{C}$ so that $\ell(\mathbf{U}) = \mathbf{C} \cdot \mathbf{U}$.

1.26. Find the cosine of the angle between the vectors $\mathbf{U} = (1,2)$ and $\mathbf{V} = (3,1)$.

1.27. Use equation (1.8) to show that for every $\mathbf{U}$ and $\mathbf{V}$ in $\mathbb{R}^2$,

$$\|\mathbf{U} + \mathbf{V}\|^2 = \|\mathbf{U}\|^2 + \|\mathbf{V}\|^2 + 2\mathbf{U} \cdot \mathbf{V}.$$

1.28. Let $\mathbf{U} = (x, y)$. Find a vector $\mathbf{C}$ to express the equation of a line $y = mx + b$ as $\mathbf{C} \cdot \mathbf{U} = b$.

1.29. If $\mathbf{C}$ and $\mathbf{D}$ are orthogonal nonzero vectors, there is a simple expression for a and b in a linear combination

$$\mathbf{U} = a\mathbf{C} + b\mathbf{D}.$$

(a) Dot both sides of this equation with $\mathbf{C}$ to show that $a = \dfrac{\mathbf{C} \cdot \mathbf{U}}{\|\mathbf{C}\|^2}$.

(b) Find a formula for b.
(c) If $(8,9) = a(\frac{3}{5}, \frac{4}{5}) + b(-\frac{4}{5}, \frac{3}{5})$, find a.

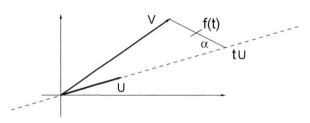

Fig. 1.12 Vectors $\mathbf{V}$ and $\mathbf{U}$ for Problem 1.30.

1.30. Let $\mathbf{U}$ be a nonzero vector and t a number. Let $f(t)$ be the distance between a point $\mathbf{V}$ and the point $t\mathbf{U}$ on the line through $\mathbf{0}$ and $\mathbf{U}$ as shown in Figure 1.12.

(a) Use calculus to find the value of t that minimizes $(f(t))^2$.
(b) Use a dot product to find the value of t that makes the angle α in the figure a right angle.
(c) Confirm that the numbers t that you found in parts (a) and (b) are the same.

1.31. Express the vectors $\mathbf{U} = (1,0)$, $\mathbf{V} = (2,2)$ in the coordinate system rotated $\frac{\pi}{4}$ counterclockwise.

1.32. A regular octagon is shown in Figure 1.13. It shows vertex $\mathbf{P} = (c,s)$ where c and s are the cosine and sine of $\frac{\pi}{8}$.

(a) Show that vertex $\mathbf{Q}$ is (s,c).
(b) Show that $\sin\left(\frac{\pi}{8}\right) = \frac{1}{2}\sqrt{2 - \sqrt{2}}$.

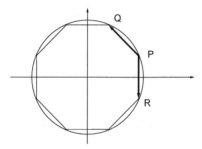

Fig. 1.13 The octagon in Problem 1.32.

1.3 Bilinear functions

Definition 1.7. A function b whose input is an ordered pair of vectors $\mathbf{U}$ and $\mathbf{V}$ and whose output is a number is called *bilinear* if, when we hold $\mathbf{U}$ fixed, $b(\mathbf{U},\mathbf{V})$ a linear function of $\mathbf{V}$, and when we hold $\mathbf{V}$ fixed, $b(\mathbf{U},\mathbf{V})$ is a linear function of $\mathbf{U}$.

As we shall see, many important functions are bilinear.

Example 1.7. Let $\mathbf{U} = (u_1,u_2)$, $\mathbf{V} = (v_1,v_2)$, $\mathbf{W} = (w_1,w_2)$, and define the function b by

$$b(\mathbf{U},\mathbf{V}) = u_1 v_1.$$

To see that b is bilinear, we first hold $\mathbf{U}$ fixed and check that $b(\mathbf{U}, \mathbf{V})$ is linear in $\mathbf{V}$. That is, check that for all numbers c and vectors $\mathbf{V}$ and $\mathbf{W}$,

$$b(\mathbf{U}, \mathbf{V} + \mathbf{W}) = b(\mathbf{U}, \mathbf{V}) + b(\mathbf{U}, \mathbf{W}), \qquad b(\mathbf{U}, c\mathbf{V}) = c\, b(\mathbf{U}, \mathbf{V}).$$

For the first we have

$$b(\mathbf{U}, \mathbf{V} + \mathbf{W}) = u_1(v_1 + w_1) = u_1 v_1 + u_1 w_1 = b(\mathbf{U}, \mathbf{V}) + b(\mathbf{U}, \mathbf{W}).$$

For the second,

$$b(\mathbf{U}, c\mathbf{V}) = u_1(cv_1) = cu_1 v_1 = c\, b(\mathbf{U}, \mathbf{V}).$$

So b is linear in $\mathbf{V}$. We ask you in Problem 1.33 to show by a similar argument that b is linear in $\mathbf{U}$ when we hold $\mathbf{V}$ fixed. $\qquad\square$

We saw in Example 1.7 that $u_1 v_1$ is a bilinear function of $(\mathbf{U}, \mathbf{V})$. Similarly $u_1 v_2$, $u_2 v_1$, and $u_2 v_2$ are bilinear. The next theorem describes all bilinear functions.

Theorem 1.6. *Every bilinear function b of $\mathbf{U} = (u_1, u_2)$ and $\mathbf{V} = (v_1, v_2)$ is of the form*

$$b(\mathbf{U}, \mathbf{V}) = eu_1 v_1 + fu_1 v_2 + gu_2 v_1 + hu_2 v_2, \qquad (1.11)$$

where e, f, g, and h are numbers.

Proof. For fixed $\mathbf{V}$, $b(\mathbf{U}, \mathbf{V})$ is a linear function of $\mathbf{U}$. According to Theorem 1.2, $b(\mathbf{U}, \mathbf{V})$ has the form (1.5):

$$b(\mathbf{U}, \mathbf{V}) = pu_1 + qu_2, \qquad (1.12)$$

where the numbers p and q depend on $\mathbf{V}$. To determine the nature of this dependence first let $\mathbf{U} = \mathbf{E}_1 = (1,0)$ and then $\mathbf{U} = \mathbf{E}_2 = (0,1)$. We get

$$b((1,0), \mathbf{V}) = (p)(1) + (q)(0) = p = p(\mathbf{V}), \qquad b((0,1), \mathbf{V}) = (p)(0) + (q)(1) = q = q(\mathbf{V}).$$

Since b is bilinear, the functions p and q are linear functions of $\mathbf{V}$. Therefore they are of the form

$$p(\mathbf{V}) = ev_1 + fv_2, \qquad q(\mathbf{V}) = gv_1 + hv_2,$$

where e, f, g, and h are numbers that do not depend on $\mathbf{V}$. Setting these formulas for p and q into formula (1.12) gives

$$b(\mathbf{U}, \mathbf{V}) = (ev_1 + fv_2)u_1 + (gv_1 + hv_2)u_2,$$

which gives

$$b(\mathbf{U}, \mathbf{V}) = eu_1 v_1 + fu_1 v_2 + gu_2 v_1 + hu_2 v_2.$$

$\qquad\square$

We ask you in Problem 1.36 to prove the following theorem.

Theorem 1.7. *A linear combination of bilinear functions is bilinear.*

Example 1.8. The dot product $\mathbf{U} \cdot \mathbf{V}$ has the properties

$$(c\mathbf{U}) \cdot \mathbf{V} = c\mathbf{U} \cdot \mathbf{V}, \quad (\mathbf{U} + \mathbf{V}) \cdot \mathbf{W} = \mathbf{U} \cdot \mathbf{W} + \mathbf{V} \cdot \mathbf{W}, \quad \mathbf{U} \cdot (\mathbf{V} + \mathbf{W}) = \mathbf{U} \cdot \mathbf{V} + \mathbf{U} \cdot \mathbf{W}.$$

This shows that $\mathbf{U} \cdot \mathbf{V}$ is a bilinear function of $\mathbf{U}$ and $\mathbf{V}$. Its formula

$$\mathbf{U} \cdot \mathbf{V} = u_1 v_1 + u_2 v_2$$

is a special case of formula (1.11) in Theorem 1.6.   □

Example 1.9. Let
$$b(\mathbf{U}, \mathbf{V}) = u_1 v_2 - u_2 v_1,$$

where $\mathbf{U} = (u_1, u_2)$, $\mathbf{V} = (v_1, v_2)$. The terms $u_1 v_2$ and $u_2 v_1$ are bilinear. By Theorem 1.7 b is a bilinear function.   □

Problems

1.33. Let $\mathbf{U} = (u_1, u_2)$, $\mathbf{V} = (v_1, v_2)$. Show that the function $b(\mathbf{U}, \mathbf{V}) = u_1 v_1$ is linear in $\mathbf{U}$ when we hold $\mathbf{V}$ fixed.

1.34. Let $\mathbf{U} = (u_1, u_2)$, $\mathbf{V} = (v_1, v_2)$. Is the function $b(\mathbf{U}, \mathbf{V}) = u_1 u_2$ bilinear?

1.35. Define $f(p, q, r, s) = qr + 3rp - sp$. Sort two of the variables p, q, r, s into a vector $\mathbf{U}$ and the other two into a vector $\mathbf{V}$ to express f as a bilinear function $b(\mathbf{U}, \mathbf{V})$.

1.36. Prove Theorem 1.7. That is, suppose $b_1(\mathbf{U}, \mathbf{V})$ and $b_2(\mathbf{U}, \mathbf{V})$ are bilinear functions and c_1, c_2 are numbers; show that the function b defined by

$$b(\mathbf{U}, \mathbf{V}) = c_1 b_1(\mathbf{U}, \mathbf{V}) + c_2 b_2(\mathbf{U}, \mathbf{V})$$

is a bilinear function.

1.4 *n*-dimensional vectors

We extend the concepts of vectors and their algebra from two dimensions to n dimensions, where n is any positive integer.

Definition 1.8. An ordered n-tuple

$$\mathbf{U} = (u_1, u_2, \ldots, u_n)$$

of numbers is called an *n-dimensional vector*. The numbers u_j are called the *components* of the vector $\mathbf{U}$, and u_j is called the j-th component of $\mathbf{U}$. The set of all n-dimensional vectors, denoted $\mathbb{R}^n$, is called n-dimensional space.

The vector all of whose components are zero is called the *zero vector* and is denoted as $\mathbf{0}$:

$$\mathbf{0} = (0, 0, \ldots, 0).$$

There is an algebra of vectors in $\mathbb{R}^n$ entirely analogous to the algebra in $\mathbb{R}^2$ described in Section 1.1.

(a) Let $\mathbf{U} = (u_1, u_2, \ldots, u_n)$ and let c be a number. The multiple $c\mathbf{U}$ is defined by multiplying each component of $\mathbf{U}$ by c:

$$c\mathbf{U} = (cu_1, cu_2, \ldots, cu_n).$$

(b) The sum of $\mathbf{U} = (u_1, u_2, \ldots, u_n)$ and $\mathbf{V} = (v_1, v_2, \ldots, v_n)$ is defined by adding the corresponding components of $\mathbf{U}$ and $\mathbf{V}$:

$$\mathbf{U} + \mathbf{V} = (u_1 + v_1, u_2 + v_2, \ldots, u_n + v_n).$$

In Problem 1.37 we ask you to verify that $\mathbb{R}^n$ has the usual algebraic properties

$$c(\mathbf{X} + \mathbf{Y}) = c\mathbf{X} + c\mathbf{Y}, \qquad \mathbf{X} + \mathbf{Y} = \mathbf{Y} + \mathbf{X}, \qquad \mathbf{X} + (\mathbf{Y} + \mathbf{Z}) = (\mathbf{X} + \mathbf{Y}) + \mathbf{Z}.$$

According to the third of these we write $\mathbf{X} + (\mathbf{Y} + \mathbf{Z})$ as $\mathbf{X} + \mathbf{Y} + \mathbf{Z}$.

Definition 1.9. Let k be a positive integer. A *linear combination* of vectors $\mathbf{U}_1, \mathbf{U}_2, \ldots, \mathbf{U}_k$ in $\mathbb{R}^n$ is a vector of the form

$$c_1\mathbf{U}_1 + c_2\mathbf{U}_2 + \cdots + c_k\mathbf{U}_k = \sum_{j=1}^{k} c_j\mathbf{U}_j,$$

where $c_1, c_2, \ldots, c_k$ are numbers. The set of all such linear combinations is called the *span* of $\mathbf{U}_1, \mathbf{U}_2, \ldots, \mathbf{U}_k$.

A linear combination is called *trivial* if all the numbers c_j are zero.

Example 1.10. Let

$$\mathbf{U} = (3, 7, 6, 9, 4),$$
$$\mathbf{V} = (2, 7, 0, 1, -5).$$

The vector

$$2\mathbf{U} + 3\mathbf{V} = (12, 35, 12, 21, -7)$$

is a linear combination of $\mathbf{U}$ and $\mathbf{V}$. In Problem 1.43 we ask you to show that the vector

$$\left(-\tfrac{1}{2}, -\tfrac{7}{2}, 3, \tfrac{7}{2}, 7\right)$$

is also a linear combination of $\mathbf{U}$ and $\mathbf{V}$. □

Definition 1.10. The vectors $\mathbf{U}_1, \ldots, \mathbf{U}_k$ in $\mathbb{R}^n$ are called *linearly independent* if the only linear combination of them that is the zero vector is the trivial one. That is,

$$\text{if } c_1\mathbf{U}_1 + c_2\mathbf{U}_2 + \cdots + c_k\mathbf{U}_k = \mathbf{0} \text{ then } c_1 = 0, \ c_2 = 0, \ \ldots, \ c_k = 0.$$

If $\mathbf{U}_1, \ldots, \mathbf{U}_k$ are not linearly independent, they are called *linearly dependent*.

Example 1.11. The vectors

$$\mathbf{E}_1 = (1,0,0,0), \quad \mathbf{E}_2 = (0,1,0,0), \quad \mathbf{E}_3 = (0,0,1,0), \quad \mathbf{E}_4 = (0,0,0,1)$$

in $\mathbb{R}^4$ are linearly independent because the linear combination

$$c_1\mathbf{E}_1 + c_2\mathbf{E}_2 + c_3\mathbf{E}_3 + c_4\mathbf{E}_4 = (c_1, c_2, c_3, c_4)$$

is the zero vector only when $c_1 = c_2 = c_3 = c_4 = 0$. □

Definition 1.11. Let $\mathbf{U}_1, \mathbf{U}_2, \ldots, \mathbf{U}_k$ be linearly independent vectors in $\mathbb{R}^n$, with $k < n$. Let $t_1, t_2, \ldots, t_k$ be numbers and let $\mathbf{U}$ be a vector in $\mathbb{R}^n$. The set of all vectors of the form

$$\mathbf{U} + t_1\mathbf{U}_1 + \cdots + t_k\mathbf{U}_k$$

is called a *k-dimensional plane* in $\mathbb{R}^n$ through $\mathbf{U}$. When $k = n - 1$ we call it a *hyperplane*. When $\mathbf{U} = \mathbf{0}$ the *k*-dimensional plane through the origin is called the *span* of $\mathbf{U}_1, \mathbf{U}_2, \ldots, \mathbf{U}_k$.

Theorem 1.8. *$n + 1$ vectors $\mathbf{U}_1, \mathbf{U}_2, \ldots, \mathbf{U}_{n+1}$ in $\mathbb{R}^n$ are linearly dependent.*

Proof. The proof is by induction on n. Take first $n = 1$, and let u and v be in $\mathbb{R}$. If both u and v are zero then $u + v = 0$ is a nontrivial linear combination. Otherwise $(v)u + (-u)v = 0$ is a nontrivial linear combination. Therefore the theorem holds for $n = 1$.

Suppose inductively that the theorem holds for $n - 1$. Look at the n-th components of the vectors $\mathbf{U}_1, \mathbf{U}_2, \ldots, \mathbf{U}_{n+1}$ in $\mathbb{R}^n$. If all of them are zero, then there are $n + 1$

vectors in $\mathbb{R}^{n-1}$ obtained by omitting the final zeros; so by the induction hypothesis they are linearly dependent.

Consider the case where one of the vectors, call it $\mathbf{U}_{n+1}$, has a nonzero n-th component. Subtract a suitable multiple c_i of $\mathbf{U}_{n+1}$ from each of the other vectors $\mathbf{U}_i$ so that the n-th component of each difference

$$\mathbf{U}_i' = \mathbf{U}_i - c_i\mathbf{U}_{n+1} \qquad (i = 1,2,\ldots,n)$$

is zero. Omitting the final zero of $\mathbf{U}_i'$ we obtain n vectors $\mathbf{V}_i$ in $\mathbb{R}^{n-1}$. According to the induction hypothesis they are linearly dependent, that is they satisfy a nontrivial linear relation

$$\sum_{i=1}^{n} k_i\mathbf{V}_i = \mathbf{0}.$$

Adjoining a final zero component to each of the vectors $\mathbf{V}_i$ gives the nontrivial linear relation

$$\sum_{i=1}^{n} k_i\mathbf{U}_i' = \mathbf{0}. \tag{1.13}$$

Setting $\mathbf{U}_i' = \mathbf{U}_i - c_i\mathbf{U}_{n+1}$ into equation (1.13) we get the nontrivial relation

$$\sum_{i=1}^{n} k_i(\mathbf{U}_i - c_i\mathbf{U}_{n+1}) = \mathbf{0}. \tag{1.14}$$

This proves the linear dependence of $\mathbf{U}_1,\ldots,\mathbf{U}_{n+1}$. This completes the proof by induction. □

Example 1.12. The vectors in $\mathbb{R}^3$,

$$\mathbf{E}_1 = (1,0,0), \quad \mathbf{E}_2 = (0,1,0), \quad \mathbf{E}_3 = (0,0,1), \quad \mathbf{X} = (2,4,3)$$

are linearly dependent: Since

$$\mathbf{X} = 2\mathbf{E}_1 + 4\mathbf{E}_2 + 3\mathbf{E}_3,$$

we have a nontrivial linear combination equal to zero:

$$2\mathbf{E}_1 + 4\mathbf{E}_2 + 3\mathbf{E}_3 - \mathbf{X} = \mathbf{0}.$$

□

Figure 1.14 illustrates other examples of linear dependence and independence in $\mathbb{R}^3$.

Theorem 1.9. *Suppose $\mathbf{U}_1,\ldots,\mathbf{U}_n$ are linearly independent in $\mathbb{R}^n$. Then each vector $\mathbf{X}$ in $\mathbb{R}^n$ can be expressed uniquely as a linear combination of the $\mathbf{U}_i$:*

$$\mathbf{X} = c_1\mathbf{U}_1 + \cdots + c_n\mathbf{U}_n. \tag{1.15}$$

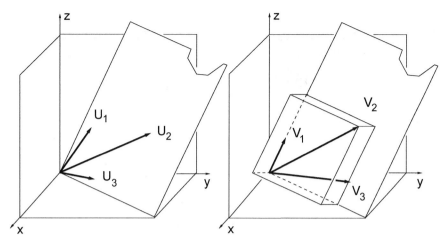

Fig. 1.14 $\mathbf{U}_1$, $\mathbf{U}_2$, $\mathbf{U}_3$ are linearly dependent. $\mathbf{V}_1$, $\mathbf{V}_2$, $\mathbf{V}_3$ are linearly independent.

Proof. According to Theorem 1.8 there is a nontrivial relation among the $n+1$ vectors $\mathbf{X}, \mathbf{U}_1, \ldots, \mathbf{U}_n$:

$$a_0\mathbf{X} + a_1\mathbf{U}_1 + \cdots + a_n\mathbf{U}_n = \mathbf{0}.$$

In this relation, a_0 is not zero. If it were zero then the linear independence of the $\mathbf{U}_i$ would imply that all the a_i are zero. Divide by a_0 to get relation (1.15) with $c_i = -a_i/a_0$.

If there were two different representations of $\mathbf{X}$ of the form (1.15), their difference would be a nontrivial linear relation among the $\mathbf{U}_i$ contrary to their linear independence. □

Linear functions. Just as for two-dimensional vectors, a function ℓ that assigns a number to each vector in $\mathbb{R}^n$ is called *linear* if for all $\mathbf{U}$ and $\mathbf{V}$ in $\mathbb{R}^n$

(a) $\ell(c\mathbf{U}) = c\ell(\mathbf{U})$ for all numbers c, and
(b) $\ell(\mathbf{U} + \mathbf{V}) = \ell(\mathbf{U}) + \ell(\mathbf{V})$.

Combining properties (a) and (b) we deduce that a function ℓ from $\mathbb{R}^n$ to $\mathbb{R}$ is linear if for all $\mathbf{U}$ and $\mathbf{V}$ in $\mathbb{R}^n$ and numbers a and b, ℓ satisfies

$$\ell(a\mathbf{U} + b\mathbf{V}) = a\ell(\mathbf{U}) + b\ell(\mathbf{V}). \tag{1.16}$$

Every function ℓ of $\mathbf{U} = (u_1, u_2, ..., u_n)$ of the form

$$\ell(\mathbf{U}) = c_1 u_1 + c_2 u_2 + \cdots + c_n u_n, \tag{1.17}$$

where $c_1, c_2, \ldots, c_n$ are numbers, has properties (a) and (b), and therefore is a linear function. (See Problem 1.45.) Conversely, we have the following theorem.

Theorem 1.10. *Let ℓ be a linear function from $\mathbb{R}^n$ to $\mathbb{R}$. Then there are numbers $c_1,\ldots,c_n$ such that*

$$\ell(\mathbf{U}) = c_1 u_1 + c_2 u_2 + \cdots + c_n u_n$$

for every vector $\mathbf{U} = (u_1, u_2, \ldots, u_n)$ in $\mathbb{R}^n$.

Proof. Let $\mathbf{E}_j$ be the vector in $\mathbb{R}^n$ whose j-th coordinate is 1 and all other coordinates are zero,

$$\mathbf{E}_1 = (1,0,0,0,\ldots,0)$$
$$\mathbf{E}_2 = (0,1,0,0,\ldots,0)$$
$$\mathbf{E}_3 = (0,0,1,0,\ldots,0)$$
$$\vdots$$
$$\mathbf{E}_n = (0,0,0,0,\ldots,1).$$

The vector $\mathbf{U} = (u_1, u_2, \ldots, u_n)$ can be expressed as the linear combination of the $\mathbf{E}_j$:

$$\mathbf{U} = u_1 \mathbf{E}_1 + u_2 \mathbf{E}_2 + \cdots + u_n \mathbf{E}_n$$

Set $c_j = \ell(\mathbf{E}_j)$, $j = 1,\ldots,n$. Since ℓ is linear,

$$\ell(\mathbf{U}) = u_1 \ell(\mathbf{E}_1) + u_2 \ell(\mathbf{E}_2) + \cdots + u_n \ell(\mathbf{E}_n)$$

$$= c_1 u_1 + c_2 u_2 + \cdots + c_n u_n.$$

$\square$

As in the two-dimensional case, we call a function b of two vectors in $\mathbb{R}^n$ *bilinear* if for every vector $\mathbf{V}$, $b(\mathbf{U}, \mathbf{V})$ is a linear function of $\mathbf{U}$, and for every vector $\mathbf{U}$ it is a linear function of $\mathbf{V}$.

Every function of the form $u_j v_k$ is bilinear, and as in Theorem 1.7, so are linear combinations of them. The following result, an extension of Theorem 1.6, characterizes all bilinear functions.

Theorem 1.11. *Let b be a bilinear function of $\mathbf{U} = (u_1,\ldots,u_n)$ and $\mathbf{V} = (v_1,\ldots,v_n)$. Then b is a linear combination of the functions f_{jk} defined by*

$$f_{jk}(\mathbf{U},\mathbf{V}) = f_{jk}(u_1,\ldots,u_n,v_1,\ldots,v_n) = u_j v_k, \qquad j = 1,\ldots,n, \quad k = 1,\ldots,n.$$

Proof. We fix the vector $\mathbf{V}$ and consider $b(\mathbf{U},\mathbf{V})$ as a linear function of $\mathbf{U}$. By Theorem 1.10 it has the form

$$b(\mathbf{U}, \mathbf{V}) = c_1 u_1 + c_2 u_2 + \cdots + c_n u_n, \tag{1.18}$$

where $c_1, c_2, \ldots, c_n$ are functions of $\mathbf{V}$. Since b is a bilinear function, the c_i are linear functions of $\mathbf{V}$. According to Theorem 1.10 each c_i is a linear combination of the v_k. Setting this into formula (1.18) we get an expression of $b(U, V)$ as a linear combination of $u_j v_k$, as asserted. □

Problems

1.37. Let $\mathbf{V} = (v_1, \ldots, v_n)$, $\mathbf{U} = (u_1, \ldots, u_n)$, and $\mathbf{W} = (w_1, \ldots, w_n)$ be vectors in $\mathbb{R}^n$, and let c and d be numbers. Show that

(a) $\mathbf{V} + \mathbf{W} = \mathbf{W} + \mathbf{V}$
(b) $(\mathbf{V} + \mathbf{U}) + \mathbf{W} = \mathbf{V} + (\mathbf{U} + \mathbf{W})$
(c) $c(\mathbf{U} + \mathbf{V}) = c\mathbf{U} + c\mathbf{V}$
(d) $(c + d)\mathbf{U} = c\mathbf{U} + d\mathbf{U}$

1.38. Express the vector $(1, 3, 5)$ as a linear combination of the vectors

$$\mathbf{U}_1 = (1, 0, 0), \quad \mathbf{U}_2 = (1, 1, 0), \quad \mathbf{U}_3 = (1, 1, 1).$$

1.39. Show that every vector in $\mathbb{R}^3$ is a linear combination of the vectors

$$\mathbf{U}_1 = (1, 0, 0), \quad \mathbf{U}_2 = (1, 1, 0), \quad \mathbf{U}_3 = (1, 1, 1).$$

1.40. Determine whether the vectors

$$\mathbf{U}_1 = (1, 0, 0), \quad \mathbf{U}_2 = (1, 1, 0), \quad \mathbf{U}_3 = (1, 1, 1).$$

are linearly independent.

1.41. Show that the vectors

$$(1, 1, 1, 1), \quad (0, 1, 1, 1), \quad (0, 0, 1, 1), \quad (0, 0, 0, 1)$$

are linearly independent in $\mathbb{R}^4$.

1.42. Are the vectors

$$(1, 2, 1), \quad (1, 2, 2), \quad (1, 2, 3), \quad (1, 2, 4)$$

linearly dependent or independent? What theorem of this section is particularly applicable to them?

1.43. Let

$$\mathbf{U} = (3,7,6,9,4),$$
$$\mathbf{V} = (2,7,0,1,-5).$$

Show that the vector

$$(-\tfrac{1}{2}, -\tfrac{7}{2}, 3, \tfrac{7}{2}, 7)$$

is a linear combination of $\mathbf{U}$ and $\mathbf{V}$.

1.44. The set of points (x,y,z) in $\mathbb{R}^3$ satisfying

$$-1 \le x \le 1, \quad -1 \le y \le 1, \quad -1 \le z \le 1,$$

is a solid cube. Write out the coordinates of the eight corner points of the cube. Is there a linear function that is equal to 8 on every corner?

1.45. A function ℓ from $\mathbb{R}^n$ to $\mathbb{R}$ is defined by

$$\ell(\mathbf{U}) = c_1 u_1 + c_2 u_2 + \cdots + c_n u_n$$

where $\mathbf{U} = (u_1, u_2, \dots, u_n)$ and $c_1, \dots, c_n$ are numbers.

(a) Show that $\ell(c\mathbf{U}) = c\ell(\mathbf{U})$ for all vectors $\mathbf{U}$ and numbers c.
(b) Show that for all vectors $\mathbf{U}$ and $\mathbf{V}$ in $\mathbb{R}^n$, $\ell(\mathbf{U}+\mathbf{V}) = \ell(\mathbf{U}) + \ell(\mathbf{V})$.

1.46. Let $\mathbf{U} = (1,3,1)$ and $\mathbf{V} = (2,2,2)$. Express $\mathbf{W} = (3,5,3)$ as a linear combination of $\mathbf{U}$ and $\mathbf{V}$.

1.47. Show that the vectors $(1,1,1)$, $(1,2,3)$, and $(3,2,1)$ are linearly dependent.

1.48. Let P be the set of all the points $(x,y,0,0)$, and let Q be the set of all points $(0,0,z,w)$ in $\mathbb{R}^4$. P and Q are two-dimensional planes in $\mathbb{R}^4$. How many points do P and Q have in common?

1.49. Let $\mathbf{X} = (x_1, x_2, \dots, x_n)$ and $\mathbf{Y} = (y_1, y_2, \dots, y_n)$ in $\mathbb{R}^n$. Which functions are bilinear? Of the bilinear functions, which are symmetric: $b(\mathbf{X}, \mathbf{Y}) = b(\mathbf{Y}, \mathbf{X})$? Which ones are antisymmetric: $b(\mathbf{X}, \mathbf{Y}) = -b(\mathbf{Y}, \mathbf{X})$?

(a) $b(\mathbf{X}, \mathbf{Y}) = x_1 y_n$
(b) $b(\mathbf{X}, \mathbf{Y}) = x_1 y_n - x_n y_1$
(c) $b(\mathbf{X}, \mathbf{Y}) = \sqrt{x_1^2 + x_2^2 + \cdots + x_n^2} \sqrt{y_1^2 + y_2^2 + \cdots + y_n^2}$
(d) $b(\mathbf{X}, \mathbf{Y}) = x_1 y_1 + x_2 y_2 + \cdots + x_n y_n$

1.50. Let $\mathbf{U} = (u_1, u_2, u_3, u_4)$, $\mathbf{V} = (v_1, v_2, v_3, v_4)$, and $\mathbf{W} = (w_1, w_2, w_3, w_4)$. Which of the following functions f have the antisymmetry property

$$f(\mathbf{U}, \mathbf{V}, \mathbf{W}) = -f(\mathbf{V}, \mathbf{U}, \mathbf{W})?$$

(a) $f(\mathbf{U}, \mathbf{V}, \mathbf{W}) = u_1 v_1 w_1$
(b) $f(\mathbf{U}, \mathbf{V}, \mathbf{W}) = u_1 w_3 - v_1 w_2$
(c) $f(\mathbf{U}, \mathbf{V}, \mathbf{W}) = (u_2 v_3 - u_3 v_2) w_4$

1.51. Let $\mathbf{U} = (u_1, u_2, u_3)$ and $\mathbf{V} = (v_1, v_2, v_3)$. Which bilinear functions b have the symmetry property

$$b(\mathbf{U}, \mathbf{V}) = b(\mathbf{V}, \mathbf{U})?$$

(a) $b(\mathbf{U}, \mathbf{V}) = 10u_1v_1$
(b) $b(\mathbf{U}, \mathbf{V}) = u_1v_2 - u_2v_1$
(c) $b(\mathbf{U}, \mathbf{V}) = u_1v_3 + u_2v_2 + 10u_3v_1$
(d) $b(\mathbf{U}, \mathbf{V}) = u_1v_3 + 10u_2v_2 + u_3v_1$

1.52. Let $\mathbf{U} = (u_1, u_2, u_3)$ and $\mathbf{V} = (v_1, v_2, v_3)$. Which bilinear functions b have the antisymmetry property

$$b(\mathbf{U}, \mathbf{V}) = -b(\mathbf{V}, \mathbf{U})?$$

(a) $b(\mathbf{U}, \mathbf{V}) = 10u_1v_1$
(b) $b(\mathbf{U}, \mathbf{V}) = u_1v_2 - u_2v_1$
(c) $b(\mathbf{U}, \mathbf{V}) = u_1v_3 + u_2v_2 + 10u_3v_1$
(d) $b(\mathbf{U}, \mathbf{V}) = u_1v_3 + 10u_2v_2 + u_3v_1$

1.5 Norm and dot product in n dimensions

We define now, in analogy with two-dimensional vectors, the norm of a vector in $\mathbb{R}^n$, and the dot product of two vectors in $\mathbb{R}^n$.

Definition 1.12. The *norm* of a vector $\mathbf{U} = (u_1, u_2, \ldots, u_n)$ is defined as

$$\|\mathbf{U}\| = \sqrt{u_1^2 + u_2^2 + \cdots + u_n^2}. \tag{1.19}$$

The zero vector $\mathbf{0}$ has norm zero, and conversely only the zero vector has norm zero. As in $\mathbb{R}^2$ we think of the norm of $\mathbf{U}$ as the length of $\mathbf{U}$ or the distance from the origin to the point $\mathbf{U}$.

Definition 1.13. We define the *dot product*, $\mathbf{U} \cdot \mathbf{V}$, of $\mathbf{U} = (u_1, u_2, \ldots, u_n)$ and $\mathbf{V} = (v_1, v_2, \ldots, v_n)$ as

$$\mathbf{U} \cdot \mathbf{V} = u_1v_1 + u_2v_2 + \cdots + u_nv_n. \tag{1.20}$$

We ask you in Problem 1.53 to verify that the dot product is distributive and commutative:

$$\mathbf{U} \cdot (\mathbf{V} + \mathbf{W}) = \mathbf{U} \cdot \mathbf{V} + \mathbf{U} \cdot \mathbf{W}, \qquad \mathbf{U} \cdot \mathbf{V} = \mathbf{V} \cdot \mathbf{U}.$$

The dot product of a vector with itself is its norm squared:

$$\mathbf{U} \cdot \mathbf{U} = \|\mathbf{U}\|^2.$$

Theorem 1.10 can be restated as follows:
Every linear function ℓ from $\mathbb{R}^n$ to $\mathbb{R}$ can be expressed as

$$\ell(\mathbf{U}) = \mathbf{C} \cdot \mathbf{U},$$

where $\mathbf{C}$ is some vector in $\mathbb{R}^n$.

Inequalities. The relation in $\mathbb{R}^2$

$$\|\mathbf{U} - \mathbf{V}\|^2 = \|\mathbf{U}\|^2 + \|\mathbf{V}\|^2 - 2\mathbf{U} \cdot \mathbf{V} \qquad (1.21)$$

was derived using only the distributive and commutative rules for the dot product. Therefore it holds for vectors in $\mathbb{R}^n$ as well. We show you in Problem 1.59 a different proof.

Next we prove a very useful inequality that compares $\mathbf{U} \cdot \mathbf{V}$ to $\|\mathbf{U}\| \|\mathbf{V}\|$.

Theorem 1.12. The Cauchy–Schwarz inequality. *Let* $\mathbf{U}$ *and* $\mathbf{V}$ *be vectors in* $\mathbb{R}^n$. *Then*

$$|\mathbf{U} \cdot \mathbf{V}| \le \|\mathbf{U}\| \|\mathbf{V}\|. \qquad (1.22)$$

Example 1.13. We have

$$(1,1,0) \cdot (0,1,1) = 1, \qquad (1,-1,0) \cdot (0,1,1) = -1,$$

and in both cases the absolute value of the dot product is less than the product of norms, $\sqrt{2}\,\sqrt{2}$.   □

Proof. If $\mathbf{U}$ is the zero vector then both sides of the inequality are zero, so it holds in that case. Since the square is nonnegative

$$0 \le \|\mathbf{V} - (\mathbf{U} \cdot \mathbf{V})\mathbf{U}\|^2. \qquad (1.23)$$

Using (1.21) rewrite the right side of (1.23) to get

$$0 \le \|\mathbf{V}\|^2 + \|(\mathbf{U} \cdot \mathbf{V})\mathbf{U}\|^2 - 2(\mathbf{U} \cdot \mathbf{V})^2$$

$$= \|\mathbf{V}\|^2 + (\mathbf{U} \cdot \mathbf{V})^2 \|\mathbf{U}\|^2 - 2(\mathbf{U} \cdot \mathbf{V})^2.$$

If $\mathbf{U}$ is a unit vector, $\|\mathbf{U}\| = 1$ and we get

$$0 \le \|\mathbf{V}\|^2 - (\mathbf{U} \cdot \mathbf{V})^2$$

so $(\mathbf{U} \cdot \mathbf{V})^2 \le \|\mathbf{V}\|^2$. Taking square roots we get that for unit vectors $\mathbf{U}$,

$$|\mathbf{U} \cdot \mathbf{V}| \le \|\mathbf{V}\|.$$

Next if $\mathbf{U}$ is a nonzero vector, $\frac{1}{\|\mathbf{U}\|}\mathbf{U}$ is a unit vector, so

$$\left|\frac{1}{\|\mathbf{U}\|}\mathbf{U} \cdot \mathbf{V}\right| \le \|\mathbf{V}\|.$$

Multiply by $\|\mathbf{U}\|$ to get (1.22).   □

We ask you in Problem 1.64 to find out when equality holds in the Cauchy–Schwarz inequality. An important consequence of the Cauchy–Schwarz inequality is the triangle inequality for vectors in $\mathbb{R}^n$.

Theorem 1.13. Triangle inequality *If* $\mathbf{U}$ *and* $\mathbf{V}$ *are vectors in* $\mathbb{R}^n$ *then*

$$\|\mathbf{U} + \mathbf{V}\| \le \|\mathbf{U}\| + \|\mathbf{V}\|.$$

Proof. We have seen in equation (1.21) that

$$\|\mathbf{U} - \mathbf{V}\|^2 = \|\mathbf{U}\|^2 + \|\mathbf{V}\|^2 - 2\mathbf{U} \cdot \mathbf{V}.$$

Replace $\mathbf{V}$ by $-\mathbf{V}$ and use the Cauchy–Schwarz inequality to get

$$\|\mathbf{U} + \mathbf{V}\|^2 = \|\mathbf{U}\|^2 + \|\mathbf{V}\|^2 + 2\mathbf{U} \cdot \mathbf{V}$$

$$\le \|\mathbf{U}\|^2 + \|\mathbf{V}\|^2 + 2\|\mathbf{U}\|\|\mathbf{V}\| = (\|\mathbf{U}\| + \|\mathbf{V}\|)^2.$$

Take the square root to get $\|\mathbf{U} + \mathbf{V}\| \le \|\mathbf{U}\| + \|\mathbf{V}\|$ as asserted.   □

Using the Cauchy–Schwarz inequality we can define the *angle* between two nonzero vectors $\mathbf{U}$ and $\mathbf{V}$ in $\mathbb{R}^n$ as follows. From the Cauchy–Schwarz inequality we see that for $\mathbf{U}$ and $\mathbf{V}$ nonzero

$$-1 \le \frac{\mathbf{U} \cdot \mathbf{V}}{\|\mathbf{U}\|\|\mathbf{V}\|} \le 1.$$

We define the angle θ between $\mathbf{U}$ and $\mathbf{V}$ to be

$$\theta = \cos^{-1}\left(\frac{\mathbf{U} \cdot \mathbf{V}}{\|\mathbf{U}\|\|\mathbf{V}\|}\right), \qquad 0 \le \theta \le \pi,$$

or $\cos\theta = \dfrac{\mathbf{U} \cdot \mathbf{V}}{\|\mathbf{U}\|\|\mathbf{V}\|}$. Using this definition we can rewrite the formula (1.21)

$$\|\mathbf{U} - \mathbf{V}\|^2 = \|\mathbf{U}\|^2 + \|\mathbf{V}\|^2 - 2\mathbf{U} \cdot \mathbf{V}$$

as

$$\|\mathbf{U} - \mathbf{V}\|^2 = \|\mathbf{U}\|^2 + \|\mathbf{V}\|^2 - 2\|\mathbf{U}\|\|\mathbf{V}\|\cos\theta.$$

$\|\mathbf{U} - \mathbf{V}\|$, $\|\mathbf{U}\|$, and $\|\mathbf{V}\|$ are the lengths of sides of the triangle $\mathbf{0UV}$ in Figure 1.15. Thus (1.21) is the Law of Cosines in $\mathbb{R}^n$.

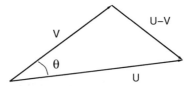

Fig. 1.15 For $\mathbf{U}$ and $\mathbf{V}$ in $\mathbb{R}^n$, $\|\mathbf{U} - \mathbf{V}\|^2 = \|\mathbf{U}\|^2 + \|\mathbf{V}\|^2 - 2\|\mathbf{U}\|\|\mathbf{V}\|\cos\theta$.

Orthonormal set. We have used the set of coordinate vectors $\mathbf{E}_j$ in $\mathbb{R}^n$ whose j-th component is 1 and all other components are zero.

$$\mathbf{E}_1 = (1,0,0,0,\ldots,0)$$
$$\mathbf{E}_2 = (0,1,0,0,\ldots,0)$$
$$\mathbf{E}_3 = (0,0,1,0,\ldots,0)$$
$$\vdots$$
$$\mathbf{E}_n = (0,0,0,0,\ldots,1)$$

Every vector $\mathbf{U} = (u_1, u_2, \ldots, u_n)$ in $\mathbb{R}^n$ can be written as a linear combination of these vectors,

$$\mathbf{U} = u_1\mathbf{E}_1 + u_2\mathbf{E}_2 + \cdots + u_n\mathbf{E}_n.$$

The list of vectors $\mathbf{E}_j$ is called the *standard basis* in $\mathbb{R}^n$.

Definition 1.14. A list of vectors $\mathbf{V}_1, \mathbf{V}_2, \ldots, \mathbf{V}_m$ in $\mathbb{R}^n$ is called an *orthonormal* set when two properties hold:

(a) The vectors $\mathbf{V}_j$ are pairwise orthogonal:

$$\mathbf{V}_j \cdot \mathbf{V}_k = 0 \text{ for all } j \neq k.$$

(a) Each vector $\mathbf{V}_j$ has norm 1:

$$\|\mathbf{V}_j\| = 1 \text{ for all } j.$$

The standard basis of $\mathbb{R}^n$ is an orthonormal set.

Example 1.14. The vectors

$$\mathbf{Q}_1 = (1,1,1,1), \quad \mathbf{Q}_2 = (1,1,-1,-1),$$

$$\mathbf{Q}_3 = (1,-1,1,-1), \quad \mathbf{Q}_4 = (-1,1,1,-1)$$

are pairwise orthogonal. For example, $\mathbf{Q}_2 \cdot \mathbf{Q}_3 = 1 - 1 - 1 + 1 = 0$. Each has norm 2. For example, $\|\mathbf{Q}_4\| = \sqrt{1 + 1 + 1 + 1} = 2$. Dividing by the norms therefore gives unit vectors $\mathbf{V}_j = \frac{1}{2}\mathbf{Q}_j$;

$$\mathbf{V}_1 = (\tfrac{1}{2}, \tfrac{1}{2}, \tfrac{1}{2}, \tfrac{1}{2}), \quad \mathbf{V}_2 = (\tfrac{1}{2}, \tfrac{1}{2}, -\tfrac{1}{2}, -\tfrac{1}{2}),$$

$$\mathbf{V}_3 = (\tfrac{1}{2}, -\tfrac{1}{2}, \tfrac{1}{2}, -\tfrac{1}{2}), \quad \mathbf{V}_4 = (-\tfrac{1}{2}, \tfrac{1}{2}, \tfrac{1}{2}, -\tfrac{1}{2})$$

are an orthonormal set in $\mathbb{R}^4$. $\square$

We show now that there are many more orthonormal sets. The basic result is the following.

Theorem 1.14. *Let $n \geq 2$ and $k < n$. Let $\mathbf{W}_1, \mathbf{W}_2, \ldots, \mathbf{W}_k$ be vectors in $\mathbb{R}^n$. Then there is a nonzero vector $\mathbf{V}$ orthogonal to each of the vectors $\mathbf{W}_i$, $i = 1, 2, \ldots, k$.*

Proof. We argue by induction on n. The case $n = 2$ is simple. If $\mathbf{W}_1$ is the zero vector we may take any vector for $\mathbf{V}$. If $\mathbf{W}_1 = (a, b) \neq \mathbf{0}$ we take $\mathbf{V} = (-b, a)$. Suppose inductively that the theorem holds for $n - 1$, where $n \geq 3$. The desired relations $\mathbf{W}_j \cdot \mathbf{V} = 0$, $j = 1, 2, \ldots, k$, form a system of k linear equations for the n components $v_1, v_2, \ldots, v_n$ of $\mathbf{V}$:

$$w_{j1}v_1 + w_{j2}v_2 + \cdots + w_{jn}v_n = 0, \quad j = 1, 2, \ldots, k. \tag{1.24}$$

We look at the last terms on the left, $w_{jn}v_n$. If all the numbers w_{jn}, $j = 1, 2, \ldots, k$ are zero, then we can satisfy these equations by choosing $v_1, v_2, \ldots, v_{n-1}$ equal to zero and v_n equal to 1. If one of the numbers w_{jn} is nonzero, we use the j-th equation in (1.24) to express v_n as a linear combination of $v_1, v_2, \ldots, v_{n-1}$. In case $k = 1$ there is no further equation and $v_1, v_2, \ldots, v_{n-1}$ may be chosen arbitrarily; otherwise set the expression for v_n into the remaining equations (1.24) we get a system of $k - 1$ equations of form (1.24) for the $n - 1$ unknowns $v_1, v_2, \ldots, v_{n-1}$. By the induction hypothesis this system has a nonzero solution. $\square$

We use Theorem 1.14 to construct many orthonormal sets $\mathbf{V}_1, \mathbf{V}_2, \ldots, \mathbf{V}_n$ in $\mathbb{R}^n$: Choose $\mathbf{V}_1$ as any vector of norm 1. According to Theorem 1.14 there is a nonzero vector, call it $\mathbf{V}_2$, orthogonal to $\mathbf{V}_1$. Again using Theorem 1.14 with $k = 2$, there is a nonzero vector $\mathbf{V}_3$ that is orthogonal to both $\mathbf{V}_1$ and $\mathbf{V}_2$. Proceeding in this way we construct a set of n nonzero vectors $\mathbf{V}_j$ that are pairwise orthogonal. Then we multiply each vector with a suitable number to make a vector of norm 1.

Theorem 1.15. *If vectors $\mathbf{V}_1, \ldots, \mathbf{V}_m$ are an orthonormal set in $\mathbb{R}^n$, then they are linearly independent.*

Proof. Suppose there is a linear relation

$$c_1 \mathbf{V}_1 + \cdots + c_m \mathbf{V}_m = \mathbf{0}.$$

Take the dot product with $\mathbf{V}_i$; since $\mathbf{V}_i \cdot \mathbf{V}_j = 0$ when $j \neq i$, and $\mathbf{V}_i \cdot \mathbf{V}_i = 1$, we get $c_i = 0$. Since this holds for each i this proves the linear independence.   □

By Theorem 1.9 every vector $\mathbf{X}$ in $\mathbb{R}^n$ can be expressed as a linear combination of the linearly independent vectors $\mathbf{V}_1, \ldots, \mathbf{V}_n$

$$\mathbf{X} = c_1 \mathbf{V}_1 + \cdots + c_n \mathbf{V}_n.$$

To find the c_j when the $\mathbf{V}_j$ are orthonormal, dot both sides of the equation with $\mathbf{V}_1$ to see that

$$\mathbf{V}_1 \cdot \mathbf{X} = \mathbf{V}_1 \cdot (c_1 \mathbf{V}_1 + \cdots + c_n \mathbf{V}_n) = c_1 \|\mathbf{V}_1\|^2 + c_2 \mathbf{V}_1 \cdot \mathbf{V}_2 + \cdots = c_1.$$

Similarly each $c_j = \mathbf{V}_j \cdot \mathbf{X}$. Thus we have proved the following theorem.

Theorem 1.16. *Let* $\mathbf{V}_1, \ldots, \mathbf{V}_n$ *be an orthonormal set of vectors in* $\mathbb{R}^n$. *Then every vector* $\mathbf{X}$ *in* $\mathbb{R}^n$ *can be expressed as the linear combination*

$$\mathbf{X} = (\mathbf{V}_1 \cdot \mathbf{X})\mathbf{V}_1 + \cdots + (\mathbf{V}_n \cdot \mathbf{X})\mathbf{V}_n. \qquad (1.25)$$

Example 1.15. To write $\mathbf{X} = (1, 2, 3, 4)$ as a linear combination of the set of orthonormal vectors $\mathbf{V}_1, \mathbf{V}_2, \mathbf{V}_3, \mathbf{V}_4$ in Example 1.14, we find

$$c_1 = \mathbf{V}_1 \cdot \mathbf{X} = \tfrac{1}{2}(1) + \tfrac{1}{2}(2) + \tfrac{1}{2}(3) + \tfrac{1}{2}(4) = 5$$
$$c_2 = \mathbf{V}_2 \cdot \mathbf{X} = -2$$
$$c_3 = \mathbf{V}_3 \cdot \mathbf{X} = -1$$
$$c_4 = \mathbf{V}_4 \cdot \mathbf{X} = 0$$

and so

$$\mathbf{X} = (1, 2, 3, 4) = \sum_{j=1}^{4} (\mathbf{V}_j \cdot \mathbf{X})\mathbf{V}_j = 5\mathbf{V}_1 - 2\mathbf{V}_2 - \mathbf{V}_3.$$

                                                                                                        □

Problems

1.53. Denote the dot product function $b(\mathbf{U}, \mathbf{V}) = \mathbf{U} \cdot \mathbf{V}$. Verify that the dot product is distributive and commutative and show that it is a symmetric bilinear function by verifying the following properties:

(a) $b(\mathbf{U}, \mathbf{V})$ is bilinear
(b) $b(\mathbf{U}, V) = b(\mathbf{V}, \mathbf{U})$

1.54. Let **U**, **V**, **W**, and **X** be the points on the rectangular box in $\mathbb{R}^3$ shown in Figure 1.16. Find the following distances:

(a) $\|\mathbf{V} - \mathbf{U}\|$,
(b) the distance between **U** and the origin,
(c) the distance between **U** and **V** + **W**,
(d) $\|\mathbf{W} - \mathbf{X}\|$.

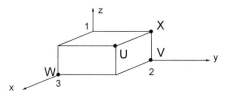

Fig. 1.16 The points used in Problem 1.54.

1.55. Find a vector **W** in $\mathbb{R}^5$ orthogonal to all three of the vectors

$$(1,2,0,0,-2), \quad (-2,1,2,0,0), \quad (0,-2,1,0,2).$$

1.56. Which of the following vectors are unit vectors?

(a) $\frac{1}{50}(3,4,5)$,
(b) $-\mathbf{U}$ if **U** is,
(c) $(-u_1, u_2, -u_3, u_4, -u_5, u_6)$, if $\mathbf{U} = (u_1, u_2, u_3, u_4, u_5, u_6)$ is.
(d) $\frac{1}{3}(1, -\sqrt{2}, \sqrt{3}, -\sqrt{3})$.

1.57. Determine whether the following pairs of vectors are orthogonal to each other.

(a) $(1,1,1,1,1)$ and $(-1,-1,-1,-1,-1)$,
(b) $(1,1,1,1)$ and $(-1,-1,-1,3)$,
(c) $(1,1,1)$ and $(-1,2,-1)$.

1.58. Show that if **X** and **Y** are orthogonal to each other in $\mathbb{R}^n$ then

$$\|\mathbf{X} + \mathbf{Y}\|^2 = \|\mathbf{X}\|^2 + \|\mathbf{Y}\|^2.$$

This is sometimes called the *Pythagorean theorem* in $\mathbb{R}^n$.

1.59. Let u and v be numbers. Use the algebraic identity

$$(u - v)^2 = u^2 - 2uv + v^2$$

n times to prove that for all vectors **U** and **V** in $\mathbb{R}^n$,

$$\|\mathbf{U} - \mathbf{V}\|^2 = \|\mathbf{U}\|^2 + \|\mathbf{V}\|^2 - 2\mathbf{U} \cdot \mathbf{V}.$$

1.60. Find the maximum norm of $\mathbf{X} = (x_1, x_2, \ldots, x_{100})$ in the unit cube in $\mathbb{R}^{100}$,

$$0 \le x_1 \le 1, \quad 0 \le x_2 \le 1, \quad \ldots \quad 0 \le x_{100} \le 1.$$

1.61. Imagine an n-dimensional cube in $\mathbb{R}^n$ with edge length $c > 0$, consisting of all the points $\mathbf{X} = (x_1, x_2, \ldots, x_n)$ with

$$0 \le x_k \le c, \qquad k = 1, \ldots, n.$$

(a) Find the point $\mathbf{P}$ in the cube that has the largest norm. Call $\mathbf{P}$ the far corner of the cube.
(b) For what value of c is the far corner $\mathbf{P}$ on the unit sphere of $\mathbb{R}^n$, $\|\mathbf{X}\| = 1$?
(c) Keeping the far corner point on the unit sphere, what happens to the edge length c as the dimension n tends to infinity?

1.62. Let $\mathbf{C} = (c_1, \ldots, c_n)$ be a given vector in $\mathbb{R}^n$ and let $\mathbf{X} = (x_1, \ldots, x_n)$. Show that the function

$$f(\mathbf{X}) = x_1 + \mathbf{C} \cdot \mathbf{X}$$

is linear, by finding $\mathbf{D} = (d_1, \ldots, d_n)$ to express f in the form $f(\mathbf{X}) = \mathbf{D} \cdot \mathbf{X}$.

1.63. Let $\mathbf{W}_1 = (1, 1, 1, 0)$ and $\mathbf{W}_2 = (0, 1, 1, 1)$. Find two linearly independent vectors that are orthogonal to both $\mathbf{W}_1$ and $\mathbf{W}_2$.

1.64. Our proof of the Cauchy–Schwarz inequality, Theorem 1.12, used that when $\mathbf{U}$ is a unit vector,
$$0 \le \|\mathbf{V} - (\mathbf{U} \cdot \mathbf{V})\mathbf{U}\|^2 = \|\mathbf{V}\|^2 - (\mathbf{U} \cdot \mathbf{V})^2.$$

(a) Show that if $\mathbf{U}$ is a unit vector and $|\mathbf{U} \cdot \mathbf{V}| = \|\mathbf{U}\| \|\mathbf{V}\|$, then $\mathbf{V} = (\mathbf{U} \cdot \mathbf{V})\mathbf{U}$.
(b) Show that equality occurs in the Cauchy–Schwarz inequality for two arbitrary vectors $\mathbf{V}$ and $\mathbf{W}$ only if one of the vectors is a multiple of the other vector.

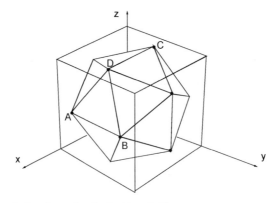

Fig. 1.17 The icosahedron in a cube, for Problem 1.65.

1.65. The regular icosahedron fits nicely into a cube, as shown in Figure 1.17. It has twenty equilateral triangle faces. In the cube, $0 \le x \le 2$, $0 \le y \le 2$, and $0 \le z \le 2$. Points **A** and **B** are located on the face of the cube where $x = 2$, and are equally spaced from the center of that face, so $\mathbf{A} = (2, 1 - h, 1)$ and $\mathbf{B} = (2, 1 + h, 1)$ for some number $h > 0$.

(a) Find the coordinates of points **C** and **D** in terms of h.
(b) Express the distance between **A** and **B** in terms of h.
(c) Express the distance between **A** and **D** in terms of h.
(d) Find h.

1.66. Let

$$\mathbf{V}_1 = (a, b, \ldots, b), \ \mathbf{V}_2 = (b, a, b, \ldots, b), \ \mathbf{V}_3 = (b, b, a, \ldots, b), \ \ldots \ \mathbf{V}_n = (b, \ldots, b, a)$$

be n vectors in $\mathbb{R}^n$, where $n > 1$. Find numbers a and b so that $\mathbf{V}_1, \ldots, \mathbf{V}_n$ is an orthonormal set.

1.67. Use the triangle inequality as needed to prove the following inequalities, where a and b are numbers and **X** and **Y** are vectors in $\mathbb{R}^n$.

(a) $|a| \le |a - b| + |b|$
(b) $|a| - |b| \le |a - b|$
(c) $\big||a| - |b|\big| \le |a - b|$
(d) $\big|\|\mathbf{X}\| - \|\mathbf{Y}\|\big| \le \|\mathbf{X} - \mathbf{Y}\|$

1.6 The determinant

The determinant of order n is a number valued function of an ordered list of n vectors $\mathbf{V}_1, \mathbf{V}_2, \ldots, \mathbf{V}_n$, each with n components. We denote it as

$$\det(\mathbf{V}_1, \mathbf{V}_2, \ldots, \mathbf{V}_n).$$

Before giving the formula for the determinant we list its algebraic properties:

(i) $\det(\mathbf{V}_1, \mathbf{V}_2, \ldots, \mathbf{V}_n)$ is a *multilinear function*, that is, a linear function of each $\mathbf{V}_i$ when all other $\mathbf{V}_j, j \ne i$ are held fixed.

(ii) If two vectors $\mathbf{V}_i$ and $\mathbf{V}_j$ of the ordered list are equal, the value of the determinant is zero:

$$\det(\ldots \mathbf{V}, \ldots, \mathbf{V}, \ldots) = 0.$$

(iii) $\det(\mathbf{E}_1, \mathbf{E}_2, \ldots, \mathbf{E}_n) = 1$, where $\mathbf{E}_i$ is the vector whose i-th component is 1 and all other components are zero.

We deduce three further properties from the properties above:

(iv) If two of the vectors in the ordered list are interchanged, the value of the determinant is multiplied by -1.

Proof. Using repeatedly properties (i) and (ii) of the determinant we get the following sequence of identities, where we only indicate the i-th and j-th vectors $\mathbf{U}$ and $\mathbf{V}$ of the ordered list:

$$0 = \det(\mathbf{U}+\mathbf{V},\mathbf{U}+\mathbf{V}) = \det(\mathbf{U},\mathbf{U})+\det(\mathbf{U},\mathbf{V})+\det(\mathbf{V},\mathbf{U})+\det(\mathbf{V},\mathbf{V}).$$

Therefore

$$0 = \det(\mathbf{U},\mathbf{V})+\det(\mathbf{V},\mathbf{U}).$$

This proves property (iv). □

(v) If $\mathbf{V}_1,\mathbf{V}_2,\ldots,\mathbf{V}_n$ are linearly dependent, then $\det(\mathbf{V}_1,\mathbf{V}_2,\ldots,\mathbf{V}_n)=0$.

Proof. If the $\mathbf{V}_i$ are linearly dependent, one of the $\mathbf{V}_i$, say $\mathbf{V}_1$, is a linear combination of the others:

$$\mathbf{V}_1 = \sum_{i=2}^{n} m_j \mathbf{V}_j$$

Using the multilinear property of the determinant we have

$$\det(\mathbf{V}_1,\mathbf{V}_2,\ldots,\mathbf{V}_n) = \det\left(\sum_{i=2}^{n} m_j \mathbf{V}_j,\mathbf{V}_2,\ldots,\mathbf{V}_n\right) = \sum_{i=2}^{n} m_j \det(\mathbf{V}_j,\mathbf{V}_2,\ldots,\mathbf{V}_n).$$

In each term of the sum on the right side of the last equation two of the vectors in det are equal. Therefore by property (ii) each term of the sum is zero. It follows that the whole sum is zero. This proves property (v). □

Next we show the converse of property (v).
(vi) If the vectors $\mathbf{V}_1,\mathbf{V}_2,\ldots,\mathbf{V}_n$ are linearly independent, $\det(\mathbf{V}_1,\mathbf{V}_2,\ldots,\mathbf{V}_n)$ is not zero.

Proof. Since the vectors $\mathbf{V}_1,\mathbf{V}_2,\ldots,\mathbf{V}_n$ are linearly independent, according to Theorem 1.9 we can express every vector as a linear combination of them. In particular the unit coordinate vectors $\mathbf{E}_i$ can be expressed:

$$\mathbf{E}_i = \sum_{j=1}^{n} b_{ji}\mathbf{V}_j.$$

Since the determinant is a multilinear function, we write

$$\det(\mathbf{E}_1,\mathbf{E}_2,\ldots,\mathbf{E}_n) = \det\left(\sum_{j=1}^{n} b_{j1}\mathbf{V}_j,\mathbf{E}_2,\ldots,\mathbf{E}_n\right) = \sum_{j=1}^{n} b_{j1}\det(\mathbf{V}_j,\mathbf{E}_2,\ldots,\mathbf{E}_n)$$

$$= \sum_{j=1}^{n} b_{j1}\det\left(\mathbf{V}_j,\sum_{k=1}^{n} b_{k2}\mathbf{V}_k,\mathbf{E}_3,\ldots,\mathbf{E}_n\right) = \sum_{j=1}^{n} b_{j1}\sum_{k=1}^{n} b_{k2}\det(\mathbf{V}_j,\mathbf{V}_k,\mathbf{E}_3,\ldots,\mathbf{E}_n)$$

$$(1.26)$$

Continuing in this fashion we get an expression of $\det(\mathbf{E}_1,\mathbf{E}_2,\ldots,\mathbf{E}_n)$ as a linear combination of the determinants $\det(\mathbf{V}_j,\mathbf{V}_k,\ldots,\mathbf{V}_z)$ where $\mathbf{V}_j,\mathbf{V}_k,\ldots,\mathbf{V}_z$ is a permutation of $\mathbf{V}_1,\mathbf{V}_2,\ldots,\mathbf{V}_n$ and therefore

$$\det(\mathbf{E}_1,\mathbf{E}_2,\ldots,\mathbf{E}_n) = \sum_{\text{permutations } jk\cdots z} (b_{j1}b_{k2}\cdots b_{zn})\det(\mathbf{V}_j,\mathbf{V}_k,\ldots,\mathbf{V}_z).$$

But for each permutation, either

$$\det(\mathbf{V}_j,\mathbf{V}_k,\ldots,\mathbf{V}_z) = \det(\mathbf{V}_1,\mathbf{V}_2,\ldots,\mathbf{V}_n)$$

or

$$\det(\mathbf{V}_j,\mathbf{V}_k,\ldots,\mathbf{V}_z) = -\det(\mathbf{V}_1,\mathbf{V}_2,\ldots,\mathbf{V}_n)$$

Therefore $\det(\mathbf{E}_1,\mathbf{E}_2,\ldots,\mathbf{E}_n)$ is a multiple of $\det(\mathbf{V}_1,\mathbf{V}_2,\ldots,\mathbf{V}_n)$. Since

$$\det(\mathbf{E}_1,\mathbf{E}_2,\ldots,\mathbf{E}_n) = 1$$

it follows that $\det(\mathbf{V}_1,\mathbf{V}_2,\ldots,\mathbf{V}_n)$ is not zero. This proves property (vi). □

We shall show that properties (i), (ii), and (iii) completely characterize the determinant. First we derive from these three properties a formula that holds for every function satisfying the properties. Then we show that a function defined by this formula satisfies the three properties.

To use properties (i), (ii), and (iii) of det to devise a formula for the determinant, we take an ordered list of n vectors $\mathbf{V}_1,\mathbf{V}_2,\ldots,\mathbf{V}_n$ in $\mathbb{R}^n$. Denote the k-th component of $\mathbf{V}_j$ by v_{kj}, $(j = 1,2,\ldots,n;\ k = 1,2,\ldots,n)$. We write each vector $\mathbf{V}_j$ as a linear combination of the unit vectors $\mathbf{E}_k$:

$$\mathbf{V}_j = \sum_{k=1}^{n} v_{kj}\mathbf{E}_k. \tag{1.27}$$

Using the above expression for $\mathbf{V}_1$ we write

$$\det(\mathbf{V}_1,\mathbf{V}_2,\ldots,\mathbf{V}_n) = \det\left(\sum_{k=1}^{n} v_{k1}\mathbf{E}_k,\mathbf{V}_2,\ldots,\mathbf{V}_n\right).$$

Using multilinearity of the determinant we can rewrite the right side as

$$\sum_{k=1}^{n} v_{k1}\det(\mathbf{E}_k,\mathbf{V}_2,\ldots,\mathbf{V}_n).$$

Next we use the expression (1.27) for $\mathbf{V}_2$ in each term of the sum above, and multilinearity to rewrite each term of this sum. We get the double sum

$$\det(\mathbf{V}_1,\mathbf{V}_2,\ldots,\mathbf{V}_n) = \sum_{k=1}^{n}\sum_{\ell=1}^{n} v_{k1}v_{\ell2}\det(\mathbf{E}_k,\mathbf{E}_\ell,\mathbf{V}_3,\ldots,\mathbf{V}_n).$$

Proceeding in this manner we get the formula as an n-tuple sum

$$\det(\mathbf{V}_1, \mathbf{V}_2, \ldots, \mathbf{V}_n) = \sum_{k,\ell,\ldots,z=1}^{n} (v_{k1} v_{\ell 2} \cdots v_{zn}) \det(\mathbf{E}_k, \mathbf{E}_\ell, \ldots, \mathbf{E}_z). \qquad (1.28)$$

Next we use properties (i), (ii), and (iii) to determine $\det(\mathbf{E}_k, \mathbf{E}_\ell, \ldots, \mathbf{E}_z)$.

According to property (ii) the determinant is zero when two of the vectors in the list are equal. This shows that in formula (1.28) we can restrict the summation to the case where $k, \ell, \ldots, z$ is a permutation of $1, 2, \ldots, n$.

Next we define the *signature* of a permutation as follows. Denote by

$$p = p_1 p_2 \cdots p_n$$

a permutation of $1, 2, \ldots, n$. That is,

$$p(1) = p_1, \qquad p(2) = p_2, \qquad \cdots \qquad p(n) = p_n.$$

Form the two products

$$\prod_{i<j} (x_{p_i} - x_{p_j}) \qquad (1.29)$$

and

$$\prod_{i<j} (x_i - x_j) \qquad (1.30)$$

Each factor in the product (1.29) is equal to one of the factors in the product (1.30) or its negative. Therefore the two products are equal or are the negatives of each other.

Definition 1.15. We define the *signature* of a permutation p of $1, 2, \ldots, n$ to be the number $s(p) = 1$ or -1, such that

$$\prod_{i<j} (x_{p_i} - x_{p_j}) = s(p) \prod_{i<j} (x_i - x_j). \qquad (1.31)$$

Signature $s(p)$ has following properties:

(a) The signature of an interchange is -1.
(b) The signature of the composite of two permutations is the product of their signatures:

$$s(pq) = s(p)s(q).$$

Proof. (a) When we interchange x_k and x_m, $(k < m)$, for every ℓ between k and m both $x_\ell - x_k$ and $x_m - x_\ell$ change sign, an even number of sign changes. In addition $x_m - x_k$ changes sign, an odd total number of sign changes.

Property (b) follows directly from the definition (1.31), as we ask you to show in Problem 1.75.   □

Every permutation p can be obtained as a composite of interchanges. Let $c(p)$ be the number of interchanges. If follows from properties (a) and (b) that

$$s(p) = (-1)^{c(p)}.$$

Example 1.16. Let $p = 312$ be a permutation of $1, 2, 3$, that is,

$$p_1 = 3, \quad p_2 = 1, \quad p_3 = 2.$$

Two interchanges bring 312 into 123: one interchange takes 312 to 132, and a second interchange takes 132 to 123. Therefore $s(p) = 1$. □

Example 1.17. One interchange takes 15342 to 12345, so $s(15342) = -1$. □

According to property (iv) a factor of -1 is introduced with each interchange of vectors, and using (iii) we get

$$\det(\mathbf{E}_{p_1}, \ldots, \mathbf{E}_{p_n}) = (-1)^{c(p)} \det(\mathbf{E}_1, \ldots, \mathbf{E}_n) = (-1)^{c(p)} = s(p).$$

Set this result into formula (1.28) for the determinant. We have shown that if a function satisfies the three properties (i), (ii), and (iii) then the value it assigns to $(\mathbf{V}_1, \ldots, \mathbf{V}_n)$ is

$$\sum_p s(p) v_{p_1 1} v_{p_2 2} \cdots v_{p_n n}$$

summed over all permutations $p = p_1 p_2 \cdots p_n$ of $1, 2, \ldots, n$. Therefore we make the following definition.

Definition 1.16. Let $\mathbf{V}_1, \ldots, \mathbf{V}_n$ be vectors in $\mathbb{R}^n$ written as columns,

$$\mathbf{V}_1 = \begin{bmatrix} v_{11} \\ v_{21} \\ \vdots \\ v_{n1} \end{bmatrix}, \quad \mathbf{V}_2 = \begin{bmatrix} v_{12} \\ v_{22} \\ \vdots \\ v_{n2} \end{bmatrix}, \quad \cdots \quad \mathbf{V}_n = \begin{bmatrix} v_{1n} \\ v_{2n} \\ \vdots \\ v_{nn} \end{bmatrix}$$

The *determinant* is defined as

$$\det(\mathbf{V}_1, \mathbf{V}_2, \ldots, \mathbf{V}_n) = \sum_p s(p) v_{p_1 1} v_{p_2 2} \cdots v_{p_n n} \tag{1.32}$$

where the sum is over all permutations $p = p_1 p_2 \cdots p_n$ of $1, 2, \ldots, n$.

The 2×2 case. For $n = 2$ let $p = 12$ and $q = 21$ be the permutations of two numbers $1, 2$. No interchanges are needed for p, so the signature $s(p) = 1$. One interchange takes 21 to 12 so $s(q) = -1$. For

$$\mathbf{V}_1 = \begin{bmatrix} v_{11} \\ v_{21} \end{bmatrix}, \quad \mathbf{V}_2 = \begin{bmatrix} v_{12} \\ v_{22} \end{bmatrix}$$

the determinant is

$$\det(\mathbf{V}_1, \mathbf{V}_2) = s(p)v_{p_11}v_{p_22} + s(q)v_{q_11}v_{q_22}$$
$$= s(12)v_{11}v_{22} + s(21)v_{21}v_{12}$$
$$= v_{11}v_{22} - v_{21}v_{12}$$

where the sum is over the two permutations of $1, 2$.

Example 1.18. $\det\left(\begin{bmatrix} 2 \\ -2 \end{bmatrix}, \begin{bmatrix} 5 \\ 1 \end{bmatrix}\right) = (2)(1) - (-2)(5) = 12.$ □

We verify now that the determinant as defined by formula (1.32) *has* the three algebraic properties listed at the beginning of this section:

(i) Each term in the sum on the right in (1.32) is a multilinear function of the $\mathbf{V}_j$. Therefore so is their sum.

(ii) In definition (1.32) interchange $\mathbf{V}_i$ and $\mathbf{V}_j$ where $i < j$. We get

$$\det(\ldots, \mathbf{V}_j, \ldots, \mathbf{V}_i, \ldots) = \sum_p s(p)(\cdots v_{p_ji} \cdots v_{p_ij} \cdots). \qquad (1.33)$$

Let r be the interchange of i and j, and denote the permutation pr as q. Since the factors v_{p_ji} and v_{p_ij} can be interchanged with no effect, we can rewrite (1.33) as a sum over all permutations q:

$$\det(\ldots, \mathbf{V}_j, \ldots, \mathbf{V}_i, \ldots) = \sum_q s(p)(v_{q_11}v_{q_22} \cdots v_{q_nn}) \qquad (1.34)$$

According to the product formula $s(q) = s(pr) = s(p)s(r)$. The signature $s(r)$ of an interchange r is -1. Therefore $s(q) = -s(p)$. Set this in formula (1.34); we get

$$\det(\ldots, \mathbf{V}_j, \ldots, \mathbf{V}_i, \ldots) = -\det(\ldots, \mathbf{V}_i, \ldots, \mathbf{V}_j, \ldots). \qquad (1.35)$$

In words: if two vectors in the list are interchanged, the value of det is multiplied by -1. It follows from (1.35) that if two of the vectors $\mathbf{V}_i$ and $\mathbf{V}_j$ are equal, $\det(\mathbf{V}_1, \ldots, \mathbf{V}_n) = 0$.

(iii) For $\mathbf{V}_j = \mathbf{E}_j$, $(j = 1, 2, \ldots, n)$, the sum on the right side of formula (1.32) for the determinant has only one nonzero term, corresponding to the trivial permutation $12 \cdots n$, and that term is equal to 1. This shows that

$$\det(\mathbf{E}_1, \mathbf{E}_2, \ldots, \mathbf{E}_n) = 1.$$

This completes the proof that the determinant defined by formula (1.32) has the three properties listed at the beginning of this section.

The components of n column vectors in $\mathbb{R}^n$ form a rectangular array of numbers called a matrix. (See also Section 1.8.) We write

$$\mathbf{M} = [\mathbf{V}_1 \; \mathbf{V}_2 \; \cdots \; \mathbf{V}_n], \qquad \det \mathbf{M} = \det(\mathbf{V}_1, \mathbf{V}_2, \ldots, \mathbf{V}_n).$$

We use Definition 1.16 to compute the determinant of $\mathbf{M}$. For example, using the vectors in Example 1.18,

$$\det \begin{bmatrix} 2 & 5 \\ -2 & 1 \end{bmatrix} = (2)(1) - (-2)(5) = 12.$$

The 3×3 case. Let $\mathbf{M} = \begin{bmatrix} m_{11} & m_{12} & m_{13} \\ m_{21} & m_{22} & m_{23} \\ m_{31} & m_{32} & m_{33} \end{bmatrix}$. For $n = 3$ there are six permutations,

$$123, \qquad 132, \qquad 213, \qquad 231, \qquad 312, \qquad 321.$$

Using (1.32)

$$
\begin{aligned}
\det \mathbf{M} &= s(123)m_{11}m_{22}m_{33} + s(132)m_{11}m_{32}m_{23} + s(213)m_{21}m_{12}m_{33} \\
&\quad + s(231)m_{21}m_{32}m_{13} + s(312)m_{31}m_{12}m_{23} + s(321)m_{31}m_{22}m_{13} \\
&= m_{11}m_{22}m_{33} - m_{11}m_{32}m_{23} - m_{21}m_{12}m_{33} \\
&\quad + m_{21}m_{32}m_{13} + m_{31}m_{12}m_{23} - m_{31}m_{22}m_{13}.
\end{aligned}
\tag{1.36}
$$

This sum can be expressed in several ways; for example we can factor out the numbers of the first column, m_{j1}, giving

$$
\begin{aligned}
\det \mathbf{M} &= m_{11}(m_{22}m_{33} - m_{32}m_{23}) \\
&\quad - m_{21}(m_{12}m_{33} - m_{32}m_{13}) + m_{31}(m_{12}m_{23} - m_{22}m_{13}).
\end{aligned}
\tag{1.37}
$$

If we factor out the numbers of the first row, m_{1k}, we get

$$\det \mathbf{M} = m_{11}(m_{22}m_{33} - m_{32}m_{23}) + m_{12}(-m_{21}m_{33} + m_{31}m_{23}) + m_{13}(m_{21}m_{32} - m_{31}m_{22})$$

and this is equal to

$$= m_{11}\det \begin{bmatrix} m_{22} & m_{23} \\ m_{32} & m_{33} \end{bmatrix} - m_{12}\det \begin{bmatrix} m_{21} & m_{23} \\ m_{31} & m_{33} \end{bmatrix} + m_{13}\det \begin{bmatrix} m_{21} & m_{22} \\ m_{31} & m_{32} \end{bmatrix}.$$

Example 1.19.

$$\det \begin{bmatrix} 0 & 1 & 2 \\ 5 & 6 & 7 \\ -1 & 3 & 2 \end{bmatrix} = 0 - 1\det \begin{bmatrix} 5 & 7 \\ -1 & 2 \end{bmatrix} + 2\det \begin{bmatrix} 5 & 6 \\ -1 & 3 \end{bmatrix} = 0 - (10+7) + 2(15+6) = 25.$$

$\square$

Example 1.20. Let

$$\mathbf{M} = \begin{bmatrix} 0 & 0 & 0 & d \\ 0 & b & 0 & 0 \\ a & 0 & 0 & 0 \\ 0 & 0 & c & 0 \end{bmatrix}.$$

The determinant must be either $abcd$ or $-abcd$, since the only nonzero term in $\det M$ is the one containing $m_{31}m_{22}m_{43}m_{14} = abcd$. In Problem 1.72 you find that the signature of 3241 is $+1$. Therefore

$$\det M = abcd.$$

□

Determinant and orientation. We turn now to the geometric meaning of the determinant. We use the notion of determinant to define algebraically the *orientation* of an ordered list of n linearly independent vectors in $\mathbb{R}^n$.

Definition 1.17. Let $V_1, V_2, \ldots, V_n$ be an ordered list of n linearly independent vectors in $\mathbb{R}^n$. The ordered list is called *positively oriented* if $\det(V_1, V_2, \ldots, V_n)$ is positive, and negatively oriented if $\det(V_1, V_2, \ldots, V_n)$ is negative. By property (vi) $\det(V_1, V_2, \ldots, V_n) \neq 0$.

Note that by property (iii) of the determinant $\det(E_1, E_2, \ldots, E_n) = 1$ and by property (i) det is multilinear. Therefore

(a) the ordered list of vectors $E_1, E_2, \ldots, E_n$ is positively oriented.
(b) the ordered list of vectors $E_1, E_2, \ldots, E_{n-1}, -E_n$ is negatively oriented.

See Figures 1.18 and 1.19.

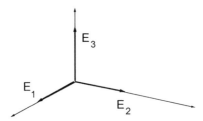

Fig. 1.18 E_1, E_2, E_3 are positively oriented in $\mathbb{R}^3$.

An n-tuple $U(t) = (u_1(t), u_2(t), \ldots, u_n(t))$ of continuous functions u_i from an interval in $\mathbb{R}$ to $\mathbb{R}$ is called a continuous vector function in $\mathbb{R}^n$. A *deformation* of an ordered list of linearly independent vectors is an ordered list of continuous vector functions $V_1(t), V_2(t), \ldots, V_n(t)$ in $\mathbb{R}^n$ such that for every t the vectors are linearly independent. The determinant of a deformation is nonzero for every t. Since $\det(V_1(t), \ldots, V_n(t))$ is a sum of products of the component functions and each component of $V_j(t)$ is a continuous function of t, $\det(V_1(t), \ldots, V_n(t))$ is continuous. It follows from the Intermediate Value Theorem that the continuous function of one variable,

$$\det(V_1(t), V_2(t), \ldots, V_n(t)) \neq 0,$$

has the same sign for all t. This shows that the ordered lists $\mathbf{V}_1(t), \mathbf{V}_2(t), \ldots, \mathbf{V}_n(t)$ have the same orientation for all t.

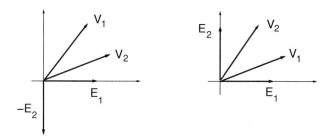

Fig. 1.19 *Left:* The ordered list $\mathbf{V}_1, \mathbf{V}_2$ is negatively oriented. *Right:* The ordered list $\mathbf{V}_1, \mathbf{V}_2$ is positively oriented.

The following is the basic result about the geometric meaning of orientation.

Theorem 1.17.*(a) Every positively oriented ordered list of n linearly independent vectors $\mathbf{V}_1, \mathbf{V}_2, \ldots, \mathbf{V}_n$ in $\mathbb{R}^n$ can be deformed into the ordered list of unit vectors $\mathbf{E}_1, \mathbf{E}_2, \ldots, \mathbf{E}_n$.*
(b) Every negatively oriented ordered list of vectors in $\mathbb{R}^n$ can be deformed into the ordered list $\mathbf{E}_1, \mathbf{E}_2, \ldots, -\mathbf{E}_n$.

Proof. We outline a proof by induction on n. See Figure 1.20.

(a) Suppose the vectors $\mathbf{V}_1, \ldots, \mathbf{V}_n$ are positively oriented. There is a rotation that takes the vector $\mathbf{V}_n$ into $p\mathbf{E}_n$, p a positive number. This rotation carries the vectors $\mathbf{V}_j$ into vectors we denote as $\mathbf{W}_j$, $(j = 1, 2, \ldots, n-1)$. We follow this by shrinking the n-th component of each of the vectors $\mathbf{W}_j$, $j < n$, to zero. This amounts to adding a multiple of the n-th vector $\mathbf{W}_n = p\mathbf{E}_n$ to each $\mathbf{W}_j$, $j < n$, which does not change the determinant of the resulting set of vectors. Therefore the resulting vectors in the hyperplane $x_n = 0$ are linearly independent.

Denote by $\mathbf{Z}_j$, $j < n$, the $(n-1)$-dimensional vectors obtained by omitting the n-th component of $\mathbf{W}_j$. Since the omitted components are zero, these are a set of $(n-1)$ linearly independent vectors with $(n-1)$ components. The determinant of the n vectors in their new position is equal to p times the $(n-1)$ by $(n-1)$ determinant of the vectors $\mathbf{Z}_1, \mathbf{Z}_2, \ldots, \mathbf{Z}_{n-1}$. Since the determinant is nonzero under a deformation, its sign doesn't change. Since p is positive, it follows that $\det(\mathbf{W}_1, \mathbf{W}_2, \ldots, \mathbf{W}_n)$ is positive. By the induction hypothesis this list of $(n-1)$ positively ordered vectors with $(n-1)$ coordinates can be deformed into the ordered list $\mathbf{E}_1, \mathbf{E}_2, \ldots, \mathbf{E}_{n-1}$. In one dimension a positively oriented vector is a positive number, which can be deformed into the number 1.

This completes the outline of the proof that the positively oriented list of vectors $\mathbf{V}_1, \mathbf{V}_2, \ldots, \mathbf{V}_n$ can be deformed into the list of vectors $\mathbf{E}_1, \mathbf{E}_2, \ldots, \mathbf{E}_n$.

(b) Analogously one shows that if the ordered list $\mathbf{V}_1, \mathbf{V}_2, \ldots, \mathbf{V}_n$ negatively oriented, it can be deformed into the ordered list $\mathbf{E}_1, \mathbf{E}_2, \ldots, -\mathbf{E}_n$.

□

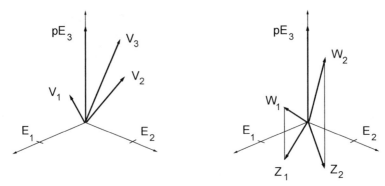

Fig. 1.20 Notation in the proof of Theorem 1.17 for $n = 3$.

Problems

1.68. Evaluate the determinants

(a) $\det \begin{bmatrix} 1 & 0 \\ 0 & 1 \end{bmatrix}$

(b) $\det \begin{bmatrix} 1 & 0 \\ 0 & -1 \end{bmatrix}$

(c) $\det \begin{bmatrix} 1 & 2 \\ 0 & -1 \end{bmatrix}$

(d) $\det \begin{bmatrix} 1 & 4 \\ 1 & 4 \end{bmatrix}$

(e) $\det[\mathbf{U}\ \mathbf{V}]$, where the columns $\mathbf{U}$ and $\mathbf{V}$ are linearly dependent in $\mathbb{R}^2$.

1.69. Show $\det[\mathbf{U}\ \mathbf{V}]$ is a bilinear function of pairs of column vectors $\mathbf{U}$, $\mathbf{V}$ in $\mathbb{R}^2$ by verifying that:

(a) $\det[\mathbf{U} + \mathbf{W}\ \mathbf{V}] = \det[\mathbf{U}\ \mathbf{V}] + \det[\mathbf{W}\ \mathbf{V}]$ and $\det[\mathbf{U}\ \mathbf{V} + \mathbf{W}] = \det[\mathbf{U}\ \mathbf{V}] + \det[\mathbf{U}\ \mathbf{W}]$,
(b) $\det[c\mathbf{U}\ \mathbf{V}] = c\det[\mathbf{U}\ \mathbf{V}]$ and $\det[\mathbf{U}\ c\mathbf{V}] = c\det[\mathbf{U}\ \mathbf{V}]$.

1.70. Use bilinearity of the determinant function to show that each expression is zero.

(a) $\det \begin{bmatrix} 5a & b \\ 5c & d \end{bmatrix} - 5\det \begin{bmatrix} a & b \\ c & d \end{bmatrix}$

(b) $\det \begin{bmatrix} x & y-z \\ v & w \end{bmatrix} - \det \begin{bmatrix} x & y \\ v & w \end{bmatrix} + \det \begin{bmatrix} x & z \\ v & 0 \end{bmatrix}$

1.71. Evaluate the determinants.

(a) $\det \begin{bmatrix} 1 & 0 & 0 \\ 0 & 1 & 0 \\ 0 & 0 & 1 \end{bmatrix}$

(b) $\det \begin{bmatrix} 1 & 0 & 0 \\ 0 & 1 & 0 \\ 0 & 0 & -1 \end{bmatrix}$

(c) $\det \begin{bmatrix} 1 & 0 & 0 \\ 0 & -1 & 0 \\ 0 & 0 & -1 \end{bmatrix}$

(d) $\det \begin{bmatrix} 1 & 0 & 0 \\ 0 & 0 & 2 \\ 0 & 3 & 0 \end{bmatrix}$

(e) $\det \begin{bmatrix} 0 & 0 & 3 \\ 0 & 2 & 0 \\ 1 & 0 & 0 \end{bmatrix}$

1.72. Find the signature $s(3241)$ from the equation

$$(x_3 - x_2)(x_3 - x_4)(x_3 - x_1)(x_2 - x_4)(x_2 - x_1)(x_4 - x_1)$$

$$= s(3241)(x_1 - x_2)(x_1 - x_3)(x_1 - x_4)(x_2 - x_3)(x_2 - x_4)(x_3 - x_4).$$

1.73. In the permutation 3241 there is an even number of cases where a larger number is to the left of a smaller one, 41, 21, 31, 32, and the signature is +1. (See Problem 1.72.)

(a) Prove in general that the signature of a permutation is +1 if there is an even number of such cases and is −1 if there is an odd number.
(b) Find $s(1237456)$.
(c) Find $s(1273456)$.

1.74. Evaluate the determinants.

(a) $\det \begin{bmatrix} 1 & 0 & 0 & \cdots & 0 \\ 0 & 2 & 0 & \cdots & 0 \\ 0 & 0 & 3 & \cdots & 0 \\ \vdots & & & \ddots & \vdots \\ 0 & 0 & 0 & \cdots & n \end{bmatrix}$

(b) $\det \begin{bmatrix} n & 1 & 1 & \cdots & 1 \\ 0 & n-1 & 1 & \cdots & 1 \\ 0 & 0 & n-2 & \cdots & 1 \\ \vdots & & & \ddots & \vdots \\ 0 & 0 & 0 & 3 & 1 \\ 0 & 0 & 0 & \cdots & 2 \end{bmatrix}$

1.75. When p and q are permutations of $1, 2, 3, \ldots, n$ we write the composition q followed by p as pq. Show that the signature of permutations has the property $s(pq) = s(p)s(q)$.

1.76. Use the result of Problem 1.75 to show that a permutation and its inverse have the same signature.

1.77. Show that the orientation of the ordered list of vectors $\mathbf{E}_1$, $\mathbf{E}_3$, $\mathbf{E}_2$ in $\mathbb{R}^3$, and the signature $s(132)$, are both negative.

1.78. Show that the orientation of the ordered list of vectors $\mathbf{E}_3$, $\mathbf{E}_1$, $\mathbf{E}_2$ in $\mathbb{R}^3$, and the signature $s(312)$, are both positive.

1.7 Signed volume

We start by defining a simplex.

Definition 1.18. Let $k \le n$ and let $\mathbf{V}_1, \ldots, \mathbf{V}_k$ be linearly independent vectors in $\mathbb{R}^n$. A k-*simplex* with vertices $\mathbf{0}, \mathbf{V}_1, \ldots, \mathbf{V}_k$ in $\mathbb{R}^n$ is the set of all points

$$\mathbf{X} = c_1 \mathbf{V}_1 + c_2 \mathbf{V}_2 + \cdots + c_k \mathbf{V}_k \qquad \text{where } 0 \le c_i \text{ and } \sum_{i=1}^{k} c_i \le 1.$$

We say the simplex is *ordered* if the order of the vertices is specified.

Example 1.21. The 2-simplex with vertices $(0,0), (1,0), (0,1)$ is the triangular region of $\mathbb{R}^2$ drawn in the left side of Figure 1.21. The 3-simplex with vertices $(0,0,0), (1,0,0), (0,1,0), (0,0,1)$ is the solid tetrahedron in $\mathbb{R}^3$ shown in the center of the figure. The 2-simplex with vertices $(0,0,0), (0,1,0), (0,0,1)$ is the triangular surface in $\mathbb{R}^3$ shown at the right of the figure.   □

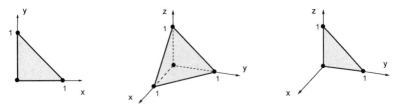

Fig. 1.21 *Left:* A 2-simplex in $\mathbb{R}^2$. *Center:* A 3-simplex in $\mathbb{R}^3$. *Right:* A 2-simplex in $\mathbb{R}^3$. See Example 1.21.

For each $j = 1, \ldots, k$ the j-face of the k-simplex is the ordered $(k-1)$-simplex whose vertices are the vectors $\mathbf{0}, \mathbf{V}_1, \ldots, \mathbf{V}_k$ omitting $\mathbf{V}_j$.

Example 1.22. The 1-face of the ordered 3-simplex with vertices

$$(0,0,0), \qquad (1,0,0), \qquad (0,1,0), \qquad (0,0,1)$$

in Example 1.21 is the ordered 2-simplex with vertices

$$(0,0,0), \qquad (0,1,0), \qquad (0,0,1).$$

See Figure 1.21.   □

We define the distance from a point $\mathbf{V}$ in $\mathbb{R}^n$ to the hyperplane spanned by $n-1$ linearly independent vectors as follows: By Theorem 1.14 there is a unit vector $\mathbf{N}$ that is orthogonal to the hyperplane. Define the distance from $\mathbf{V}$ to the hyperplane to be $|\mathbf{V} \cdot \mathbf{N}|$.

The *volume* of an n-simplex is defined to be the product of the $(n-1)$-dimensional volume of the j-face times the distance from $\mathbf{V}_j$ to the hyperplane spanned by the $\mathbf{V}_i, (i \neq j)$ that contains the j-face, divided by n. As we shall see, this number is the same for all j. To complete this definition we need the volume of the face. In the case of a 3-simplex in $\mathbb{R}^3$, the face is a triangular region that lies in a two-dimensional plane. The volume of the face is then the area of that triangle. In higher dimensions it is possible to introduce a coordinate system with $n-1$ coordinates in the hyperplane containing the face, so that the volume of a face can be defined inductively by dimension.

Definition 1.19. The *signed volume*, denoted $S(\mathbf{V}_1,\ldots,\mathbf{V}_n)$, of the ordered n-simplex with vertices $\mathbf{0}, \mathbf{V}_1, \ldots, \mathbf{V}_n$ is defined as the n-dimensional volume of the simplex in case the ordered collection $\mathbf{V}_1, \ldots, \mathbf{V}_n$ is positively oriented, and as minus the volume in case the ordered collection $\mathbf{V}_1, \ldots, \mathbf{V}_n$ is negatively oriented,

We shall derive a formula for the signed volume and show that it is a multilinear function of $\mathbf{V}_1, \ldots, \mathbf{V}_n$.

Let $\mathbf{V}_1, \ldots, \mathbf{V}_n$ be an ordered set of linearly independent vectors. We define the *signed distance* $s(\mathbf{V}_j)$ of a vector $\mathbf{V}_j$ from the hyperplane spanned by the vectors $\mathbf{V}_1, \ldots, \mathbf{V}_i, \ldots, \mathbf{V}_n, (i \neq j)$ as follows:

(a) $s(\mathbf{V}_j)$ is the distance of $\mathbf{V}_j$ from the hyperplane spanned by the vectors $\mathbf{V}_i, (i \neq j)$, in case the n-simplex $S(\mathbf{V}_1, \ldots, \mathbf{V}_n)$ is positively oriented.
(b) $s(\mathbf{V}_j)$ is -1 times the distance of $\mathbf{V}_j$ from the hyperplane spanned by the vectors $\mathbf{V}_i, (i \neq j)$, in case the n-simplex $S(\mathbf{V}_1, \ldots, \mathbf{V}_n)$ is negatively oriented.
(c) In case the vectors $\mathbf{V}_i$ are linearly dependent, we define $s(\mathbf{V}_j)$ as zero.

In Problem 1.83 we ask you to verify that the function $s(\mathbf{V})$ is a linear function in each half-space outside the hyperplane. Let $\mathbf{W}$ be the reflection of $\mathbf{V}$ across the hyperplane. It follows from the definition above of signed distance that $s(\mathbf{W}) = -s(\mathbf{V})$. This shows that $s(\mathbf{V})$ is a linear function in the whole space.

The signed volume of the n-simplex is equal to the product of $s(\mathbf{V}_n)$ times the $(n-1)$-dimensional volume of the face with vertices $\mathbf{0}, \mathbf{V}_1, \mathbf{V}_2, \ldots, \mathbf{V}_{n-1}$. By the induction hypothesis the $(n-1)$-volume of this face is a multilinear function of the variables $\mathbf{V}_i$, $(i < n)$. This proves that the signed volume $S(\mathbf{V}_1, \mathbf{V}_2, \ldots, \mathbf{V}_n)$ is a multilinear function. That is property (i) of the determinant as defined in Section 1.6.

If two of the $\mathbf{V}_j$ are the same, the volume is zero. That is property (ii) of the determinant.

The volume of the simplex whose vertices are the unit vectors $\mathbf{E}_1, \mathbf{E}_2, \ldots, \mathbf{E}_n$ is $\frac{1}{n!}$. This shows that

$$n! S(\mathbf{V}_1, \mathbf{V}_2, \ldots, \mathbf{V}_n)$$

has the property (iii) of the determinant. We have seen that the three properties characterize the determinant. Therefore it follows that $n! S(\mathbf{V}_1, \mathbf{V}_2, \ldots, \mathbf{V}_n)$ is the determinant. Dividing by $n!$ we get the formula for the signed volume of an ordered simplex:

$$S(\mathbf{V}_1, \mathbf{V}_2, \ldots, \mathbf{V}_n) = \frac{1}{n!} \det(\mathbf{V}_1, \mathbf{V}_2, \ldots, \mathbf{V}_n).$$

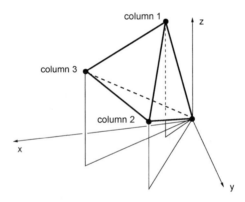

Fig. 1.22 The tetrahedron in Example 1.23.

Example 1.23. The tetrahedron with ordered vertices $(0,0,0)$, $(1,1,3)$, $(2,4,1)$, $(5,2,2)$ (see Figure 1.22) has signed volume

$$\frac{1}{3!} \det \begin{bmatrix} 1 & 2 & 5 \\ 1 & 4 & 2 \\ 3 & 1 & 2 \end{bmatrix} = \frac{1}{3!} (1(4(2) - 1(2)) - 2(1(2) - 3(2)) + 5(1(1) - 3(4)))$$

$$= \frac{6+8-55}{6} = -\frac{41}{6}.$$

□

The *parallelopiped* in $\mathbb{R}^n$ determined by vectors $\mathbf{V}_1, \ldots, \mathbf{V}_n$ is the set of points

$$c_1 \mathbf{V}_1 + \cdots + c_n \mathbf{V}_n, \qquad 0 \le c_i \le 1.$$

The signed volume of the parallelopiped determined by $\mathbf{V}_1, \ldots, \mathbf{V}_n$ is $n!$ times the signed volume of the n-simplex and is equal to

$$\det(\mathbf{V}_1, \ldots, \mathbf{V}_n).$$

Example 1.24. The signed volume of the parallelopiped determined by vertices $(1,1,3)$, $(2,4,1)$, $(5,2,2)$ is

$$\det \begin{bmatrix} 1 & 2 & 5 \\ 1 & 4 & 2 \\ 3 & 1 & 2 \end{bmatrix} = -41.$$

See Example 1.23 and Figure 1.23.   □

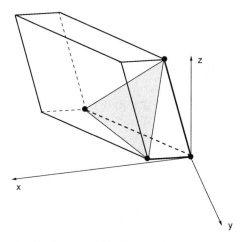

Fig. 1.23 The parallelopiped in Example 1.24. Compare the darkened vertices with Figure 1.22.

Problems

1.79. Justify the following items to prove that the area of the triangle with vertices $\mathbf{0}$, $\mathbf{U} = (u_1, u_2)$ and $\mathbf{V} = (v_1, v_2)$ is $\frac{1}{2}|u_1 v_2 - u_2 v_1|$. (See Figure 1.24.)

(a) The area of the triangle is $\frac{1}{2}\|\mathbf{U}\|(\|\mathbf{V}\|\sin\theta)$.

(b) $\sin\theta = \sqrt{1 - \left(\frac{\mathbf{U}\cdot\mathbf{V}}{\|\mathbf{U}\|\|\mathbf{V}\|}\right)^2}$.

(c) The area is equal to $\frac{1}{2}\sqrt{\|\mathbf{U}\|^2\|\mathbf{V}\|^2 - (\mathbf{U}\cdot\mathbf{V})^2}$.

(d) The expression $\|\mathbf{U}\|^2\|\mathbf{V}\|^2 - (\mathbf{U}\cdot\mathbf{V})^2$ is equal to the perfect square $(u_1 v_2 - u_2 v_1)^2$.

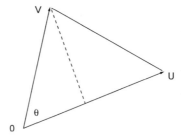

Fig. 1.24 The triangle in Problem 1.79.

1.80. Show that the signed area of the ordered triangle with vertices $\mathbf{0}$, $\mathbf{U} = (u_1, u_2)$ and $\mathbf{V} = (v_1, v_2)$ is $\frac{1}{2}(u_1 v_2 - u_2 v_1)$. (See Figure 1.24 and Problem 1.79.)

1.81. Find the area of the parallelogram with vertices $(0,0)$, $(1,3)$, $(2,1)$, and $(3,4)$. (See Problem 1.79.)

1.82. Draw the ordered tetrahedra (ordered simplices in $\mathbb{R}^3$) with the following vertices and find their signed volume.

(a) $\mathbf{0}$, $\mathbf{U} = (2,1,0)$, $\mathbf{V} = (1,2,0)$ and $\mathbf{W} = (0,0,7)$.
(b) $\mathbf{0}$, $\mathbf{U} = (2,1,0)$, $\mathbf{V} = (1,2,0)$ and $\mathbf{W} = (7,7,7)$.
(c) $\mathbf{0}$, $\mathbf{U} = (2,1,0)$, $\mathbf{W} = (7,7,7)$ and $\mathbf{V} = (1,2,0)$.

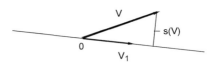

Fig. 1.25 Signed distance $s(\mathbf{V})$ for positively oriented vectors $\mathbf{V}_1, \mathbf{V}$. See Problem 1.83.

1.83. Let $\mathbf{V}_1 \neq \mathbf{0}$, $\mathbf{V}$ and $\mathbf{W}$ be vectors in $\mathbb{R}^2$ with $\mathbf{V}_1, \mathbf{V}$ and $\mathbf{V}_1, \mathbf{W}$ positively oriented. See Figure 1.25. Let $s(\mathbf{V})$ be the distance from $\mathbf{V}$ to the line containing $\mathbf{V}_1$. Use a sketch to illustrate the following.

(a) $s(c\mathbf{V}) = c\, s(\mathbf{V})$ for all numbers $c \geq 0$,
(b) $s(\mathbf{V} + \mathbf{W}) = s(\mathbf{V}) + s(\mathbf{W})$.

1.8 Linear functions and their representation by matrices

The notion of *linear function* is basic.

Definition 1.20. A function $\mathbf{F}$ from $\mathbb{R}^n$ to $\mathbb{R}^k$ is called *linear* when it has the following properties:

(a) For all numbers c and vectors $\mathbf{V}$ in $\mathbb{R}^n$, $\mathbf{F}(c\mathbf{V}) = c\mathbf{F}(\mathbf{V})$.
(b) For all vectors $\mathbf{V}$ and $\mathbf{W}$ in $\mathbb{R}^n$, $\mathbf{F}(\mathbf{V}+\mathbf{W}) = \mathbf{F}(\mathbf{V})+\mathbf{F}(\mathbf{W})$.

Combining properties (a) and (b) we deduce that for all vectors $\mathbf{V}_1, \mathbf{V}_2, \ldots, \mathbf{V}_\ell$ and numbers $c_1, c_2, \ldots, c_\ell$

$$\mathbf{F}(c_1\mathbf{V}_1 + c_2\mathbf{V}_2 + \cdots + c_\ell\mathbf{V}_\ell) = c_1\mathbf{F}(\mathbf{V}_1) + c_2\mathbf{F}(\mathbf{V}_2) + \cdots + c_\ell\mathbf{F}(\mathbf{V}_\ell). \tag{1.38}$$

We describe now all linear functions from $\mathbb{R}^n$ to $\mathbb{R}^k$.

We express vector $\mathbf{V} = (v_1, \ldots, v_n)$ as a linear combination of the standard basis vectors $\mathbf{E}_1, \ldots, \mathbf{E}_n$

$$\mathbf{V} = v_1\mathbf{E}_1 + v_2\mathbf{E}_2 + \cdots + v_n\mathbf{E}_n.$$

Using equation (1.38) we get

$$\mathbf{F}(\mathbf{V}) = v_1\mathbf{F}(\mathbf{E}_1) + v_2\mathbf{F}(\mathbf{E}_2) + \cdots + v_n\mathbf{F}(\mathbf{E}_n). \tag{1.39}$$

This shows that the linear function $\mathbf{F}$ is determined by the vectors $\mathbf{F}(\mathbf{E}_1), \ldots, \mathbf{F}(\mathbf{E}_n)$. Next we use formula (1.39) to characterize all linear functions.

Choose vectors $\mathbf{M}_1, \mathbf{M}_2, \ldots, \mathbf{M}_n$ in $\mathbb{R}^k$. We define the function $\mathbf{F}$ by

$$\mathbf{F}(\mathbf{V}) = v_1\mathbf{M}_1 + v_2\mathbf{M}_2 + \cdots + v_n\mathbf{M}_n \tag{1.40}$$

for every $\mathbf{V} = (v_1, v_2, \ldots, v_n)$ in $\mathbb{R}^n$. It is easy to verify, and we ask you to show in Problem 1.87 that the function $\mathbf{F}$ defined by formula (1.40) is linear. According to (1.39) all linear functions are of this form. So we have proved the following theorem.

Theorem 1.18. *A function $\mathbf{F}$ from $\mathbb{R}^n$ to $\mathbb{R}^k$ is linear if and only if there is a list of vectors $\mathbf{M}_1, \ldots, \mathbf{M}_n$ in $\mathbb{R}^k$ so that for every $\mathbf{V} = (v_1, \ldots, v_n)$ in $\mathbb{R}^n$*

$$\mathbf{F}(\mathbf{V}) = v_1\mathbf{M}_1 + v_2\mathbf{M}_2 + \cdots + v_n\mathbf{M}_n.$$

Matrix notation. If $\mathbf{F}$ is a linear function denote the components of the vectors $\mathbf{M}_i$ as $m_{1i}, m_{2i}, \ldots, m_{ki}$, $(i = 1, 2, \ldots, n)$. We arrange these numbers in a rectangular array:

$$\mathbf{M} = \begin{bmatrix} m_{11} & m_{12} & \cdots & m_{1n} \\ m_{21} & m_{22} & \cdots & m_{2n} \\ \vdots & \vdots & \cdots & \vdots \\ m_{k1} & m_{k2} & \cdots & m_{kn} \end{bmatrix}. \tag{1.41}$$

The j-th column consists of the components of the vector $\mathbf{M}_j$.

Definition 1.21. A rectangular array of numbers $\mathbf{M}$ in (1.41), with k rows and n columns is called a k by n (or $k \times n$) *matrix*. The number in the i-th row and j-th column is denoted as m_{ij}.

Definition 1.22. The *product* $\mathbf{MV}$ of a k by n matrix $\mathbf{M}$ and a column vector $\mathbf{V}$ whose components are $v_1, v_2, \ldots, v_n$, is the column vector whose i-th component is

$$m_{i1}v_1 + m_{i2}v_2 + \cdots + m_{in}v_n \qquad (i = 1, 2, \ldots, k)$$

In words: the i-th component of $\mathbf{MV}$ is the dot product of the i-th row of the matrix $\mathbf{M}$ and the vector $\mathbf{V}$.

Example 1.25. Here are three examples:

$$\begin{bmatrix} 1 & 2 \\ 3 & 4 \end{bmatrix}\begin{bmatrix} -1 \\ 1 \end{bmatrix} = \begin{bmatrix} 1 \\ 1 \end{bmatrix}, \quad \begin{bmatrix} 1 & 2 & 3 \\ 4 & 5 & 6 \\ 7 & 8 & 9 \end{bmatrix}\begin{bmatrix} 1 \\ 1 \\ -1 \end{bmatrix} = \begin{bmatrix} 0 \\ 3 \\ 6 \end{bmatrix}, \quad \begin{bmatrix} 1 & 4 \\ 2 & 1 \\ -3 & 3 \end{bmatrix}\begin{bmatrix} -1 \\ 1 \end{bmatrix} = \begin{bmatrix} 3 \\ -1 \\ 6 \end{bmatrix}.$$

$\square$

Matrices give us an alternative way to represent linear functions.

Theorem 1.19. *Every linear function* $\mathbf{F}$ *from* $\mathbb{R}^n$ *to* $\mathbb{R}^k$ *can be written in matrix form*

$$\mathbf{F(V)} = \mathbf{MV}$$

for some $k \times n$ *matrix* $\mathbf{M}$.

Linear functions can be multiplied by a number, added and composed, and the result is a linear function; we formulate these operations as an "algebra" of linear functions.

(i) For a linear function $\mathbf{F}$ from $\mathbb{R}^n$ to $\mathbb{R}^k$ and a number c the product $c\mathbf{F}$ is defined by

$$(c\mathbf{F})(\mathbf{V}) = c\,\mathbf{F(V)}.$$

(ii) For a pair of linear functions $\mathbf{F}$ and $\mathbf{G}$ both mapping $\mathbb{R}^n$ to $\mathbb{R}^k$, their sum $\mathbf{F} + \mathbf{G}$ is defined by

$$(\mathbf{F} + \mathbf{G})(\mathbf{V}) = \mathbf{F(V)} + \mathbf{G(V)}.$$

(iii) Denote by $\mathbf{F}$ a linear function from $\mathbb{R}^n$ to $\mathbb{R}^k$ and by $\mathbf{G}$ a linear function from $\mathbb{R}^k$ to $\mathbb{R}^m$. Their composition is defined by

$$\mathbf{G} \circ \mathbf{F(V)} = \mathbf{G(F(V))}.$$

It is a function from $\mathbb{R}^n$ to $\mathbb{R}^m$.

We ask you in Problem 1.90 to verify that constant multiples, sums, and composites of linear functions are linear.

Since linear functions are represented by matrices, these operations can be expressed as an algebra of matrices:

(i)' For a matrix $\mathbf{M}$ and number c, $c\mathbf{M}$ is defined to be the matrix whose elements are c times the elements of $\mathbf{M}$.

$$(c\mathbf{M})_{ij} = cm_{ij}.$$

(ii)' For a pair of k by n matrices $\mathbf{M}$ and $\mathbf{N}$, their sum $\mathbf{M}+\mathbf{N}$ is defined to be the k by n matrix whose elements are the sums of the elements of $\mathbf{M}$ and $\mathbf{N}$:

$$m_{ij} + n_{ij}.$$

(iii)' For a k by n matrix $\mathbf{M}$ and a m by k matrix $\mathbf{N}$ the matrix product $\mathbf{NM}$ is an m by n matrix whose ij-th element is the sum

$$\sum_{h=1}^{k} n_{ih}m_{hj}.$$

One can think of this sum as the dot product of the i-th row of the matrix $\mathbf{N}$ and j-th column of the matrix $\mathbf{M}$.

Formulas (i)' and (ii)' clearly express rules (i) and (ii) for linear functions. We ask you in Problem 1.91 to verify that formula (iii)' expresses rule (iii).

Example 1.26. Here are some examples of matrix products:

$$\begin{bmatrix} 1&2&3 \\ 4&5&6 \\ 7&8&9 \end{bmatrix}\begin{bmatrix} 1&-1 \\ 2&-1 \\ 3&1 \end{bmatrix} = \begin{bmatrix} 14&0 \\ 32&-3 \\ 50&-6 \end{bmatrix}, \qquad \begin{bmatrix} 1&3 \\ 4&5 \end{bmatrix}\begin{bmatrix} 2&1 \\ -1&3 \end{bmatrix} = \begin{bmatrix} -1&10 \\ 3&19 \end{bmatrix}.$$

□

Matrix multiplication is not commutative; that is, $\mathbf{KL}$ and $\mathbf{LK}$ are in general not equal.

Example 1.27. Let

$$\mathbf{K} = \begin{bmatrix} 1&2 \\ 3&4 \end{bmatrix}, \qquad \mathbf{L} = \begin{bmatrix} 1&-1 \\ -3&4 \end{bmatrix}.$$

then

$$\mathbf{KL} = \begin{bmatrix} 1&2 \\ 3&4 \end{bmatrix}\begin{bmatrix} 1&-1 \\ -3&4 \end{bmatrix} = \begin{bmatrix} -5&7 \\ -9&13 \end{bmatrix}, \qquad \mathbf{LK} = \begin{bmatrix} 1&-1 \\ -3&4 \end{bmatrix}\begin{bmatrix} 1&2 \\ 3&4 \end{bmatrix} = \begin{bmatrix} -2&-2 \\ 9&10 \end{bmatrix}.$$

□

If **K** and **L** are matrices, it is possible that one of the two products **KL** and **LK** is defined and the other not defined.

We turn now to square matrices; they represent linear functions from $\mathbb{R}^n$ to $\mathbb{R}^n$. Denote by $\mathbf{I}_n$ the $n \times n$ matrix whose diagonal elements are 1 and off-diagonal elements are zero. For $n = 3$

$$\mathbf{I}_3 = \begin{bmatrix} 1 & 0 & 0 \\ 0 & 1 & 0 \\ 0 & 0 & 1 \end{bmatrix}.$$

$\mathbf{I}_n$ is called the $n \times n$ *identity matrix*. Multiplication by $\mathbf{I}_n$ is the identity function:

$$\mathbf{I}_n \mathbf{V} = \mathbf{V}$$

for every vector **V** in $\mathbb{R}^n$.

The basic result about square matrices is the following.

Theorem 1.20. *Let* **M** *be an* $n \times n$ *matrix, and let* **U**, **V** *and* **W** *be vectors in* $\mathbb{R}^n$.

(a) *If* **MV** = **0** *only for* **V** = **0**, *then every vector* **W** *can be represented as* **W** = **MU** *for some* **U**.

(b) *If every vector* **W** *can be represented as* **MU** *for some* **U** *then* **MV** = **0** *only for* **V** = **0**.

We say that a function **F** is *one to one* if $\mathbf{F(U)} = \mathbf{F(V)}$ only when **U** = **V**. Before presenting the proof of Theorem 1.20 we draw some of its consequences. In case (a) the function **F** defined by $\mathbf{F(V)} = \mathbf{MV}$ is one to one. For if

$$\mathbf{F(V)} = \mathbf{F(U)},$$

then by linearity $\mathbf{F(V - U)} = \mathbf{F(V)} - \mathbf{F(U)} = \mathbf{0}$. Since the only vector mapped into zero by multiplication by **M** is the zero vector, it follows that **V** − **U** = **0** and so **V** = **U**. Therefore every vector **W** can be represented as **W** = **MU** = **F(U)** for some **U**, and **U** is uniquely determined because **F** is one to one. That is, **F** has an inverse function that we denote $\mathbf{F}^{-1}$.

We show now that $\mathbf{F}^{-1}$ is linear. Let $\mathbf{F(U)} = \mathbf{W}$ and $\mathbf{F(V)} = \mathbf{Z}$. Then by linearity

$$\mathbf{F(U + V)} = \mathbf{F(U)} + \mathbf{F(V)} = \mathbf{W} + \mathbf{Z}.$$

By definition of inverse

$$\mathbf{U} = \mathbf{F}^{-1}(\mathbf{W}), \qquad \mathbf{V} = \mathbf{F}^{-1}(\mathbf{Z}) \quad \text{and} \quad \mathbf{U} + \mathbf{V} = \mathbf{F}^{-1}(\mathbf{W} + \mathbf{Z}).$$

Therefore

$$\mathbf{F}^{-1}(\mathbf{W}) + \mathbf{F}^{-1}(\mathbf{Z}) = \mathbf{F}^{-1}(\mathbf{W} + \mathbf{Z}).$$

One can verify similarly that

$$\mathbf{F}^{-1}(c\mathbf{W}) = c\,\mathbf{F}^{-1}(\mathbf{W}).$$

According to Theorem 1.19 the inverse function $\mathbf{F}^{-1}$ can be represented by a matrix. We make the following definition.

Definition 1.23. Let $\mathbf{M}$ be a square matrix, and denote the corresponding linear function from $\mathbb{R}^n$ to $\mathbb{R}^n$ as $\mathbf{F}$:

$$\mathbf{F}(\mathbf{V}) = \mathbf{M}\mathbf{V}.$$

If $\mathbf{F}$ has an inverse function $\mathbf{F}^{-1}$ then we say $\mathbf{M}$ is *invertible* and denote by

$$\mathbf{M}^{-1}$$

the matrix that represents $\mathbf{F}^{-1}$. We call $\mathbf{M}^{-1}$ the *inverse* of the matrix $\mathbf{M}$.

We present now a proof of Theorem 1.20.

Proof. (a) Take the case that $\mathbf{M}\mathbf{V} = \mathbf{0}$ only for the vector $\mathbf{V} = \mathbf{0}$. We show that the n vectors $\mathbf{M}\mathbf{E}_1, \mathbf{M}\mathbf{E}_2, \dots, \mathbf{M}\mathbf{E}_n$ are linearly independent. For suppose there were a nontrivial linear relation

$$c_1\mathbf{M}\mathbf{E}_1 + c_2\mathbf{M}\mathbf{E}_2 + \cdots + c_n\mathbf{M}\mathbf{E}_n = \mathbf{0}.$$

By properties of matrices (i)'–(iii)' we can write this relation as

$$\mathbf{M}(c_1\mathbf{E}_1 + c_2\mathbf{E}_2 + \cdots + c_n\mathbf{E}_n) = \mathbf{0}.$$

Since $\mathbf{M}\mathbf{V} = \mathbf{0}$ only for $\mathbf{V} = \mathbf{0}$, it follows that $c_1 = c_2 = \cdots = c_n = 0$. This proves the linear independence of the vectors $\mathbf{M}\mathbf{E}_j$. According to Theorem 1.9, every vector $\mathbf{W}$ can be represented uniquely as

$$\mathbf{W} = a_1\mathbf{M}\mathbf{E}_1 + a_2\mathbf{M}\mathbf{E}_2 + \cdots + a_n\mathbf{M}\mathbf{E}_n.$$

By linearity we can write this relation as

$$\mathbf{W} = \mathbf{M}(a_1\mathbf{E}_1 + a_2\mathbf{E}_2 + \cdots + a_n\mathbf{E}_n).$$

This proves that every vector $\mathbf{W}$ can be represented as $\mathbf{M}\mathbf{U}$.

(b) Take the case that every vector can be represented as $\mathbf{M}\mathbf{U}$. Then the unit vectors $\mathbf{E}_i$ can be represented as

$$\mathbf{E}_i = \mathbf{M}\mathbf{U}_i. \qquad\qquad (i = 1, 2, \dots, n)$$

By linearity, for all numbers $c_1, c_2, \dots, c_n$ we have

$$\mathbf{M}(c_1\mathbf{U}_1 + c_2\mathbf{U}_2 + \cdots + c_n\mathbf{U}_n) = c_1\mathbf{M}\mathbf{U}_1 + c_2\mathbf{M}\mathbf{U}_2 + \cdots + c_n\mathbf{M}\mathbf{U}_n$$

$$= c_1 \mathbf{E}_1 + c_2 \mathbf{E}_2 + \cdots + c_n \mathbf{E}_n. \tag{1.42}$$

The right side is nonzero unless all the numbers c_i are zero. Therefore so is the left side. It follows that $\mathbf{M}(c_1 \mathbf{U}_1 + c_2 \mathbf{U}_2 + \cdots + c_n \mathbf{U}_n)$ is nonzero unless all numbers c_i are zero. Consequently $c_1 \mathbf{U}_1 + c_2 \mathbf{U}_2 + \cdots + c_n \mathbf{U}_n$ is nonzero unless all the numbers c_i are zero. This shows that the vectors $\mathbf{U}_i$ are linearly independent. By Theorem 1.9, every vector $\mathbf{W}$ is some linear combination $c_1 \mathbf{U}_1 + c_2 \mathbf{U}_2 + \cdots + c_n \mathbf{U}_n$ of the linearly independent vectors $\mathbf{U}_1, \ldots, \mathbf{U}_n$. It follows from equation (1.42) that $\mathbf{MW} = \mathbf{0}$ only if all the numbers c_i are zero. But then $\mathbf{W} = \mathbf{0}$. This proves part (b).   □

In Problems 1.92, 1.93, and 1.94 we ask you to justify some additional results regarding matrices and determinants:

(a) For square matrices $\mathbf{A}$ and $\mathbf{B}$ of the same size,

$$\det(\mathbf{AB}) = \det(\mathbf{A})\det(\mathbf{B}).$$

(b) If the $n \times n$ matrix $\mathbf{A}$ has an inverse matrix $\mathbf{A}^{-1}$, then

$$\mathbf{AA}^{-1} = \mathbf{I}_n = \mathbf{A}^{-1}\mathbf{A}$$

and

$$\det(\mathbf{A}^{-1}) = (\det(\mathbf{A}))^{-1}.$$

Problems

1.84. Evaluate the products.

(a) $\begin{bmatrix} 1 & 0 & 0 \\ 0 & 2 & 0 \\ 0 & 0 & 6 \end{bmatrix} \begin{bmatrix} x_1 \\ x_2 \\ x_3 \end{bmatrix}$

(b) $\begin{bmatrix} 3 & -4 \\ 4 & 3 \end{bmatrix} \begin{bmatrix} 6 \\ 6 \end{bmatrix}$

(c) $\begin{bmatrix} \cos\theta & -\sin\theta \\ \sin\theta & \cos\theta \end{bmatrix} \begin{bmatrix} 1 \\ 0 \end{bmatrix}$

(d) $\begin{bmatrix} \cos\theta & -\sin\theta \\ \sin\theta & \cos\theta \end{bmatrix}^2 = \begin{bmatrix} \cos\theta & -\sin\theta \\ \sin\theta & \cos\theta \end{bmatrix} \begin{bmatrix} \cos\theta & -\sin\theta \\ \sin\theta & \cos\theta \end{bmatrix}$

1.85. Evaluate the products, where

$$\mathbf{X} = \begin{bmatrix} 1 \\ 3 \end{bmatrix}, \quad \mathbf{Y} = \begin{bmatrix} -3 \\ 1 \end{bmatrix}, \quad \mathbf{A} = \begin{bmatrix} 0 & 1 \\ -1 & 0 \end{bmatrix}, \quad \mathbf{B} = \begin{bmatrix} b_{11} & b_{12} \\ b_{21} & b_{22} \end{bmatrix}.$$

(a) $\mathbf{AX}$ (b) $\mathbf{X} \cdot \mathbf{Y}$ (c) $\mathbf{Y} \cdot (\mathbf{AX})$ (d) $\mathbf{E}_i \cdot \mathbf{BE}_j, \ i, j = 1, 2$

1.86. Evaluate the products, where

$$X = \begin{bmatrix} 1 \\ 2 \\ 3 \end{bmatrix}, \quad Y = \begin{bmatrix} -3 \\ 0 \\ 1 \end{bmatrix}, \quad A = \begin{bmatrix} 0 & 1 & 0 \\ 1 & 0 & 0 \\ 0 & 0 & 7 \end{bmatrix}, \quad B = \begin{bmatrix} b_{11} & b_{12} & b_{13} \\ b_{21} & b_{22} & b_{23} \\ b_{31} & b_{32} & b_{33} \end{bmatrix}.$$

(a) AX (b) $X \cdot Y$ (c) $Y \cdot (AX)$ (d) $E_i \cdot BE_j$, $i, j = 1, 2, 3$

1.87. Show that the function defined in equation (1.40) is linear.

1.88. Express these formulas as a product AX of a matrix A and a column vector X.

(a) $\begin{bmatrix} x_1 + x_2 \\ x_1 - x_2 \end{bmatrix}$ where X is in $\mathbb{R}^2$

(b) $x_2 - 4x_3$, where X is in $\mathbb{R}^3$

(c) $x_2 - 4x_3$, where X is in $\mathbb{R}^4$

(d) $\begin{bmatrix} x_2 - 4x_3 \\ x_1 + x_4 \\ x_1 + x_2 + x_3 + x_4 \end{bmatrix}$ where X is in $\mathbb{R}^4$.

1.89. Express these formulas as $X \cdot (AY)$, where X and Y are column vectors of $\mathbb{R}^2$ and A is a matrix.

(a) $x_1 y_1 + 2x_2 y_2$

(b) $x_1 y_2 - x_2 y_1$

(c) $x_1(y_1 + 3y_2) + x_2(y_1 - y_2)$

1.90. Let F and G be linear functions from $\mathbb{R}^n$ to $\mathbb{R}^m$, let H be a linear function from $\mathbb{R}^m$ to $\mathbb{R}^k$, and let c be a number. Show the following.

(a) $F + G$ is a linear function.

(b) cF is a linear function.

(c) $H \circ F$ is a linear function.

1.91. Let N be an m by k matrix, M a k by n matrix, and X in $\mathbb{R}^n$. Show that the product $N(MX)$ can be found by multiplying X by the matrix with ij entry

$$\sum_{h=1}^{k} n_{ih} m_{hj},$$

the dot product of row i of N with column j of M. That is,

$$N(MX) = (NM)X.$$

1.92. Assume A is an $n \times n$ matrix with $\det A \neq 0$, and consider the function of $n \times n$ matrices B defined by

$$f(B) = \frac{\det(AB)}{\det A}.$$

Show that f satisfies the three basic properties that characterize the determinant. Thus by uniqueness $f(B)$ must be equal to $\det B$. Conclude that

$$\det(AB) = \det(A)\det(B).$$

1.93. Assume **A** and **B** are $n \times n$ matrices with $\det \mathbf{A} = 0$. Show the following:

(a) There is a vector **W** such that for all **V**, $\mathbf{AV} \neq \mathbf{W}$. (See Theorem 1.20.)
(b) There is a vector **W** such that for all **U**, $\mathbf{ABU} \neq \mathbf{W}$.
(c) $\det(\mathbf{AB}) = 0$.

Therefore $\det(\mathbf{AB}) = \det(\mathbf{A})\det(\mathbf{B})$ holds when $\det \mathbf{A} = 0$.

1.94. Prove the following.

(a) If the $n \times n$ matrix **A** has an inverse, then $\mathbf{AA}^{-1} = \mathbf{A}^{-1}\mathbf{A} = \mathbf{I}_n$.
(b) Use the formula $\det(\mathbf{AB}) = \det(\mathbf{A})\det(\mathbf{B})$ to prove that $\det(\mathbf{A}^{-1}) = (\det \mathbf{A})^{-1}$ for every invertible matrix **A**.

1.9 Geometric applications

Lines, planes, and hyperplanes. We have noted that for two vectors **U** and **V** in $\mathbb{R}^2$, and all numbers c, the points

$$\mathbf{U} + c\mathbf{V}$$

lie along a line through **U** and parallel to **V**. If **U** and **V** are vectors in $\mathbb{R}^n$, the same expression describes a *line* in $\mathbb{R}^n$. See Figure 1.26 for the case $n = 3$.

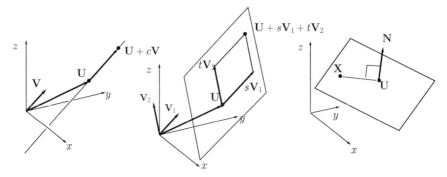

Fig. 1.26 *Left:* A line in $\mathbb{R}^3$. *Center:* A two-dimensional plane in $\mathbb{R}^3$. *Right:* A plane specified as in Example 1.28.

Let **U**, $\mathbf{V}_1$, and $\mathbf{V}_2$ be three vectors in $\mathbb{R}^3$, $\mathbf{V}_1$ and $\mathbf{V}_2$ linearly independent, and let s and t be numbers. For each of the points $\mathbf{U} + s\mathbf{V}_1$ on the line though **U** parallel to $\mathbf{V}_1$, there is a line $\mathbf{U} + s\mathbf{V}_1 + t\mathbf{V}_2$ through $\mathbf{U} + s\mathbf{V}_1$ parallel to $\mathbf{V}_2$. The set of points

$$\mathbf{U} + s\mathbf{V}_1 + t\mathbf{V}_2$$

is a plane in $\mathbb{R}^3$. The same expression represents a two-dimensional plane in any number of dimensions greater than three.

In the case of $\mathbb{R}^3$ there is an alternate description for a plane, given by specifying a *normal* vector **N** perpendicular to the plane.

Example 1.28. Let **U** = $(1,2,3)$ and **N** = $(2,6,-3)$, and consider the plane containing **U** and perpendicular to **N**. That is the set of points **X** = (x,y,z) that satisfy

$$\mathbf{N} \cdot (\mathbf{X} - \mathbf{U}) = 0.$$

For **U** and **N** specified above this equation can be written as

$$2(x-1) + 6(y-2) - 3(z-3) = 0$$

or as $z = -\frac{5}{3} + \frac{2}{3}x + 2y$. See Figure 1.26. Since $\mathbf{N} \cdot (\mathbf{U} - \mathbf{U}) = \mathbf{N} \cdot \mathbf{0} = 0$, **U** satisfies the equation. If **X** and **Y** satisfy the equation, then the vector **X** − **Y** is orthogonal to **N**,

$$\mathbf{N} \cdot (\mathbf{X} - \mathbf{Y}) = \mathbf{N} \cdot (\mathbf{X} - \mathbf{U} - (\mathbf{Y} - \mathbf{U})) = \mathbf{N} \cdot (\mathbf{X} - \mathbf{U}) - \mathbf{N} \cdot (\mathbf{Y} - \mathbf{U}) = 0 - 0 = 0.$$

□

When **U** and $\mathbf{V}_1, \mathbf{V}_2, \ldots, \mathbf{V}_{n-1}$ are vectors in $\mathbb{R}^n$ and the $\mathbf{V}_j$ are linearly independent, we have called the set of all vectors of the form

$$\mathbf{U} + t_1 \mathbf{V}_1 + \cdots + t_{n-1} \mathbf{V}_{n-1}$$

a hyperplane. A hyperplane in $\mathbb{R}^n$ can also be defined, given a point **U** in the hyperplane and a nonzero vector **N** orthogonal to the plane, by the equation

$$\mathbf{N} \cdot (\mathbf{X} - \mathbf{U}) = 0.$$

The distance from a point **V** in $\mathbb{R}^n$ to that hyperplane is $\|\mathbf{V} - \mathbf{U}\| \cos\theta$ where θ is the angle between **N** and **V** − **U**. (See Figure 1.27.) In the case where **N** is a unit vector that is

$$\text{distance } = \mathbf{N} \cdot (\mathbf{V} - \mathbf{U}), \qquad (\|\mathbf{N}\| = 1). \qquad (1.43)$$

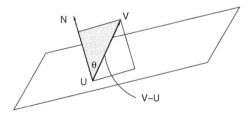

Fig. 1.27 Distance from point to hyperplane in equation (1.43).

Cross product in $\mathbb{R}^3$. For a pair of vectors **V** and **W** in $\mathbb{R}^3$ we introduce the *cross product*, denoted as $\mathbf{V} \times \mathbf{W}$. For all column vectors **U**, **V**, and **W** in $\mathbb{R}^3$, the array

$$[\mathbf{U}\ \mathbf{V}\ \mathbf{W}]$$

is a 3×3 matrix. For given vectors $\mathbf{V}$ and $\mathbf{W}$ let

$$\ell(\mathbf{U}) = \det[\mathbf{U}\ \mathbf{V}\ \mathbf{W}]. \tag{1.44}$$

Since the determinant is a multilinear function of its columns, $\ell(\mathbf{U})$ is a linear function of $\mathbf{U}$. According to the representation theorem for linear functions, $\ell(\mathbf{U})$ is of the form

$$\ell(\mathbf{U}) = \mathbf{U} \cdot \mathbf{Z} \tag{1.45}$$

$\mathbf{Z}$ some vector. Since the linear function $\ell(\mathbf{U})$ is determined by the vectors $\mathbf{V}$ and $\mathbf{W}$, so is the vector $\mathbf{Z}$. We call this vector $\mathbf{Z}$ the cross product of $\mathbf{V}$ and $\mathbf{W}$,

$$\mathbf{Z} = \mathbf{V} \times \mathbf{W}. \tag{1.46}$$

Combining (1.44), (1.45), and (1.46), we get that for all vectors $\mathbf{U}$, $\mathbf{V}$, and $\mathbf{W}$

$$\det[\mathbf{U}\ \mathbf{V}\ \mathbf{W}] = \mathbf{U} \cdot (\mathbf{V} \times \mathbf{W}). \tag{1.47}$$

Since the determinant on the left side of (1.47) is a multilinear function of $\mathbf{U}$, $\mathbf{V}$, and $\mathbf{W}$, so is the right side. It follows that

The cross product $\mathbf{V} \times \mathbf{W}$ as defined by (1.47) is a bilinear function of $\mathbf{V}$ and $\mathbf{W}$.

We ask you in Problem 1.103 to derive from (1.47) the following formula for the cross product.

Definition 1.24. The *cross product* of two vectors

$$\mathbf{V} = (v_1, v_2, v_3), \qquad \mathbf{W} = (w_1, w_2, w_3)$$

in $\mathbb{R}^3$ is defined as

$$\mathbf{V} \times \mathbf{W} = (v_2 w_3 - v_3 w_2, -(v_1 w_3 - v_3 w_1), v_1 w_2 - v_2 w_1).$$

Example 1.29. Let $\mathbf{i} = (1,0,0)$, $\mathbf{j} = (0,1,0)$, and $\mathbf{k} = (0,0,1)$. Then

$$\mathbf{i} \times \mathbf{j} = \mathbf{k}, \qquad \mathbf{j} \times \mathbf{k} = \mathbf{i}, \qquad \mathbf{k} \times \mathbf{i} = \mathbf{j}.$$

$\square$

The formula for $\mathbf{V} \times \mathbf{W}$ in Definition 1.24 can be found by formally computing $\det[\mathbf{U}, \mathbf{V}, \mathbf{W}]$ where $\mathbf{U}$ is the symbolic vector $[\mathbf{i}, \mathbf{j}, \mathbf{k}]$.

The cross product has the following properties.

Theorem 1.21.(a) $\mathbf{V} \times \mathbf{W}$ is orthogonal to $\mathbf{V}$ and to $\mathbf{W}$,

$$\mathbf{V} \cdot (\mathbf{V} \times \mathbf{W}) = 0, \qquad \mathbf{W} \cdot (\mathbf{V} \times \mathbf{W}) = 0.$$

(b) The signed volume of the ordered tetrahedron with vertices $\mathbf{0}, \mathbf{U}, \mathbf{V}, \mathbf{W}$ is $\frac{1}{6} \mathbf{U} \cdot (\mathbf{V} \times \mathbf{W})$.

(c) $\mathbf{W} \times \mathbf{V} = -\mathbf{V} \times \mathbf{W}$

Proof. (a) If we set $\mathbf{U} = \mathbf{V}$ or $\mathbf{U} = \mathbf{W}$ in (1.47), the left side is zero, because the determinant of a matrix that has two equal columns is zero. Therefore the right side is zero too.

(b) We have shown in Section 1.7 that $\frac{1}{6} \det[\mathbf{U}\ \mathbf{V}\ \mathbf{W}]$ is the signed volume of the ordered tetrahedron $\mathbf{0UVW}$. Since $\det[\mathbf{U}\ \mathbf{V}\ \mathbf{W}] = \mathbf{U} \cdot (\mathbf{V} \times \mathbf{W})$, part (b) follows.

(c) If we interchange $\mathbf{V}$ and $\mathbf{W}$ on the left side of (1.47) the sign changes by properties of determinant. Part (c) follows from this observation. $\qquad\square$

We can give now four related uses for cross products and determinant. See Figure 1.28. Let $\mathbf{U}$, $\mathbf{V}$, and $\mathbf{W}$ be vectors in $\mathbb{R}^3$.

- The volume of the parallelopiped determined by $\mathbf{U}, \mathbf{V}, \mathbf{W}$ is $|\mathbf{U} \cdot (\mathbf{V} \times \mathbf{W})|$.
- The area of the parallelogram determined by linearly independent vectors $\mathbf{V}$ and $\mathbf{W}$ is $\|\mathbf{V} \times \mathbf{W}\|$. To see this let $\mathbf{U}$ be $\dfrac{\mathbf{V} \times \mathbf{W}}{\|\mathbf{V} \times \mathbf{W}\|}$. The volume of the parallelopiped determined by $\mathbf{U}, \mathbf{V}, \mathbf{W}$ is $\|\mathbf{V} \times \mathbf{W}\|$ and its height is 1. Therefore the area of its base is $\|\mathbf{V} \times \mathbf{W}\|$.
- The area of the triangle with vertices $\mathbf{0}$, $\mathbf{V}$, and $\mathbf{W}$ is $\frac{1}{2}\|\mathbf{V} \times \mathbf{W}\|$.
- Imagine that at each point in space the velocity of a fluid is $\mathbf{U} = (u_1, u_2, u_3)$. Consider a parallelogram determined by vectors $\mathbf{V} = (v_1, v_2, v_3)$ and $\mathbf{W} = (w_1, w_2, w_3)$ as shown in Figure 1.28. Then $\mathbf{U} \cdot (\mathbf{V} \times \mathbf{W})$ is the volume of fluid that flows across the parallelogram in one time unit. This is called the *volumetric flow rate* or *flux*. Define a unit vector orthogonal to the parallelogram as $\mathbf{N} = \frac{\mathbf{V} \times \mathbf{W}}{\|\mathbf{V} \times \mathbf{W}\|}$; we get

$$\text{flux } = \mathbf{U} \cdot (\mathbf{V} \times \mathbf{W}) = \mathbf{U} \cdot \mathbf{N} \|\mathbf{V} \times \mathbf{W}\|,$$

Example 1.30. Find a plane containing three noncollinear points $\mathbf{U}, \mathbf{V}$, and $\mathbf{W}$.

(a) We can express the plane as the set of all points

$$\mathbf{X} = \mathbf{U} + s(\mathbf{V} - \mathbf{U}) + t(\mathbf{W} - U)$$

as s and t vary over all real numbers.

(b) A normal vector to the plane is $\mathbf{N} = (\mathbf{V} - \mathbf{U}) \times (\mathbf{W} - \mathbf{U})$, so we get an equation for the plane,

$$\mathbf{N} \cdot (\mathbf{X} - \mathbf{U}) = 0.$$

$\qquad\square$

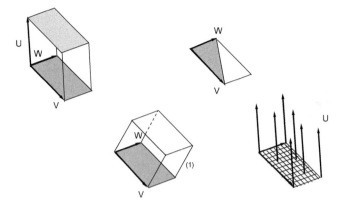

Fig. 1.28 One volume, two areas, and a flux, all based on the cross product.

Example 1.31. Let $\mathbf{U} = (1,1,-2)$ [length/time] be the velocity of a fluid flow. Find the volumetric flow rate across the oriented triangles $\mathbf{0VW}$, where $\mathbf{V}$ and $\mathbf{W}$ have length dimensions.

(a) $\mathbf{V} = (2,2,-4)$, $\mathbf{W} = (3,2,1)$
(b) $\mathbf{V} = (1,0,0)$, $\mathbf{W} = (3,2,1)$

The flow rate through the triangle is half that through the parallelogram, so it is $\frac{1}{2}\mathbf{U}\cdot(\mathbf{V}\times\mathbf{W})$. In (a), $\mathbf{V}$ is parallel to $\mathbf{U}$, and

$$\tfrac{1}{2}\mathbf{U}\cdot(\mathbf{V}\times\mathbf{W}) = \tfrac{1}{2}\det[\mathbf{U}\ 2\mathbf{U}\ \mathbf{W}] = 0.$$

No fluid flows across the triangle because the fluid velocity is parallel to it. In (b) the flow rate is

$$\tfrac{1}{2}\mathbf{U}\cdot(\mathbf{V}\times\mathbf{W}) = \tfrac{1}{2}\det\begin{bmatrix} 1 & 1 & 3 \\ 1 & 0 & 2 \\ -2 & 0 & 1 \end{bmatrix}. = -\tfrac{5}{2}.$$

The minus sign means that the angle between $\mathbf{U}$ and $\mathbf{V}\times\mathbf{W}$ is greater than $\pi/2$.   □

We ask you in Problem 1.106 to explore other properties of the cross product.

Problems

1.95. Find equations in the form $ax + by + cz = d$ for the following planes in $\mathbb{R}^3$.

(a) The plane through the origin with normal $(1,0,0)$.
(b) The plane through $(0,0,0)$, $(0,1,1)$, and $(-3,0,0)$.

(c) The plane containing the point $(1,1,1)$ and parallel to the plane with equation $x - 3y + 5z = 60$.

1.96. Which of these points are on the line $2x_1 + x_2 = 0$ in $\mathbb{R}^2$?
   (a) $(0,0)$ (b) $(-1,2)$ (c) $c\mathbf{X}$ if $\mathbf{X}$ is.   (d) $\mathbf{X} + \mathbf{Y}$ if $\mathbf{X}$ and $\mathbf{Y}$ are.

1.97. Which of these points are on the line $10x_1 + 5x_2 = 0$ in $\mathbb{R}^2$?
   (a) $(0,0)$ (b) $(-1,2)$ (c) $c\mathbf{X}$ if $\mathbf{X}$ is.   (d) $\mathbf{X} + \mathbf{Y}$ if $\mathbf{X}$ and $\mathbf{Y}$ are.

1.98. Which of these points are on the line $2x_1 + x_2 = 100$ in $\mathbb{R}^2$?
   (a) $(0,0)$ (b) $(50,0)$ (c) $(50,0) + \mathbf{Y}$ if $2y_1 + y_2 = 0$ (d) $c\mathbf{X}$ if $\mathbf{X}$ is.

1.99. Which of these points are on the plane in $\mathbb{R}^3$ given by $20x_1 + 10x_2 - 50x_3 = 0$?

(a) $(0,0,0)$
(b) $(0,5,1)$
(c) $(-1,2,0)$
(d) $c\mathbf{X}$ if $\mathbf{X}$ is.
(e) $\mathbf{X} + \mathbf{Y}$ if $\mathbf{X}$ and $\mathbf{Y}$ are.

1.100. Which of these points are on the plane in $\mathbb{R}^3$ given by $2x_1 + x_2 - 5x_3 = 100$?

(a) $(0,0,0)$
(b) $(50,0,0)$
(c) $(0,100,0)$
(d) $(0,100,0) + \mathbf{Y}$ if $2y_1 + y_2 - 5y_3 = 0$.
(e) $\mathbf{X} + \mathbf{Y}$ if $\mathbf{X}$ and $\mathbf{Y}$ are.

1.101. Let $\mathbf{U} = \mathbf{0}$, $\mathbf{V}_1 = (0,1,1)$, and $\mathbf{V}_2 = (-3,0,0)$.

(a) Find an equation of the line through $\mathbf{U}$ and $\mathbf{V}_1$ in the form $\mathbf{X}(s) = \mathbf{U} + s\mathbf{V}_1$.
(b) Find an equation of the line through $\mathbf{U}$ and $\mathbf{V}_2$ in the form $\mathbf{X}(t) = \mathbf{U} + t\mathbf{V}_2$.
(c) Find an equation of the plane through $\mathbf{U}$, $\mathbf{V}_1$, and $\mathbf{V}_2$ in the form

$$\mathbf{X}(s,t) = \mathbf{U} + s\mathbf{V}_1 + t\mathbf{V}_2.$$

1.102. Which of these points are on the hyperplane in $\mathbb{R}^4$ where $x_1 + x_2 + x_3 + x_4 = 0$?
   (a) $(\frac{1}{5}, -\frac{1}{5}, \frac{1}{5}, -\frac{1}{5})$ (b) $(a,a,a,-3a)$ (c) $(-1,1)$ (d) $c\mathbf{X}$ if $\mathbf{X}$ is.   (e) $(1,-3,2)$

1.103. Let

$$\mathbf{U} = \begin{bmatrix} u_1 \\ u_2 \\ u_3 \end{bmatrix}, \qquad \mathbf{V} = \begin{bmatrix} v_1 \\ v_2 \\ v_3 \end{bmatrix}, \qquad \mathbf{W} = \begin{bmatrix} w_1 \\ w_2 \\ w_3 \end{bmatrix}.$$

Write expressions a, b, and c in terms of the components of $\mathbf{V}$ and $\mathbf{W}$ so that

$$\det[\mathbf{U} \ \mathbf{V} \ \mathbf{W}] = au_1 + bu_2 + cu_3.$$

Compare your results to formula (1.47), $\det[\mathbf{U} \ \mathbf{V} \ \mathbf{W}] = \mathbf{U} \cdot \mathbf{V} \times \mathbf{W}$, to derive a formula for $\mathbf{V} \times \mathbf{W}$.

1.104. Evaluate the cross products using the formula

$$\mathbf{V} \times \mathbf{W} = (v_2 w_3 - v_3 w_2, v_3 w_1 - v_1 w_3, v_1 w_2 - v_2 w_1).$$

(a) $(1,0,0) \times (0,1,0)$
(b) $(0,1,0) \times (1,0,0)$
(c) $(0,0,1) \times (a,b,c)$

1.105. Using $\mathbf{i} \times \mathbf{j} = \mathbf{k}$, $\mathbf{j} \times \mathbf{k} = \mathbf{i}$, $\mathbf{k} \times \mathbf{i} = \mathbf{j}$, and the distributive and anticommutative laws $\mathbf{U} \times (\mathbf{V} + \mathbf{W}) = \mathbf{U} \times \mathbf{V} + \mathbf{U} \times \mathbf{W}$, $\mathbf{V} \times \mathbf{W} = -\mathbf{W} \times \mathbf{V}$, evaluate the following cross products.

(a) $(1,0,0) \times (0,1,0)$
(b) $\mathbf{j} \times (\mathbf{i} + \mathbf{k})$
(c) $(2\mathbf{i} + 3\mathbf{k}) \times (a\mathbf{i} + b\mathbf{j} + c\mathbf{k})$

1.106. Verify the following properties of the cross product in $\mathbb{R}^3$.

(a) the product is anticommutative: $\mathbf{U} \times \mathbf{V} = -\mathbf{V} \times \mathbf{U}$
(b) the "back-cab" rule: $\mathbf{A} \times (\mathbf{B} \times \mathbf{C}) = \mathbf{B}(\mathbf{A} \cdot \mathbf{C}) - \mathbf{C}(\mathbf{A} \cdot \mathbf{B})$,
(c) the product is not associative: $\mathbf{A} \times (\mathbf{B} \times \mathbf{C}) \neq (\mathbf{A} \times \mathbf{B}) \times \mathbf{C}$. In fact there is a product rule similar to the product rule for derivatives:

$$\mathbf{A} \times (\mathbf{B} \times \mathbf{C}) = (\mathbf{A} \times \mathbf{B}) \times \mathbf{C} + \mathbf{B} \times (\mathbf{A} \times \mathbf{C}).$$

(d) As a function of $\mathbf{V}$, the product $\mathbf{U} \times \mathbf{V}$ is linear, so can be written as $\mathbf{U} \times \mathbf{V} = \mathbf{M} \mathbf{V}$ for some matrix $\mathbf{M}$. Find $\mathbf{M}$.
(e) $\|\mathbf{U} \times \mathbf{V}\|^2 + (\mathbf{U} \cdot \mathbf{V})^2 = \|\mathbf{U}\|^2 \|\mathbf{V}\|^2$.
(f) Use part (e) and $\mathbf{U} \cdot \mathbf{V} = \|\mathbf{U}\| \|\mathbf{V}\| \cos\theta$, where θ is the angle between the vectors, to show that $\|\mathbf{U} \times \mathbf{V}\| = \|\mathbf{U}\| \|\mathbf{V}\| \sin\theta$.

1.107. Find the volumetric flow rate $\mathbf{U} \cdot (\mathbf{V} \times \mathbf{W})$ of a fluid of velocity $\mathbf{U}$ through the parallelogram determined by $\mathbf{V}$ and $\mathbf{W}$. Sketch the parallelogram, and the parallelepiped determined by $\mathbf{U}, \mathbf{V}$, and $\mathbf{W}$.

(a) $\mathbf{U} = (2,0,0)$, $\mathbf{V} = (0,2,0)$, $\mathbf{W} = (0,0,7)$.
(b) $\mathbf{U} = (-2,0,0)$, $\mathbf{V} = (0,2,0)$, $\mathbf{W} = (0,0,7)$.
(c) $\mathbf{U} = (2,1,0)$, $\mathbf{V} = (1,2,0)$, $\mathbf{W} = (7,7,7)$.

Chapter 2
Functions

Abstract The concept of function is central to mathematics. In single variable calculus we studied functions that assign to each number in their domain a number. In multivariable calculus we study functions that assign to each vector with n components in their domain a vector with m components.

2.1 Functions of several variables

We use the notation $\mathbf{F} : D \subset \mathbb{R}^n \to \mathbb{R}^m$ for a function $\mathbf{F}$ that assigns a vector $\mathbf{F}(\mathbf{X})$ in $\mathbb{R}^m$ to each vector $\mathbf{X}$ in a subset D of $\mathbb{R}^n$, and say that $\mathbf{F}$ is a function from $\mathbb{R}^n$ to $\mathbb{R}^m$. When the *domain* D of a function $\mathbf{F}$ of n variables is not specified we assume, as we have with functions of a single variable, that the domain is the largest set for which the definition makes sense. We call the set of outputs $\mathbf{F}(D)$ the *range* of $\mathbf{F}$ or the *image* of D. We call $\mathbf{F}$ *one to one* if $\mathbf{F}(\mathbf{U}) = \mathbf{F}(\mathbf{V})$ only when $\mathbf{U} = \mathbf{V}$. We say $\mathbf{F}$ maps D *onto* a set B in $\mathbb{R}^m$ if $\mathbf{F}(D) = B$. We usually denote a function whose output is a vector by a bold capital letter. A function whose output is a number is called a scalar valued, or real valued, function and we usually denote it by a lower case letter.

If f and $\mathbf{F}$ are functions then $f(\mathbf{X})$ and $\mathbf{F}(\mathbf{X})$ are values assigned to $\mathbf{X}$. It is sometimes convenient to indicate names of domain and range variables by speaking of "a function" $r = f(\theta)$, $\mathbf{U} = \mathbf{F}(\mathbf{X})$, $\mathbf{V} = (u(x,y,t), v(x,y,t))$, and so on.

Definition 2.1. A *function* $\mathbf{F} : D \subset \mathbb{R}^n \to \mathbb{R}^m$ assigns a vector $\mathbf{F}(\mathbf{X})$ in $\mathbb{R}^m$ to each $\mathbf{X}$ in D, denoted

$$\mathbf{F}(\mathbf{X}) = (f_1(\mathbf{X}), f_2(\mathbf{X}), \dots, f_m(\mathbf{X})).$$

The function f_j is called the j-th *component function* of $\mathbf{F}$.

Next we look at some examples of functions.

© Springer International Publishing AG 2017
P. D. Lax and M. S. Terrell, *Multivariable Calculus with Applications*,
Undergraduate Texts in Mathematics, https://doi.org/10.1007/978-3-319-74073-7_2

Definition 2.2. A function that assigns the same vector $\mathbf{C}$ in $\mathbb{R}^m$ to each vector $\mathbf{X}$ in $\mathbb{R}^n$ is called a *constant* function.

Example 2.1. The functions $f(x,y) = 7$ and $\mathbf{G}(x,y) = (8,3,2)$ are examples of constant functions. f is a constant function from $\mathbb{R}^2$ to $\mathbb{R}$, and $\mathbf{G}$ is a constant function from $\mathbb{R}^2$ to $\mathbb{R}^3$.   □

Linear functions In Chapter 1 we defined linear functions from $\mathbb{R}^n$ to $\mathbb{R}^m$. We review that definition and Theorem 1.19 about representing linear functions.

Definition 2.3. A function $\mathbf{L}$ from $\mathbb{R}^n$ to $\mathbb{R}^m$ is *linear* if

$$a\mathbf{L}(\mathbf{U}) = \mathbf{L}(a\mathbf{U}), \quad \text{and} \quad \mathbf{L}(\mathbf{U}+\mathbf{V}) = \mathbf{L}(\mathbf{U})+\mathbf{L}(\mathbf{V}) \qquad (2.1)$$

for all numbers a and all vectors $\mathbf{U}$ and $\mathbf{V}$ in $\mathbb{R}^n$.

Example 2.2. Is $\mathbf{L}(x,y) = (2x-3y,x,5y)$ a linear function from $\mathbb{R}^2$ to $\mathbb{R}^3$? Let's check to see whether (2.1) holds.

Let a be a number, and let (x,y) and (u,v) be vectors in $\mathbb{R}^2$.

$$a\mathbf{L}(x,y) = a(2x-3y,x,5y) = (2ax-3ay,ax,5ay) = \mathbf{L}(ax,ay) = \mathbf{L}(a(x,y)).$$

Also,

$$\mathbf{L}(x,y)+\mathbf{L}(u,v) = (2x-3y,x,5y)+(2u-3v,u,5v)$$
$$= (2(x+u)-3(y+v),x+u,5(y+v)) = \mathbf{L}(x+u,y+v).$$

So $\mathbf{L}$ is linear.   □

In Problem 2.4 we ask you to show that a constant function $\mathbf{F}$ from $\mathbb{R}^n$ to $\mathbb{R}^m$ is linear if and only if $\mathbf{F}(x_1,x_2,\ldots,x_n) = \mathbf{0}$.

Recall that according to Theorem 1.10 every linear function ℓ from $\mathbb{R}^n$ to $\mathbb{R}$ is of the form

$$\ell(x_1,x_2,\ldots,x_n) = c_1x_1 + c_2x_2 + c_3x_3 + \ldots + c_nx_n,$$

which we can write as

$$\ell(\mathbf{X}) = \mathbf{C}\cdot\mathbf{X},$$

for some $\mathbf{C} = (c_1,c_2,\ldots,c_n)$ in $\mathbb{R}^n$.

Let's look again at why a linear function $\mathbf{L}$ from $\mathbb{R}^n$ to $\mathbb{R}^m$ can be represented using matrix multiplication. Let ℓ_k be the k-th component function of $\mathbf{L}$, $k = 1,2,\ldots,m$. Since $\mathbf{L}$ is linear it follows that each component $\ell_k(\mathbf{X})$ is a linear function. So by Theorem 1.10 there is a vector $\mathbf{C}_k$ in $\mathbb{R}^n$ so that $\ell_k(\mathbf{X}) = \mathbf{C}_k\cdot\mathbf{X}$. Denote the vectors $\mathbf{C}_k = (c_{k1},c_{k2},\ldots c_{kn})$ as row vectors and denote $\mathbf{X}$ and $\mathbf{L}(\mathbf{X})$ as column vectors. Let $\mathbf{C}$ be the matrix whose k-th row is $\mathbf{C}_k$. The relations $\ell_k(\mathbf{X}) = \mathbf{C}_k\cdot\mathbf{X}$, $k = 1,2,\ldots,m$ can be expressed as the product of the matrix $\mathbf{C}$ and the vector $\mathbf{X}$:

$$\mathbf{L}(\mathbf{X}) = \mathbf{CX}.$$

Theorem 2.1. *Every linear function* **L** *from* $\mathbb{R}^n$ *to* $\mathbb{R}^m$ *can be written in matrix form*

$$\mathbf{L}(\mathbf{X}) = \begin{bmatrix} c_{11} & c_{12} & \cdots & c_{1n} \\ c_{21} & c_{22} & \cdots & c_{2n} \\ \vdots & \vdots & \cdots & \vdots \\ c_{m1} & c_{m2} & \cdots & c_{mn} \end{bmatrix} \begin{bmatrix} x_1 \\ x_2 \\ \vdots \\ x_n \end{bmatrix} = \mathbf{CX}. \tag{2.2}$$

C *is called the matrix of* **L** *or the matrix that represents* **L**. *(See also Theorem 1.19.)*

We use the following definition to quantify the size of a matrix.

Definition 2.4. For an $m \times n$ matrix $\mathbf{C} = [c_{ij}]$, the *norm* $\|\mathbf{C}\|$ of $\mathbf{C}$ is defined by

$$\|\mathbf{C}\| = \sqrt{\sum_{i=1}^{m} \sum_{j=1}^{n} c_{ij}^2}. \tag{2.3}$$

There is an important relation between the norms of $\mathbf{X}$, $\mathbf{CX}$, and $\mathbf{C}$.

Theorem 2.2. *Let* **C** *be an* $m \times n$ *matrix. Then*

$$\|\mathbf{CX}\| \le \|\mathbf{C}\| \|\mathbf{X}\|$$

for every vector **X** *in* $\mathbb{R}^n$.

Proof. The k-th component of $\mathbf{CX}$ is the dot product of the k-th row of $\mathbf{C}$,

$$\mathbf{C}_k = (c_{k1}, c_{k2}, \dots, c_{kn})$$

with $\mathbf{X} = (x_1, x_2, \dots, x_n)$. By the definition of norm,

$$\|\mathbf{CX}\| = \sqrt{(\mathbf{C}_1 \cdot \mathbf{X})^2 + (\mathbf{C}_2 \cdot \mathbf{X})^2 + \cdots + (\mathbf{C}_m \cdot \mathbf{X})^2}.$$

According to the Cauchy–Schwarz inequality, Theorem 1.12, applied to each component this is

$$\le \sqrt{\|\mathbf{C}_1\|^2 \|\mathbf{X}\|^2 + \|\mathbf{C}_2\|^2 \|\mathbf{X}\|^2 + \cdots + \|\mathbf{C}_m\|^2 \|\mathbf{X}\|^2}$$

$$= \|\mathbf{X}\| \sqrt{\|\mathbf{C}_1\|^2 + \|\mathbf{C}_2\|^2 + \cdots + \|\mathbf{C}_m\|^2} = \|\mathbf{X}\| \sqrt{\sum_{i=1}^{m} \sum_{j=1}^{n} c_{ij}^2} = \|\mathbf{C}\| \|\mathbf{X}\|. \qquad \square$$

Combining Theorems 2.1 and 2.2 we see that for every linear function **L** from $\mathbb{R}^n$ to $\mathbb{R}^m$

$$\|\mathbf{LX}\| \le \|\mathbf{C}\|\|\mathbf{X}\|.$$

That is, the norm of the output of a linear function is *not greater* than $\|\mathbf{C}\|$ times the norm of the input.

Functions from $\mathbb{R}^n$ to $\mathbb{R}$ Next we look at some examples of functions from $\mathbb{R}^n$ to $\mathbb{R}$.

Example 2.3. The area a of a rectangle that has length x, width y is given by $a(x,y) = xy$. The domain of a is the set D of all ordered pairs (x,y) with $x > 0$ and $y > 0$, $a : D \subset \mathbb{R}^2 \to \mathbb{R}$. Notice that even though the rule for a makes sense for all ordered pairs (x,y), we restricted the domain based on the context of the problem. □

Example 2.4. The volume v of a rectangular box that has length x, width y, and height z is given by the function $v(x,y,z) = xyz$. The domain of v, D, is the set of all ordered triples (x,y,z) with $x > 0$, $y > 0$, and $z > 0$. The function $v : D \subset \mathbb{R}^3 \to \mathbb{R}$. □

Example 2.5. Let $g(x,y) = \sqrt{x^2 + y^2}$, $g : \mathbb{R}^2 \to \mathbb{R}$. $g(x,y)$ is the norm of (x,y), $g(x,y) = \|(x,y)\|$. □

Visualizing functions; graphs of functions from $\mathbb{R}^2$ to $\mathbb{R}$. In single variable calculus we visualized a function f by sketching its graph, the set of all points $(x,f(x))$. For a function from $\mathbb{R}^2$ to $\mathbb{R}$, the *graph* of f is the set of ordered triples $(x,y,f(x,y))$, and often looks like a surface in (x,y,z) space, as in Figure 2.1.

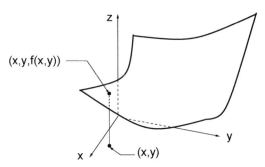

Fig. 2.1 The graph of a function $f : \mathbb{R}^2 \to \mathbb{R}$ is a subset of $\mathbb{R}^3$.

Next we sketch the graph of a few functions from $\mathbb{R}^2$ to $\mathbb{R}$.

Example 2.6. Figure 2.2 shows the graph of the constant function $f(x,y) = 7$. □

Example 2.7. To sketch the graph of $f(x,y) = x^2 + y^2$ we notice that the points in the domain where $f(x,y) = c$, $c \ge 0$, lie on a circle of radius $\sqrt{c}$. The circle of radius zero is one point. On the left side of Figure 2.3. we sketched the

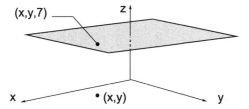

Fig. 2.2 The graph of the constant function in Example 2.6.

points in the domain satisfying $x^2 + y^2 = c$ for $c = 0, 1, 4, 9$. On the right hand side of Figure 2.3. we sketched the corresponding points on the graph of f. $\square$

Definition 2.5. Let $f : D \subset \mathbb{R}^n \to \mathbb{R}$ be a real valued function, and c a number. The set of all points $(x_1, x_2, \ldots, x_n)$ in the domain D where $f(x_1, x_2, \ldots, x_n) = c$ is called the c *level set* of f.

For functions of two variables we can draw level sets in the domain and plot corresponding points (x, y, c) on the graph. This gives an idea what the graph looks like, as we did in Example 2.7. The set of points on the graph that correspond to a level set $f(x, y) = c$ is called the *contour* curve of the graph at $z = c$.

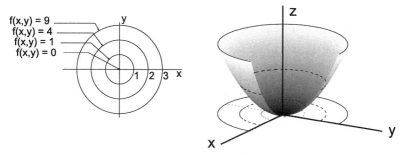

Fig. 2.3 *Left:* level sets $c = 0, 1, 4, 9$ for $f(x, y) = x^2 + y^2$, *Right:* the graph (cut away for clarity) of $f(x, y)$ with contour curves shown at $z = 1, 4$ and 9. See Example 2.7.

Example 2.8. Let $f(x, y) = \dfrac{x^2 - y^2}{x^2 + y^2}$, $(x, y) \neq (0, 0)$. Find level set $c = \dfrac{x^2 - y^2}{x^2 + y^2}$, or $x^2 - y^2 = c(x^2 + y^2)$ with $x^2 + y^2 \neq 0$. First we consider some simple choices for c. When $c = 1$ we find that $x^2 - y^2 = x^2 + y^2$. Thus the c=1 level set is the set of all points $y = 0$, the x axis, with the origin removed. Similarly, the level set for $c = -1$ satisfies $x^2 - y^2 = -(x^2 + y^2)$ with $x^2 + y^2 \neq 0$. So this level set is the set of points $x = 0$, the y axis, with the origin removed. For $c \neq -1$ and $x \neq 0$ we can divide by x^2 to get

$$1 - \frac{y^2}{x^2} = c\left(1 + \frac{y^2}{x^2}\right).$$

Solving for $\left(\frac{y}{x}\right)^2$ we get $\left(\frac{y}{x}\right)^2 = \frac{1-c}{1+c}$. Thus, each c level set is a pair of lines $y = \sqrt{\frac{1-c}{1+c}}x$ and $y = -\sqrt{\frac{1-c}{1+c}}x$. So for example, the level set for $c = -\frac{1}{2}$ is the pair of lines $y = \sqrt{3}x$, $y = -\sqrt{3}x$, origin deleted. In Figure 2.4 we sketched the level sets for $c = -\frac{1}{2}, 0, 1$ and the graph of f with the corresponding contour lines on the graph.   □

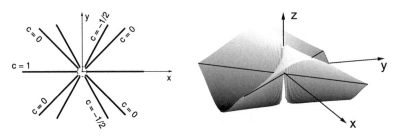

Fig. 2.4 *Left:* level sets $c = -\frac{1}{2}, 0, 1$ of $f(x,y) = \frac{x^2 - y^2}{x^2 + y^2}$. *Right:* the graph of f with contour lines at $z = 0$ shown. See Example 2.8.

Here are some further examples of level sets of a function.

Example 2.9. Figure 2.5 shows a circular disk, D, and all the points in D where the value of a function $t(x,y)$ is 2, 3, 4, 5, 6, 8, 10 or 10.5. The level sets $t = 2$ and $t = 10.5$ are points. The level sets $t = 3$, $t = 4$, $t = 6$, and $t = 8$ are curves. The level set $t = 5$ consists of all the points of the shaded band.   □

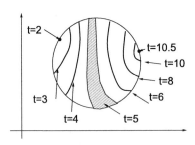

Fig. 2.5 Level sets in Example 2.9.

Example 2.10. Let $f(x,y,z) = \dfrac{1}{\sqrt{x^2 + y^2 + z^2}}$. The domain of f is the set of all points $(x,y,z) \neq (0,0,0)$. The *graph* of f is the set of points $(x,y,z,f(x,y,z))$.

The c level set of f for $c > 0$ is the set of points that satisfy $\dfrac{1}{\sqrt{x^2 + y^2 + z^2}} = c$,

or $x^2 + y^2 + z^2 = \dfrac{1}{c^2}$. It is the sphere with radius $\dfrac{1}{c}$ centered at the origin of $\mathbb{R}^3$.
□

Functions from $\mathbb{R}$ to $\mathbb{R}^n$ Next we look at examples of functions from $\mathbb{R}$ to $\mathbb{R}^n$.

Example 2.11. Let I be an interval $[a,b]$. Let x, y, and z be functions from I to $\mathbb{R}$ and let
$$\mathbf{P}(t) = (x(t), y(t), z(t)).$$

Figure 2.6 shows a, t_1, t_2 and b on the interval I and the corresponding image positions in space $\mathbf{P}(a)$, $\mathbf{P}(t_1)$, $\mathbf{P}(t_2)$, and $\mathbf{P}(b)$.
□

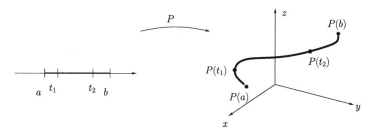

Fig. 2.6 Position $\mathbf{P}(t)$ in Example 2.11.

Example 2.12. Let $\mathbf{F}(t) = (\cos t, \sin t, t)$, $0 \leq t \leq 4\pi$. Figure 2.7 shows solid dots as points $\mathbf{F}(t)$ where $t = 0, \frac{\pi}{2}, \pi, 2\pi, \frac{5\pi}{2}, 3\pi$, and 4π, and a curve for the other points in the range of $\mathbf{F}$. The resulting curve is called a *helix*.
□

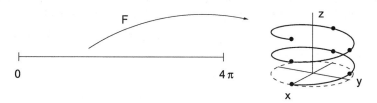

Fig. 2.7 The helix in Example 2.12 and its projection (dotted) on the x, y plane.

Example 2.13. Let $\mathbf{F}(t) = t(2,3,4) = (2t,3t,4t)$. Figure 2.8 shows points in the range where $t = -.5, -.25, 0, 1.1$. The range of $\mathbf{F}$ is the line that goes through the origin and $(2,3,4)$.
□

Example 2.14. Let $\mathbf{G}(t) = (1,2,0) + t(2,3,4) = (1 + 2t, 2 + 3t, 4t)$. The range of $\mathbf{G}$ is the line in Figure 2.9 that goes through $(1,2,0)$ and $(3,5,4)$.
□

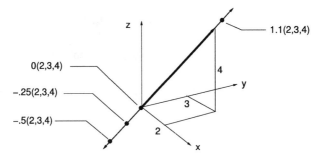

Fig. 2.8 The line in Example 2.13.

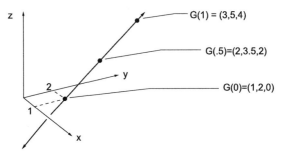

Fig. 2.9 The line in Example 2.14.

In Examples 2.11–2.14 we sketched the range of the functions. In Examples 2.11 and 2.12 we sketched the domain of each function showing how a point gets mapped to a point in $\mathbb{R}^n$.

Functions from $\mathbb{R}^n$ to $\mathbb{R}^n$ Next we consider functions from $\mathbb{R}^n$ to $\mathbb{R}^n$, called *vector fields*. One way to visualize a vector field from $\mathbb{R}^2$ to $\mathbb{R}^2$ is to draw an arrow representing $\mathbf{F}(x,y)$ starting at the point (x,y). Let's look at an example.

Example 2.15. Describe the vector field $\mathbf{F}(x,y) = (-y,x)$ by sketching a few vectors. We first make a list.

(x,y)	$(1,0)$	$(0,1)$	$(-1,0)$	$(0,-1)$	$(2,2)$	$(-2,2)$	$(-2,-2)$	$(2,-2)$
$\mathbf{F}(x,y)$	$(0,1)$	$(-1,0)$	$(0,-1)$	$(1,0)$	$(-2,2)$	$(-2,-2)$	$(2,-2)$	$(2,2)$

See Figure 2.10. The vectors (x,y) and $\mathbf{F}(x,y)$ are perpendicular since the dot product $(x,y) \cdot (-y,x) = 0$. In fact if we take (x,y) on a circle centered at $(0,0)$ we see that $\mathbf{F}(x,y)$ has magnitude equal to the radius of the circle, and direction tangent to the circle.   □

Example 2.16. Let

$$\mathbf{F}(x,y,z) = \left(\frac{x}{\sqrt{x^2 + y^2 + z^2}}, \frac{y}{\sqrt{x^2 + y^2 + z^2}}, \frac{z}{\sqrt{x^2 + y^2 + z^2}} \right).$$

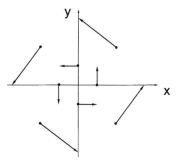

Fig. 2.10 A sketch of $\mathbf{F}(x,y) = (-y,x)$ in Example 2.15.

Let $\mathbf{X} = (x,y,z)$, then $\|\mathbf{X}\| = \sqrt{x^2 + y^2 + z^2}$ and we can express $\mathbf{F}$ as

$$\mathbf{F}(\mathbf{X}) = \frac{\mathbf{X}}{\|\mathbf{X}\|}.$$

The domain of $\mathbf{F}$ contains every point in $\mathbb{R}^3$ except the origin. Every value of $\mathbf{F}$ is a unit vector. The range of $\mathbf{F}$ is the unit sphere centered at the origin. Figure 2.11 shows a sketch of this vector field. □

Fig. 2.11 The vector field in Example 2.16. *Left*: a sketch of two points and the vectors assigned by the vector field $\mathbf{F}$. *Right*: a display of 45 points in the first octant and the vectors assigned by $\mathbf{F}$.

Example 2.17. Let $\mathbf{F}$ be the vector field

$$\mathbf{F}(x.y.z) = -\frac{1}{(\sqrt{x^2 + y^2 + z^2})^3}(x,y,z).$$

that can also be expressed as

$$\mathbf{F}(\mathbf{X}) = -\frac{\mathbf{X}}{\|\mathbf{X}\|^3} = -\frac{\mathbf{X}}{\|\mathbf{X}\|}\frac{1}{\|\mathbf{X}\|^2}.$$

To visualize $\mathbf{F}$, plot a point $\mathbf{X}$ and then sketch the unit vector $-\dfrac{\mathbf{X}}{\|\mathbf{X}\|}$ whose direction is opposite $\mathbf{X}$. Adjust the length by the factor $\dfrac{1}{\|\mathbf{X}\|^2}$. We call $\mathbf{F}$ the *inverse square* vector field. See Figure 2.12.   □

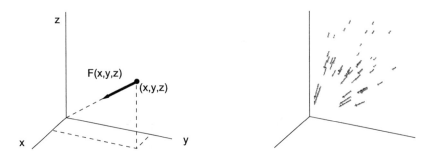

Fig. 2.12 The inverse square vector field in Example 2.17. *Left*: a sketch of a single point (x,y,z) and the vector $\mathbf{F}(x,y,z)$ assigned to it by $\mathbf{F}$. *Right*: a display of 45 points in the first octant and the vectors assigned to them.

For a function $\mathbf{F}$ from $\mathbb{R}^3$ to $\mathbb{R}^3$ physicists employ the notation

$$\mathbf{F}(x,y,z) = (f_1(x,y,z),f_2(x,y,z),f_3(x,y,z)) = f_1(x,y,z)\mathbf{i}+f_2(x,y,z)\mathbf{j}+f_3(x,y,z)\mathbf{k}$$

where
$$\mathbf{i} = (1,0,0), \quad \mathbf{j} = (0,1,0), \quad \mathbf{k} = (0,0,1).$$

Similarly, a function from $\mathbb{R}^2$ to $\mathbb{R}^2$ can be written using $\mathbf{i} = (1,0)$, $\mathbf{j} = (0,1)$ as

$$\mathbf{F}(x,y) = (f_1(x,y),f_2(x,y)) = f_1(x,y)\mathbf{i}+f_2(x,y)\mathbf{j}.$$

Example 2.18. Let $\mathbf{F}(x,y,z) = (2,3,4) = 2\mathbf{i}+3\mathbf{j}+4\mathbf{k}$. $\mathbf{F}$ is a constant vector field. A sketch of $\mathbf{F}$ looks like a field of arrows of the same length all pointing in the same direction. See Figure 2.13.   □

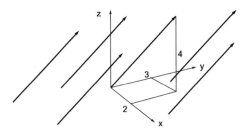

Fig. 2.13 The constant vector field $\mathbf{F}(x,y,z) = (2,3,4) = 2\mathbf{i}+3\mathbf{j}+4\mathbf{k}$ of Example 2.18.

Definition 2.6. Let $\mathbf{F}$ be a function from $\mathbb{R}^m$ to $\mathbb{R}^n$ and let $\mathbf{G}$ be a function from $\mathbb{R}^k$ to $\mathbb{R}^m$. Suppose the range of $\mathbf{G}$ is contained in the domain of $\mathbf{F}$. We define the *composite function* $\mathbf{F} \circ \mathbf{G}$ from $\mathbb{R}^k$ to $\mathbb{R}^n$ by

$$\mathbf{F} \circ \mathbf{G}(\mathbf{X}) = \mathbf{F}(\mathbf{G}(\mathbf{X})).$$

Example 2.19. Let $\mathbf{G}(x,y,z) = (x^2, y^2 + z^2)$ and let $f(u,v) = uv$. Then

$$f \circ \mathbf{G}(x,y,z) = x^2(y^2 + z^2).$$

$\square$

Problems

2.1. Express each linear function as $f(\mathbf{X}) = \mathbf{C} \cdot \mathbf{X}$ for some vector $\mathbf{C}$ or $\mathbf{F}(\mathbf{X}) = \mathbf{C}\mathbf{X}$ for some matrix $\mathbf{C}$.

(a) $f(x_1, x_2) = x_1 + 2x_2$
(b) $f(x_1, x_2, x_3) = x_1 + 2x_2$
(c) $\mathbf{F}(x_1, x_2) = (x_1, x_1 + 2x_2)$
(d) $\mathbf{F}(x_1, x_2, x_3) = (x_1, x_1 + 2x_2)$
(e) $\mathbf{F}(x_1, x_2, x_3, x_4) = (x_4 - x_2, x_3 - x_1, x_2 + 5x_1, x_1 + x_2 + x_3 + x_4)$
(f) $\mathbf{F}(x_1, x_2) = (x_1, 5x_1, -x_2, -2x_1, x_2)$
(g) $\mathbf{F}(x_1, x_2, x_3, x_4, x_5) = (x_1, 5x_1, -x_2, -2x_1, x_2)$

2.2. Let $f(x,y)$ be equal to 1 when (x,y) is inside the unit disk centered at the origin, and 0 when $x^2 + y^2 \geq 1$. Describe the level sets of f. Sketch a graph of f.

2.3. Let $f(\mathbf{X})$ be equal to $\|\mathbf{X}\|^2$ when $\mathbf{X}$ is inside the unit ball $\|\mathbf{X}\| < 1$, and 0 when $\|\mathbf{X}\| \geq 1$. Describe the level sets of f when the domain is

(a) $\mathbb{R}$,
(b) $\mathbb{R}^3$,
(c) $\mathbb{R}^5$.

2.4. Show that a constant function $\mathbf{F} : \mathbb{R}^n \to \mathbb{R}^m$ is linear if and only if

$$\mathbf{F}(x_1, x_2, \dots, x_n) = \mathbf{0}.$$

2.5. A plane in $\mathbb{R}^3$ with equation $z = 2x + 3y$ is the 0 level set of a linear function ℓ from $\mathbb{R}^3$ to $\mathbb{R}$. Find ℓ.

2.6. The function

$$f(x,y) = 5 + x + x^2 + y^2,$$

is the sum of the constant function 5, the linear function x, and the degree 2 polynomial $x^2 + y^2$.

(a) Give two examples of points (x,y) near $(0,0)$ for which the approximation by the first two parts,

$$f(x,y) \approx 5 + x$$

differs from $f(x,y)$ less than $\frac{1}{100}$.

(b) Show that

$$g(u,v) = f(1 + u, 2 + v)$$

is the sum of a constant function, a linear function of (u,v), and a degree 2 polynomial function of u and v.

2.7. Let $f(\mathbf{X}) = \|\mathbf{X}\|^2$, $\mathbf{X}$ in $\mathbb{R}^4$. Let $\mathbf{A}$ be in $\mathbb{R}^4$ and define

$$g(\mathbf{X}) = \|\mathbf{A}\|^2 + 2\mathbf{A} \cdot (\mathbf{X} - \mathbf{A}).$$

(a) Let $\mathbf{U} = \mathbf{X} - \mathbf{A}$. Use the formula

$$\|\mathbf{A} + \mathbf{U}\|^2 = \|\mathbf{A}\|^2 + 2\mathbf{A} \cdot \mathbf{U} + \|\mathbf{U}\|^2$$

to show that the difference between $f(\mathbf{X})$ and $g(\mathbf{X})$ is $\|\mathbf{U}\|^2$.

(b) Show that the difference between $f(\mathbf{X})$ and $g(\mathbf{X})$ does not exceed 10^{-4} when $\|\mathbf{X} - \mathbf{A}\| < 10^{-2}$.

2.8. Let $\mathbf{L}$ be a linear function from $\mathbb{R}^n$ to $\mathbb{R}^m$. Show that

(a) If $\mathbf{L}(\mathbf{X}) = \mathbf{0}$ and $\mathbf{L}(\mathbf{Y}) = \mathbf{0}$, then $\mathbf{L}(\mathbf{X} + \mathbf{Y})$ is also $\mathbf{0}$.

(b) If $\mathbf{L}(\mathbf{X}) = \mathbf{0}$ and c is a number, then $\mathbf{L}(c\mathbf{X})$ is also $\mathbf{0}$.

2.9. Rework Example 2.8 by assuming $y \neq 0$ rather than $x \neq 0$.

2.10. Sketch or describe the c level sets, $f(x,y) = c$ in $\mathbb{R}^2$, for the following functions f and values c. Use the level sets to help sketch the graph of the function.

(a) $f(x,y) = x + 2y$, $c = -1, 0, 1, 2$

(b) $f(x,y) = xy$, $c = -1, 0, 1, 2$

(c) $f(x,y) = x^2 - y$, $c = -1, 0, 1, 2$

(d) $f(x,y) = \sqrt{1 - x^2 - y^2}$, $c = 0, \frac{1}{2}, 1$

2.11. Sketch or describe the c level sets, $f(x,y) = c$ in $\mathbb{R}^2$, for the following functions f from $\mathbb{R}^2$ to $\mathbb{R}$ and values c.

(a) $f(x,y) = x^2 + y^2$, $c = 0, 1, 2$

(b) $f(x,y) = \sqrt{x^2 + y^2}$, $c = 0, 1, 2$

(c) $f(x,y) = \dfrac{1}{x^2 + y^2}$, $c = 0, 1, 2$. Which one is empty?

2.12. Justify the claim that Theorem 1.20 can be restated as follows: "A linear function from $\mathbb{R}^n$ to $\mathbb{R}^n$ is onto if and only if it is one to one."

2.13. Sketch or describe the c level sets, $f(\mathbf{X}) = c$ in $\mathbb{R}^3$, for the following functions f from $\mathbb{R}^3$ to $\mathbb{R}$ and values c.

(a) $f(\mathbf{X}) = \|\mathbf{X}\|^2$, $c = \frac{1}{2}, 1, 2$

(b) $f(\mathbf{X}) = \|\mathbf{X}\|$, $c = \frac{1}{2}, 1, 2$

(c) $f(\mathbf{X}) = \dfrac{1}{\|\mathbf{X}\|^2}$, $c = \frac{1}{2}, 1, 2$

(d) $f(\mathbf{X}) = \mathbf{U} \cdot \mathbf{X}$, $c = \frac{1}{2}, 1, 2$ where $\mathbf{U}$ is a unit vector. Hint: write $\mathbf{X}$ as a vector parallel to $\mathbf{U}$ plus a vector orthogonal to $\mathbf{U}$.

2.14. Let $\mathbf{C} = \begin{bmatrix} 3 & 1 \\ 2 & 4 \end{bmatrix}$, and $\mathbf{X} = \begin{bmatrix} 1 \\ 2 \end{bmatrix}$. verify the inequality

$$\|\mathbf{C}\mathbf{X}\| \le \|\mathbf{C}\|\,\|\mathbf{X}\|.$$

2.15. Show that for a linear function $\mathbf{L}$ from $\mathbb{R}^n$ to $\mathbb{R}^m$, the value $\mathbf{L}(\mathbf{X})$ is a linear combination of the columns $\mathbf{V}_j$ of the matrix of $\mathbf{L}$:

$$\mathbf{L}(\mathbf{X}) = x_1 \mathbf{V}_1 + \cdots + x_n \mathbf{V}_n.$$

2.16. Suppose $\mathbf{C}$ is an n by n matrix with orthonormal columns. Use Theorem 2.2 to show that for every $\mathbf{X}$ in $\mathbb{R}^n$,

$$\|\mathbf{C}\mathbf{X}\| \le \sqrt{n}\|\mathbf{X}\|.$$

Use the Pythagorean theorem and the result of Problem 2.15 to show that in fact for every $\mathbf{X}$ in $\mathbb{R}^n$,

$$\|\mathbf{C}\mathbf{X}\| = \|\mathbf{X}\|$$

for such a matrix $\mathbf{C}$.

2.17. For every $\mathbf{X} \ne \mathbf{0}$ in $\mathbb{R}^n$ let $\mathbf{F}(\mathbf{X}) = \dfrac{\mathbf{X}}{\|\mathbf{X}\|}$ and let $\mathbf{G}(\mathbf{X}) = -\dfrac{\mathbf{X}}{\|\mathbf{X}\|^3}$ and denote their norms by $f(\mathbf{X}) = \|\mathbf{F}(\mathbf{X})\|$ and $g(\mathbf{X}) = \|\mathbf{G}(\mathbf{X})\|$. Describe the level sets

$$f(\mathbf{X}) = 1, \quad g(\mathbf{X}) = 1, \quad g(\mathbf{X}) = 2, \quad \text{and} \quad g(\mathbf{X}) = 4$$

in $\mathbb{R}^n$. Are there any points in $\mathbb{R}^n$ where $f(\mathbf{X}) = 2$?

2.18. Consider a function $\mathbf{F}(t) = (1-t)\mathbf{A} + t\mathbf{B}$, where $\mathbf{A}$ and $\mathbf{B}$ are in $\mathbb{R}^2$.

(a) Express $\mathbf{F}(t)$ as the sum of $\mathbf{A}$ and a multiple of $\mathbf{B} - \mathbf{A}$.

(b) For what value of t is $\mathbf{F}(t) = \mathbf{A}$? $\mathbf{B}$? the midpoint $\frac{1}{2}(\mathbf{A} + \mathbf{B})$?

(c) For what interval of t are the points $\mathbf{F}(t)$ on the line segment from $\mathbf{A}$ to $\mathbf{B}$?

2.19. Consider a function given by $\mathbf{F}(t,\theta) = (x(t,\theta), y(t,\theta))$ from $\mathbb{R}^2$ to $\mathbb{R}^2$ such that for each fixed θ, as t varies from 0 to 1, $\mathbf{F}(t,\theta)$ runs along the radius of the unit circle centered at the origin from $(0,0)$ to $(\cos\theta, \sin\theta)$.

(a) Write a rule for $\mathbf{F}(t,\theta)$.

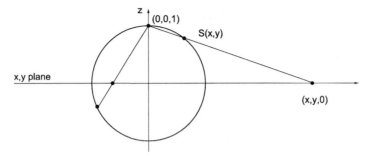

Fig. 2.14 The plane and sphere in Problem 2.20.

(b) What is the image of the rectangle $0 \le t \le 1$, $|\theta| \le 1$?

2.20. Figure 2.14 shows the unit sphere centered at the origin

$$x^2 + y^2 + z^2 = 1$$

in $\mathbb{R}^3$, with the x,y plane viewed edgewise. In this problem we introduce a correspondence between the x,y plane and the sphere with the North Pole $(0,0,1)$ deleted, known as *stereographic projection*.

(a) The line segment from the North Pole to $(x,y,0)$ in the x,y plane can be parametrized as
$$(1-t)(0,0,1) + t(x,y,0),$$

with $0 \le t \le 1$. Find t in terms of x and y, so that this point is on the sphere.
(b) Conclude that the function

$$\mathbf{S}(x,y) = \frac{1}{1+x^2+y^2}(2x, 2y, x^2+y^2-1)$$

maps the plane onto the sphere, missing the North Pole.
(c) Which points of the x,y plane correspond to the upper hemisphere? the lower?
(d) Which points of the x,y plane correspond to the equator? the South Pole?
(e) Show that the function $\mathbf{S}^{-1}$ given by

$$\mathbf{S}^{-1}(s_1, s_2, s_3) = \frac{1}{1-s_3}(s_1, s_2),$$

defined for points (s_1, s_2, s_3) of the sphere other than the North Pole, is the inverse function of $\mathbf{S}$.

2.21. Show that the linear function

$$\mathbf{L}(x, y, z) = (x, z, -y)$$

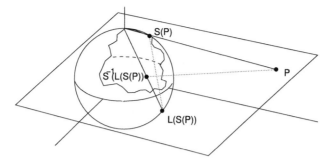

Fig. 2.15 The rotation L and stereographic projection S in Problem 2.21.

maps the unit sphere centered at the origin into itself. If $\mathbf{S}$ is the stereographic projection defined in Problem 2.20, conclude that the composition

$$\mathbf{S}^{-1} \circ \mathbf{L} \circ \mathbf{S}$$

maps the x, y plane to itself. See Figure 2.15. Where does the right half-plane $(x > 0)$ go?

2.22. The gravity force on a particle at point $\mathbf{X}$ in $\mathbb{R}^3$, due to a mass at the origin, is some negative multiple of

$$\mathbf{G}(\mathbf{X}) = \frac{\mathbf{X}}{||\mathbf{X}||^3}.$$

For points $\mathbf{X} = \mathbf{A} + \mathbf{U}$ near a given nonzero point $\mathbf{A}$, we compare two approximations of $\mathbf{G}$ given by

$$\mathbf{G}_1(\mathbf{X}) = \frac{\mathbf{X}}{||\mathbf{A}||^3} = \mathbf{G}(\mathbf{A}) + \frac{\mathbf{U}}{||\mathbf{A}||^3} = \mathbf{G}(\mathbf{A}) + \mathbf{L}_1(\mathbf{U})$$

and

$$\mathbf{G}_2(\mathbf{X}) = \mathbf{G}(\mathbf{A}) + \left(\frac{\mathbf{U}}{||\mathbf{A}||^3} - 3\frac{\mathbf{A} \cdot \mathbf{U}\mathbf{A}}{||\mathbf{A}||^5} \right) = \mathbf{G}(\mathbf{A}) + \mathbf{L}_2(\mathbf{U}).$$

(a) Show that $\mathbf{L}_1$ and $\mathbf{L}_2$ are linear functions of $\mathbf{U}$.
(b) Take $\mathbf{A} = (1,0,0)$ and $\mathbf{U} = (\frac{1}{10},0,0)$. Show that the relative errors in these approximations are about

$$\frac{||\mathbf{G}(\mathbf{X}) - \mathbf{G}_1(\mathbf{X})||}{||\mathbf{G}(\mathbf{X})||} \approx .33, \qquad \frac{||\mathbf{G}(\mathbf{X}) - \mathbf{G}_2(\mathbf{X})||}{||\mathbf{G}(\mathbf{X})||} \approx .03.$$

(c) Take $\mathbf{A} = (1,0,0)$ and $\mathbf{U} = (\frac{1}{100},0,0)$. Show that the relative errors are about

$$\frac{||\mathbf{G}(\mathbf{X}) - \mathbf{G}_1(\mathbf{X})||}{||\mathbf{G}(\mathbf{X})||} \approx .03, \qquad \frac{||\mathbf{G}(\mathbf{X}) - \mathbf{G}_2(\mathbf{X})||}{||\mathbf{G}(\mathbf{X})||} \approx .0003.$$

2.23. In this problem we use single variable calculus to derive the linear approximations for

$$\frac{\mathbf{X}}{\|\mathbf{X}\|^3} = \frac{\mathbf{A}+\mathbf{U}}{(\|\mathbf{A}\|^2 + 2\mathbf{A}\cdot\mathbf{U} + \|\mathbf{U}\|^2)^{3/2}}$$

given in Problem 2.22.

(a) Suppose a and u are positive numbers. Use Taylor's Theorem to show there are numbers θ_1 and θ_2 between zero and u with

$$(a^2+u)^{-3/2} = a^{-3} - \tfrac{3}{2}(a^2+\theta_1)^{-5/2}u, \qquad \text{and}$$

$$= a^{-3} - \tfrac{3}{2}(a^2+u)^{-5/2}u + \tfrac{15}{8}(a^2+\theta_2)^{-7/2}u^2.$$

(b) Conclude that

$$\mathbf{G}(\mathbf{A}+\mathbf{U}) = \left(\|\mathbf{A}\|^{-3} - \tfrac{3}{2}(\|\mathbf{A}\|^2+\theta_1)^{-5/2}(2\mathbf{A}\cdot\mathbf{U} + \|\mathbf{U}\|^2)\right)(\mathbf{A}+\mathbf{U}), \qquad \text{and}$$

$$= \left(\|\mathbf{A}\|^{-3} - \tfrac{3}{2}(\|\mathbf{A}\|^2)^{-5/2}(2\mathbf{A}\cdot\mathbf{U} + \|\mathbf{U}\|^2)\right.$$

$$\left. + \tfrac{15}{8}(\|\mathbf{A}\|^2+\theta_2)^{-7/2}(2\mathbf{A}\cdot\mathbf{U} + \|\mathbf{U}\|^2)^2\right)(\mathbf{A}+\mathbf{U}).$$

(c) Sort those terms into the form

$$\mathbf{G}(\mathbf{A}+\mathbf{U}) = \mathbf{G}(\mathbf{A}) + \mathbf{L}_1(\mathbf{U}) + (\text{large}) = \mathbf{G}(\mathbf{A}) + \mathbf{L}_2(\mathbf{U}) + (\text{small})$$

where "large" is of order $\|\mathbf{U}\|$ and "small" is of order $\|\mathbf{U}\|^2$.

2.2 Continuity

In single variable calculus we motivated the definition of continuity of f at x by asking whether approximate knowledge of x is sufficient to give approximate knowledge of $f(x)$. This is a very practical question, because we almost always round off or approximate the input of a function. We said f is continuous at x means that for every tolerance $\epsilon > 0$ for error in the output, we can find a level of precision $\delta > 0$ for the input, so that

$$\text{if} \quad |x-y| < \delta \quad \text{then} \quad |f(x)-f(y)| < \epsilon.$$

This is also what continuity means for a function $\mathbf{F}: D \subset \mathbb{R}^n \to \mathbb{R}^m$.

Definition 2.7. A function $\mathbf{F} : D \subset \mathbb{R}^n \to \mathbb{R}^m$ is continuous at $\mathbf{X}$ in D if for every tolerance $\epsilon > 0$, there exists a precision $\delta > 0$, that depends on ϵ, so that

$$\text{if} \quad \|\mathbf{X} - \mathbf{Y}\| < \delta \quad \text{then} \quad \|\mathbf{F}(\mathbf{X}) - \mathbf{F}(\mathbf{Y})\| < \epsilon. \qquad (\mathbf{Y} \text{ in } D)$$

Let $\mathbf{L} : \mathbb{R}^m \to \mathbb{R}^n$ be a linear function. We show that $\mathbf{L}$ is continuous at every $\mathbf{X}$. We need to show that for every tolerance $\epsilon > 0$, there is a precision $\delta > 0$ so that if $\|\mathbf{Y} - \mathbf{X}\| < \delta$ then $\|\mathbf{L}(\mathbf{Y}) - \mathbf{L}(\mathbf{X})\| < \epsilon$. Denote $\mathbf{Y}$ as $\mathbf{X} + \mathbf{H}$. Then

$$\|\mathbf{L}(\mathbf{Y}) - \mathbf{L}(\mathbf{X})\| = \|\mathbf{L}(\mathbf{Y} - \mathbf{X})\| = \|\mathbf{L}(\mathbf{H})\|.$$

By Theorem 2.1 there is a matrix $\mathbf{C}$ so that $\mathbf{L}(\mathbf{H}) = \mathbf{C}\mathbf{H}$, and by Theorem 2.2

$$\|\mathbf{L}(\mathbf{H})\| = \|\mathbf{C}\mathbf{H}\| \leq \|\mathbf{C}\| \, \|\mathbf{H}\|.$$

If $\|\mathbf{C}\|$ is zero then $\mathbf{L}$ is the constant function $\mathbf{0}$. Constant functions $\mathbf{F}$ are continuous at each $\mathbf{X}$ because $\|\mathbf{F}(\mathbf{X}) - \mathbf{F}(\mathbf{Y})\| = 0$. So we assume $\|\mathbf{C}\| \neq 0$. Given a tolerance $\epsilon > 0$, take $\|\mathbf{H}\| < \delta = \dfrac{\epsilon}{\|\mathbf{C}\|}$. We get

$$\|\mathbf{L}(\mathbf{Y}) - \mathbf{L}(\mathbf{X})\| = \|\mathbf{L}(\mathbf{H})\| \leq \|\mathbf{C}\| \, \|\mathbf{H}\| < \|\mathbf{C}\| \frac{\epsilon}{\|\mathbf{C}\|} = \epsilon.$$

Therefore $\mathbf{L}$ is continuous at $\mathbf{X}$.

Definition 2.8. A function $\mathbf{F} : D \subset \mathbb{R}^n \to \mathbb{R}^m$ is *continuous on D* if $\mathbf{F}$ is continuous at every $\mathbf{X}$ in D.

Definition 2.9. A sequence of points $\mathbf{X}_1, \mathbf{X}_2, \ldots, \mathbf{X}_k, \ldots$ in $\mathbb{R}^n$ *converges* to $\mathbf{X}$ if for every $\epsilon > 0$ there is a whole number N so that if $k > N$, then $\|\mathbf{X}_k - \mathbf{X}\| < \epsilon$.

As with functions from $\mathbb{R}$ to $\mathbb{R}$, a continuous function takes a convergent sequence of points in the domain to a convergent sequence of points in the range.

Theorem 2.3. *If* $\mathbf{F} : D \subset \mathbb{R}^n \to \mathbb{R}^m$ *is continuous on D, then for every sequence*

$$\mathbf{X}_1, \ \mathbf{X}_2, \ \ldots, \ \mathbf{X}_k, \ \ldots$$

of points in D that converges to a point $\mathbf{X}$ *in D, the sequence*

$$\mathbf{F}(\mathbf{X}_1), \ \mathbf{F}(\mathbf{X}_2), \ \ldots, \ \mathbf{F}(\mathbf{X}_k), \ \ldots$$

converges to $\mathbf{F}(\mathbf{X})$.

Proof. Take $\epsilon > 0$. Since $\mathbf{F}$ is continuous at $\mathbf{X}$, there is $\delta > 0$ so that

$$\text{if } \|\mathbf{X} - \mathbf{Y}\| < \delta, \text{ then } \|\mathbf{F}(\mathbf{X}) - \mathbf{F}(\mathbf{Y})\| < \epsilon.$$

Since the $\mathbf{X}_k$ converge to $\mathbf{X}$, given this δ there is a whole number N so that

$$\text{if } k > N, \text{ then } \|\mathbf{X} - \mathbf{X}_k\| < \delta.$$

Therefore

$$\text{If } k > N, \text{ then } \|\mathbf{F}(\mathbf{X}) - \mathbf{F}(\mathbf{X}_k)\| < \epsilon.$$

This proves that the sequence $\mathbf{F}(\mathbf{X}_k)$ converges to $\mathbf{F}(\mathbf{X})$.   □

In Problem 2.44 we ask you to prove the converse of Theorem 2.3.

The next theorem gives us a tool to reduce the question of continuity of $\mathbf{F}$ to checking the continuity of the component functions.

Theorem 2.4. *A function* $\mathbf{F} : D \subset \mathbb{R}^n \to \mathbb{R}^m$, *denoted*

$$\mathbf{F}(\mathbf{X}) = (f_1(\mathbf{X}), f_2(\mathbf{X}), \dots, f_m(\mathbf{X})),$$

is continuous on D if and only if each component function $f_j : D \subset \mathbb{R}^n \to \mathbb{R}$ *is continuous on D.*

Proof. Suppose $\mathbf{F}$ is continuous at $\mathbf{X}$. For every tolerance $\epsilon > 0$ there is a precision $\delta > 0$ so that if $\|\mathbf{Y} - \mathbf{X}\| < \delta$ then $\|\mathbf{F}(\mathbf{Y}) - \mathbf{F}(\mathbf{X})\| < \epsilon$. Let f_i be one of the component functions. Each component of a vector has absolute value less than or equal to the norm of the vector. Therefore

$$|f_i(\mathbf{Y}) - f_i(\mathbf{X})| \leq \|\mathbf{F}(\mathbf{Y}) - \mathbf{F}(\mathbf{X})\| < \epsilon.$$

So continuity of $\mathbf{F}$ at $\mathbf{X}$ implies continuity of each component function f_i at $\mathbf{X}$.

Now suppose each component function f_i is continuous at $\mathbf{X}$. We show that $\mathbf{F}$ is continuous at $\mathbf{X}$. For every tolerance $\epsilon > 0$, there are m precisions $\delta_1, \delta_2, \dots, \delta_m$, one for each component function, so that if $\|\mathbf{Y} - \mathbf{X}\| < \delta_i$ then $|f_i(\mathbf{Y}) - f_i(\mathbf{X})| < \epsilon$.

Take δ to be the smallest of the δ_i. If $\|\mathbf{Y} - \mathbf{X}\| < \delta$ then $|f_i(\mathbf{Y}) - f_i(\mathbf{X})| < \epsilon$ for all $i = 1, 2, \dots, m$, and

$$\|\mathbf{F}(\mathbf{Y}) - \mathbf{F}(\mathbf{X})\| = \sqrt{(f_1(\mathbf{Y}) - f_1(\mathbf{X}))^2 + (f_2(\mathbf{Y}) - f_2(\mathbf{X}))^2 + \cdots + (f_m(\mathbf{Y}) - f_m(\mathbf{X}))^2}$$

$$\leq \sqrt{\epsilon^2 + \epsilon^2 + \cdots + \epsilon^2} = \sqrt{m}\,\epsilon.$$

Since ϵ can be chosen as small as we like, this shows that $\|\mathbf{F}(\mathbf{Y}) - \mathbf{F}(\mathbf{X})\|$ can be made as small as we like by taking δ small enough; this shows that $\mathbf{F}$ is continuous at $\mathbf{X}$.   □

Taking $n = 1$ in Theorem 2.4 shows that $\mathbf{F}(t) = (f_1(t), f_2(t), \ldots, f_m(t))$ is continuous on an interval if and only if each component function f_i is continuous there.

> *Example 2.20.* $\mathbf{F}(t) = (\cos t, \sin t, t)$ is continuous at t because each component function is continuous at t. $\qquad\qquad\qquad\qquad\qquad\qquad\qquad\qquad\qquad\qquad\square$

The next two theorems help us find continuous functions from $\mathbb{R}^n$ to $\mathbb{R}$.

Theorem 2.5. *If $f : D \subset \mathbb{R}^n \to \mathbb{R}$ and $g : D \subset \mathbb{R}^n \to \mathbb{R}$ are continuous on D then*

(a) $f + g$ is continuous on D,
(b) fg is continuous on D,
(c) $\frac{1}{g}$ is continuous at every point $\mathbf{X}$ where $g(\mathbf{X}) \neq 0$.

Proof. (a) Let $\mathbf{X}$ be in D and let $\epsilon > 0$ be a tolerance. There are two precisions, δ_f and δ_g, so that if $\|\mathbf{X} - \mathbf{Y}\| < \delta_f$ then $|f(\mathbf{X}) - f(\mathbf{Y})| < \epsilon$, and if $\|\mathbf{X} - \mathbf{Y}\| < \delta_g$ then $|g(\mathbf{X}) - g(\mathbf{Y})| < \epsilon$.

Let δ be the smaller of δ_f and δ_g. Now if $\|\mathbf{X} - \mathbf{Y}\| < \delta$ then $|f(\mathbf{X}) - f(\mathbf{Y}| < \epsilon$, and $|g(\mathbf{X}) - g(\mathbf{Y})| < \epsilon$. According to the triangle inequality

$$|f(\mathbf{X}) - f(\mathbf{Y}) + g(\mathbf{X}) - g(\mathbf{Y})| \leq |f(\mathbf{X}) - f(\mathbf{Y})| + |g(\mathbf{X}) - g(\mathbf{Y})| < 2\epsilon.$$

Regrouping terms gives $|(f + g)(\mathbf{X}) - (f + g)(\mathbf{Y})| < 2\epsilon$. Since 2ϵ can be made as small as we like, we are done.

(b) Let ϵ be a tolerance and let δ be as in the proof of part (a). By algebra we have

$$|f(\mathbf{X})g(\mathbf{X}) - f(\mathbf{Y})g(\mathbf{Y})| = |f(\mathbf{X})g(\mathbf{X}) - f(\mathbf{X})g(\mathbf{Y}) + f(\mathbf{X})g(\mathbf{Y}) - f(\mathbf{Y})g(\mathbf{Y})|.$$

By the triangle inequality and properties of absolute values,

$$|f(\mathbf{X})g(\mathbf{X}) - f(\mathbf{Y})g(\mathbf{Y})| \leq |f(\mathbf{X})||g(\mathbf{X}) - g(\mathbf{Y})| + |g(\mathbf{Y})||f(\mathbf{X}) - f(\mathbf{Y})|.$$

If $\|\mathbf{X} - \mathbf{Y}\| < \delta$ we have by continuity of f and g that

$$|f(\mathbf{X}) - f(\mathbf{Y})| < \epsilon \text{ and } |g(\mathbf{X}) - g(\mathbf{Y})| < \epsilon.$$

Therefore

$$|f(\mathbf{X})g(\mathbf{X}) - f(\mathbf{Y})g(\mathbf{Y})| < \epsilon(|f(\mathbf{X})| + |g(\mathbf{Y})|).$$

We also know by the triangle inequality (see Problem 1.67) that

$$\big||g(\mathbf{X})| - |g(\mathbf{Y})|\big| \leq |g(\mathbf{X}) - g(\mathbf{Y})| < \epsilon$$

so $|g(\mathbf{Y})| < |g(\mathbf{X})| + \epsilon$. Therefore

$$|f(\mathbf{X})g(\mathbf{X}) - f(\mathbf{Y})g(\mathbf{Y})| < \epsilon(|f(\mathbf{X})| + |g(\mathbf{X})| + \epsilon).$$

For a given $\mathbf{X}$, $f(\mathbf{X})$ and $g(\mathbf{X})$ are fixed numbers. Therefore the expression on the right side can be made as small as we like by taking ϵ small enough.

(c) Let $g(\mathbf{X}) = k \neq 0$. Since g is continuous at $\mathbf{X}$ there is a $\gamma > 0$ so that for all $\mathbf{Y}$ with $\|\mathbf{X} - \mathbf{Y}\| < \gamma$, $|g(\mathbf{Y}) - g(\mathbf{X})| < |\frac{1}{2}k|$. Since $g(\mathbf{X}) = k$, $|g(\mathbf{Y})| > |\frac{1}{2}k|$. Now let $\epsilon > 0$. Since g is continuous at $\mathbf{X}$ there is a $\delta > 0$ so that if $\mathbf{Y}$ is within δ of $\mathbf{X}$, $|g(\mathbf{X}) - g(\mathbf{Y})| < \epsilon$. For $\|\mathbf{X} - \mathbf{Y}\|$ less than the smaller of γ and δ,

$$\left| \frac{1}{g(\mathbf{X})} - \frac{1}{g(\mathbf{Y})} \right| = \left| \frac{g(\mathbf{X}) - g(\mathbf{Y})}{g(\mathbf{X})g(\mathbf{Y})} \right| = \frac{|g(\mathbf{X}) - g(\mathbf{Y})|}{|g(\mathbf{X})||g(\mathbf{Y})|} < \frac{\epsilon}{\left| \frac{1}{2}k \right| |k|}.$$

Since k is fixed and ϵ can be taken as small as we like, we have that $\frac{1}{g}$ is continuous at $\mathbf{X}$.

$\square$

Example 2.21. Consider the function from $\mathbb{R}^3$ to $\mathbb{R}$,

$$f(x,y,z) = \frac{x^2 + xy - 2x + z + 7}{x^2 - y^3}.$$

The numerator is continuous at every (x,y,z) because it is a sum of functions that are constants, or linear, or are products of constants and linear functions. The denominator is a continuous function for the same reason. According to Theorem 2.5 then f is continuous at every point (x,y,z) where the denominator $x^2 - y^3 \neq 0$ and z is arbitrary.

$\square$

Theorem 2.6. *If $\mathbf{F} : D \subset \mathbb{R}^n \to \mathbb{R}^m$ is continuous on D, $g : A \subset \mathbb{R}^m \to \mathbb{R}$ is continuous on A, and the range of $\mathbf{F}$ is contained in A, then the composition $g \circ \mathbf{F} : D \to \mathbb{R}$ is continuous on D.*

Proof. Let $\mathbf{X}$ be in D. Since g is continuous at $\mathbf{F}(\mathbf{X})$, for every $\epsilon > 0$ there is a $\delta > 0$ so that if $\|\mathbf{F}(\mathbf{X}) - \mathbf{Y}\| < \delta$, then $|g(\mathbf{F}(\mathbf{X})) - g(\mathbf{Y})| < \epsilon$. Since $\mathbf{F}$ is continuous at $\mathbf{X}$, we can find a precision $\gamma > 0$ so that if $\|\mathbf{X} - \mathbf{Z}\| < \gamma$, $\|\mathbf{F}(\mathbf{X}) - \mathbf{F}(\mathbf{Z})\| < \delta$. It follows that for $\|\mathbf{X} - \mathbf{Z}\| < \gamma$, $\left| g(\mathbf{F}(\mathbf{X})) - g(\mathbf{F}(\mathbf{Z})) \right| < \epsilon$.

$\square$

Theorem 2.7. *If $\mathbf{F} : D \subset \mathbb{R}^n \to \mathbb{R}^m$ is continuous on D, $\mathbf{G} : A \subset \mathbb{R}^m \to \mathbb{R}^k$ is continuous on A and the range of $\mathbf{F}$ is contained in A, then the composition $\mathbf{G} \circ \mathbf{F} : D \subset \mathbb{R}^n \to \mathbb{R}^k$ is continuous on D.*

Proof. Denote $\mathbf{G}(y_1, \ldots, y_m) = (g_1(y_1, \ldots, y_m), \ldots, g_k(y_1, \ldots, y_m))$. By Theorem 2.4, g_i is continuous on A, and by Theorem 2.6, $g_i \circ \mathbf{F}$ is continuous on D. By Theorem 2.4, a vector function whose components are continuous is continuous. Therefore $\mathbf{G} \circ \mathbf{F}$ is continuous on D.

$\square$

Example 2.22. Let **F** be the function

$$\mathbf{F}(\mathbf{X}) = -\frac{\mathbf{X}}{\|\mathbf{X}\|^3}, \qquad \mathbf{X} \neq \mathbf{0}.$$

Rewriting **F** in coordinate notation,

$$\mathbf{F}(x,y,z) = \left(\frac{-x}{(x^2+y^2+z^2)^{3/2}}, \frac{-y}{(x^2+y^2+z^2)^{3/2}}, \frac{-z}{(x^2+y^2+z^2)^{3/2}} \right).$$

According to Theorem 2.6 the function $(x^2+y^2+z^2)^{3/2}$ is continuous. Then according to Theorem 2.5 the component

$$f_1(x,y,z) = \frac{-x}{(x^2+y^2+z^2)^{3/2}}$$

is a continuous function at every (x,y,z) in $\mathbb{R}^3$ except the origin, so f_1 is continuous on $\mathbb{R}^3 - \{(0,0,0)\}$. Similarly the components f_2 and f_3 are continuous on $\mathbb{R}^3 - \{(0,0,0)\}$. So by Theorem 2.4, **F** is continuous on $\mathbb{R}^3 - \{(0,0,0)\}$. $\quad\square$

Example 2.23. Consider

$$\mathbf{F}(x,y,z) = (\sin(x+y), e^{xz+y^2}, \log(xz)).$$

The component functions $f_1(x,y,z) = \sin(x+y)$ and $f_2(x,y,z) = e^{xz+y^2}$ are continuous at every (x,y,z), and the component $f_3(x,y,z) = \log(xz)$ is continuous at each point (x,y,z) where $xz > 0$. By Theorem 2.4, **F** is continuous at each point (x,y,z) where $xz > 0$. Alternatively, let

$$\mathbf{H}(x,y,z) = (x+y, xz+y^2, xz), \qquad \mathbf{G}(u,v,w) = (\sin u, e^v, \log w).$$

Then $\mathbf{G} \circ \mathbf{H} = \mathbf{F}$. By Theorem 2.7, **F** is continuous at all points (x,y,z) where $xz > 0$. $\quad\square$

Definition 2.10. A *curve* is the range of a continuous function $\mathbf{X} : I \subset \mathbb{R} \to \mathbb{R}^n$ from an interval I to $\mathbb{R}^n$,

$$\mathbf{X}(t) = (x_1(t), x_2(t), \ldots, x_n(t)), \qquad t \text{ in } I.$$

The function **X** is called a *parametrization* of the curve. If I is a closed interval $[a,b]$ then $\mathbf{X}(a)$ and $\mathbf{X}(b)$ are called the *endpoints* of the curve. If $\mathbf{X}(a) = \mathbf{X}(b)$ we say the curve is *closed* and call the curve a *loop*.

Definition 2.11. A subset A of $\mathbb{R}^n$ is *connected* if for all points **P** and **Q** in A there is a curve in A with endpoints **P** and **Q**.

Example 2.24. $\mathbf{X}(t) = (2\cos t, 3\sin t, t)$ for t in $[0, 4\pi]$ is sketched in Figure 2.16, part of an elliptical helix.   □

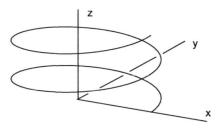

Fig. 2.16 The curve in Example 2.24.

Example 2.25. The curve given by the function

$$\mathbf{X}(t) = (a_1 + c_1 t, a_2 + c_2 t, a_3 + c_3 t)$$

is a straight line in $\mathbb{R}^3$. See Figure 2.17. Let

$$\mathbf{A} = (a_1, a_2, a_3), \qquad \mathbf{C} = (c_1, c_2, c_3)$$

and rewrite the function as $\mathbf{X}(t) = \mathbf{A} + t\mathbf{C}$. Since $\mathbf{X}(0) = \mathbf{A}$, the line goes through point $\mathbf{A}$.   □

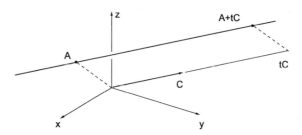

Fig. 2.17 The curve $\mathbf{X}(t) = \mathbf{A} + t\mathbf{C}$ in Example 2.25.

We close this section with some useful notions about the geometry of $\mathbb{R}^n$ and important theorems about continuous functions.

Definition 2.12. An *open ball* of radius $r > 0$ in $\mathbb{R}^n$ centered at $\mathbf{A}$ is the set of all $\mathbf{X}$ in $\mathbb{R}^n$ with

$$\|\mathbf{X} - \mathbf{A}\| < r.$$

Example 2.26. The open ball of radius 2 centered at $(4,5,6)$ is shown in Figure 2.18. It includes all the points inside the sphere surface, but not the points on the surface, as we indicate using dotted lines.   □

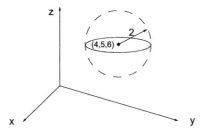

Fig. 2.18 A sketch of the open ball in Example 2.26.

Definition 2.13. A point **A** in $D \subset \mathbb{R}^n$ is called an *interior point* of D if there is an open ball centered at **A** that is contained in D. The *interior* of D is the set of interior points of D.

Example 2.27. Take S to be the square region in $\mathbb{R}^2$ consisting of all points (x,y) where $0 < x < 1$ and $0 < y < 1$. See Figure 2.19. We show that every point of S is an interior point of S. Let r be the smallest of

$$x, \quad y, \quad 1-x, \quad 1-y.$$

Then the open disk of radius r centered at (x,y) is contained in S, so (x,y) is an interior point.   □

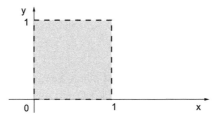

Fig. 2.19 The region S in Example 2.27 shown in gray. The dotted points are not included in S.

Example 2.28. Take D to be the unit disk $x^2 + y^2 \leq 1$ centered at the origin in the x,y plane. We show that every point **P** whose distance from the origin is less than 1 is an interior point of D: Let $r = 1 - \|\mathbf{P}\|$ and let **Q** be a point in the open disk of radius r centered at **P**. By the triangle inequality

$$\|Q\| = \|Q - P + P\| \le \|Q - P\| + \|P\|$$

$$< r + \|P\| = 1.$$

The points that are within r of P are all in D. So P is an interior point. See Figure 2.20.   □

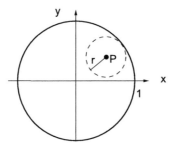

Fig. 2.20 P is an interior point in Example 2.28.

Definition 2.14. A set D in $\mathbb{R}^n$ is said to be *open* if every point in D is an interior point.

Definition 2.15. A point B of a set D in $\mathbb{R}^n$ is a *boundary point* of D if every ball centered at B contains points that are in D and also points that are not in D. The *boundary* of D is the set of boundary points of D, denoted ∂D.

It follows from this definition that a set and its complement have the same boundary points.

We denote the complement of D as $\mathbb{R}^n - D$, and more generally write $A - B$ for the set of points in a set A that are not in B.

Definition 2.16. A set D is called *closed* if it contains all its boundary points. The *closure* of a set D is the union of the set D and its boundary points. The closure of D is denoted as $\overline{D}$.

An open set B that contains the closure $\overline{C}$ of a set C is called a *neighborhood* of C.

Example 2.29. The set S in Example 2.27, where $0 < x < 1$, $0 < y < 1$ is an open set since every point of S is an interior point. S is a neighborhood of the rectangle R defined by $.2 \le x \le .4$ and $.1 \le y \le .5$. See Figure 2.21.   □

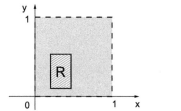 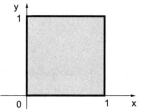

Fig. 2.21 *Left:* The open set in Example 2.29 is a neighborhood of rectangle *R*. *Right:* The closed set in Example 2.30. The boundary points are drawn solid.

Example 2.30. Take *S* to be the square region

$$0 \leq x \leq 1, \qquad 0 \leq y \leq 1,$$

shown in Figure 2.21. Every point with $x = 0$ or $x = 1$ or $y = 0$ or $y = 1$ is a boundary point. Because *S* contains all its boundary points, *S* is a closed set. ☐

Theorem 2.8. *The closure $\overline{D}$ of a set D is a closed set.*

Proof. We claim that a boundary point **B** of $\overline{D}$ is a boundary point of *D*. To see this take any ball Σ centered at **B**. The ball Σ contains a point that is not in $\overline{D}$ and therefore not in *D*. We must show that Σ contains a point of *D*. We know that Σ contains a point **A** that is in $\overline{D}$. Take a ball centered at **A** of radius so small that the ball is contained in Σ. Since **A** belongs to $\overline{D}$ the small ball is either contained in *D* or contains points of both *D* and the complement of *D*. This shows that Σ contains points of both *D* and the complement *D*. Therefore **B** is a boundary point of *D*.   ☐

The following result is basic in the geometry of $\mathbb{R}^n$.

Theorem 2.9. *The complement of an open set is a closed set and conversely.*

Proof. It follows from the definition that an open set *D* contains none of its boundary points. So all the boundary points of *D* belong the complement of *D*. Since a set and its complement have the same boundary points, this proves the theorem.   ☐

Definition 2.17. A set *D* in $\mathbb{R}^n$ is *bounded* if there is a number *b* so that

$$\|\mathbf{X}\| < b$$

for every **X** in *D*.

Example 2.31. The set U of points in $\mathbb{R}^2$ that satisfy

$$x^2 + y^2 < 1$$

is bounded. It is called the open unit disk.   □

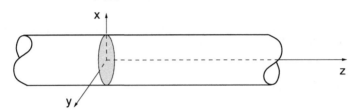

Fig. 2.22 The cylinder surface, boundary of the solid cylinders in Example 2.32.

Example 2.32. The set S of points in $\mathbb{R}^3$ that satisfy

$$x^2 + y^2 < 1$$

is the solid circular *cylinder* of radius 1 centered about the z axis. We ask you in Problem 2.35 to show that S is an open set. Let T be the set of points in $\mathbb{R}^3$ that satisfy

$$x^2 + y^2 \leq 1.$$

The boundary of S and of T is the cylindrical surface of radius 1 centered on the z axis. See Figure 2.22. T is closed. Neither S nor T is bounded because the z coordinate of points can be arbitrarily large positive or negative.   □

Theorem 2.10. *If a sequence of points* $\mathbf{X}_1, \ldots, \mathbf{X}_k, \ldots$ *of a closed set C in $\mathbb{R}^n$ converges to a point* $\mathbf{X}$, *then* $\mathbf{X}$ *is in C.*

Proof. Suppose $\mathbf{X}$ is not in C. Then $\mathbf{X}$ is in the complement of C, which is open. So there is an open ball of some radius r centered at $\mathbf{X}$, that contains no point of C. But the points $\mathbf{X}_k$ are in C and for k large enough,

$$\|\mathbf{X}_k - \mathbf{X}\| < r.$$

This is a contradiction. Therefore $\mathbf{X}$ is in C.   □

The following result is basic.

Theorem 2.11. Extreme Value Theorem *A continuous function*

$$f : C \subset \mathbb{R}^n \to \mathbb{R}$$

defined on a closed and bounded set C attains its maximum value and its minimum at some points in C.

Proof. We show first that a continuous function f defined on a closed, bounded subset C of $\mathbb{R}^n$ is *bounded*, that is, there is a number b so that $|f(\mathbf{X})| \leq b$ for all $\mathbf{X}$ in C. We argue indirectly and suppose on the contrary that f is unbounded. That means that for every whole number k there is a point $\mathbf{X}_k$ in C such that

$$|f(\mathbf{X}_k)| > k. \tag{2.4}$$

Divide the space $\mathbb{R}^n$ into n-dimensional cubes of unit edge. Since C is bounded, it is contained in a finite number of these cubes. Therefore there is a unit cube C_1 that contains infinitely many of these points $\mathbf{X}_k$.

Divide C_1 into n-dimensional cubes of edge $\frac{1}{2}$. Since the number of these cubes is finite, one of them, call it C_2, will contain infinitely many of the points $\mathbf{X}_k$.

Continuing in this fashion we obtain a sequence of cubes C_k of edge length 2^{-k}, each contained in the previous one, each containing infinitely many of the points $\mathbf{X}_k$.

We choose now a sequence of points $\mathbf{Y}_m$ as follows: $\mathbf{Y}_m$ is one of the points $\mathbf{X}_k$ contained in the cube C_m that is not one of the points $\mathbf{Y}_k$, $k < m$ previously chosen. Since we have infinitely many points to choose from, such a choice is possible. The nested sequence of cubes C_n have exactly one point in common; call it $\mathbf{Y}$. The sequence of points $\mathbf{Y}_m$ converges to $\mathbf{Y}$. Since the points $\mathbf{Y}_m$ belong to C, and since C is closed, it follows that $\mathbf{Y}$ belongs to C. Since f is a continuous function,

$$\lim_{m \to \infty} f(\mathbf{Y}_m) = f(\mathbf{Y}). \tag{2.5}$$

The sequence $\mathbf{Y}_m$ is a subsequence of $\mathbf{X}_n$. Since according to (2.4) the sequence $|f(\mathbf{X}_n)|$ tends to infinity, it follows that also $|f(\mathbf{Y}_m)|$ tends to infinity. This contradicts equation (2.5).

We have arrived at this contradiction by assuming in (2.4) that the function f is unbounded. Since this assumption led to a contradiction, we conclude that f is bounded on C. That is, the values of f are a bounded set. Therefore, by the Least Upper Bound Theorem referenced in the Preface, there is a least upper bound.

Let M be the least upper bound of the values of f on C. Then for each whole number $k > 0$ the number $M - \frac{1}{k}$ is not an upper bound for the values of f. Therefore there is a point $\mathbf{Z}_k$ where

$$M \geq f(\mathbf{Z}_k) \geq M - \frac{1}{k}.$$

This shows that

$$\lim_{m \to \infty} f(\mathbf{Z}_m) = M. \tag{2.6}$$

Arguing as above we can show that this sequence has a subsequence that converges to a point; call this limit point $\mathbf{Z}$. Since C is closed, $\mathbf{Z}$ belongs to C. Since f is continuous, it follows from (2.6) that

$$f(\mathbf{Z}) = M.$$

This proves that f attains its maximum value.

Since every f has a maximum, it follows that $-f$ has a maximum, so f has a minimum. This completes the proof of the Extreme Value Theorem.   □

Uniform continuity. The concept of *uniform continuity* is basic:

Definition 2.18. Denote by S a subset of $\mathbb{R}^n$. A function $\mathbf{F} : S \to \mathbb{R}^m$ is *uniformly continuous* on S if for every tolerance $\epsilon > 0$, there is a precision $\delta > 0$ so that if $\mathbf{X}$ and $\mathbf{Z}$ in S are within δ of each other, $\mathbf{F}(\mathbf{X})$ and $\mathbf{F}(\mathbf{Z})$ are within ϵ of each other. That is,

if $\|\mathbf{X} - \mathbf{Z}\| < \delta$ then $\|\mathbf{F}(\mathbf{X}) - \mathbf{F}(\mathbf{Z})\| < \epsilon.$

Uniform continuity on S implies continuity at every point of S. Surprisingly a converse is true if S is closed and bounded.

Theorem 2.12. *A continuous function* $\mathbf{F} : C \subset \mathbb{R}^n \to \mathbb{R}^m$ *on a closed and bounded set* C *is uniformly continuous on* C.

A proof of this theorem is outlined in Problem 2.40.

Example 2.33. Let $f(x,y) = xy$ on the square $0 \le x \le 1$, $0 \le y \le 1$. According to Theorem 2.12 f is uniformly continuous.   □

Problems

2.24. Rewrite the proof of Theorem 2.5, part (a), by justifying the following steps.

(a) Let $\mathbf{X}$ be in D and let $\epsilon > 0$ be a tolerance. Show that there is a $\delta > 0$ so that if $\|\mathbf{X} - \mathbf{Y}\| < \delta$ then $|f(\mathbf{X}) - f(\mathbf{Y})| < \frac{1}{2}\epsilon$ and $|g(\mathbf{X}) - g(\mathbf{Y})| < \frac{1}{2}\epsilon$.

(b) Show that

$$|f(\mathbf{X}) - f(\mathbf{Y}) + g(\mathbf{X}) - g(\mathbf{Y})| \le |f(\mathbf{X}) - f(\mathbf{Y})| + |g(\mathbf{X}) - g(\mathbf{Y})| < \epsilon.$$

(c) Show that this proves $f + g$ is continuous at $\mathbf{X}$.

2.25. Suppose $\mathbf{F}(x_1, \ldots, x_n) = (f_1(x_1, \ldots, x_n), f_2(x_1, \ldots, x_n))$ is a continuous function from $\mathbb{R}^n$ to $\mathbb{R}^2$, and g is a continuous function from $\mathbb{R}^2$ to $\mathbb{R}$. Prove the following. (This is an alternate way to prove parts (a) and (b) of Theorem 2.5).

(a) The composite $g \circ \mathbf{F}$ is continuous.
(b) The function $g(x,y) = x + y$ is continuous.
(c) Suppose f_1 and f_2 are continuous functions from $\mathbb{R}^n$ to $\mathbb{R}$. Use parts (a) and (b) to show that $f_1 + f_2$ is continuous.
(d) Suppose f_1 and f_2 are continuous functions from $\mathbb{R}^n$ to $\mathbb{R}$. Use part (a) and some function g to show that the product $f_1 f_2$ is continuous.

2.26. Show that the function

$$f(x,y,z) = \frac{\sin(x^2 + y^2)}{e^{z+y}}$$

is continuous at all points (x,y,z).

2.27. The slope of the graph of $\cos(cx)$ lies in the interval $[-c,c]$. Fill in the missing numbers.

(a) if $|x - a| < (?)$ then $|\cos(2x) - \cos(2a)| < \epsilon$.
(b) if $|y - b| < \delta$ then $|\cos(3y) - \cos(3b)| < (?)$.
(c) if $\|(x,y) - (a,b)\| < (?)$ then $|\cos(2x)\cos(3y) - \cos(2a)\cos(3b)| < \epsilon$.

2.28. According to the single variable Intermediate Value Theorem a continuous function f on a closed interval assumes every value between its values at the end-points. Suppose now $D \subset \mathbb{R}^n$ is a set in which any two points can be joined by a curve (D is *connected*) and let $f : D \to \mathbb{R}^n$ be continuous. Justify the following steps to show that if y is a number between two values of f then y is a value of f.
    Suppose $\mathbf{A}$ and $\mathbf{B}$ are points of D and y a number with

$$f(\mathbf{A}) < y < f(\mathbf{B}).$$

(a) Show there is a curve $\mathbf{X}(t)$, $a \le t \le b$, in D with $\mathbf{X}(a) = \mathbf{A}$, $\mathbf{X}(b) = \mathbf{B}$.
(b) The composite $f \circ \mathbf{X}$ is continuous on $[a,b]$.
(c) $f(\mathbf{X}(a)) < y < f(\mathbf{X}(b))$.
(d) There is a number t_1 in $[a,b]$ such that $f(\mathbf{X}(t_1)) = y$.

2.29. A function f from $\mathbb{R}^2$ to $\mathbb{R}$ is continuous in the disk $x^2 + y^2 \le 1$, the maximum value of f is 10 and $f(1,0) = 10$, $f(0,\frac{1}{4}) = -10$. Which are true?

(a) $f(x,y) = 0$ at some point of the disk.
(b) f has a minimum value on the disk.
(c) -10 is the minimum value of f on the disk.
(d) If $x^2 + (y - \frac{1}{4})^2$ is small enough then $f(x,y) < -9.98$.

2.30. A function f from $\mathbb{R}^3$ to $\mathbb{R}$ is continuous on an open set that contains the cube where $0 \le x \le 1, 0 \le y \le 1, 0 \le z \le 1$, the maximum value of f on the cube is 10 and $f(0,\frac{1}{2},1) = 5$. Which are true?

(a) f has a minimum value on the cube.
(b) $f(0,0,0)$ is the minimum value of f on the cube.

(c) $f(x,y,z) = 2\pi$ at some point in the cube.
(d) f could be 10 at two points.
(e) If $x^2 + (y - \frac{1}{2})^2 + (z - 1)^2$ is small enough then $f(x,y,z) > 4.98$.

2.31. A function f from $\mathbb{R}^3$ to $\mathbb{R}$ is continuous on $\mathbb{R}^3$. Show that these functions are continuous.

(a) $10 + xf(x,y,z)$ on $\mathbb{R}^3$
(b) $f(x,x,y)$ on $\mathbb{R}^2$
(c) $f(x_1x_2, x_2x_3, x_3x_4)$ on $\mathbb{R}^4$

2.32. A continuous function $f : (a,b) \to \mathbb{R}$ on an open interval does not necessarily have a maximum or minimum value.

(a) Give an example of a continuous function $f : (0,1) \to \mathbb{R}$ with arbitrarily large values.
(b) Give an example of a continuous function $g : (0,1) \to \mathbb{R}$ that is bounded but does not attain a maximum or minimum value.

2.33. Sketch the graph of $f(x_1,x_2) = x_2$ on $x_1^2 + x_2^2 \le 2$ and find the maximum value of f.

2.34. Which of the following sets are bounded?

(a) The points of $\mathbb{R}^3$ where $x^2 + y^2 + z^2 = 25$.
(b) The points of $\mathbb{R}^3$ where $x^2 + y^2 - z^2 = 1$.
(c) The points of $\mathbb{R}^2$ where $x < 1$ and $y < 1$.

2.35. Show that the set $x^2 + y^2 < 1$ in $\mathbb{R}^3$ is an open set.

2.36. Let S be the set $\mathbb{R}^2$ with the origin removed. Show that $\mathbf{0}$ is a boundary point of S.

2.37. Let T be the triangular region in $\mathbb{R}^2$ defined by $x \ge 0$, $y \ge 0$, and $x + y \le 1$.

(a) Describe the boundary of T.
(b) Show that the point $(.0001, .9998)$ is an interior point of T.

2.38. State the domain of each function. Is the domain closed? bounded? Is f continuous? Does f have a maximum? a minimum?

(a) $f(\mathbf{X}) = e^{-\|\mathbf{X}\|^2}$ where $\mathbf{X}$ is in $\mathbb{R}^2$
(b) $f(x,t) = (4\pi t)^{-1/2} e^{-x^2/4t}$
(c) $f(\mathbf{X},t) = (4\pi t)^{-n/2} e^{-\|\mathbf{X}\|^2/4t}$ where $\mathbf{X}$ is in $\mathbb{R}^n$.
(d) $f(x_1,x_2,x_3,x_4,x_5) = \dfrac{1}{\sqrt{x_2^2 + x_3^2 + x_4^2}}$

2.39. Consider the linear function

$$\mathbf{F}(\mathbf{X}) = \begin{bmatrix} -1 & 5 \\ 5 & -1 \end{bmatrix} \begin{bmatrix} x_1 \\ x_2 \end{bmatrix}.$$

(a) Find a number c so that $\|\mathbf{F}(\mathbf{X})\| \le c\|\mathbf{X}\|$.
(b) Find a number d so that $\|\mathbf{F}(\mathbf{X}) - \mathbf{F}(\mathbf{Y})\| \le d\|\mathbf{X} - \mathbf{Y}\|$.
(c) Is $\mathbf{F}$ uniformly continuous?

2.40. In this problem we prove Theorem 2.12. Suppose $\mathbf{F} : C \subset \mathbb{R}^n \to \mathbb{R}^m$ is continuous on a closed and bounded set C. Either $\mathbf{F}$ is uniformly continuous on C, or it is not. Justify the following statements that show that the statement that $\mathbf{F}$ is not uniformly continuous on C leads to a contradiction, hence $\mathbf{F}$ is uniformly continuous.

(a) Since f is not uniformly continuous there is some tolerance ϵ and a sequence of pairs of points $\mathbf{X}_k$, $\mathbf{Y}_k$ in C, $k = 1, 2, 3, \dots$ such that $\|\mathbf{X}_k - \mathbf{Y}_k\| < \frac{1}{k}$ and

$$\|\mathbf{F}(\mathbf{X}_k) - \mathbf{F}(\mathbf{Y}_k)\| \ge \epsilon.$$

(b) As we saw in the proof of Theorem 2.11, the sequence $\mathbf{X}_k$ must have a subsequence $\mathbf{X}_{k_i}$ that converges to a point $\mathbf{X}$ in C. Then $\|\mathbf{X}_{k_i} - \mathbf{Y}_{k_i}\| < \frac{1}{k_i}$ and since $\mathbf{F}$ is continuous

$$\lim_{k_i \to \infty} \mathbf{F}(\mathbf{X}_{k_i}) = \mathbf{F}(\mathbf{X}).$$

(c) Use the triangle inequality

$$\|\mathbf{X} - \mathbf{Y}_{k_i}\| \le \|\mathbf{X} - \mathbf{X}_{k_i}\| + \|\mathbf{X}_{k_i} - \mathbf{Y}_{k_i}\|$$

to show that the $\mathbf{Y}_{k_i}$ also converge to $\mathbf{X}$.
(d) The sequences $\mathbf{F}(\mathbf{X}_{k_i})$ and $\mathbf{F}(\mathbf{Y}_{k_i})$ both converge to $\mathbf{F}(\mathbf{X})$.
(e) That contradicts $\|\mathbf{F}(\mathbf{X}_{k_i}) - \mathbf{F}(\mathbf{Y}_{k_i})\| \ge \epsilon$.
(f) $\mathbf{F}$ is uniformly continuous on C.

2.41. Let $\mathbf{C}$ be a vector in $\mathbb{R}^n$ and let $\mathbf{X}$ and $\mathbf{Y}$ be vectors in $\mathbb{R}^n$. Use the Cauchy–Schwarz inequality
$$|\mathbf{A} \cdot \mathbf{B}| \le \|\mathbf{A}\|\|\mathbf{B}\|$$

to prove:

(a) the function $f(\mathbf{X}) = \mathbf{C} \cdot \mathbf{X}$ from $\mathbb{R}^n$ to $\mathbb{R}$ is uniformly continuous,
(b) the function $g(\mathbf{X}, \mathbf{Y}) = \mathbf{X} \cdot \mathbf{Y}$ from $\mathbb{R}^{2n}$ to $\mathbb{R}$ is continuous.

2.42. Let $f(\mathbf{X}) = \|\mathbf{X}\|$ and $g(\mathbf{X}) = \|\mathbf{X}\|^2$, for $\mathbf{X}$ in $\mathbb{R}^n$.

(a) Give examples of points $\mathbf{X}$ and $\mathbf{Y}$ one unit apart from each other, such that

$$\left|g(\mathbf{Y}) - g(\mathbf{X})\right| > 10^{60}.$$

(b) Show that f is uniformly continuous.
(c) Show that g is not uniformly continuous.

2.43. Consider the function $f(\mathbf{X}) = \dfrac{1}{\|\mathbf{X}\|}$ in the set D in $\mathbb{R}^n$ where $\|\mathbf{X}\| \ge 2$. Use the identity

$$\frac{1}{\|\mathbf{X}\|} - \frac{1}{\|\mathbf{Y}\|} = \frac{\|\mathbf{Y}\| - \|\mathbf{X}\|}{\|\mathbf{X}\|\|\mathbf{Y}\|}$$

to show that f is uniformly continuous on D.

2.44. Suppose $\mathbf{F}$ is a function from $\mathbb{R}^n$ to $\mathbb{R}^m$ with the following property. For every sequence $\mathbf{X}_1, \mathbf{X}_2, \mathbf{X}_3, \ldots$ that converges to $\mathbf{X}$ in the domain of $\mathbf{F}$, the sequence $\mathbf{F}(\mathbf{X}_1), \mathbf{F}(\mathbf{X}_2), \mathbf{F}(\mathbf{X}_3), \ldots$ converges to $\mathbf{F}(\mathbf{X})$. Justify the following steps to show that $\mathbf{F}$ is continuous.

(a) If $\mathbf{F}$ is not continuous at a point $\mathbf{A}$, then there is some $\epsilon > 0$ so that for every δ there is a point $\mathbf{B}$ with

$$\|\mathbf{A} - \mathbf{B}\| < \delta \qquad \text{and} \quad \|\mathbf{F}(\mathbf{A}) - \mathbf{F}(\mathbf{B})\| > \epsilon.$$

(b) If $\mathbf{F}$ is not continuous at a point $\mathbf{A}$, then there is some $\epsilon > 0$ so that for every integer $k > 0$ there is a point $\mathbf{X}_k$ with

$$\|\mathbf{A} - \mathbf{X}_k\| < \frac{1}{k} \qquad \text{and} \quad \|\mathbf{F}(\mathbf{A}) - \mathbf{F}(\mathbf{X}_k)\| > \epsilon.$$

(c) If $\mathbf{F}$ is not continuous at a point $\mathbf{A}$, then there is a sequence $\mathbf{X}_1, \mathbf{X}_2, \mathbf{X}_3, \ldots$ converging to $\mathbf{A}$ such that the sequence $\mathbf{F}(\mathbf{X}_1), \mathbf{F}(\mathbf{X}_2), \mathbf{F}(\mathbf{X}_3), \ldots$ does not converge to $\mathbf{F}(\mathbf{A})$.

2.3 Other coordinate systems

Polar coordinates. It is often convenient to use polar coordinates (r, θ) rather than rectangular coordinates (x, y) to describe a curve or a region in the plane. Figure 2.23 shows the coordinates of a point in a plane given in terms of rectangular coordinates and polar coordinates, where we usually take $r \geq 0$ and $0 \leq \theta \leq 2\pi$.

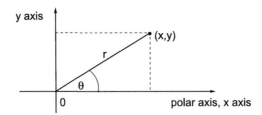

Fig. 2.23 Polar and rectangular coordinates

The coordinates are related by

$$x = r\cos\theta$$
$$y = r\sin\theta, \tag{2.7}$$

and

$$r = \sqrt{x^2 + y^2} \tag{2.8}$$

Let $\mathbf{F} : \mathbb{R}^2 \to \mathbb{R}^2$ be the function defined by

$$\mathbf{F}(r,\theta) = (r\cos\theta, r\sin\theta).$$

The component functions of $\mathbf{F}$ are $x = r\cos\theta$, $y = r\sin\theta$. $\mathbf{F}$ is called the polar coordinate mapping. Let's see how regions in the (r,θ) plane are mapped to the (x,y) plane. See Figure 2.24. If we restrict $\mathbf{F}$ to a rectangular region $0 < a \le r \le b$, $0 < \alpha \le \theta \le \beta < 2\pi$, then $\mathbf{F}$ is one to one. Three such regions are indicated in the figure.

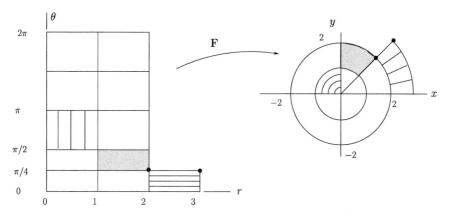

Fig. 2.24 The polar coordinate mapping.

Cylindrical coordinates. An alternative to Cartesian coordinates (x,y,z) in space uses polar coordinates in the x,y plane and retains the z coordinate.

Definition 2.19. The *cylindrical coordinates* (r,θ,z) that correspond to Cartesian coordinates (x,y,z) are related by

$$x = r\cos\theta$$
$$y = r\sin\theta$$
$$z = z$$

$r \ge 0$ and $0 \le \theta \le 2\pi$.

Example 2.34. The point with Cartesian coordinates $(x,y,z) = (\sqrt{2}, \sqrt{2}, 3)$ has cylindrical coordinates $(r,\theta,z) = (2, \frac{\pi}{4}, 3)$.   □

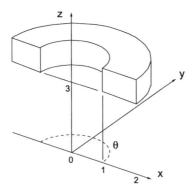

Fig. 2.25 The region in Example 2.35.

Cylindrical coordinates can simplify the description of some regions in space.

Example 2.35. Let D be the region described in cylindrical coordinates by

$$1 \le r \le 2, \quad 0 \le \theta \le \pi, \quad 3 \le z \le 4.$$

A sketch of this region is shown in Figure 2.25. Using Cartesian coordinates D is described by

$$0 \le y, \quad 1 \le x^2 + y^2 \le 4, \quad 3 \le z \le 4.$$

□

If we think of the function

$$\mathbf{F}(r,\theta,z) = (r\cos\theta, r\sin\theta, z)$$

from $\mathbb{R}^3$ to $\mathbb{R}^3$, we can see that a solid rectangular bar in (r,θ,z) space, described by the inequalities in Example 2.35, is mapped to the region D there. If we extend the bar to $0 \le \theta \le 2\pi$, then we get a full ring. Describing a region as the range of some mapping will be useful later when we study the integral.

Spherical coordinates. Another way to describe the location of a point (x,y,z) is to denote by ρ the distance between the point and the origin, let ϕ be the angle between the positive z axis and the line through $(0,0,0)$ and (x,y,z), and let θ be the angle between the plane containing (x,y,z) and the z axis, and the x,z plane. See Figure 2.26.

Definition 2.20. The *spherical coordinates* (ρ, ϕ, θ) that correspond to the Cartesian coordinates (x, y, z) are related by

$$x = \rho \sin \phi \cos \theta$$
$$y = \rho \sin \phi \sin \theta$$
$$z = \rho \cos \phi.$$

$0 \le \rho, \; 0 \le \phi \le \pi, \; 0 \le \theta \le 2\pi.$

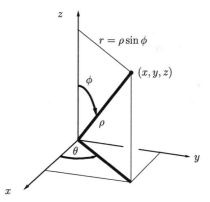

Fig. 2.26 The spherical and cylindrical coordinate systems.

Example 2.36. The set of points in space that lie on a sphere of radius 3 centered at the origin satisfy

$$\sqrt{x^2 + y^2 + z^2} = 3.$$

Using spherical coordinates we describe these points by

$$\rho = 3$$

and ϕ and θ take their full range: $0 \le \theta \le 2\pi, 0 \le \phi \le \pi.$ □

Example 2.37. To describe the set of points on the circular cone surface

$$z = \sqrt{x^2 + y^2}$$

shown in Figure 2.27 we note that every point on the cone has the same angle ϕ. To find ϕ we look at the intersection of the cone with the plane $y = 0$.

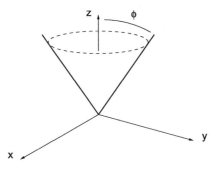

Fig. 2.27 The cone surface in Example 2.37.

At such points we get $z = \sqrt{x^2} = |x|$, so $\phi = \frac{\pi}{4}$, and ρ and θ are unrestricted. Another approach is to substitute the formulas $x = \rho \sin\phi \cos\theta$, $y = \rho \sin\phi \sin\theta$ and $z = \rho \cos\phi$ into the equation for the cone:

$$\rho \cos\phi = \sqrt{\rho^2 \sin^2\phi \cos^2\theta + \rho^2 \sin^2\phi \sin^2\theta}$$
$$= \rho \sin\phi.$$

Since $0 \le \phi \le \pi$ this is satisfied only by $\phi = \frac{\pi}{4}$. $\square$

Example 2.38. Let D be the solid region of points that satisfy

$$x^2 + y^2 + z^2 \le 9, \qquad x^2 + y^2 + z^2 > 1, \qquad z \ge 0$$

shown in Figure 2.28. The surface of the inner hemisphere is sketched with dots to indicate that the points on that surface are not included. Region D is described in spherical coordinates as

$$1 < \rho \le 3, \qquad 0 \le \phi \le \frac{\pi}{2}.$$

                                                                                $\square$

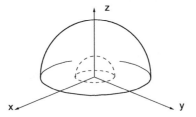

Fig. 2.28 The region in Example 2.38.

Problems

2.45. Consider the polar coordinate mapping illustrated in Figure 2.24. Sketch a region in the r, θ plane that corresponds to the upper half plane $y > 0$.

2.46. Sketch the region in the x, y plane that corresponds to the polar coordinate rectangle $1 \le r \le 2$, $0 \le \theta \le \pi$. Find the polar coordinates of the point $(x, y) = (0, 1.5)$.

2.47. Use equations and inequalities to describe the following sets in polar coordinates.

(a) The open unit disk $x^2 + y^2 < 1$.
(b) The first quadrant $x > 0$, $y > 0$.

2.48. Consider the set of points in the x, y plane whose polar coordinates satisfy

$$0 \le \theta \le 2\pi, \quad r = \frac{1}{2 + \sin \theta}.$$

Show that this is a bounded set in the x, y plane.

2.49. Let $0 < b < a$ and consider the set of points in the plane whose polar coordinates r, θ satisfy

$$r = \frac{1}{a + b \cos \theta}.$$

(a) Show that the x, y coordinates of the points satisfy

$$1 = a \sqrt{x^2 + y^2} + bx.$$

(b) Show that the equation $1 = a \sqrt{x^2 + y^2} + bx$ is the equation of an ellipse by completing a square and getting the equation into the form

$$1 = \frac{(x - \alpha)^2}{c^2} + \frac{(y - \beta)^2}{d^2}.$$

2.50. Make a sketch to show the image of a rectangle $r_0 < r < r_0 + h$, $0 \le \theta \le 2\pi$ under the polar coordinate mapping. Find the area of the image rectangle. Show that the ratio

$$\frac{\text{area(image)}}{\text{area(rectangle)}}$$

tends to r_0 as h tends to 0.

2.51. Match the formulas (i–iv) in spherical coordinates with the descriptions (a–d) of sets in $\mathbb{R}^3$. Note that if we don't specify any restrictions on a coordinate, then the usual ones apply: $\rho \ge 0$, $0 \le \phi \le \pi$, $0 \le \theta \le 2\pi$.

(i) $\rho < 1$ (ii) $\phi = \frac{2\pi}{3}$ (iii) $0 < \theta < \pi$ (iv) $\rho > 1$ and $\frac{\pi}{2} < \phi$

(a) a half-space

(b) an open ball

(c) a cone surface

(d) a half-space with points removed that were within one unit of the origin

2.52. Let D be the region in $\mathbb{R}^3$ where the spherical coordinate ρ is restricted:

$$2 \leq \rho \leq 4,$$

and let $\mathbf{P} = (0,0,1)$, $\mathbf{Q} = (7,0,0)$, $\mathbf{R} = (3,0,0)$ in Cartesian coordinates.

(a) Describe the region D.

(b) Is $\mathbf{R}$ in D?

(c) There is a point $\mathbf{A}$ in D that is closest to $\mathbf{P}$, and a point $\mathbf{B}$ in D that is farthest from $\mathbf{Q}$. Use a sketch to find $\mathbf{A}$ and $\mathbf{B}$.

2.53. Consider functions

$$s_1(\rho,\phi,\theta) = e^{-\rho}, \qquad s_2(\rho,\phi,\theta) = (1-\rho)\rho e^{-\rho}, \qquad d_3(\rho,\phi,\theta) = (3\cos^2\phi - 1)\rho^2 e^{-\rho},$$

of the type used to describe orbitals of an electron in an atom. (See also Section 9.5.)

(a) What is the maximum value of s_1, and what do the level sets of s_1 look like for values less than the maximum?

(b) Which of the three functions is zero on some sphere centered at the origin?

(c) What are the limits of s_1, s_2, and d_3 as ρ tends to infinity?

(d) d_3 is nonnegative on the z axis and in a region enclosed by a double cone surrounding the z axis. Sketch the region.

(e) Is d_3 positive or negative outside the region of part (d)?

2.54. For a function that is a product $f(x,y) = g(x)h(y)$ of nonnegative functions g and h, show that the maximum of f is the product of the maximum values of g and of h, if these exist. Use this idea in spherical coordinates to find the locations of the maximum values of

$$d_3^2(\rho,\phi,\theta) = (3\cos^2\phi - 1)^2 \rho^4 e^{-2\rho}.$$

2.55. We outline a proof of the Fundamental Theorem of Algebra: For every polynomial with complex coefficients

$$p(z) = p_0 + p_1 z + p_2 z^2 + \cdots + p_n z^n, \qquad (n > 0)$$

there is a complex number z where $p(z) = 0$.

Recall that complex numbers $z = x + iy$ can be identified with points (x,y) in $\mathbb{R}^2$, together with the multiplication

$$(x+iy)(u+iv) = xu - vy + i(yu + xv).$$

Thus polynomials with complex inputs can be viewed as functions from $\mathbb{R}^2$ to $\mathbb{R}^2$. Prove the following statements.

(a) Multiplication is continuous, so z^2, z^3, etc., are continuous functions, so polynomials are continuous functions.

(b) Complex numbers have square roots, cube roots, fourth roots, etc: use the fact that $z = (x, y)$ can be written $r(\cos\theta + i\sin\theta)$ in polar coordinates, and that multiplication is given by

$$r_1(\cos\theta_1 + i\sin\theta_1)r_2(\cos\theta_2 + i\sin\theta_2) = r_1 r_2(\cos(\theta_1 + \theta_2) + i\sin(\theta_1 + \theta_2)).$$

(c) The absolute value $|z| = \|(x, y)\|$ is a continuous function of x and y and, using part (b), $|zw| = |z||w|$.

(d) $|p(z)|$ tends to infinity as $|z|$ tends to infinity. To see this, first use the triangle inequality to show that for $|z| > 1$

$$|p_{n-1}z^{n-1} + \cdots + p_0| < P|z|^{n-1}$$

for some number P; second use the triangle inequality

$$|p_0 + \cdots + p_n z^n| \geq |p_n z^n| - |p_{n-1}z^{n-1} + \cdots + p_0|$$

to show that for $|z| > 1$, $|p(z)| \geq |p_n||z|^n - P|z|^{n-1}$, that tends to infinity as $|z|$ tends to infinity.

(e) Now a proof by contradiction: If a polynomial p does not have a root then the function

$$f(z) = \frac{1}{|p(z)|}$$

is a continuous function from $\mathbb{R}^2$ to $\mathbb{R}$ that tends to zero as $|z|$ tends to infinity. Such a function f has a maximum value at some number a. Therefore $|p(z)|$ has a minimum value $|p(a)| = m \neq 0$.

(f) For every polynomial $q(z)$ of degree n and every number a, $q(z)$ can be expressed as a polynomial in $z - a$,

$$q(z) = q(a) + q'(a)(z - a) + q''(a)\frac{(z - a)^2}{2!} + \cdots + q^{(n)}(a)\frac{(z - a)^n}{n!}.$$

(g) Use part (f) to express

$$p(z) = p(a) + c(z - a)^k + \cdots$$

where $c \neq 0$ and the dots are powers of $z - a$ greater than k. According to part (b) there is a k-th root h:

$$h^k = -\frac{p(a)}{c}.$$

Then use

$$p(a + \epsilon h) = p(a)(1 - \epsilon^k) + \cdots$$

where the dots are powers of ϵ greater then k. This shows that

$$|p(a + \epsilon h)| \le m(1 - \epsilon^k) + \cdots < m$$

for ϵ small, a contradiction.

2.56. Suppose two points in $\mathbb{R}^3$ have cylindrical coordinates (r_1, θ_1, z_1), (r_2, θ_2, z_2). Show that the distance between the points is given by

$$\sqrt{r_1^2 + r_2^2 - 2r_1 r_2 \cos(\theta_2 - \theta_1) + (z_2 - z_1)^2}.$$

2.57. Suppose two points on the unit sphere centered at the origin in $\mathbb{R}^3$ have spherical coordinates $(1, \phi_1, \theta_1)$, $(1, \phi_2, \theta_2)$. Show that the dot product of the points is given by

$$\cos(\phi_2 - \phi_1) - \sin\phi_2 \sin\phi_1 (1 - \cos(\theta_2 - \theta_1)).$$

Chapter 3
Differentiation

Abstract In this chapter we introduce the notion of derivative of functions of several variables. We start with functions of two variables and then extend to several variables. By using vector and matrix notation we find that many of the concepts and results look familiar.

3.1 Differentiable functions

Recall that for a function f of a single variable x, we say that f is differentiable at a if f is *locally linear* at a. That is, there is a constant m so that for all h sufficiently close to zero the change in f at a,

$$f(a+h) - f(a),$$

is well approximated by mh. By "well approximated" we mean that the difference between $f(a+h) - f(a)$ and mh is small compared to h when h is small. That is,

$$\frac{(f(a+h) - f(a)) - mh}{h}$$

tends to zero as h tends to zero. In this case we say f is differentiable at a and we write

$$\lim_{h \to 0} \frac{f(a+h) - f(a)}{h} = m.$$

The number m is called the derivative of f at a, denoted $f'(a)$, and the linear approximation of f at a is

$$f(a) + f'(a)(x - a).$$

Now let's extend the notion of "local linearity" to functions from $\mathbb{R}^2$ to $\mathbb{R}$. First we recall from Theorem 1.2 that a linear function ℓ from $\mathbb{R}^2$ to $\mathbb{R}$ is a function of the form

© Springer International Publishing AG 2017
P. D. Lax and M. S. Terrell, *Multivariable Calculus with Applications*,
Undergraduate Texts in Mathematics, https://doi.org/10.1007/978-3-319-74073-7_3

$$\ell(h,k) = ph + qk, \tag{3.1}$$

where p and q are some numbers. We recall Definition 1.5 of the norm of a vector $\|(h,k)\| = \sqrt{h^2 + k^2}$.

Definition 3.1. A function f defined in an open disk in $\mathbb{R}^2$ centered at (a,b) is *differentiable* at (a,b) if

$$f(a+h, b+k) - f(a,b)$$

can be well approximated by a linear function ℓ in the following sense:

$$\frac{(f(a+h,b+k) - f(a,b)) - \ell(h,k)}{\|(h,k)\|} \tag{3.2}$$

tends to zero as $\|(h,k)\|$ tends to zero.

We call

$$L(x,y) = f(a,b) + \ell(x-a, y-b)$$

the *linear approximation* of $f(x,y)$ at (a,b).

Definition 3.1 can be rewritten in vector language. Let $\mathbf{A} = (a,b)$ and $\mathbf{H} = (h,k)$.

Definition 3.2. (Vector notation version) A function f from $\mathbb{R}^2$ to $\mathbb{R}$ defined on an open disk centered at $\mathbf{A}$ is *differentiable* at $\mathbf{A}$ if $f(\mathbf{A}+\mathbf{H}) - f(\mathbf{A})$ can be well approximated by a linear function ℓ in the following sense:

$$\frac{(f(\mathbf{A}+\mathbf{H}) - f(\mathbf{A})) - \ell(\mathbf{H})}{\|\mathbf{H}\|}$$

tends to zero as $\|\mathbf{H}\|$ tends to zero.

Theorem 3.1. *If a function f from $\mathbb{R}^2$ to $\mathbb{R}$ is differentiable at $\mathbf{A}$ then f is continuous at $\mathbf{A}$.*

Proof. By Definition 3.2,

$$f(\mathbf{A}+\mathbf{H}) - f(\mathbf{A}) - \ell(\mathbf{H})$$

tends to zero as $\|\mathbf{H}\|$ tends to zero. Since $\ell(\mathbf{H})$ tends to zero as $\|\mathbf{H}\|$ tends to zero, so does $f(\mathbf{A}+\mathbf{H}) - f(\mathbf{A})$.   □

Next we relate the numbers p and q in the linear function $\ell(h,k) = ph + qk$ in formula (3.2) to the function f.

Suppose f is differentiable at (a,b) and $\ell(h,k) = ph + qk$. Set $k = 0$ in Definition 3.1; then

$$\lim_{h \to 0} \frac{f(a+h,b) - f(a,b) - ph}{h} = 0,$$

so $f(x,b)$ is a differentiable function of a single variable x at a, and

$$p = \lim_{h \to 0} \frac{f(a+h,b) - f(a,b)}{h}.$$

The number p is called the *partial derivative* of f with respect to x at (a,b) and is denoted

$$\frac{\partial f}{\partial x}(a,b) \quad \text{or} \quad f_x(a,b).$$

Thus $\frac{\partial f}{\partial x}(a,b)$ is found by holding y equal to the constant b and differentiating $f(x,b)$ with respect to x at a. Similarly if we let $h = 0$ in Definition 3.1 we see that if f is differentiable at (a,b) then $f(a,y)$ is a differentiable function of y at b and

$$\lim_{k \to 0} \frac{f(a,b+k) - f(a,b) - qk}{k} = 0$$

so that

$$q = \lim_{k \to 0} \frac{f(a,b+k) - f(a,b)}{k}.$$

The number q is called the partial derivative of f with respect to y at (a,b) and is denoted

$$\frac{\partial f}{\partial y}(a,b) \quad \text{or} \quad f_y(a,b).$$

This shows that if f is differentiable at (a,b) then it has partial derivatives $f_x(a,b)$ and $f_y(a,b)$ there, and the linear approximation of f at (a,b) is

$$L(x,y) = f(a,b) + f_x(a,b)(x-a) + f_y(a,b)(y-b).$$

The rules for finding partial derivatives follow from the rules for ordinary differentiation.

(a) $(f+g)_x = f_x + g_x$ and $(f+g)_y = f_y + g_y$
(b) $(fg)_x = f_x g + f g_x$ and $(fg)_y = f_y g + f g_y$
(c) $\left(\frac{1}{f}\right)_x = -\frac{f_x}{f^2}$ and $\left(\frac{1}{f}\right)_y = -\frac{f_y}{f^2}$

Example 3.1. We show that $f(x,y) = xy^2$ is differentiable at $(1,3)$. First we find $f_x(1,3)$ and $f_y(1,3)$. Holding y fixed and differentiating f with respect to x we get $f_x = y^2$, $f_x(1,3) = 9$. Holding x fixed and differentiating f with respect to y we get $f_y = 2xy$, $f_y(1,3) = 6$. Now that we have p and q we check that f is locally linear at $(1,3)$, where $\ell(h,k) = 9h + 6k$.

$$\frac{f(1+h,3+k) - f(1,3) - \ell(h,k)}{\|(h,k)\|} = \frac{(1+h)(3^2+6k+k^2) - (1)(3)^2 - (9h+6k)}{\|(h,k)\|}$$

$$= \frac{k^2 + 6hk + hk^2}{\|(h,k)\|}.$$

By the triangle inequality $|k^2 + 6hk + hk^2| \le k^2 + 6|hk| + |h|k^2$. Since we are taking the limit as $\|(h,k)\|$ tends to zero we can restrict attention to where $\|(h,k)\| \le 1$, and there $|h| \le 1$. We get

$$|k^2 + 6hk + hk^2| \le (1 + |h|)k^2 + 6|hk| \le 2k^2 + 6|hk|.$$

Since $(h \pm k)^2 = h^2 + k^2 \pm 2hk \ge 0$ it follows that $2|hk| \le h^2 + k^2$. Therefore if $\|(h,k)\| \le 1$ we have

$$\frac{|k^2+6hk+hk^2|}{\|(h,k)\|} \le \frac{2k^2+6|hk|}{\|(h,k)\|} \le \frac{2k^2+3h^2+3k^2}{\|(h,k)\|} \le \frac{5(h^2+k^2)}{\|(h,k)\|} = 5\sqrt{h^2+k^2}.$$

Therefore as $\|(h,k)\|$ tends to zero

$$\frac{f(1+h,3+k) - f(1,3) - \ell(h,k)}{\|(h,k)\|}$$

tends to zero, and $f(x,y) = xy^2$ is differentiable at $(1,3)$.   □

The next example shows that the existence of partial derivatives, while necessary for differentiability at a point, is not sufficient.

Example 3.2. Define a function f as

$$f(x,y) = |x+y| - |x-y|. \tag{3.3}$$

The single variable functions,

$$f(x,0) = |x+0| - |x-0| = 0, \qquad f(0,y) = |0+y| - |0-y| = 0,$$

are both constant and therefore differentiable at $(0,0)$, and

$$f_x(0,0) = 0, \qquad f_y(0,0) = 0.$$

Next we show that

$$\frac{f(0+h,0+k) - f(0,0) - \ell(h,k)}{\|(h,k)\|}$$

does not tend to zero as $\|(h,k)\|$ tends to zero. Take $k = h$; then

$$\frac{f(0+h,0+h) - f(0,0) - (0h+0h)}{\sqrt{h^2+h^2}} = \frac{|2h| - |0| - 0 - (0h+0k)}{\sqrt{2h^2}} = \frac{2}{\sqrt{2}}$$

does not tend to zero as $\|(h,h)\|$ tends to zero. Therefore f is not differentiable at $(0,0)$. □

Next we show that if f has *continuous* partial derivatives on an open set containing (a,b) then f is differentiable at (a,b). We use the following theorem.

Theorem 3.2. A Mean Value Theorem. *Let f be a function from $\mathbb{R}^2$ to $\mathbb{R}$ whose partial derivatives f_x and f_y exist on an open set containing (a,b). Then for each (h,k) with $\|(h,k)\|$ sufficiently small there are numbers h' and k' where $a+h'$ lies between a and $a+h$, and $b+k'$ lies between b and $b+k$, such that*

$$f(a+h,b+k) - f(a,b) = hf_x(a+h',b+k) + kf_y(a,b+k'). \qquad (3.4)$$

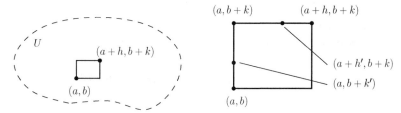

Fig. 3.1 *Left:* a small rectangle in U, in Theorem 3.2. *Right:* points used in the proof.

Proof. Write

$$f(a+h,b+k) - f(a,b) = f(a+h,b+k) - f(a,b+k) + f(a,b+k) - f(a,b).$$

For $|h|$ and $|k|$ sufficiently small all the points on the sides of a small rectangle (see Figure 3.1) are in U and $f(x,b+k)$ is differentiable on the closed interval from a to $a+h$. We apply the Mean Value Theorem for the single variable function $f(x,b+k)$ and conclude that there is a number $a+h'$ between a and $a+h$ for which

$$f(a+h,b+k) - f(a,b+k) = hf_x(a+h',b+k).$$

Similarly $f(a,y)$ is differentiable on the interval between b and $b+k$, and by the Mean Value Theorem for single variable functions there is a number $b+k'$ between b and $b+k$ for which

$$f(a,b+k) - f(a,b) = kf_y(a,b+k').$$

Add these two equations, and we obtain formula (3.4) to complete the proof. □

Theorem 3.3. *If the partial derivatives of f are continuous in an open set containing (a,b) then f is differentiable at (a,b).*

Proof. Using formula (3.4) of Theorem 3.2

$$\frac{f(a+h,b+k) - f(a,b) - (hf_x(a,b) + kf_y(a,b))}{\sqrt{h^2 + k^2}}$$

$$= \frac{hf_x(a+h',b+k) + kf_y(a,b+k')) - (hf_x(a,b) + kf_y(a,b))}{\sqrt{h^2 + k^2}}$$

$$= \frac{h(f_x(a+h',b+k) - f_x(a,b)) + k(f_y(a,b+k') - f_y(a,b))}{\sqrt{h^2 + k^2}} \tag{3.5}$$

where h' is between 0 and h and k' is between 0 and k. By the triangle inequality the absolute value of the last expression is less than or equal to

$$\frac{|h|}{\sqrt{h^2 + k^2}}\left|f_x(a+h',b+k) - f_x(a,b)\right| + \frac{|k|}{\sqrt{h^2 + k^2}}\left|f_y(a,b+k') - f_y(a,b)\right|.$$

Since $\dfrac{|h|}{\sqrt{h^2 + k^2}}$ and $\dfrac{|k|}{\sqrt{h^2 + k^2}}$ are less than or equal to 1, and f_x and f_y are continuous, each of the terms in this expression tends to zero as $\|(h,k)\|$ tends to zero. This proves that (3.5) tends to zero as $\|(h,k)\|$ tends to zero. This shows that f is differentiable at (a,b). $\square$

Example 3.3. Let $f(x,y) = y + \sin(xy) + \sinh x$. Use the linear approximation of f at $(0,1)$ to approximate $f(.1,.9)$. The linear approximation of f at $(0,1)$ is

$$L(x,y) = f(0,1) + f_x(0,1)(x-0) + f_y(0,1)(y-1).$$

We calculate

$$f(0,1) = 1 + \sin(0) + \sinh(0) = 1 + 0 + \tfrac{1}{2}(e^0 - e^{-0}) = 1$$

$$f_x(x,y) = \cos(xy)y + \cosh x, \qquad f_x(0,1) = \cos(0)1 + \tfrac{1}{2}(e^0 + e^{-0}) = 2$$

$$f_y(x,y) = 1 + \cos(xy)x, \qquad f_y(0,1) = 1 + \cos(0)0 = 1.$$

Therefore $L(x,y) = 1 + 2(x-0) + 1(y-1)$ and

$$L(.1,.9) = 1 + 2(.1) + 1(.9 - 1) = 1 + .2 - .1 = 1.1$$

is a good approximation of $f(.1,.9) - 1.09005\ldots$. $\square$

Partial derivatives and differentiability, $\mathbf{F} : \mathbb{R}^n \to \mathbb{R}^m$. We extend the definitions of differentiability, local linearity, and partial derivative. For functions $f : \mathbb{R}^n \to \mathbb{R}$

we define partial derivatives by differentiating with respect to the i-th variable while holding the others fixed:

$$\frac{\partial f}{\partial x_i} = \lim_{h \to 0} \frac{f(x_1,\ldots,x_i+h,\ldots,x_n) - f(x_1,\ldots,x_n)}{h}.$$

Let $\mathbf{F}$ be a function $\mathbb{R}^n \to \mathbb{R}^m$ whose component functions

$$\mathbf{F}(x_1,\ldots,x_n) = (f_1(x_1,\ldots,x_n), f_2(x_1,\ldots,x_n),\ldots,f_m(x_1,\ldots,x_n))$$

are differentiable. Each function f_i from $\mathbb{R}^n \to \mathbb{R}$ has n partial derivatives denoted $\dfrac{\partial f_i}{\partial x_j}$ or f_{i,x_j}. We arrange the partial derivatives in an m by n matrix

$$DF(\mathbf{A}) = \begin{bmatrix} \frac{\partial f_1}{\partial x_1}(\mathbf{A}) & \frac{\partial f_1}{\partial x_2}(\mathbf{A}) & \cdots & \frac{\partial f_1}{\partial x_n}(\mathbf{A}) \\ \frac{\partial f_2}{\partial x_1}(\mathbf{A}) & \frac{\partial f_2}{\partial x_2}(\mathbf{A}) & \cdots & \frac{\partial f_2}{\partial x_n}(\mathbf{A}) \\ \vdots & \vdots & \cdots & \vdots \\ \frac{\partial f_m}{\partial x_1}(\mathbf{A}) & \frac{\partial f_m}{\partial x_2}(\mathbf{A}) & \cdots & \frac{\partial f_m}{\partial x_n}(\mathbf{A}) \end{bmatrix}, \tag{3.6}$$

called the *matrix derivative* of $\mathbf{F}$ at $\mathbf{A}$.

Using vectors and matrices we can express the definition of differentiability of $\mathbf{F}$ at $\mathbf{A}$.

Definition 3.3. A function $\mathbf{F}$ from $\mathbb{R}^n$ to $\mathbb{R}^m$ defined on an open set U containing $\mathbf{A}$ is *differentiable* at $\mathbf{A}$ if $\mathbf{F}(\mathbf{A}+\mathbf{H}) - \mathbf{F}(\mathbf{A})$ can be well approximated by a linear function $\mathbf{L_A}$ in the sense that

$$\frac{\|\mathbf{F}(\mathbf{A}+\mathbf{H}) - \mathbf{F}(\mathbf{A}) - \mathbf{L_A}(\mathbf{H})\|}{\|\mathbf{H}\|}$$

tends to zero as $\|\mathbf{H}\|$ tends to zero.

By an argument similar to the one we gave for $f : \mathbb{R}^2 \to \mathbb{R}$ we can show that if $\mathbf{F}$ is differentiable at $\mathbf{A}$ then $\mathbf{F}$ is continuous at $\mathbf{A}$. We can also show that the component functions have partial derivatives, and that

$$\mathbf{L_A}(\mathbf{H}) = DF(\mathbf{A})\mathbf{H}$$

where $DF(\mathbf{A})$ is the matrix (3.6) of partial derivatives. We ask you to justify the steps to show this in Problem 3.9.

Example 3.4. Let $f(x,y,z) = x^2 \sin(yz)$. The partial derivatives of f are

$$f_x = 2x\sin(yz), \quad f_y = zx^2\cos(yz), \quad f_z = yx^2\cos(yz),$$

and

$$f_x(1, \tfrac{\pi}{2}, 2) = 0, \quad f_y(1, \tfrac{\pi}{2}, 2) = -2, \quad f_z(1, \tfrac{\pi}{2}, 2) = -\tfrac{\pi}{2}.$$

Therefore

$$Df(1, \tfrac{\pi}{2}, 2)) = \begin{bmatrix} 0 & -2 & -\tfrac{\pi}{2} \end{bmatrix}.$$

□

Example 3.5. Let $\mathbf{F}(x, y) = (x^2 + y^2, x, -y^3)$. Find $D\mathbf{F}(1, -2)$.

$$\begin{array}{ll} \frac{\partial f_1}{\partial x} = 2x, & \frac{\partial f_1}{\partial y} = 2y, \\ \frac{\partial f_2}{\partial x} = 1, & \frac{\partial f_2}{\partial y} = 0, \\ \frac{\partial f_3}{\partial x} = 0, & \frac{\partial f_3}{\partial y} = -3y^2. \end{array}$$

At the point $(1, -2)$ we have

$$D\mathbf{F}(1, -2) = \begin{bmatrix} 2 & -4 \\ 1 & 0 \\ 0 & -12 \end{bmatrix}.$$

□

Definition 3.4. For $n = m$ the matrix derivative (3.6) of $\mathbf{F}$ at $\mathbf{A}$ is a square matrix. Its determinant is called the *Jacobian* of $\mathbf{F}$ at $\mathbf{A}$ and is denoted as $J\mathbf{F}(\mathbf{A})$:

$$J\mathbf{F}(\mathbf{A}) = \det D\mathbf{F}(\mathbf{A}) \tag{3.7}$$

Just as the derivative $f'(a)$ can be thought of as a local stretching factor by f, $\frac{f(x) - f(a)}{x - a}$ as $x - a$ tends to zero, the geometric meaning of the Jacobian is local magnification of volume by $\mathbf{F}$. That is, denote by $B_r(\mathbf{A})$ the ball of radius r centered at $\mathbf{A}$, and by $C_r(\mathbf{A})$ its image under the mapping $\mathbf{F}$. The ratio

$$\frac{\text{Vol}(C_r(\mathbf{A}))}{\text{Vol}(B_r(\mathbf{A}))}$$

tends to $|J\mathbf{F}(\mathbf{A})|$ as r tends to zero.

Example 3.6. Let $\mathbf{F}(x, y) = (x^2 + y, y^3 + xy)$. Find the Jacobian of $\mathbf{F}$ at $(1, 2)$, and interpret it as a local magnification of area.

$$D\mathbf{F}(x, y) = \begin{bmatrix} 2x & 1 \\ y & 3y^2 + x \end{bmatrix}, \quad D\mathbf{F}(1, 2) = \begin{bmatrix} 2 & 1 \\ 2 & 13 \end{bmatrix}, \quad J\mathbf{F}(1, 2) = \det \begin{bmatrix} 2 & 1 \\ 2 & 13 \end{bmatrix} = 24.$$

The area of the image $\mathbf{F}(B_r)$ of a small disk B_r of radius r centered at $(1, 2)$ is about 24 times as large as the area of B_r.

□

If a function $\mathbf{X}$ from $\mathbb{R}$ to $\mathbb{R}^n$ is differentiable at t, then by definition

$$\frac{\mathbf{X}(t + h) - \mathbf{X}(t)}{h} - \mathbf{X}'(t)$$

tends to zero as h tends to zero. That means $\mathbf{X}'(t)$ is the limit of secant vectors divided by h. Figure 3.2 illustrates that $\mathbf{X}'(t)$ is tangent to the curve at $\mathbf{X}(t)$.

If we think of $\mathbf{X}(t)$ as the position of a particle at time t, then $\mathbf{X}'(t)$ is the *velocity* of the particle at time t and $\|\mathbf{X}'(t)\|$ is its speed.

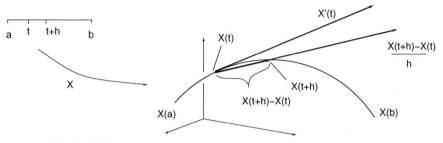

Fig. 3.2 $\dfrac{\mathbf{X}(t+h)-\mathbf{X}(t)}{h}$ tends to $\mathbf{X}'(t)$ as h tends to 0.

Example 3.7. Let $\mathbf{X}(t) = (\cos t, \sin t, t)$. The matrix derivative of $\mathbf{X}$ at t is

$$D\mathbf{X}(t) = \mathbf{X}'(t) = \begin{bmatrix} -\sin t \\ \cos t \\ 1 \end{bmatrix}.$$

Thinking of $\mathbf{X}'(t)$ as the velocity vector at time t we may write

$$\mathbf{X}'(t) = (-\sin t, \cos t, 1).$$

Its speed is $\|\mathbf{X}'(t)\| = \sqrt{(-\sin t)^2 + (\cos t)^2 + 1^2} = \sqrt{2}.$ □

Definition 3.5. The vector of partial derivatives of a function $f : \mathbb{R}^n \to \mathbb{R}$ is denoted
$$\nabla f = (f_{x_1}, \ldots, f_{x_n}) \quad \text{or} \quad \text{grad} f$$
and is called the *gradient* of f.

Example 3.8. Let $f(x,y,z) = x^2 \sin(yz)$. By the calculations in Example 3.4,

$$\nabla f(x,y,z) = (2x \sin(yz), zx^2 \cos(yz), yx^2 \cos(yz))$$

and $\nabla f(1, \frac{\pi}{2}, 2) = (0, -2, -\frac{\pi}{2})$. □

Using the gradient notation the linear approximation of $f(\mathbf{A} + \mathbf{H})$ at $\mathbf{A}$ can be written using the dot product

$$f(\mathbf{A} + \mathbf{H}) \approx f(\mathbf{A}) + \nabla f(\mathbf{A}) \cdot \mathbf{H}.$$

Definition 3.6. A function $\mathbf{F}$ from an open set U of $\mathbb{R}^n$ to $\mathbb{R}^m$ is called *continuously differentiable* if it has partial derivatives that are continuous functions on U. A continuously differentiable function is called a C^1 function.

In Theorem 3.3 we have shown that a function of two variables that has continuous partial derivatives is differentiable; the analogous theorem holds for functions of several variables.

Theorem 3.4. *If $\mathbf{F}$ from $\mathbb{R}^n$ to $\mathbb{R}^m$ is C^1 on an open set U then $\mathbf{F}$ is differentiable at each point of U.*

In Problem 3.10 we show you how to extend the proof of Theorem 3.3 to prove Theorem 3.4.

Problems

3.1. Find the partial derivatives.

(a) $\displaystyle\lim_{h\to 0} \frac{(3(x+h)^2 + 4y) - (3x^2 + 4y)}{h}$

(b) $\displaystyle\lim_{k\to 0} \frac{(3x^2 + 4(y+k)) - (3x^2 + 4y)}{k}$

3.2. Let $f(x,y) = x^2 + 3y$. Find the linear approximation of $f(x,y)$ near $(2,4)$, and use it to estimate $f(2.01, 4.03)$.

3.3. Find the indicated partial derivatives of the functions.

(a) $f_x(x,y)$ and $f_y(2,0)$ if $f(x,y) = e^{-x^2 - y^2}$

(b) $\dfrac{\partial}{\partial y}(xe^y + ye^x)$

(c) $\dfrac{\partial}{\partial x}\left(\cos(xy) + \dfrac{\partial}{\partial y}(\sin(xy))\right)$

3.4. Let $f(x,y) = x^2 y^3$. Let $(x,y) = (a+u, b+v)$. Use binomial expansion on the right side of

$$x^2 y^3 = (a+u)^2(b+v)^3$$

to find the numbers c_1, c_2, c_3 in

$$f(x,y) = c_1 + c_2(x-a) + c_3(y-b) + \cdots$$

where the dots represent polynomials in $(x-a)$ and $(y-b)$ of degree 2 or more.

(a) Express c_1 in terms of f, a, and b.

(b) Express c_2 and c_3 in terms of partial derivatives of f at (a,b).

(c) Find functions ℓ and s so that

$$f(a+u,b+v) = f(a,b) + \ell(u,v) + s(u,v),$$

and ℓ is linear, and $s(u,v)$ is small compared to (u,v) as (u,v) tends to $(0,0)$.

3.5. Two mathematics students are discussing the values of $(1+x+3y)^2$ for small x and y. They find two linear functions ℓ_1 and ℓ_2 to help estimate the values,

$$
\begin{aligned}
(1+x+3y)^2 &= (1+x+3y)(1+x+3y) \\
&\approx (1)(1+x+3y) = 1 + \underbrace{x+3y}_{\ell_1(x,y)}, \quad \text{and}
\end{aligned}
$$

$$
\begin{aligned}
(1+x+3y)^2 &= 1 + 2x + 6y + 6xy + x^2 + 9y^2 \\
&\approx 1 + \underbrace{2x + 6y}_{\ell_2(x,y)}.
\end{aligned}
$$

(x,y)	$(.1,.2)$	$(.01,.02)$
$(1+x+3y)^2$		
$1+x+3y$		
$1+2x+6y$		

Fill in the table of values, and observe that some linear functions track small changes better than others.

3.6. Consider a linear function $\ell(x,y,z) = ax + by + cz$. Show that $\nabla\ell$ is constant. Show that when a, b, c are not all zero $\nabla\ell$ is normal to the level set $\ell = 0$.

3.7. Consider two linear functions $\ell(x,y) = x+2y$ and $m(x,y) = -3\ell(x,y)$. Sketch level sets $\ell = -1,0,1$ and $m = -1,0,1$. For which function are these more closely spaced? Determine the gradient vectors $\nabla\ell(x,y)$ and $\nabla m(x,y)$. For which function are these vectors longer?

3.8. Let $\mathbf{X}$ be in $\mathbb{R}^n$. Find the gradients.

(a) $\nabla\left(2\|\mathbf{X}\|^{1/2}\right)$

(b) $\nabla\left(-\|\mathbf{X}\|^{-1}\right)$

(c) $\nabla\left(\dfrac{1}{r}\|\mathbf{X}\|^r\right), \quad r \neq 0$

3.9. Suppose a function $\mathbf{F}$ from $\mathbb{R}^n$ to $\mathbb{R}^m$ is differentiable at $\mathbf{A}$. Justify the following statements that prove

$$\mathbf{L}_{\mathbf{A}}\mathbf{H} = D\mathbf{F}(\mathbf{A})\mathbf{H},$$

that is, the linear function $\mathbf{L}_{\mathbf{A}}$ in Definition 3.3 is given by the matrix of partial derivatives $D\mathbf{F}(\mathbf{A})$.

(a) There is a matrix $\mathbf{C}$ such that $\mathbf{L_A(H)} = \mathbf{CH}$ for all $\mathbf{H}$.

(b) Let $\mathbf{C}_i$ be the i-th row of $\mathbf{C}$. The fraction

$$\frac{\|\mathbf{F(A+H)} - \mathbf{F(A)} - \mathbf{L_A(H)}\|}{\|\mathbf{H}\|} = \frac{\|\mathbf{F(A+H)} - \mathbf{F(A)} - \mathbf{CH}\|}{\|\mathbf{H}\|}$$

tends to zero as $\|\mathbf{H}\|$ tends to zero if and only if each component

$$\frac{f_i(\mathbf{A+H}) - f_i(\mathbf{A}) - \mathbf{C}_i\mathbf{H}}{\|\mathbf{H}\|}$$

tends to zero as $\|\mathbf{H}\|$ tends to zero.

(c) Set $\mathbf{H} = h\mathbf{E}_j$ in the i-th component of the numerator to show that the partial derivative $f_{i,x_j}(\mathbf{A})$ exists and is equal to the (i,j) entry of $\mathbf{C}$.

3.10. Justify the following steps to prove Theorem 3.4, that a function with continuous first partial derivatives is differentiable. In parts (a)-(d) we suppose $f : \mathbb{R}^n \to \mathbb{R}$ has continuous first partial derivatives at all points in a ball of radius r centered at point $\mathbf{P}$. In parts (e)-(f) we assume the components f_i of $\mathbf{F} : \mathbb{R}^n \to \mathbb{R}^m$ have continuous first partial derivatives at all points in a ball of radius r centered at point $\mathbf{P}$. Let $\mathbf{H}$ be a vector with $\|\mathbf{H}\| < r$.

(a)

$$\begin{aligned}
f(\mathbf{P+H}) - f(\mathbf{P}) &= f(p_1+h_1,\ldots,p_n+h_n) - f(p_1,\ldots,p_n) \\
&= f(p_1+h_1,p_2+h_2\ldots,p_n+h_n) - f(p_1,p_2+h_2,\ldots,p_n+h_n) \\
&\quad + f(p_1,p_2+h_2,p_3+h_3,\ldots,p_n+h_n) - f(p_1,p_2,p_3+h_3,\ldots,p_n+h_n) \\
&\quad + \cdots \\
&\quad + f(p_1,p_2,\ldots,p_{n-1},p_n+h_n) - f(p_1,p_2,\ldots,p_{n-1},p_n).
\end{aligned}$$

(b) There are numbers $0 \le h_i' \le h_i$ such that

$$f(p_1,\ldots,p_{i-1},p_i+h_i,p_{i+1}+h_{i+1},\ldots,p_n+h_n)$$
$$- f(p_1,\ldots,p_{i-1},p_i,p_{i+1}+h_{i+1},\ldots,p_n+h_n)$$
$$= h_i f_{x_i}(p_1,\ldots,p_{i-1},p_i+h_i',p_{i+1}+h_{i+1},\ldots,p_n+h_n).$$

(c) $f(\mathbf{P+H}) - f(\mathbf{P}) = \displaystyle\sum_{i=1}^n h_i f_{x_i}(p_1,\ldots,p_{i-1},p_i+h_i',p_{i+1}+h_{i+1},\ldots,p_n+h_n).$

(d) $\dfrac{f(\mathbf{P+H}) - f(\mathbf{P}) - \mathbf{H}\cdot\nabla f(\mathbf{P})}{\|\mathbf{H}\|}$ tends to zero as $\mathbf{H}$ tends to $\mathbf{0}$.

(e) Given $\epsilon > 0$ there are numbers r_i so that if $\|\mathbf{H}\| < r_i$, then

$$\frac{|f_i(\mathbf{P+H}) - f(\mathbf{P}) - \nabla f_i(\mathbf{P})\cdot\mathbf{H}|}{\|\mathbf{H}\|} < \epsilon.$$

(f) Let r be the smallest of $r_1,\ldots,r_m$. Then if $\|\mathbf{H}\| < r$,

$$\frac{\|\mathbf{F}(\mathbf{P}+\mathbf{H}) - \mathbf{F}(\mathbf{P}) - D\mathbf{F}(\mathbf{P})\mathbf{H}\|}{\|\mathbf{H}\|} < \epsilon m.$$

3.11. Define functions f and g from $\mathbb{R}^2$ to $\mathbb{R}$ by

$$f(x,y) = \cos(x+y), \qquad g(x,y) = \sin(2x-y).$$

Find the gradients ∇f and ∇g, and show that

$$f_x - f_y = 0, \qquad g_x + 2g_y = 0.$$

3.12. Let $f_1(x,y) = e^x \cos y$, $f_2(x,y) = x^2 - y^2$. Find ∇f_1 and ∇f_2, and show that

$$\frac{\partial}{\partial x}(f_x) + \frac{\partial}{\partial y}(f_y) = 0$$

for each of f_1 and f_2.

3.13. Let $g(x,y) = e^{ax+by}$, where a and b are some numbers. Find the gradient of g.

3.14. Let a, b, and c be some numbers, and define

$$f(x,y,z) = \sin(ax + by + cz).$$

Let $\mathbf{C} = (p,q,r)$ be a vector such that $ap + bq + cr = 0$. Show that $\mathbf{C} \cdot \nabla f = 0$, that is,

$$pf_x + qf_y + rf_z = 0.$$

3.2 The tangent plane and partial derivatives

The geometric interpretation of the derivative $f'(a)$ of a function f of a single variable is the slope of the line

$$y = f(a) + f'(a)(x-a)$$

tangent to the graph of the function. There is a similar geometric interpretation of the partial derivatives of a function of two variables. Suppose f is differentiable at (a,b), with partial derivatives $f_x(a,b)$ and $f_y(a,b)$ at (a,b). The geometric interpretation of linear approximation is that the graph of

$$z = L(x,y) = f(a,b) + f_x(a,b)(x-a) + f_y(a,b)(y-b)$$

is the plane tangent to the graph of f at the point $(a,b,f(a,b))$. Rewriting the equation

$$f_x(a,b)(x-a) + f_y(a,b)(y-b) + (-1)(z - f(a,b)) = 0 \qquad (3.8)$$

we see that a normal to the tangent plane is the vector

$$\mathbf{N} = (f_x(a,b), f_y(a,b), -1).$$

If f is differentiable at (a,b) we say $\mathbf{N}$ is normal to the graph of f at $(a,b,f(a,b))$.

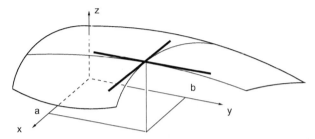

Fig. 3.3 A graph partly cut away for clarity at its intersections with planes $x = a$ and $y = b$.

Another way to obtain an equation for the plane tangent to the graph of f at (a,b) is to intersect the graph of f with the planes $x = a$ and $y = b$ (see Figure 3.3), and use the lines tangent to the resulting curves to determine the tangent plane.

In the plane $x = a$ an equation for the line tangent to the curve at $(a,b,f(a,b))$ is

$$c_1(t) = (a,b,f(a,b)) + t(0,1,f_y(a,b)).$$

Similarly, an equation for the line tangent to the intersection curve in the plane $y = b$ is

$$c_2(s) = (a,b,f(a,b)) + s(1,0,f_x(a,b)).$$

As we saw in Section 1.9 a parametric equation for the plane determined by these two lines is

$$\mathbf{P}(s,t) = (a,b,f(a,b)) + s(1,0,f_x(a,b)) + t(0,1,f_y(a,b)).$$

A normal to this plane is the cross product of the two tangent vectors,

$$(1,0,f_x(a,b)) \times (0,1,f_y(a,b)) = (-f_x(a,b), -f_y(a,b), 1)$$

which is a normal to the tangent plane given by equation (3.8).

Example 3.9. Let $f(x,y) = x^{1/2}y^{1/3}$.

(a) Find an equation of the plane tangent to the graph of f at $(1,-1)$, and
(b) Use the linear approximation of f at $(1,-1)$ to estimate $f(.9,-1.1)$.

We have $f_y(x,y) = \frac{1}{3}x^{1/2}y^{-2/3}$ and $f_x(x,y) = \frac{1}{2}x^{-1/2}y^{1/3}$. Since f_x and f_y are continuous near $(1,-1)$, f is differentiable at $(1,-1)$.

$$f(1,-1) = -1, \qquad f_y(1,-1) = \tfrac{1}{3}, \qquad f_x(1,-1) = -\tfrac{1}{2}.$$

(a) An equation for the plane tangent to the graph of f at $(1,-1)$ is

$$z = f(1,-1) + f_x(1,-1)(x-1) + f_y(1,-1)(y-(-1)) = -1 - \tfrac{1}{2}(x-1) + \tfrac{1}{3}(y+1),$$

or $3x - 2y + 6z + 1 = 0$.

(b) To approximate $f(.9,-1.1)$ we use the linear approximation

$$L(x,y) = f(1,-1) + f_x(1,-1)(x-1) + f_y(1,-1)(y-(-1)).$$

$$L(.9,-1.1) = -1 - \tfrac{1}{2}(.9-1) + \tfrac{1}{3}(-1.1+1) = -1 + \tfrac{1}{20} - \tfrac{1}{30} = -.9833\ldots,$$

a good approximation of $f(.9,-1.1) = -.9793\ldots$.

□

Example 3.10. Find all points on the graph of

$$f(x,y) = x^2 - 2xy - y^2 + 6x - 6y$$

where the tangent plane is horizontal. The function f has continuous partial derivatives $f_x = 2x - 2y + 6$ and $f_y = -2x - 2y - 6$, so f is differentiable at every point. A tangent plane to the graph of f at (x,y) is horizontal if $f_x(x,y) = 0$ and $f_y(x,y) = 0$. The equations $2x - 2y + 6 = 0$ and $-2x - 2y - 6 = 0$ have the solution $y = 0$ and $x = -3$. There is one point on the graph with horizontal tangent plane, $(-3,0,-9)$.

□

If two C^1 functions f and g have the same value $f(a,b) = g(a,b)$ and the same partial derivatives at (a,b), then their graphs have the same tangent plane at $(a,b,f(a,b))$, and we say that the graphs are tangent at that point.

Example 3.11. To find points of tangency of the graphs of

$$f(x,y) = x^2 - 2xy - y^2$$
$$g(x,y) = x^2 - 3xy + 4x - 16$$

and the common tangent plane we look for points where the normal vectors, $(2x-2y, -2x-2y, -1)$ and $(2x-3y+4, -3x, -1)$, are multiples of each other. Since the third components are equal this only happens when

$$2x - 2y = 2x - 3y + 4$$
$$-2x - 2y = -3x,$$

which are satisfied by $y = 4$ and $x = 8$. We verify that the point $(8,4,-16)$ is on both graphs. Therefore there is a common tangent plane

$$z = -16 + 8(x-8) - 24(y-4)$$

at the point $(8,4,-16)$.

□

Problems

3.15. Find an equation for the plane tangent to the graph of $f(x,y) = x+y^2$,

(a) at $(x,y) = (0,0)$,
(b) at $(x,y) = (1,2)$.

3.16. Let $f(x,y) = \sqrt{1-x^2-y^2}$.

(a) Sketch the graph of f.
(b) Find an equation of the plane tangent to the graph of f at $(x,y) = (\sqrt{.4}, \sqrt{.5})$.

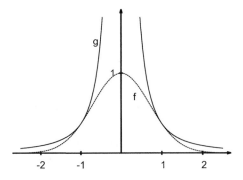

Fig. 3.4 Cross sections of graphs of f and g in Problem 3.17.

3.17. Let $f(x,y) = e^{-(x^2+y^2)}$ and $g(x,y) = \dfrac{e^{-1}}{x^2+y^2}$.

(a) Show that $f(a,b) = g(a,b)$ at every point (a,b) where $a^2+b^2 = 1$.
(b) Find the gradients ∇f and ∇g.
(c) Show that the graph of f is tangent to the graph of g at every point (a,b) where

$$a^2 + b^2 = 1.$$

See Figure 3.4 for cross sections of graphs of f and g.

3.3 The Chain Rule

Composition. There are many types of functions of several variables so the formula for the Chain Rule takes several forms. Here is one of them.

Theorem 3.5. Chain Rule 1 (for curves). *Let f from $\mathbb{R}^2$ to $\mathbb{R}$ be continuously differentiable on an open set U, and let $\mathbf{X}(t) = (x(t), y(t))$ be a differentiable function from an open interval I to U. Then the composition $f(x(t), y(t))$ is a differentiable function from I to $\mathbb{R}$ and*

$$\frac{d}{dt} f(x(t), y(t)) = \frac{\partial f}{\partial x} \frac{dx}{dt} + \frac{\partial f}{\partial y} \frac{dy}{dt}.$$

The right side can be written as the dot product

$$= \nabla f(x(t), y(t)) \cdot \mathbf{X}'(t).$$

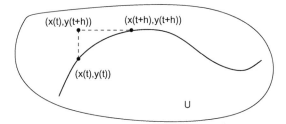

Fig. 3.5 Line segments used in the proof of Theorem 3.5.

Proof. Because $\mathbf{X}$ is continuous, we know that for h sufficiently small, $h \neq 0$, the line segments from $(x(t), y(t))$ to $(x(t), y(t+h))$ and from $(x(t), y(t+h))$ to $(x(t+h), y(t+h))$ in Figure 3.5 are in U. Rewrite

$$f(x(t+h), y(t+h)) - f(x(t), y(t))$$

as

$$= \Big(f(x(t+h), y(t+h)) - f(x(t), y(t+h))\Big) + \Big(f(x(t), y(t+h)) - f(x(t), y(t))\Big).$$

By the Mean Value Theorem for single variable functions there is a number x_* between $x(t+h)$ and $x(t)$ for which

$$f(x(t+h), y(t+h)) - f(x(t), y(t+h)) = f_x(x_*, y(t+h))(x(t+h) - x(t)).$$

Similarly there is a number y_* between $y(t)$ and $y(t+h)$ so that

$$f(x(t), y(t+h)) - f(x(t), y(t)) = f_y(x(t), y_*)(y(t+h) - y(t)).$$

Now divide by h; we get

$$\frac{f(x(t+h),y(t+h)) - f(x(t),y(t))}{h}$$

$$= \frac{f(x(t+h),y(t+h)) - f(x(t),y(t+h))}{h} + \frac{f(x(t),y(t+h)) - f(x(t),y(t))}{h}$$

$$= \frac{f_x(x_*,y(t+h))(x(t+h) - x(t))}{h} + \frac{f_y(x(t),y_*)(y(t+h) - y(t))}{h}.$$

As h tends to zero $\dfrac{x(t+h) - x(t)}{h}$ and $\dfrac{y(t+h) - y(t)}{h}$ tend to $x'(t)$ and $y'(t)$. Since f_x and f_y are continuous, $f_x(x_*,y(t+h))$ tends to $f_x(x(t),y(t))$ and $f_y(x(t),y_*)$ tends to $f_y(x(t),y(t))$ as h tends to zero. Hence

$$\frac{\mathrm{d}}{\mathrm{d}t} f(x(t),y(t)) = f_x(x(t),y(t))x'(t) + f_y(x(t),y(t))y'(t) = \nabla f(x(t),y(t)) \cdot \mathbf{X}'(t).$$

$\square$

Example 3.12. For $0 \le t \le 1$, $\mathbf{X}(t) = (x(t),y(t)) = (t,2t)$ is the straight path from the origin to point $(1,2)$. Let

$$f(x,y) = x^2 + y^4.$$

The derivatives of $\mathbf{X}$ and f are $\mathbf{X}'(t) = (1,2),$, $\nabla f = (2x,4y^3)$. By the Chain Rule,

$$\frac{\mathrm{d}}{\mathrm{d}t} f(\mathbf{X}(t)) = \nabla f(x(t),y(t)) \cdot \mathbf{X}'(t) = (2(t),4(2t)^3) \cdot (1,2) = 2t + 64t^3.$$

Alternatively, the composite of f and $\mathbf{X}$ is

$$f(\mathbf{X}(t)) = t^2 + 16t^4,$$

and the derivative is $\dfrac{\mathrm{d}}{\mathrm{d}t}(t^2 + 16t^4) = 2t + 64t^3$ as we found using the Chain Rule.

$\square$

We proved the Chain Rule for curves in $\mathbb{R}^2$ and functions from $\mathbb{R}^2$ to $\mathbb{R}$. The analogous theorem can be proved in n dimensions.

Theorem 3.6. Chain Rule 1 (for curves). *Let f from $\mathbb{R}^n$ to $\mathbb{R}$ have continuous partial derivatives on an open set U, and let $\mathbf{X}(t) = (x_1(t),\ldots,x_n(t))$ be a differentiable function from an open interval I in $\mathbb{R}$ into U. Then the composite $f(\mathbf{X}(t))$ is differentiable on I and*

$$\frac{\mathrm{d}}{\mathrm{d}t} f(\mathbf{X}(t)) = \frac{\partial f}{\partial x_1}\frac{\mathrm{d}x_1}{\mathrm{d}t} + \cdots + \frac{\partial f}{\partial x_n}\frac{\mathrm{d}x_n}{\mathrm{d}t} = \nabla f(\mathbf{X}(t)) \cdot \mathbf{X}'(t) = Df(\mathbf{X}(t))D\mathbf{X}(t).$$

Example 3.13. Let $\mathbf{X}(t) = (\cos t, \sin t, t)$, $-\infty < t < \infty$ be a curve in $\mathbb{R}^3$ that gives the position of a particle at time t. Let

$$f(x, y, z) = e^{-x^2 - y^2 - z^2}$$

represent the temperature of the particle at point (x, y, z). Then the composite

$$f(\mathbf{X}(t)) = e^{-(1+t^2)}$$

is the temperature of the particle at time t. The rate at which the temperature of the particle changes with respect to time is

$$\frac{d}{dt} f(\mathbf{X}(t)) = -2t e^{-(1+t^2)}.$$

We calculate derivatives: the velocity of the particle at time t,

$$\mathbf{X}'(t) = (-\sin t, \cos t, 1),$$

and the gradient of the temperature at (x, y, z)

$$\nabla f(x, y, z) = -2e^{-x^2 - y^2 - z^2} (x, y, z).$$

Using the Chain Rule we have that the rate at which the temperature of the particle changes with respect to time is

$$\frac{d}{dt} f(\mathbf{X}(t)) = \nabla f(\mathbf{X}(t)) \cdot \mathbf{X}'(t)$$

$$= -2e^{-(1+t^2)}(\cos t, \sin t, t) \cdot (-\sin t, \cos t, 1) = -2t e^{-(1+t^2)}.$$

□

Directional derivative. The rate at which a function f from $\mathbb{R}^n$ to $\mathbb{R}$ changes as we move from a point $\mathbf{P}$ along the straight line $\mathbf{X}(t) = \mathbf{P} + t\mathbf{V}$, $-h \le t \le h$ is

$$\lim_{t \to 0} \frac{f(\mathbf{X}(t)) - f(\mathbf{X}(0))}{t - 0} = \frac{d}{dt} f(\mathbf{X}(t))\Big|_{t=0}.$$

By the Chain Rule for curves

$$\frac{d}{dt} f(\mathbf{X}(t)) = \nabla f(\mathbf{X}(t)) \cdot \mathbf{X}'(t).$$

At $t = 0$

$$\nabla f(\mathbf{X}(0)) \cdot \mathbf{X}'(0) = \nabla f(\mathbf{P}) \cdot \mathbf{V}.$$

Definition 3.7. Let f be a C^1 function from an open set in $\mathbb{R}^n$ containing $\mathbf{P}$ to $\mathbb{R}$, and let $\mathbf{V}$ be a unit vector in $\mathbb{R}^n$. We call

$$\nabla f(\mathbf{P}) \cdot \mathbf{V}$$

the *directional derivative* of f at $\mathbf{P}$ in the direction of $\mathbf{V}$, and denote it $D_{\mathbf{V}} f(\mathbf{P})$.

The directional derivative gives insight into the gradient of f at $\mathbf{P}$. Since $\|\mathbf{V}\| = 1$,

$$D_{\mathbf{V}} f(\mathbf{P}) = \nabla f(\mathbf{P}) \cdot \mathbf{V} = \|\nabla f(\mathbf{P})\| \cos \theta$$

where θ is the angle between $\nabla f(\mathbf{P})$ and $\mathbf{V}$. We see that the directional derivative is greatest when $\cos \theta = 1$, i.e., when $\mathbf{V}$ is in the direction of $\nabla f(\mathbf{P})$. The unit vector in the direction of the gradient of f at $\mathbf{P}$ is

$$\frac{\nabla f(\mathbf{P})}{\|\nabla f(\mathbf{P})\|}.$$

Therefore the greatest directional rate of change in f at $\mathbf{P}$ is in the direction of the gradient and is

$$\nabla f(\mathbf{P}) \cdot \frac{\nabla f(\mathbf{P})}{\|\nabla f(\mathbf{P})\|} = \|\nabla f(\mathbf{P})\|.$$

Example 3.14. Let $f(x,y,z) = x + y^2 + z^4$. Find the direction and magnitude of the greatest directional derivative of f at $(3,2,1)$.

$$\nabla f = (1, 2y, 4z^3), \qquad \nabla f(3,2,1) = (1,4,4).$$

The greatest rate of change in f at $(3,2,1)$ is

$$\|\nabla f(3,2,1)\| = \sqrt{1^2 + 4^2 + 4^2} = \sqrt{33} = 5.744\ldots$$

and this occurs in the direction of $(1,4,4)$. $\qquad\qquad\qquad\qquad\qquad\qquad\square$

Example 3.15. Let $f(x,y,z) = x + y^2 + z^4$. In which directions is the rate of change of f at $(3,2,1)$ zero? From Example 3.14,

$$\nabla f(3,2,1) = (1,4,4).$$

The directional derivative $D_{\mathbf{V}} f(3,2,1)$ is zero if $(1,4,4) \cdot \mathbf{V} = 0$. That is, if we move from $(3,2,1)$ in any direction orthogonal to the gradient, the rate of change is zero. $\qquad\qquad\qquad\qquad\qquad\qquad\qquad\qquad\qquad\qquad\square$

The gradient and level sets. Let k be a number, and consider the level set S given by $f(x,y,z) = k$. We assume

- a point (a,b,c) is on S, so $f(a,b,c) = k$
- f is C^1 on S

- $\nabla f(a,b,c) \neq \mathbf{0}$

We show that $\nabla f(a,b,c)$ is *normal* to S.

Suppose $\mathbf{X}(t)$ is a differentiable curve in S that goes through (a,b,c) at $t = t_0$. Then

$$f(\mathbf{X}(t)) = k, \qquad \text{and} \quad \frac{d}{dt} f(\mathbf{X}(t))\Big|_{t=t_0} = \frac{d}{dt} k = 0.$$

By the Chain Rule,

$$\frac{d}{dt} f(\mathbf{X}(t))\Big|_{t=t_0} = \nabla f(\mathbf{X}(t_0)) \cdot \mathbf{X}'(t_0).$$

Combining these two expressions we get

$$0 = \nabla f(\mathbf{X}(t_0)) \cdot \mathbf{X}'(t_0).$$

Therefore $\nabla f(a,b,c)$ is orthogonal to the tangent vector of *every* differentiable curve in S through (a,b,c). This is what we mean when we say that $\nabla f(a,b,c)$ is normal to S at (a,b,c). See Figure 3.6. The Implicit Function Theorem in Section 3.4 will show that there is a tangent plane to S at (a,b,c) and that it is the set of points (x,y,z) that satisfy

$$\nabla f(a,b,c) \cdot (x-a, y-b, z-c) = 0. \tag{3.9}$$

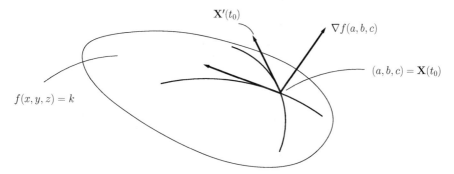

Fig. 3.6 $\nabla f(a,b,c)$ is normal to the k level set $f(x,y,z) = k$ at (a,b,c).

Example 3.16. Find the plane tangent to the level set

$$xyz + z^3 = 3$$

at the point $(1,2,1)$. Let $f(x,y,z) = xyz + z^3$. Then $\nabla f = (yz, xz, xy + 3z^2)$ and $\nabla f(1,2,1) = (2,1,5)$. The tangent plane is given by

$$(2,1,5) \cdot (x-1, y-2, z-1) = 0.$$

$\square$

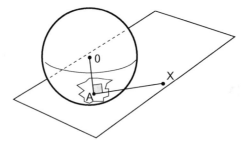

Fig. 3.7 A sphere and its tangent plane at **A**, $\mathbf{A} \cdot (\mathbf{X} - \mathbf{A}) = 0$. See Example 3.17.

Example 3.17. Let S be the sphere of radius R in $\mathbb{R}^3$ centered at the origin, satisfying

$$x^2 + y^2 + z^2 = R^2.$$

S is the R^2 level set of the function $f(x,y,z) = x^2 + y^2 + z^2$. The gradient of f is

$$\nabla f(x,y,z) = (2x, 2y, 2z).$$

Using (3.9) the tangent plane at a point (a,b,c) on S has equation

$$2(a,b,c) \cdot (x-a, y-b, z-c) = 0.$$

Let $\mathbf{A} = (a,b,c)$ and $\mathbf{X} = (x,y,z)$. Then the equation of the tangent plane (see Figure 3.7) to the sphere S can also be written

$$\mathbf{A} \cdot (\mathbf{X} - \mathbf{A}) = 0.$$

$\square$

Let S be the graph of a function f from $\mathbb{R}^2$ to $\mathbb{R}$. We can define a new function from $\mathbb{R}^3$ to $\mathbb{R}$ by

$$g(x,y,z) = f(x,y) - z.$$

The level surface $g(x,y,z) = 0$ is the same set of points in $\mathbb{R}^3$ as the graph of f. The gradient of g is

$$\nabla g = \left(\frac{\partial g}{\partial x}, \frac{\partial g}{\partial y}, \frac{\partial g}{\partial z}\right) = \left(\frac{\partial f}{\partial x}, \frac{\partial f}{\partial y}, -1\right).$$

Using (3.9) the equation of the plane tangent to the level surface $g(x,y,z) = 0$ at the point $(a,b,f(a,b))$ is

$$\nabla g(a,b,f(a,b)) \cdot (x-a, y-b, z-f(a,b))$$

$$= (f_x(a,b), f_y(a,b), -1) \cdot (x-a, y-b, z-f(a,b)) = 0,$$

which we rewrite as $f_x(a,b)(x-a) + f_y(a,b)(y-b) - (z-f(a,b)) = 0$ as we found earlier.

The Chain Rule also holds if the first function in the composition is a function of n variables and the second maps $\mathbb{R}$ to $\mathbb{R}$.

Theorem 3.7. Chain Rule 2. *Suppose $y = f(x_1,\ldots,x_n)$ is continuously differentiable on an open set U in $\mathbb{R}^n$ and $z = g(y)$ is continuously differentiable on an open interval V in $\mathbb{R}$ that contains the range of f. Then the composite $g(f(x_1,\ldots,x_n))$ is continuously differentiable on U and for each x_i*

$$\frac{\partial}{\partial x_i}(g(f(x_1,\ldots,x_n))) = \frac{dg}{dy}\frac{\partial f}{\partial x_i}, \qquad (i = 1,\ldots,n).$$

Therefore

$$D(g \circ f)(\mathbf{X}) = \frac{dg}{dy}\nabla f(\mathbf{X}).$$

Proof. Since f and g are continuously differentiable on U and V their partial derivatives exist and are continuous there. By the Chain Rule for single variable functions

$$\frac{dz}{dy}\frac{\partial y}{\partial x_i} = \frac{\partial z}{\partial x_i}.$$

Since $\dfrac{\partial z}{\partial x_i}$ is a product of continuous functions these partial derivatives are continuous on U. $\qquad\square$

The most general form of the Chain Rule can be proved for compositions of continuously differentiable functions $\mathbf{X}$ from $\mathbb{R}^k$ to $\mathbb{R}^m$ and $\mathbf{F}$ from $\mathbb{R}^m$ to $\mathbb{R}^n$.

Theorem 3.8. Chain Rule. *Let U be an open set in $\mathbb{R}^k$, and let V be an open set in $\mathbb{R}^m$. Suppose that $\mathbf{X}(\mathbf{T}) = (x_1(t_1,\ldots,t_k),\ldots,x_m(t_1,\ldots,t_k))$ is continuously differentiable on U and $\mathbf{F}(\mathbf{X}) = (y_1(x_1,\ldots,x_m),\ldots,y_n(x_1,\ldots,x_m))$ is continuously differentiable on V, and that the range of $\mathbf{X}$ is contained in V. Then the composite $\mathbf{F}(\mathbf{X}(\mathbf{T}))$ is continuously differentiable on U and the derivative of the composite is the product of the derivative matrices:*

$$D(\mathbf{F} \circ \mathbf{X})(\mathbf{T}) = D\mathbf{F}(\mathbf{X}(\mathbf{T}))D\mathbf{X}(\mathbf{T}).$$

Componentwise we can write this relation as

$$\frac{\partial y_i}{\partial t_j} = \frac{\partial y_i}{\partial x_1}\frac{\partial x_1}{\partial t_j} + \frac{\partial y_i}{\partial x_2}\frac{\partial x_2}{\partial t_j} + \cdots + \frac{\partial y_i}{\partial x_m}\frac{\partial x_m}{\partial t_j}.$$

Figure 3.8 illustrates the composition of a function from $\mathbb{R}^3$ to $\mathbb{R}^2$ and a function from $\mathbb{R}^2$ to $\mathbb{R}$.

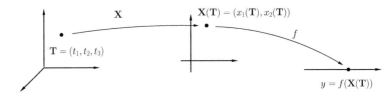

Fig. 3.8 A composition of functions $\mathbb{R}^3 \to \mathbb{R}^2 \to \mathbb{R}$.

Proof. Consider each component function $f_i(\mathbf{X}(t)) = y_i$ as a function of a single variable t_j by holding the other t_k constant. We can use the Chain Rule for curves, Theorem 3.6, to find the partial derivative of the composite $f_i(\mathbf{X}(\mathbf{T}))$ with respect to t_j. This gives

$$\frac{\partial y_i}{\partial t_j} = \frac{\partial y_i}{\partial x_1}\frac{\partial x_1}{\partial t_j} + \frac{\partial y_i}{\partial x_2}\frac{\partial x_2}{\partial t_j} + \cdots + \frac{\partial y_i}{\partial x_m}\frac{\partial x_m}{\partial t_j}. \tag{3.10}$$

Since we have assumed that the partial derivatives of $\mathbf{X}$ and $\mathbf{F}$ are continuous, each such partial derivative is continuous on U. Hence $\mathbf{F} \circ \mathbf{X}$ is C^1 on U. By Theorem 3.4 $\mathbf{F} \circ \mathbf{X}$ is differentiable on U.

The partial derivative $\dfrac{\partial y_i}{\partial t_j}$ is by definition the i, j entry of the matrix

$$D(\mathbf{F} \circ \mathbf{X})(\mathbf{T}).$$

According to (3.10) it is also the i, j entry of the product

$$DF(\mathbf{X}(\mathbf{T}))D\mathbf{X}(\mathbf{T}),$$

where

$$DF(\mathbf{X}) = \begin{bmatrix} \frac{\partial y_1}{\partial x_1} & \frac{\partial y_1}{\partial x_2} & \cdots & \cdots \\ \frac{\partial y_2}{\partial x_1} & \frac{\partial y_2}{\partial x_2} & \cdots & \cdots \\ \vdots & \vdots & \vdots & \vdots \\ \frac{\partial y_n}{\partial x_1} & \cdots & \cdots & \frac{\partial y_n}{\partial x_m} \end{bmatrix}, \qquad D\mathbf{X}(\mathbf{T}) = \begin{bmatrix} \frac{\partial x_1}{\partial t_1} & \frac{\partial x_1}{\partial t_2} & \cdots & \cdots \\ \frac{\partial x_2}{\partial t_1} & \frac{\partial x_2}{\partial t_2} & \cdots & \cdots \\ \vdots & \vdots & \vdots & \vdots \\ \frac{\partial x_m}{\partial t_1} & \cdots & \cdots & \frac{\partial x_m}{\partial t_k} \end{bmatrix}.$$

$\square$

We can use the Chain Rule to get another version of the Mean Value Theorem.

Theorem 3.9. Mean Value Theorem *Let* $\mathbf{F} = (f_1, f_2, \ldots, f_m)$ *be a* C^1 *map from* $\mathbb{R}^n$ *to* $\mathbb{R}^m$, *on an open set in* $\mathbb{R}^n$ *that contains points* $\mathbf{A}$, $\mathbf{A} + \mathbf{H}$, *and the line segment*

$$\mathbf{A} + t\mathbf{H}, \qquad 0 \leq t \leq 1$$

joining them. Then there exist numbers $\theta_1, \theta_2, \ldots, \theta_m$, *with* $0 < \theta_i < 1$ *such that*

$$\mathbf{F}(\mathbf{A} + \mathbf{H}) - \mathbf{F}(\mathbf{A}) = \mathbf{MH}$$

where $\mathbf{M}$ *is the m by n matrix whose i-th row is* $\nabla f_i(\mathbf{A} + \theta_i \mathbf{H})$.

Proof. Let $\phi_i(t) = f_i(\mathbf{A} + t\mathbf{H})$, $0 \leq t \leq 1$. Then

$$\phi_i(0) = f_i(\mathbf{A}), \qquad \phi_i(1) = f_i(\mathbf{A} + \mathbf{H}).$$

By the Mean Value Theorem for single variable functions, there exists a number θ_i between 0 and 1 so that

$$\phi_i'(\theta_i) = \phi_i(1) - \phi_i(0) = f_i(\mathbf{A} + \mathbf{H}) - f_i(\mathbf{A}).$$

By the Chain Rule,

$$\phi_i'(t) = \nabla f_i(\mathbf{A} + t\mathbf{H}) \cdot \mathbf{H}.$$

Therefore for each component of $\mathbf{F}(\mathbf{A} + \mathbf{H}) - \mathbf{F}(\mathbf{A})$ we have

$$f_i(\mathbf{A} + \mathbf{H}) - f_i(\mathbf{A}) = \nabla f_i(\mathbf{A} + \theta_i \mathbf{H}) \cdot \mathbf{H}.$$

This shows that $\mathbf{F}(\mathbf{A} + \mathbf{H}) - \mathbf{F}(\mathbf{A}) = \mathbf{MH}$. $\qquad\qquad\square$

Second derivatives. We turn now to second partial derivatives.

Definition 3.8. A function f defined in an open disk in the x, y plane is *twice continuously differentiable* in the disk if its partial derivatives f_x and f_y exist and have continuous partial derivatives. The partial derivatives of f_x and f_y are denoted as f_{xx}, f_{xy} and f_{yx}, f_{yy}. They are called the *second partial derivatives* of f.

A twice continuously differentiable function is called a C^2 function.
The following result is basic:

Theorem 3.10. *For a twice continuously differentiable function f on an open disk in the x, y plane, the mixed second partial derivatives are equal:*

$$f_{xy} = f_{yx}.$$

Proof. Let (a,b) be a point in the open disk where f is twice continuously differentiable, and consider the following combination of translates of f in the x and y directions: Define

$$C(h,k) = f(a+h,b+k) - f(a,b+k) - \big(f(a+h,b) - f(a,b)\big) \qquad (3.11)$$

and let p and q be the functions

$$p(x) = f(x,b+k) - f(x,b), \qquad q(y) = f(a+h,y) - f(a,y).$$

We can write $C(h,k)$ in two different ways as

$$C(h,k) = p(a+h) - p(a)$$
$$= q(b+k) - q(b).$$

Applying the Mean Value Theorem twice for single variable functions to $C(h,k)$ we get that there is a number h_1 between 0 and h and a number k_1 between 0 and k such that

$$C(h,k) = hp'(a+h_1) = h(f_x(a+h_1,b+k) - f_x(a+h_1,b)) = hkf_{xy}(a+h_1,b+k_1).$$

Similarly there are h_2 and k_2 such that

$$C(h,k) = kq'(b+k_2) = k(f_y(a+h,b+k_2) - f_y(a,b+k_2)) = hkf_{yx}(a+h_2,b+k_2).$$

This shows that
$$f_{xy}(a+h_1,b+k_1) = f_{yx}(a+h_2,b+k_2).$$

Since f_{xy} and f_{yx} are assumed to be continuous functions, the left side tends to $f_{xy}(a,b)$, and the right to $f_{yx}(a,b)$ as h and k tend to zero. This proves that the mixed second partial derivatives are equal. $\qquad \square$

Problems

3.18. Find the following derivatives or partial derivatives.

(a) $\dfrac{d}{dx}(f(x))^3$, where f is a differentiable function from $\mathbb{R}$ to $\mathbb{R}$

(b) $\dfrac{\partial}{\partial x}(y+x^2)^3$

(c) $\dfrac{\partial}{\partial x}g(y+x^2)$, where g is a differentiable function from $\mathbb{R}$ to $\mathbb{R}$

(d) $\dfrac{\partial}{\partial y}(f(x,y))^3$, where f is a differentiable function from $\mathbb{R}$ to $\mathbb{R}$.

3.19. Find the derivatives f_{xx}, f_{xy}, and f_{yy}.

(a) $f(x,y) = x^2 - y^2$

(b) $f(x,y) = x^2 + y^2$
(c) $f(x,y) = (x+y)^2$
(d) $f(x,y) = e^{-x}\cos y$
(e) $f(x,y) = e^{-ay}\sin(ax)$, a constant

3.20. Express the indicated derivatives of f as a constant multiple of f if this is possible. For example

$$f(x,y) = ye^{ax}, \quad f_{xx} = a^2 f.$$

(a) $f(x,y) = e^{-x}\cos y$, $f_{xx} + 2f_{yy} =?$ (constant)f
(b) $f(x,y) = e^{-ay}\sin(ax)$, $f_{xx} =?$ (constant)f
(c) $f(x,t) = \sin(x - 3t)$, $f_x - 2f_t =?$ (constant)f
(d) $f(x,t) = \cos(x + ct)$, $f_{xx} - af_{tt} =?$ (constant)f

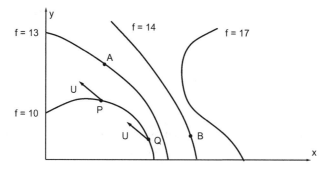

Fig. 3.9 Level sets for Problem 3.21.

3.21. Some level sets of a differentiable function f from $\mathbb{R}^2$ to $\mathbb{R}$ are sketched in Figure 3.9. Which of the following derivatives are positive?

(a) $f_x(\mathbf{A})$
(b) $f_y(\mathbf{B})$
(c) $D_{\mathbf{U}}f(\mathbf{P}) = \mathbf{U} \cdot \nabla f(\mathbf{P})$
(d) $D_{\mathbf{U}}f(\mathbf{Q})$

3.22. Recall the addition formulas

$$\cos(u+v) = \cos u \cos v - \sin u \sin v, \qquad \sin(u+v) = \sin u \cos v + \cos u \sin v.$$

(a) Use the addition formulas and the change of coordinates $x = r\cos\theta$, $y = r\sin\theta$ to express the following functions, given in polar coordinates, in terms of x and y:

$$f_1 = r^2\cos(2\theta) \qquad\qquad f_2 = r^{-2}\sin(2\theta) \qquad\qquad f_3 = r^3\sin(3\theta)$$

(b) Show that each function in part (a) satisfies the equation $f_{xx} + f_{yy} = 0$.

3.23. Let $f(x,y) = 4 + 3x + 2y + xy$ and $\mathbf{A} = (1,1)$. Find the directional derivatives $D_{\mathbf{V}} f(\mathbf{A}) = \mathbf{V} \cdot \nabla f(\mathbf{A})$ in each direction.

$$\mathbf{V} = (1,0), \qquad \mathbf{V} = (\tfrac{4}{5}, \tfrac{3}{5}), \qquad \mathbf{V} = (0,1), \qquad \mathbf{V} = (-\tfrac{3}{5}, \tfrac{4}{5}), \qquad \mathbf{V} = (-1,0).$$

Why is one of these derivatives equal to $\|\nabla f(\mathbf{A})\|$?

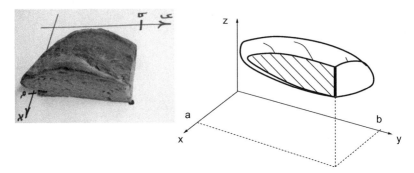

Fig. 3.10 A loaf of bread illustrates a mixed derivative f_{xy} in Problem 3.24.

3.24. Imagine a region in space occupied by a loaf of bread, and let $f(a,b)$ be the volume of bread in the region where $x < a$ and $y < b$, as in Figure 3.10. The following arguments illustrate $f_{xy} = f_{yx}$.

(a) Note the difference quotient $\dfrac{f(a+h,b) - f(a,b)}{h}$; the numerator is the volume of a certain slice of bread, and the denominator is the thickness.

(b) Conclude that $f_x(a,b)$ is the area of the shaded $x = a$ cross section.

(c) Explain similarly why $f_y(a,b)$ is the area of the unshaded $y = b$ cross section.

(d) Note the difference quotient $\dfrac{f_x(a,b+k) - f_x(a,b)}{k}$; the numerator is the area of a certain object, and the denominator is its width.

(e) Conclude that $f_{xy}(a,b)$ is the length of the vertical segment at the corner where $x = a$, $y = b$.

(f) Conclude by a similar argument that $f_{yx}(a,b)$ is also the length of the same segment.

3.25. Consider the partial differential equation

$$u_t + 3u_x = 0 \tag{3.12}$$

for a differentiable function $u(x,t)$.

(a) Suppose f is a differentiable function of one variable. Show that the function $u(x,t) = f(x - 3t)$ satisfies (3.12).

(b) Let $\mathbf{V} = \frac{1}{\sqrt{10}}(3,1)$. Show that if $u(x,t)$ satisfies (3.12) then the directional derivative

$$D_{\mathbf{V}}u = 0.$$

(Here coordinates in $\mathbb{R}^2$ are (x,t).) Therefore u is constant on lines that are parallel to $\mathbf{V}$.

(c) Show that the line that goes through point (x,t) and that is parallel to $\mathbf{V}$ also goes through the point $(x-3t,0)$.

(d) Show that every solution $u(x,t)$ has the property $u(x,t) = u(x-3t,0)$, i.e., is a function of $x-3t$.

3.26. Let $a(x,y)$ be a continuously differentiable function of x and y, and define $b(x) = a(x,x)$. Show that b is a differentiable function.

3.27. Consider a function that depends only on the distance to the origin, that is,

$$f(x,y) = g(r), \quad r = \sqrt{x^2 + y^2},$$

for some function g of one variable.

(a) Find the function $g(r)$ in the case $f(x,y) = (x^2 + y^2)^3$, and verify the relation

$$xf_x + yf_y = rg_r. \tag{3.13}$$

(b) Use the Chain Rule to show that relation (3.13) holds for all differentiable functions f that depend only on r as $f(x,y) = g(r)$.

(c) Show that for twice differentiable functions of the form $f(x,y) = g(r)$,

$$f_{xx} + f_{yy} = g_{rr} + \frac{1}{r}g_r.$$

3.28. For functions f of two variables let $\Delta f = f_{xx} + f_{yy}$. Let a,b,p be numbers.

(a) Find examples of polynomials $f(x,y) = ax^2 + by^2$ for which Δf is positive, zero, or negative.

(b) Evaluate Δf for the functions $f(x,y) = x^4 + y^4$, $f(x,y) = (x^2 + y^2)^p$.

(c) Show that $\Delta(\log(x^2 + y^2)) = 0$. What is the domain of $\log(x^2 + y^2)$?

3.29. Suppose $u(x,y)$ is a C^2 function from $\mathbb{R}^2$ to $\mathbb{R}$. After a change to polar coordinates $u(x,y) = u(r\cos\theta, r\sin\theta)$ we have

$$u_r = \frac{\partial u}{\partial x}\frac{\partial x}{\partial r} + \frac{\partial u}{\partial y}\frac{\partial y}{\partial r}, \qquad u_\theta = \frac{\partial u}{\partial x}\frac{\partial x}{\partial \theta} + \frac{\partial u}{\partial y}\frac{\partial y}{\partial \theta}.$$

Use the Chain Rule to show that

$$u_{xx} + u_{yy} = u_{rr} + \frac{1}{r}u_r + \frac{1}{r^2}u_{\theta\theta}.$$

3.30. Show that the function $u(x,t) = t^{-1/2}e^{-x^2/4t}$ $(t > 0)$ satisfies $u_t = u_{xx}$.

3.31. Let $u(x,y) = x^2 - y^2$ and $v(x,y) = 2xy$.

(a) Show that $u_x = v_y$ and $u_y = -v_x$.
(b) Show that $u_{xx} + u_{yy} = 0$.
(c) Define $w(x,y) = u(u(x,y), v(x,y))$. Show that $w_{xx} + w_{yy} = 0$.
(d) Suppose p, q, and r are a C^2 functions such that

$$p_x = q_y, \qquad p_y = -q_x, \qquad r_{xx} + r_{yy} = 0.$$

Define $w(x,y) = r(p(x,y), q(x,y))$. Show that $w_{xx} + w_{yy} = 0$.

3.32. Given twice differentiable functions f and g, use the Chain Rule to express the following derivatives of

$$u(x,t) = f(x + 4t) + g(x - 4t)$$

in terms of f, g, f', g', f'', g''.

(a) u_x and u_t
(b) u_{xx} and u_{tt}
(c) $u_t + 4u_x$.
(d) $u_{tt} - 16u_{xx}$.

3.33. Let $z = x + iy$ be a complex number, and $f(z) = u(x,y) + iv(x,y)$ a complex valued function.

(a) For $f(z) = z^2$, find $u(x,y)$ and $v(x,y)$.
(b) Define f to be differentiable at z if there is a complex number m so that for all complex h near 0 the change

$$f(z+h) - f(z)$$

is well approximated by mh. Thus

$$f(z+h) = f(z) + mh + r(h)$$

where r is some function that is small in the sense that $\dfrac{r(h)}{h}$ tends to zero as the complex number h tends to zero. Define $f'(z) = m$. Show that $(z^2)' = 2z$.
(c) Suppose $f = u + iv$ is differentiable. Take h real in part (b) to show that f' is equal to the partial derivative $u_x + iv_x$.
(d) Take h imaginary, and show that f' is equal to $-iu_y + v_y$.
(e) If $f = u + iv$ is differentiable conclude that

$$u_x = v_y$$
$$u_y = -v_x$$

(f) Assume that u and v have continuous second partial derivatives. Deduce that

$$u_{xx} + u_{yy} = 0, \qquad v_{xx} + v_{yy} = 0,$$

and verify these for the cases $f(z) = z^2$ and $f(z) = z^3$.

3.34. Justify the following steps to prove

$$\frac{d}{dt} \int_{y_1(t)}^{y_2(t)} g(x,t)\,dx = \int_{y_1(t)}^{y_2(t)} g_t(x,t)\,dx + g(y_2(t),t)y_2'(t) - g(y_1(t),t)y_1'(t)$$

where $g(x,t)$ is continuously differentiable in (x,t) and $y_1(t)$ and $y_2(t)$ are continuously differentiable functions of one variable.

(a) Define $f(u,v,w) = \int_u^v g(x,w)\,dx$. Use the Fundamental Theorem of Calculus to show that

$$f_u(u,v,w) = -g(u,w), \qquad f_v(u,v,w) = g(v,w).$$

(b) Use differentiation under the integral to show that

$$f_w(u,v,w) = \int_u^v \frac{dg}{dw}(x,w)\,dx.$$

(c) Show that $\dfrac{d}{dt} f(y_1(t), y_2(t), t) = f_u y_1'(t) + f_v y_2'(t) + f_w.$

3.4 Inverse functions

We recall the following result about differentiable functions of a single variable:

Let f be a continuously differentiable function of a single variable x on an open interval, and suppose that at some point a in the interval, $f'(a)$ is nonzero. Then f maps a sufficiently small interval around the point a one to one onto an interval around the point $f(a)$. Denote the inverse function of f by g, then g is differentiable at $f(a)$, and $g'(f(a)) = \dfrac{1}{f'(a)}$.

In this section we state and prove the analogous theorem for functions from $\mathbb{R}^2$ to $\mathbb{R}^2$. That is, if a C^1 function $\mathbf{F}$ has an invertible derivative matrix at $\mathbf{A}$ then there is a disk about $\mathbf{A}$ where $\mathbf{F}$ has a differentiable inverse function. First we compare linear approximations of $\mathbf{F}$ at different points near $\mathbf{A}$.

Suppose $\mathbf{F}(x,y) = (f(x,y), g(x,y)) = (u,v)$ is a continuously differentiable function on an open set O in $\mathbb{R}^2$ and that $\mathbf{A} = (a,b)$ is a point in O. By the definition of differentiability f and g can be well approximated near (a,b) by a constant plus linear function in the sense

$$u = f(x,y) = f(a,b) + f_x(a,b)(x-a) + f_y(a,b)(y-b) + r_1(x-a,y-b)$$
$$v = g(x,y) = g(a,b) + g_x(a,b)(x-a) + g_y(a,b)(y-b) + r_2(x-a,y-b) \qquad (3.14)$$

where

$$\frac{|r_1(x-a,y-b)|}{\sqrt{(x-a)^2+(y-b)^2}} \quad \text{and} \quad \frac{|r_2(x-a,y-b)|}{\sqrt{(x-a)^2+(y-b)^2}}$$

tend to zero as $\sqrt{(x-a)^2+(y-b)^2}$ tends to zero.

Vector and matrix notation simplifies the description. Denote the vectors

$$\mathbf{A} = \begin{bmatrix} a \\ b \end{bmatrix}, \qquad \mathbf{X} = \begin{bmatrix} x \\ y \end{bmatrix}, \qquad \mathbf{P} = \mathbf{X} - \mathbf{A} = \begin{bmatrix} x-a \\ y-b \end{bmatrix}, \qquad \mathbf{U} = \begin{bmatrix} u \\ v \end{bmatrix}.$$

Denote the functions $\mathbf{F}(\mathbf{X}) = \begin{bmatrix} f(x,y) \\ g(x,y) \end{bmatrix}$, the linear function

$$\mathbf{L_A}(\mathbf{P}) = \begin{bmatrix} f_x(a,b) & f_y(a,b) \\ g_x(a,b) & g_y(a,b) \end{bmatrix} \mathbf{P} = D\mathbf{F}(\mathbf{A})\mathbf{P}$$

and the remainder function $\mathbf{R}(\mathbf{P}) = \begin{bmatrix} r_1(x-a,y-b) \\ r_2(x-a,y-b) \end{bmatrix}$. Then the vector notation for (3.14) is

$$\mathbf{U} = \mathbf{F}(\mathbf{X}) = \mathbf{F}(\mathbf{A}+\mathbf{P}) = \mathbf{F}(\mathbf{A}) + \mathbf{L_A}(\mathbf{P}) + \mathbf{R}(\mathbf{P}). \qquad (3.15)$$

The next lemma compares the remainders $\mathbf{R}(\mathbf{P})$ and $\mathbf{R}(\mathbf{Q})$ at different points near $\mathbf{A}$.

Lemma 3.1. *Let $\mathbf{F} = (f,g)$ be a continuously differentiable function defined on an open set O in $\mathbb{R}^2$. Then for every $\mathbf{A}$ in O there is a disk N_r centered at $\mathbf{A}$ of radius $r > 0$ so that*

$$\|\mathbf{R}(\mathbf{P}) - \mathbf{R}(\mathbf{Q})\| \le s(\mathbf{P},\mathbf{Q})\|\mathbf{P} - \mathbf{Q}\| \qquad (3.16)$$

for all $\mathbf{A}+\mathbf{P}$ and $\mathbf{A}+\mathbf{Q}$ in N_r, and $s(\mathbf{P},\mathbf{Q})$ tends to zero as r tends to zero.

Proof. Let N_r be an open ball centered at $\mathbf{A}$ that is contained in O. For $\|\mathbf{P}\|$ and $\|\mathbf{Q}\|$ less than r, $\mathbf{A}+\mathbf{P}$ and $\mathbf{A}+\mathbf{Q}$ are in O. By (3.15)

$$\mathbf{R}(\mathbf{P}) = \mathbf{F}(\mathbf{A}+\mathbf{P}) - \mathbf{F}(\mathbf{A}) - \mathbf{L_A}(\mathbf{P}),$$
$$\mathbf{R}(\mathbf{Q}) = \mathbf{F}(\mathbf{A}+\mathbf{Q}) - \mathbf{F}(\mathbf{A}) - \mathbf{L_A}(\mathbf{Q}).$$

Subtract $\mathbf{R}(\mathbf{Q})$ from $\mathbf{R}(\mathbf{P})$. By the linearity of $\mathbf{L_A}$ we get

$$\mathbf{R}(\mathbf{P}) - \mathbf{R}(\mathbf{Q}) = \mathbf{F}(\mathbf{A}+\mathbf{P}) - \mathbf{F}(\mathbf{A}+\mathbf{Q}) - \mathbf{L_A}(\mathbf{P}-\mathbf{Q}).$$

Since $\mathbf{A}+\mathbf{P}$ and $\mathbf{A}+\mathbf{Q}$ are in N_r the segment from $\mathbf{A}+\mathbf{Q}$ to $\mathbf{A}+\mathbf{P}$ is in N_r. By the Mean Value Theorem 3.9 there is a matrix $\mathbf{M}$ with rows

$$\nabla f_i(\mathbf{A}+\mathbf{Q}+\theta_i(\mathbf{P}-\mathbf{Q})), \quad i=1,2, \quad 0<\theta_i<1,$$

for which

$$\mathbf{F}(\mathbf{A}+\mathbf{P})-\mathbf{F}(\mathbf{A}+\mathbf{Q})=\mathbf{M}(\mathbf{P}-\mathbf{Q}).$$

Therefore

$$\mathbf{R}(\mathbf{P})-\mathbf{R}(\mathbf{Q})=\mathbf{M}(\mathbf{P}-\mathbf{Q})-\mathbf{L_A}(\mathbf{P}-\mathbf{Q})$$

$$=\mathbf{M}(\mathbf{P}-\mathbf{Q})-DF(\mathbf{A})(\mathbf{P}-\mathbf{Q})=(\mathbf{M}-DF(\mathbf{A}))(\mathbf{P}-\mathbf{Q}).$$

Taking the norm of both sides and applying Theorem 2.2 we get

$$\|\mathbf{R}(\mathbf{P})-\mathbf{R}(\mathbf{Q})\|\le\|\mathbf{M}-DF(\mathbf{A})\|\|\mathbf{P}-\mathbf{Q}\|.$$

This is inequality (3.16) with $s(\mathbf{P},\mathbf{Q})=\|\mathbf{M}-DF(\mathbf{A})\|$. By the continuity of the partial derivatives, as r tends to zero, $\mathbf{M}$ tends to $DF(\mathbf{A})$ so $\|\mathbf{M}-DF(\mathbf{A})\|$ tends to zero. □

Theorem 3.11. Inverse function. *Let $\mathbf{U}=\mathbf{F}(\mathbf{X})$ be a continuously differentiable function from $\mathbb{R}^2$ to $\mathbb{R}^2$ defined on an open set containing $\mathbf{A}$. If $DF(\mathbf{A})$ is invertible, then $\mathbf{F}$ maps in a one to one way a sufficiently small disk around $\mathbf{A}$ onto a set of points in the $\mathbf{U}$ plane that includes all points in some circular disk around the point $\mathbf{F}(\mathbf{A})$. That is, on a small enough disk centered at $\mathbf{A}$, $\mathbf{F}$ has an inverse, $\mathbf{G}$, that is differentiable at $\mathbf{F}(\mathbf{A})$ and $DG(\mathbf{F}(\mathbf{A}))=DF(\mathbf{A})^{-1}$.*

Proof. (i) We show first that the function $\mathbf{U}=\mathbf{F}(\mathbf{X})$ is one to one for $\mathbf{X}$ near $\mathbf{A}$. Suppose that $\mathbf{F}(\mathbf{X})=\mathbf{F}(\mathbf{Y})$, where

$$\mathbf{X}=\mathbf{A}+\mathbf{P}, \quad \mathbf{Y}=\mathbf{A}+\mathbf{Q}.$$

We use formula (3.15) to express both $\mathbf{F}(\mathbf{A}+\mathbf{P})$ and $\mathbf{F}(\mathbf{A}+\mathbf{Q})$:

$$\mathbf{F}(\mathbf{A}+\mathbf{P})=\mathbf{F}(\mathbf{A})+\mathbf{L_A}(\mathbf{P})+\mathbf{R}(\mathbf{P}), \qquad \mathbf{F}(\mathbf{A}+\mathbf{Q})=\mathbf{F}(\mathbf{A})+\mathbf{L_A}(\mathbf{Q})+\mathbf{R}(\mathbf{Q}).$$

If $\mathbf{F}(\mathbf{A}+\mathbf{P})$ is equal to $\mathbf{F}(\mathbf{A}+\mathbf{Q})$, then

$$\mathbf{L_A}(\mathbf{P})+\mathbf{R}(\mathbf{P})=\mathbf{L_A}(\mathbf{Q})+\mathbf{R}(\mathbf{Q}),$$

which implies that

$$\mathbf{L_A}(\mathbf{P}-\mathbf{Q})=\mathbf{R}(\mathbf{Q})-\mathbf{R}(\mathbf{P}). \tag{3.17}$$

The matrix $DF(\mathbf{A})$ representing $\mathbf{L_A}$ was assumed to be invertible. Multiply both sides of (3.17) by $DF(\mathbf{A})^{-1}$. We get

$$\mathbf{P}-\mathbf{Q}=DF(\mathbf{A})^{-1}(\mathbf{R}(\mathbf{P})-\mathbf{R}(\mathbf{Q})).$$

Since the two sides are equal, so are their norms:

$$\|\mathbf{P} - \mathbf{Q}\| = \|DF(\mathbf{A})^{-1}(\mathbf{R}(\mathbf{P}) - \mathbf{R}(\mathbf{Q}))\| \le \|DF(\mathbf{A})^{-1}\| \|\mathbf{R}(\mathbf{P}) - \mathbf{R}(\mathbf{Q})\|. \tag{3.18}$$

By Lemma 3.1 we can choose r so small that $|s(\mathbf{P}, \mathbf{Q})| < \frac{1}{2} \left\| DF(\mathbf{A})^{-1} \right\|^{-1}$. Then by (3.16)

$$\|\mathbf{R}(\mathbf{P}) - \mathbf{R}(\mathbf{Q})\| \le \frac{1}{2\|DF(\mathbf{A})^{-1}\|} \|\mathbf{P} - \mathbf{Q}\| \qquad \text{for } \mathbf{A} + \mathbf{P} \text{ and } \mathbf{A} + \mathbf{Q} \text{ in } N_r \tag{3.19}$$

where N_r is the disk of radius r centered at $\mathbf{A}$. Using (3.19) to estimate the right side of (3.18) we get

$$\|\mathbf{P} - \mathbf{Q}\| \le \|DF(\mathbf{A})^{-1}\| \frac{1}{2\|DF(\mathbf{A})^{-1}\|} \|\mathbf{P} - \mathbf{Q}\| = \frac{1}{2} \|\mathbf{P} - \mathbf{Q}\|. \tag{3.20}$$

From this it follows that $\|\mathbf{P} - \mathbf{Q}\| = 0$; but then $\mathbf{P} = \mathbf{Q}$. This proves that the mapping $\mathbf{U} = \mathbf{F}(\mathbf{X})$ is one to one for $\|\mathbf{X} - \mathbf{A}\| < r$. Therefore $\mathbf{F}$ has an inverse function. Call it $\mathbf{G}$.

(ii) Next we show that the range of $\mathbf{F}$ includes all points in some small disk around $\mathbf{F}(\mathbf{A})$. Let $\mathbf{U}$ be such a vector. We shall construct a vector $\mathbf{X} = \mathbf{A} + \mathbf{P}$ that is in N_r and satisfies

$$\mathbf{U} = \mathbf{F}(\mathbf{X}) = \mathbf{F}(\mathbf{A}) + \mathbf{L_A}(\mathbf{P}) + \mathbf{R}(\mathbf{P}). \tag{3.21}$$

Here r is the number determined in the proof of part (i). We construct $\mathbf{P}$ as the limit of a sequence of approximations $\mathbf{P}_n$ defined recursively as follows:

$$\begin{aligned} \mathbf{P}_0 &= \mathbf{0}, \\ \mathbf{U} &= \mathbf{F}(\mathbf{A}) + DF(\mathbf{A})\mathbf{P}_n + \mathbf{R}(\mathbf{P}_{n-1}). \end{aligned} \tag{3.22}$$

Multiply equation (3.22) by $DF(\mathbf{A})^{-1}$ to express $\mathbf{P}_n$ as

$$\mathbf{P}_n = DF(\mathbf{A})^{-1}(\mathbf{U} - \mathbf{F}(\mathbf{A}) - \mathbf{R}(\mathbf{P}_{n-1})). \tag{3.23}$$

By Lemma 3.1, $\mathbf{R}$ is continuous.

If the sequence $\mathbf{P}_n$ converges to a point $\mathbf{P}$ then (3.21) follows from (3.22).

To prove that the sequence $\mathbf{P}_n$ converges to a point $\mathbf{P}$ with $\mathbf{A} + \mathbf{P}$ in N_r, we first show by induction that for all j

$$\|\mathbf{P}_j - \mathbf{P}_{j-1}\| < \frac{r}{2^{j+1}}. \tag{3.24}$$

To start the induction we need the $n = 1$ case of (3.24). We have $\mathbf{P}_0 = \mathbf{0}$ and

$$\mathbf{P}_1 = DF(\mathbf{A})^{-1}(\mathbf{U} - \mathbf{F}(\mathbf{A})),$$

so

$$\|\mathbf{P}_1 - \mathbf{P}_0\| \le \|DF(\mathbf{A})^{-1}\| \|\mathbf{U} - \mathbf{F}(\mathbf{A})\|.$$

We assume $\mathbf{U}$ is close enough to $\mathbf{F}(\mathbf{A})$ so that

$$\|DF(\mathbf{A})^{-1}\|\,\|\mathbf{U} - \mathbf{F}(\mathbf{A})\| < \frac{r}{2^2}.$$

This makes $\|\mathbf{P}_1\| < \frac{r}{2^2}$ and specifies the small disk

$$\|\mathbf{U} - \mathbf{F}(\mathbf{A})\| < \frac{r}{2^2\|DF(\mathbf{A})^{-1}\|}$$

of $\mathbf{U}$ such that the sequence $\mathbf{P}_n$ converges to a point $\mathbf{P}$ that satisfies $\mathbf{F}(\mathbf{A} + \mathbf{P}) = \mathbf{U}$.
   As n-th step of the induction we use the inductive assumption that

$$\|\mathbf{P}_j - \mathbf{P}_{j-1}\| < \frac{r}{2^{j+1}}$$

for $j = 1, 2, \dots, n$ to estimate the norm of $\mathbf{P}_j$, $j = 1, 2, \dots, n$. We write

$$\mathbf{P}_j = \mathbf{P}_1 + (\mathbf{P}_2 - \mathbf{P}_1) + \cdots + (\mathbf{P}_j - \mathbf{P}_{j-1}).$$

We deduce from this and (3.24) by the triangle inequality that

$$\|\mathbf{P}_j\| \leq \|\mathbf{P}_1\| + \|\mathbf{P}_2 - \mathbf{P}_1\| + \cdots + \|\mathbf{P}_j - \mathbf{P}_{j-1}\| < \frac{r}{2^2} + \frac{r}{2^3} + \cdots + \frac{r}{2^{j+1}}. \qquad (3.25)$$

The sum of this geometric series is less than $\frac{r}{2}$; this gives $\|\mathbf{P}_j\| < \frac{r}{2}$.
   By definition (3.23) of $\mathbf{P}_n$ the difference

$$\mathbf{P}_{n+1} - \mathbf{P}_n = DF(\mathbf{A})^{-1}(\mathbf{U} - \mathbf{F}(\mathbf{A}) - R(\mathbf{P}_n)) - DF(\mathbf{A})^{-1}(\mathbf{U} - \mathbf{F}(\mathbf{A}) - R(\mathbf{P}_{n-1}))$$

$$= DF(\mathbf{A})^{-1}(R(\mathbf{P}_{n-1}) - R(\mathbf{P}_n)).$$

Therefore by (3.19)

$$\|\mathbf{P}_{n+1} - \mathbf{P}_n\| \leq \tfrac{1}{2}\|\mathbf{P}_n - \mathbf{P}_{n-1}\|.$$

This completes the inductive proof of inequality (3.24) which shows that the sequence $\mathbf{P}_j$ converges to a point $\mathbf{P}$ with $\|\mathbf{P}\| \leq \frac{r}{2} < r$. Hence we have determined a point $\mathbf{A} + \mathbf{P}$ in N_r that is mapped by $\mathbf{F}$ to $\mathbf{U}$. So the domain of $\mathbf{G}$ contains a small disk centered at $\mathbf{F}(\mathbf{A})$.
   (iii) To prove that $\mathbf{G}$, the inverse of $\mathbf{F}$, is differentiable at $\mathbf{B} = \mathbf{F}(\mathbf{A})$, and to prove that its derivative is $(DF(\mathbf{A}))^{-1}$ we show that

$$\frac{\|\mathbf{G}(\mathbf{B} + \mathbf{K}) - \mathbf{G}(\mathbf{B}) - (DF(\mathbf{A}))^{-1}\mathbf{K}\|}{\|\mathbf{K}\|} \qquad (3.26)$$

tends to zero as $\|\mathbf{K}\|$ tends to zero. Let

$$\mathbf{H} = \mathbf{G}(\mathbf{B} + \mathbf{K}) - \mathbf{G}(\mathbf{B}).$$

Since $\mathbf{G}(\mathbf{B}) = \mathbf{A}$, we can rewrite this as

$$\mathbf{A} + \mathbf{H} = \mathbf{G}(\mathbf{B} + \mathbf{K}).$$

Applying $\mathbf{F}$ to both sides we get $\mathbf{F}(\mathbf{A} + \mathbf{H}) = \mathbf{B} + \mathbf{K}$. Since $\mathbf{B} = \mathbf{F}(\mathbf{A})$ it follows that

$$\mathbf{K} = \mathbf{F}(\mathbf{A} + \mathbf{H}) - \mathbf{F}(\mathbf{A}).$$

Using these relations and factoring out $DF(\mathbf{A})^{-1}$ we see that (3.26) is equal to

$$\frac{\|\mathbf{H} - (DF(\mathbf{A}))^{-1}\mathbf{K}\|}{\|\mathbf{K}\|} = \frac{\|(DF(\mathbf{A}))^{-1}(DF(\mathbf{A})\mathbf{H} - \mathbf{K})\|}{\|\mathbf{K}\|}$$

$$= \frac{\|(DF(\mathbf{A}))^{-1}\big(DF(\mathbf{A})\mathbf{H} - (\mathbf{F}(\mathbf{A} + \mathbf{H}) - \mathbf{F}(\mathbf{A}))\big)\|}{\|\mathbf{K}\|}.$$

Using the matrix norm (Theorem 2.2) we get that

$$\frac{\|\mathbf{G}(\mathbf{B} + \mathbf{K}) - \mathbf{G}(\mathbf{B}) - (DF(\mathbf{A}))^{-1}\mathbf{K}\|}{\|\mathbf{K}\|} \leq \frac{\|(DF(\mathbf{A}))^{-1}\|\|\big(DF(\mathbf{A})\mathbf{H} - (\mathbf{F}(\mathbf{A} + \mathbf{H}) - \mathbf{F}(\mathbf{A}))\big)\|}{\|\mathbf{K}\|}.$$

$$(3.27)$$

We next show that

$$\|\mathbf{K}\| \geq \frac{\|\mathbf{H}\|}{2\|(DF(\mathbf{A}))^{-1}\|} = \frac{\|\mathbf{G}(\mathbf{B} + \mathbf{K}) - \mathbf{G}(\mathbf{B})\|}{2\|(DF(\mathbf{A}))^{-1}\|}. \qquad (3.28)$$

To prove this, multiply $\mathbf{K} = \mathbf{F}(\mathbf{A} + \mathbf{H}) - \mathbf{F}(\mathbf{A}) = DF(\mathbf{A})\mathbf{H} + R(\mathbf{H})$ by $DF(\mathbf{A})^{-1}$ to get

$$DF(\mathbf{A})^{-1}\mathbf{K} = \mathbf{H} + DF(\mathbf{A})^{-1}R(\mathbf{H}).$$

Theorem 2.2 and the triangle inequality then give

$$\|DF(\mathbf{A})^{-1}\|\|\mathbf{K}\| \geq \|DF(\mathbf{A})^{-1}\mathbf{K}\|$$
$$= \|\mathbf{H} + DF(\mathbf{A})^{-1}R(\mathbf{H})\| \geq \|\mathbf{H}\| - \|DF(\mathbf{A})^{-1}R(\mathbf{H})\|. \qquad (3.29)$$

In inequality (3.19) set $\mathbf{P} = \mathbf{H}$ and $\mathbf{Q} = \mathbf{0}$; we get

$$\|R(\mathbf{H})\| \leq \frac{\|\mathbf{H}\|}{2\|DF(\mathbf{A})^{-1}\|}$$

which implies

$$\|DF(\mathbf{A})^{-1}R(\mathbf{H})\| \leq \|DF(\mathbf{A})^{-1}\|\|R(\mathbf{H})\| < \tfrac{1}{2}\|\mathbf{H}\|.$$

Replacing $\|DF(\mathbf{A})^{-1}R(\mathbf{H})\|$ in (3.29) by the larger $\tfrac{1}{2}\|\mathbf{H}\|$ we get a new inequality

$$\|DF(\mathbf{A})^{-1}\|\|\mathbf{K}\| \geq \|\mathbf{H}\| - \tfrac{1}{2}\|\mathbf{H}\| = \tfrac{1}{2}\|\mathbf{H}\|.$$

Dividing by $\|DF(\mathbf{A})^{-1}\|$ completes the proof of (3.28).

Inequality (3.28) implies that $\mathbf{G}$ is continuous at $\mathbf{B}$. If we replace $\|\mathbf{K}\|$ in the denominator of (3.26) by the smaller $\dfrac{\|\mathbf{H}\|}{2\|(D\mathbf{F(A)})^{-1}\|}$ and use inequality (3.27) we get

$$0 \le \frac{\|\mathbf{G(B+K)} - \mathbf{G(B)} - (D\mathbf{F(A)})^{-1}\mathbf{K}\|}{\|\mathbf{K}\|}.$$

$$\le 2\|(D\mathbf{F(A)})^{-1}\|^2 \frac{\|\mathbf{F(A+H)} - \mathbf{F(A)} - D\mathbf{F(A)H}\|}{\|\mathbf{H}\|}. \tag{3.30}$$

It follows from the continuity of $\mathbf{G}$ at $\mathbf{B}$ that as $\|\mathbf{K}\|$ tends to zero, $\|\mathbf{H}\|$ tends to zero. Since $\mathbf{F}$ is differentiable at $\mathbf{A}$,

$$\frac{\|\mathbf{F(A+H)} - \mathbf{F(A)} - D\mathbf{F(A)H}\|}{\|\mathbf{H}\|}$$

tends to zero as $\|\mathbf{H}\|$ tends to zero. Using inequality (3.30) we get

$$\frac{\|\mathbf{G(B+K)} - \mathbf{G(B)} - (D\mathbf{F(A)})^{-1}\mathbf{K}\|}{\|\mathbf{K}\|}$$

tends to zero. This shows that $\mathbf{G}$ is differentiable at $\mathbf{B}$, and its matrix derivative at $\mathbf{B}$ is $(D\mathbf{F(A)})^{-1}$. That is,

$$D\mathbf{G(F(A))} = (D\mathbf{F(A)})^{-1}.$$

This concludes the proof! □

Example 3.18. The function $\mathbf{F}(x,y) = (2-y+x^2y, 3x+2y+xy)$ maps $\mathbb{R}^2$ to $\mathbb{R}^2$. It is continuously differentiable.

$$\mathbf{F}(1,1) = (2,6), \qquad D\mathbf{F}(x,y) = \begin{bmatrix} 2xy & -1+x^2 \\ 3+y & 2+x \end{bmatrix}, \qquad D\mathbf{F}(1,1) = \begin{bmatrix} 2 & 0 \\ 4 & 3 \end{bmatrix}.$$

Since $\det D\mathbf{F}(1,1) \ne 0$, by Theorem 3.11 $\mathbf{F}$ has an inverse defined in an open set about $(2,6)$. The inverse is differentiable and

$$D\mathbf{F}^{-1}(2,6) = (D\mathbf{F}(1,1))^{-1}.$$

Since the inverse of a 2 by 2 matrix is given by $\begin{bmatrix} a & b \\ c & d \end{bmatrix}^{-1} = \dfrac{1}{ad-bc}\begin{bmatrix} d & -b \\ -c & a \end{bmatrix}$ the derivative of $\mathbf{F}^{-1}$ at $(2,6)$ is

$$D\mathbf{F}^{-1}(2,6) = \frac{1}{6}\begin{bmatrix} 3 & 0 \\ -4 & 2 \end{bmatrix} = \begin{bmatrix} \frac{1}{2} & 0 \\ -\frac{2}{3} & \frac{1}{3} \end{bmatrix}.$$

To estimate $\mathbf{F}^{-1}(2.1, 5.8)$ we can use the linear approximation

$$\mathbf{F}^{-1}(2+.1,6+(-.2)) \approx \mathbf{F}^{-1}(2,6) + D\mathbf{F}^{-1}(2,6)\begin{bmatrix} .1 \\ -.2 \end{bmatrix}$$

$$= \begin{bmatrix} 1 \\ 1 \end{bmatrix} + \begin{bmatrix} \frac{1}{2} & 0 \\ -\frac{2}{3} & \frac{1}{3} \end{bmatrix}\begin{bmatrix} .1 \\ -.2 \end{bmatrix} = \begin{bmatrix} 1.05 \\ .86... \end{bmatrix}.$$

$\square$

Example 3.19. Given the system of equations

$$2 - y + x^2 y = 2$$
$$3x + 2y + xy = 6$$

we see that $(x,y) = (1,1)$ is one solution. Let $\mathbf{F}(x,y) = (2 - y + x^2 y, 3x + 2y + xy)$. We've seen in Example 3.18 that $\mathbf{F}$ has an inverse defined in an open set about $(2,6)$. That means for each (u,v) in a disk centered at $(2,6)$ of some small enough radius, the system of equations

$$2 - y + x^2 y = u$$
$$3x + 2y + xy = v$$

has a unique solution (x,y) and it is close to $(1,1)$. $\square$

The statement and the proof of the Inverse Function Theorem make no use the of the fact that the number variables is two. Therefore the n-dimensional analogue of Theorem 3.11 holds for differentiable functions of any number of variables.

Theorem 3.12. Inverse function. *Let* $\mathbf{F} = (f_1, f_2, \ldots, f_n)$ *be a continuously differentiable function on an open set* O *in* $\mathbb{R}^n$. *Assume* $D\mathbf{F}(\mathbf{A})$ *is invertible at a point* $\mathbf{A}$ *of* O. *Then* $\mathbf{F}$ *maps a sufficiently small open ball centered at* $\mathbf{A}$ *one to one onto a set that includes an open ball centered at* $\mathbf{F}(\mathbf{A})$. $\mathbf{F}$ *has an inverse* $\mathbf{G}$ *on this set.* $\mathbf{G}$ *is differentiable at* $\mathbf{F}(\mathbf{A})$ *and* $D\mathbf{G}(\mathbf{F}(\mathbf{A})) = (D\mathbf{F}(\mathbf{A}))^{-1}$.

Implicitly defined functions. A consequence of the Inverse Function Theorem is the following theorem for functions from $\mathbb{R}^3$ to $\mathbb{R}$.

Theorem 3.13. Implicit Function. *Let* f *be a continuously differentiable function from a ball centered at a point* $\mathbf{P}$ *in* x, y, z *space* $\mathbb{R}^3$, *to* $\mathbb{R}$. *Suppose* $f_z(\mathbf{P}) \neq 0$. *Then all points* $\mathbf{X}$ *sufficiently close to* $\mathbf{P}$ *that satisfy* $f(\mathbf{X}) = f(\mathbf{P})$ *are of the form* $\mathbf{X} = (x, y, g(x,y))$ *where* g *is a continuously differentiable function. The partial derivatives of* g *are related to the partial derivatives of* f *by*

$$g_x = -\frac{f_x}{f_z}, \quad g_y = -\frac{f_y}{f_z}.$$

Proof. Define the vector function $\mathbf{F}(x,y,z) = (x,y,f(x,y,z))$. The matrix derivative of $\mathbf{F}$ is

$$DF(x,y,z) = \begin{bmatrix} 1 & 0 & 0 \\ 0 & 1 & 0 \\ f_x & f_y & f_z \end{bmatrix}.$$

Its determinant is $f_z(x,y,z)$. Since this is not zero at $\mathbf{P} = (p_1,p_2,p_3)$, the Inverse Function Theorem gives us that $\mathbf{F}$ is invertible near $\mathbf{P}$, with differentiable inverse. Let $f(\mathbf{P}) = c$; then $\mathbf{F}(\mathbf{P}) = (p_1,p_2,c)$. $\mathbf{F}$ maps a small ball centered at $\mathbf{P}$ one to one onto a set in $\mathbb{R}^3$ that includes a ball centered at $\mathbf{F}(\mathbf{P})$. See Figure 3.11. The inverse of $\mathbf{F}$ is of the form

$$\mathbf{F}^{-1}(x,y,w) = (x,y,h(x,y,w))$$

where h is a differentiable function. Define $g(x,y) = h(x,y,c)$. Then g is differentiable and the relations

$$\mathbf{F}^{-1}(x,y,c) = (x,y,g(x,y)), \qquad (x,y,c) = \mathbf{F}(x,y,g(x,y)) = (x,y,f(x,y,g(x,y))),$$

give

$$f(x,y,g(x,y)) = f(\mathbf{P}) = c. \tag{3.31}$$

Differentiating both sides with respect to x by the Chain Rule gives

$$\frac{\partial f}{\partial x}\frac{\partial(x)}{\partial x} + \frac{\partial f}{\partial y}\frac{\partial(y)}{\partial x} + \frac{\partial f}{\partial z}\frac{\partial g}{\partial x} = \frac{\partial(c)}{\partial x} = 0$$

or

$$\frac{\partial f}{\partial x} + 0 + \frac{\partial f}{\partial z}\frac{\partial g}{\partial x} = 0.$$

Since $f_z(\mathbf{P}) \neq 0$ and f_z is continuous, it follows that f_z is not zero for points sufficiently near $\mathbf{P}$. Then dividing by f_z we get $g_x = -\dfrac{f_x}{f_z}$. Similarly differentiating both sides of (3.31) with respect to y by the Chain Rule gives

$$\frac{\partial f}{\partial x}\frac{\partial(x)}{\partial y} + \frac{\partial f}{\partial y}\frac{\partial(y)}{\partial y} + \frac{\partial f}{\partial z}\frac{\partial g}{\partial y} = \frac{\partial(c)}{\partial y} = 0$$

so

$$\frac{\partial f}{\partial y} + 0 + \frac{\partial f}{\partial z}\frac{\partial g}{\partial y} = 0.$$

Therefore $g_y = -\dfrac{f_y}{f_z}$. □

Example 3.20. Consider the point $(1,2,1)$ on the level surface

$$xyz + z^3 = 3$$

of the function $f(x,y,z) = xyz + z^3$. We compute $f_z(x,y,z) = xy + 3z^2$ and $f_z(1,2,1) = 5 \neq 0$. By the Implicit Function Theorem we can solve for z as

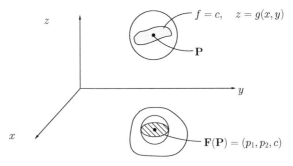

Fig. 3.11 A ball centered at **P** is mapped to a set that includes a ball centered at **F(P)**.

a function of x and y near $(1,2)$. That is, there is a differentiable function g defined in an open disk centered at $(1,2)$, with $g(1,2) = 1$, whose graph $(x,y,g(x,y))$ is in the 3-level set of f,

$$f(x,y,g(x,y)) = xyg(x,y) + g(x,y)^3 = 3.$$

The partial derivatives of g at $(1,2)$ are

$$g_x(1,2) = -\frac{f_x(1,2,1)}{f_z(1,2,1)} = -\frac{yz}{xy+3z^2}\Big|_{(1,2,1)} = -\frac{2}{5},$$

$$g_y(1,2) = -\frac{f_y(1,2,1)}{f_z(1,2,1)} = -\frac{xz}{xy+3z^2}\Big|_{(1,2,1)} = -\frac{1}{5}.$$

The equation of the plane tangent to the graph of g at $(1,2,1)$ is

$$\left(-\tfrac{2}{5},-\tfrac{1}{5},-1\right)\cdot(x-1,y-2,z-1) = 0,$$

or $(2,1,5)\cdot(x-1,y-2,z-1) = 0$, the same equation we found in Example 3.16. □

The Implicit Function Theorem says that a level surface S, $f(x,y,z) = k$, of a C^1 function f is locally the graph of a differentiable function, say $z = g(x,y)$. If (a,b,c) is on S and $f_z(a,b,c) \neq 0$ then

$$(g_x(a,b),g_y(a,b),-1)\cdot(x-a,y-b,z-c) = 0 \qquad (3.32)$$

is the equation of the tangent plane at (a,b,c). Since

$$g_x(a,b) = -\frac{f_x(a,b,c)}{f_z(a,b,c)} \quad \text{and} \quad g_y(a,b) = -\frac{f_y(a,b,c)}{f_z(a,b,c)},$$

equation (3.32) is the same as

$$(f_x(a,b,c), f_y(a,b,c), f_z(a,b,c)) \cdot (x-a, y-b, z-c) = 0$$

or

$$\nabla f(a,b,c) \cdot (x-a, y-b, z-c) = 0$$

which is equation (3.9) in Section 3.3.

Example 3.21. The pressure P, temperature T, and density ρ of a gas are related by

$$P + a\rho^2 = \frac{R\rho T}{1-b\rho}$$

where a, b, and R are nonzero constants. Find an expression for the rate at which the density changes with respect to a change in temperature. Rather than trying to solve for ρ explicitly in terms of T and P we use the Implicit Function Theorem. Define

$$f(P,T,\rho) = P + a\rho^2 - \frac{R\rho T}{1-b\rho}.$$

Then

$$f_\rho = 2a\rho - \frac{RT}{1-b\rho} - \frac{bR\rho T}{(1-b\rho)^2}$$

$$= \left(2a - \frac{RT}{(1-b\rho)^2}\right)\rho$$

is nonzero at a given P and T provided that $2a - \dfrac{RT}{(1-b\rho)^2} \neq 0$. In that case there is a function g defined near (P,T), $\rho = g(P,T)$. The partial derivative

$$g_T(P,T) = -\frac{f_T}{f_\rho} = \frac{\frac{Rg(P,T)}{1-bg(P,T)}}{\left(2a - \frac{RT}{(1-bg(P,T))^2}\right)g(P,T)}$$

$$= \frac{R}{2a(1-bg(P,T)) - \frac{RT}{(1-bg(P,T))}}.$$

$\square$

Remark: The physics notation for g_T in Example 3.21 is $\left(\dfrac{\partial \rho}{\partial T}\right)_P$ as a reminder of what variable was held constant.

Next we state a more general version of Theorem 3.13.

Theorem 3.14. Implicit Function. *Let*

$$\mathbf{F}(\mathbf{X},\mathbf{Y}) = (f_1(x_1,\ldots,x_n,y_1,\ldots,y_m),\ldots,f_m(x_1,\ldots,x_n,y_1,\ldots,y_m))$$

be a C^1 function from $\mathbb{R}^{n+m}$ to $\mathbb{R}^m$ on an open set that contains $(\mathbf{A},\mathbf{B})$, and consider the set S in $\mathbb{R}^{n+m}$ of points $(\mathbf{X},\mathbf{Y})$ that satisfy the system of equations

$$f_1(x_1,\ldots,x_n,y_1,\ldots,y_m) = c_1$$
$$f_2(x_1,\ldots,x_n,y_1,\ldots,y_m) = c_2$$
$$\vdots$$
$$f_m(x_1,\ldots,x_n,y_1,\ldots,y_m) = c_m.$$

Suppose $(\mathbf{A},\mathbf{B}) = (a_1,\ldots,a_n,b_1,\ldots,b_m)$ is in S and that the determinant of the matrix of partial derivatives

$$\left[\frac{\partial f_i}{\partial y_j}(\mathbf{A},\mathbf{B}) \right]$$

is not zero. Then there is a C^1 function $\mathbf{G}$ from $\mathbb{R}^n$ to $\mathbb{R}^m$ such that every point $(\mathbf{X},\mathbf{Y})$ in S sufficiently close to $(\mathbf{A},\mathbf{B})$ can be expressed as $(\mathbf{X},\mathbf{Y}) = (\mathbf{X},\mathbf{G}(\mathbf{X}))$, that is,

$$y_1 = g_1(x_1,\ldots,x_n)$$
$$y_2 = g_2(x_1,\ldots,x_n)$$
$$\vdots$$
$$y_m = g_m(x_1,\ldots,x_n).$$

The partial derivatives of the f_i and g_j are related by

$$\begin{bmatrix} \frac{\partial f_1}{\partial x_j} \\ \vdots \\ \frac{\partial f_m}{\partial x_j} \end{bmatrix} + \begin{bmatrix} \frac{\partial f_1}{\partial y_1} & \cdots & \frac{\partial f_1}{\partial y_m} \\ \vdots & & \vdots \\ \frac{\partial f_m}{\partial y_1} & \cdots & \frac{\partial f_m}{\partial y_m} \end{bmatrix} \begin{bmatrix} \frac{\partial g_1}{\partial x_j} \\ \vdots \\ \frac{\partial g_m}{\partial x_j} \end{bmatrix} = \begin{bmatrix} 0 \\ \vdots \\ 0 \end{bmatrix}, \qquad j = 1,\ldots,n. \qquad (3.33)$$

Proof. We will prove just the last part of this theorem. Assume such C^1 functions g_i exist. Apply the Chain Rule to each function f_i,

$$\frac{\partial}{\partial x_j}(f_i(x_1,\ldots,x_n,g_1(x_1,\ldots,x_n),\ldots,g_m(x_1,\ldots,x_n))) = \frac{\partial}{\partial x_j}(c_i) = 0.$$

That gives

$$\frac{\partial f_i}{\partial x_j} + \frac{\partial f_i}{\partial y_1}\frac{\partial g_1}{\partial x_j} + \frac{\partial f_i}{\partial y_2}\frac{\partial g_2}{\partial x_j} + \cdots + \frac{\partial f_i}{\partial y_m}\frac{\partial g_m}{\partial x_j} = 0$$

or

$$\frac{\partial f_i}{\partial x_j} + \begin{bmatrix} \frac{\partial f_i}{\partial y_1} & \frac{\partial f_i}{\partial y_2} & \cdots & \frac{\partial f_i}{\partial y_m} \end{bmatrix} \begin{bmatrix} \frac{\partial g_1}{\partial x_j} \\ \frac{\partial g_2}{\partial x_j} \\ \cdots \\ \frac{\partial g_m}{\partial x_j} \end{bmatrix} = 0$$

as we see in the matrix equation (3.33). □

Example 3.22. The equations

$$f_1(x_1, x_2, x_3, y_1, y_2) = 2x_1 y_1 + x_2 y_2 = 4$$
$$f_2(x_1, x_2, x_3, y_1, y_2) = x_1^2 x_3 y_1^4 + x_2 y_2^2 = 3$$

are satisfied by

$$(x_1, x_2, x_3, y_1, y_2) = (1, 1, -1, 1, 2).$$

Find $\dfrac{\partial y_1}{\partial x_2}$ and $\dfrac{\partial y_2}{\partial x_2}$ at $(1, 1, -1, 1, 2)$.

To see whether y_1 and y_2 are functions of (x_1, x_2, x_3) near $(1, 1, -1, 1, 2)$ we check the determinant there:

$$\det \begin{bmatrix} \frac{\partial f_1}{\partial y_1} & \frac{\partial f_1}{\partial y_2} \\ \frac{\partial f_2}{\partial y_1} & \frac{\partial f_2}{\partial y_2} \end{bmatrix} = \det \begin{bmatrix} 2x_1 & x_2 \\ 4x_1^2 x_3 y_1^3 & 2x_2 y_2 \end{bmatrix} = \det \begin{bmatrix} 2 & 1 \\ -4 & 4 \end{bmatrix} = 12.$$

That is not zero so the Implicit Function Theorem applies. Equation (3.33) gives at $(1, 1, -1, 1, 2)$,

$$\begin{bmatrix} \frac{\partial f_1}{\partial x_2} \\ \frac{\partial f_2}{\partial x_2} \end{bmatrix} + \begin{bmatrix} 2 & 1 \\ -4 & 4 \end{bmatrix}\begin{bmatrix} \frac{\partial y_1}{\partial x_2} \\ \frac{\partial y_2}{\partial x_2} \end{bmatrix} = \begin{bmatrix} 2 \\ 2^2 \end{bmatrix} + \begin{bmatrix} 2 & 1 \\ -4 & 4 \end{bmatrix}\begin{bmatrix} \frac{\partial y_1}{\partial x_2} \\ \frac{\partial y_2}{\partial x_2} \end{bmatrix} = \begin{bmatrix} 0 \\ 0 \end{bmatrix},$$

Multiply by the inverse matrix to get

$$\begin{bmatrix} \frac{\partial y_1}{\partial x_2} \\ \frac{\partial y_2}{\partial x_2} \end{bmatrix} = -\frac{1}{12}\begin{bmatrix} 4 & -1 \\ 4 & 2 \end{bmatrix}\begin{bmatrix} 2 \\ 4 \end{bmatrix} = \begin{bmatrix} -\frac{1}{3} \\ -\frac{4}{3} \end{bmatrix}.$$

□

Problems

3.35. Let $F(x, y) = (e^x \cos y, e^x \sin y)$.

(a) Find the derivative matrix DF. That is, writing $u = e^x \cos y$, $v = e^x \sin y$ find

$$\begin{bmatrix} \frac{\partial u}{\partial x} & \frac{\partial u}{\partial y} \\ \frac{\partial v}{\partial x} & \frac{\partial v}{\partial y} \end{bmatrix}.$$

(b) Show that $DF(A)^{-1}$ exists at every point $A = (a,b)$ so F is locally invertible at A.
(c) Show that F is not invertible globally; that is, find different points of the domain that are mapped to the same value in the range.

3.36. Let $F(x,y) = (x \cos y, x \sin y)$ for $x > 0$.

(a) Find the derivative matrix $DF(A)$ at point $A = (a,b)$.
(b) Show that $DF(A)^{-1}$ exists at every point with $a > 0$ so F is locally invertible at A.

3.37. Let $F(x,y) = (x^4 + 2xy + 1, y)$, and consider the equations

$$x^4 + 2xy + 1 = u$$
$$y = v$$

that is, $F(x,y) = (u,v)$.

(a) Show that for $(u,v) = (0,1)$, one solution is $(x,y) = (-1,1)$.
(b) Find the derivative matrix $DF(x,y)$.
(c) Show that $DF(-1,1)$ is invertible, and hence F^{-1} exists on a small enough disk centered at $(0,1)$.
(d) Use the linear approximation to estimate $F^{-1}(.2, 1.01)$.

3.38. Let $F(x,y) = (x + y^{-1}, x^{-1} + y)$, and consider the system of equations

$$x + y^{-1} = 1.5$$
$$x^{-1} + y = 3$$

(a) Show that $(x,y) = (1,2)$ is a solution.
(b) Show that the derivative matrix $DF(1,2)$ is invertible.
(c) Find a linear approximation for $F^{-1}(1.49, 2.9)$.

3.39. Suppose f is a C^1 function from $\mathbb{R}^3$ to $\mathbb{R}$, with

$$f(1,0,3) = f(1,2,3) = f(1,2,-3) = 5$$

and

$$\nabla f(1,0,3) = (0,1,0), \quad \nabla f(1,2,3) = (4,1,-\tfrac{1}{2}), \quad \nabla f(1,2,-3) = (0,0,0).$$

(a) Can we solve the equation $f(x,y,z) = 5$ for z as a function of (x,y) near $(1,2,-3)$? That means: does the Implicit Function Theorem guarantee existence of a function g from a disk around $(1,2)$ into $\mathbb{R}$ so that $g(1,2) = -3$ and $f(x,y,g(x,y)) = 5$?
(b) Can we solve $f(x,y,z) = 5$ for y in terms of (x,z) near $(1,0,3)$?
(c) Can we solve $f(x,y,z) = 5$ for y in terms of (x,z) near $(1,2,3)$?

(d) Can a function $y = g(x,z)$ that satisfies part (b) also satisfy part (c)?
(e) Can we solve $f(x,y,z) = 5$ for x in terms of (y,z) near $(1,0,3)$?

3.40. Let $f(x,y,z) = z^4 + 2yz + x$, and consider the level set

$$z^4 + 2yz + x = 0.$$

(a) Show that $(0,2,0)$ is on the level set.
(b) Find the partial derivatives of f.
(c) Show that there is a function $z = g(x,y)$ defined near $(0,2)$ so that

$$f(x,y,g(x,y)) = 0.$$

(d) Find the partial derivatives $g_x(x,y)$ and $g_y(x,y)$, and use them to find the second
derivatives $g_{xx}(0,2)$, $g_{xy}(0,2)$, $g_{yy}(0,2)$.

3.41. Let

$$f_1(x,y_1,y_2) = 3y_1 + y_2^2 + 4x = 0$$
$$f_2(x,y_1,y_2) = 4y_1^3 + y_2 + x = 0$$

(a) Verify that $(x,y_1,y_2) = (-4,0,4)$ is a solution.
(b) Show that there is a function $\mathbf{G} = (g_1,g_2)$ such that $g_1(x) = y_1$, $g_2(x) = y_2$ for all
solutions near $(-4,0,4)$.
(c) Show that at $x = -4$,

$$\frac{dg_1}{dx} = \frac{4}{3}, \qquad \frac{dg_2}{dx} = -1.$$

3.42. Let $\mathbf{F}(u,v,w) = (u+v+w, uv+vw+wu)$, and consider the system of equations
$\mathbf{F}(u,v,w) = (2,-4)$:

$$u+v+w = 2$$
$$uv+vw+wu = -4.$$

(a) Verify that $(2,2,-2)$ is a solution.
(b) Writing $\mathbf{F}(u,v,w) = (f_1(u,v,w), f_2(u,v,w))$, find all the first partial derivatives of
f_1 and f_2.
(c) Verify that $DF(2,2,-2) = \begin{bmatrix} 1 & 1 & 1 \\ 0 & 0 & 4 \end{bmatrix}$.
(d) Can the equations be solved for v and w in terms of u near $(2,2,-2)$? That
means: does the Implicit Function Theorem guarantee existence of a func-
tion $\mathbf{G} = (g_1,g_2)$ defined on some interval around 2 so that $\mathbf{G}(2) = (2,-2)$ and
$\mathbf{F}(u,g_1(u),g_2(u)) = (2,-4)$?
(e) Can the equations be solved for u and v in terms of w near $(2,2,-2)$?

3.43. Let T be the set of all points (x,y,u,v) in $\mathbb{R}^4$ that satisfy

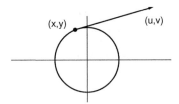

Fig. 3.12 A tangent vector at a point of the unit circle in Problem 3.43.

$$x^2 + y^2 = 1,$$
$$ux + vy = 0.$$

T is the set of all vectors tangent to the unit circle; see Figure 3.12.

(a) We relabel coordinates to conform to the Implicit Function Theorem: Define
$\mathbf{F}(x_1, x_2, y_1, y_2) = (x_1^2 + y_1^2, x_2 x_1 + y_2 y_1)$ where $x_1 = x$, $y_1 = y$, $x_2 = u$ and $y_2 = v$.
Then T is the set defined by

$$\mathbf{F}(x_1, x_2, y_1, y_2) = (1, 0).$$

Calculate the matrix $\left[\frac{\partial f_i}{\partial y_j} \right]$.

(b) Use the Implicit Function Theorem to deduce that near any point where $y \neq 0$,
the equations $x^2 + y^2 = 1$, $ux + vy = 0$ can be solved for y and v in terms of x
and u.

(c) Find formulas for y and v in terms of x and u valid when $y \neq 0$.

(d) Find formulas for x and u in terms of y and v valid when $x \neq 0$.

3.5 Divergence and curl

In this section we introduce two combinations of partial derivatives for functions
from $\mathbb{R}^2$ to $\mathbb{R}^2$ or from $\mathbb{R}^3$ to $\mathbb{R}^3$. Their significance for integration will be discussed
in Chapter 8. Given

$$\mathbf{G}(x, y) = (g_1(x, y), g_2(x, y)) \quad \text{and} \quad \mathbf{F}(x, y, z) = (f_1(x, y, z), f_2(x, y, z), f_3(x, y, z)),$$

the derivative matrices of $\mathbf{G}$ at (x, y) and of $\mathbf{F}$ at (x, y, z) are

$$DG(x, y) = \begin{bmatrix} \frac{\partial g_1}{\partial x} & \frac{\partial g_1}{\partial y} \\ \frac{\partial g_2}{\partial x} & \frac{\partial g_2}{\partial y} \end{bmatrix} = \begin{bmatrix} g_{1x} & g_{1y} \\ g_{2x} & g_{2y} \end{bmatrix}, \quad DF(x, y, z) = \begin{bmatrix} \frac{\partial f_1}{\partial x} & \frac{\partial f_1}{\partial y} & \frac{\partial f_1}{\partial z} \\ \frac{\partial f_2}{\partial x} & \frac{\partial f_2}{\partial y} & \frac{\partial f_2}{\partial z} \\ \frac{\partial f_3}{\partial x} & \frac{\partial f_3}{\partial y} & \frac{\partial f_3}{\partial z} \end{bmatrix} = \begin{bmatrix} f_{1x} & f_{1y} & f_{1z} \\ f_{2x} & f_{2y} & f_{2z} \\ f_{3x} & f_{3y} & f_{3z} \end{bmatrix}.$$

Two special combinations of the partial derivatives of $\mathbf{G}$ and of $\mathbf{F}$ are the *divergence*,

$$\text{div}\,\mathbf{G} = \frac{\partial g_1}{\partial x} + \frac{\partial g_2}{\partial y}, \qquad \text{div}\,\mathbf{F} = \frac{\partial f_1}{\partial x} + \frac{\partial f_2}{\partial y} + \frac{\partial f_3}{\partial z},$$

and the *curl*,

$$\text{curl}\,\mathbf{G} = \frac{\partial g_2}{\partial x} - \frac{\partial g_1}{\partial y}, \qquad \text{curl}\,\mathbf{F} = \left(\frac{\partial f_3}{\partial y} - \frac{\partial f_2}{\partial z}, \ -\left(\frac{\partial f_3}{\partial x} - \frac{\partial f_1}{\partial z}\right), \frac{\partial f_2}{\partial x} - \frac{\partial f_1}{\partial y}\right).$$

We use the notation $\nabla = (\frac{\partial}{\partial x}, \frac{\partial}{\partial y}, \frac{\partial}{\partial z})$ and the formulas for computing dot product, cross product, and multiplication by a scalar to express the formulas for $\text{div}\,\mathbf{F}$, $\text{curl}\,\mathbf{F}$, and the gradient of f. For a differentiable vector field $\mathbf{F} = (f_1, f_2, f_3)$,

$$\text{div}\,\mathbf{F} = \nabla \cdot \mathbf{F} = \frac{\partial}{\partial x} f_1 + \frac{\partial}{\partial y} f_2 + \frac{\partial}{\partial z} f_3,$$

and

$$\text{curl}\,\mathbf{F} = \nabla \times \mathbf{F} = \det \begin{bmatrix} \mathbf{i} & \mathbf{j} & \mathbf{k} \\ \frac{\partial}{\partial x} & \frac{\partial}{\partial y} & \frac{\partial}{\partial z} \\ f_1 & f_2 & f_3 \end{bmatrix}$$

$$= \left(\frac{\partial f_3}{\partial y} - \frac{\partial f_2}{\partial z}\right)\mathbf{i} - \left(\frac{\partial f_3}{\partial x} - \frac{\partial f_1}{\partial z}\right)\mathbf{j} + \left(\frac{\partial f_2}{\partial x} - \frac{\partial f_1}{\partial y}\right)\mathbf{k}.$$

For a differentiable scalar valued function f from $\mathbb{R}^3$ to $\mathbb{R}$

$$\text{grad}\,f = \nabla f = \left(\frac{\partial f}{\partial x}, \frac{\partial f}{\partial y}, \frac{\partial f}{\partial z}\right).$$

Example 3.23. Let $\mathbf{F}(x,y,z) = (x^2, xy, zy)$. Then

$$\text{div}\,\mathbf{F} = 2x + x + y = 3x + y$$

and

$$\text{curl}\,\mathbf{F} = (z - 0, -(0 - 0), y - 0) = (z, 0, y).$$

$\square$

Example 3.24. Let $\mathbf{F}(x,y,z) = \dfrac{(x,y,z)}{(x^2 + y^2 + z^2)^{3/2}}$ for $(x,y,z) \neq (0,0,0)$. Then

$$\frac{\partial}{\partial x}\left(\frac{x}{(x^2 + y^2 + z^2)^{3/2}}\right) = \frac{(x^2 + y^2 + z^2)^{3/2} - x\frac{3}{2}(x^2 + y^2 + z^2)^{1/2}(2x)}{((x^2 + y^2 + z^2)^{3/2})^2}.$$

Canceling a factor of $(x^2 + y^2 + z^2)^{1/2}$ we get

$$f_{1x} = \frac{x^2 + y^2 + z^2 - 3x^2}{(x^2 + y^2 + z^2)^{5/2}}.$$

Similarly for f_{2y} and f_{3z} we get

$$f_{2y} = \frac{x^2 + y^2 + z^2 - 3y^2}{(x^2 + y^2 + z^2)^{5/2}},$$

$$f_{3z} = \frac{x^2 + y^2 + z^2 - 3z^2}{(x^2 + y^2 + z^2)^{5/2}}.$$

Therefore $\text{div}\,\mathbf{F}(x,y,z) = \dfrac{3(x^2 + y^2 + z^2) - 3x^2 - 3y^2 - 3z^2}{(x^2 + y^2 + z^2)^{5/2}} = 0.$ □

Example 3.25. Let **F** be the inverse square vector field in Example 3.24, and let

$$\mathbf{H}(\mathbf{X}) = -\mathbf{F}(\mathbf{X}).$$

Then

$$\text{div}\,\mathbf{H}(\mathbf{X}) = \text{div}\,(-\mathbf{F}(\mathbf{X})) = -\text{div}\,\mathbf{F}(\mathbf{X}) = 0.$$

□

Example 3.26. Let $\mathbf{G}(x,y) = (-y, x)$. Then

$$\text{div}\,\mathbf{G}(x,y) = \frac{\partial}{\partial x}(-y) + \frac{\partial}{\partial y}(x) = 0$$

and

$$\text{curl}\,\mathbf{G}(x,y) = \frac{\partial}{\partial x}(x) - \frac{\partial}{\partial y}(-y) = 1 - (-1) = 2.$$

□

Example 3.27. Let $\mathbf{G}(x,y) = \left(\dfrac{-y}{x^2 + y^2}, \dfrac{x}{x^2 + y^2}\right)$, $(x,y) \neq (0,0)$. Then

$$\text{div}\,\mathbf{G} = \frac{\partial}{\partial x}\left(\frac{-y}{x^2 + y^2}\right) + \frac{\partial}{\partial y}\left(\frac{x}{x^2 + y^2}\right) = \frac{-(-y)2x}{(x^2 + y^2)^2} + \frac{-(x)2y}{(x^2 + y^2)^2} = 0,$$

and

$$\text{curl}\,\mathbf{G}(x,y) = \frac{\partial}{\partial x}\left(\frac{x}{x^2 + y^2}\right) - \frac{\partial}{\partial y}\left(\frac{-y}{x^2 + y^2}\right)$$

$$= \frac{(x^2 + y^2)1 - x(2x)}{(x^2 + y^2)^2} - \frac{(x^2 + y^2)(-1) - (-y)2y}{(x^2 + y^2)^2}$$

$$= \frac{2(x^2 + y^2) - 2x^2 - 2y^2}{(x^2 + y^2)^2} = 0.$$

□

Next we investigate what the divergence and curl tell us about the vector field itself.

Divergence. Suppose that $\mathbf{F}(x,y,z) = (f_1(x,y,z), f_2(x,y,z), f_3(x,y,z))$ represents the velocity of a fluid at point (x,y,z) at some moment in time. Imagine a small box with one vertex at the point $\mathbf{P} = (x_0, y_0, z_0)$ and the edge lengths $\Delta x, \Delta y, \Delta z$, as illustrated in Figure 3.13. Let **N** be the outward unit normal vector at points on the surface of the box.

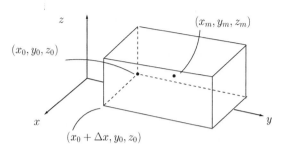

Fig. 3.13 Fluid flows through a box.

We estimate the rate of flow out of this box across its faces. Denote the midpoint of the box as (x_m, y_m, z_m). For a C^1 vector field **F** on a small box, at points of the bottom face **F** is approximately $\mathbf{F}(x_m, y_m, z_0)$, and on the top face is approximately $\mathbf{F}(x_m, y_m, z_0 + \Delta z)$. The combined rate at which fluid flows out of the box through these two faces is approximately

$$\mathbf{F}(x_m, y_m, z_0 + \Delta z) \cdot (0,0,1)\Delta x \Delta y + \mathbf{F}(x_m, y_m, z_0) \cdot (0,0,-1)\Delta x \Delta y$$

$$= (f_3(x_m, y_m, z_0 + \Delta z) - f_3(x_m, y_m, z_0))\Delta x \Delta y.$$

See Figure 3.14. Using the Mean Value Theorem of calculus this is equal to

$$\frac{\partial f_3}{\partial z}(x_m, y_m, \overline{z})\Delta z \Delta x \Delta y$$

where $\overline{z}$ lies between z_0 and $z_0 + \Delta z$.

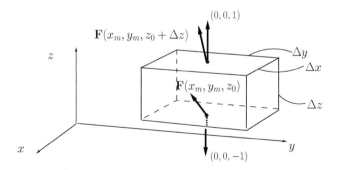

Fig. 3.14 Approximate flow across the top and bottom of the box.

By similar arguments the approximate outward flow rate of **F** across the faces of the box parallel to the x, z plane is

$$\frac{\partial f_2}{\partial y}(x_m, \bar{y}, z_m) \Delta y \Delta x \Delta z$$

for some $\bar{y}$ between y_0 and $y_0 + \Delta y$, and across the faces of the box parallel to the y, z plane is

$$\frac{\partial f_1}{\partial x}(\bar{x}, y_m, z_m) \Delta x \Delta y \Delta z$$

for some $\bar{x}$ between x_0 and $x_0 + \Delta x$.

The total approximate outward flow rate across the surface of the box is

$$\left(\frac{\partial f_1}{\partial x}(\bar{x}, y_m, z_m) + \frac{\partial f_2}{\partial y}(x_m, \bar{y}, z_m) + \frac{\partial f_3}{\partial z}(x_m, y_m, \bar{z}) \right) \Delta x \Delta y \Delta z.$$

We define the *average flux density* to be the net outward flow rate across the surface divided by the volume of the box. Then the average flux density of **F** across the surface of the box is

$$\frac{\partial f_1}{\partial x}(\bar{x}, y_m, z_m) + \frac{\partial f_2}{\partial y}(x_m, \bar{y}, z_m) + \frac{\partial f_3}{\partial z}(x_m, y_m, \bar{z}).$$

Since the partial derivatives are continuous, as Δx, Δy, and Δz tend to zero, $\bar{x}$ and x_m tend to x_0, $\bar{y}$ and y_m tend to y_0, and $\bar{z}$ and z_m tend to z_0, and the average flux density tends to

$$\frac{\partial f_1}{\partial x}(x_0, y_0, z_0) + \frac{\partial f_2}{\partial y}(x_0, y_0, z_0) + \frac{\partial f_3}{\partial z}(x_0, y_0, z_0) = \operatorname{div} \mathbf{F}(\mathbf{P}).$$

$\operatorname{div} \mathbf{F}(\mathbf{P})$ is called the *flux density* of **F** at **P**.

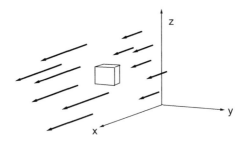

Fig. 3.15 Outward flux is positive, in Example 3.28.

Example 3.28. Let $\mathbf{F}(x, y, z) = (x, 0, 0)$. Then

$$\operatorname{div} \mathbf{F} = \frac{\partial x}{\partial x} + \frac{\partial 0}{\partial y} + \frac{\partial 0}{\partial z} = 1.$$

See Figure 3.15.   □

Curl in $\mathbb{R}^2$**.** Imagine that $\mathbf{G}(x,y)$ is the velocity of a fluid in the x,y plane at some moment in time, and suppose $\mathbf{G}$ is continuously differentiable. Consider the tangential component of $\mathbf{G}$ as we traverse the boundary of a small rectangle as shown in Figure 3.16.

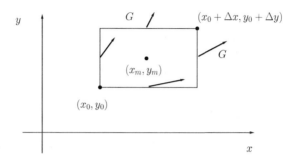

Fig. 3.16 We consider the tangential component of $\mathbf{G}$ at the boundary.

Because $\mathbf{G}$ is continuous and the rectangle is small, $\mathbf{G}(x,y)$ for points along the lower edge is approximately $\mathbf{G}(x_m,y_0)$, along the right edge approximately $\mathbf{G}(x_0+\Delta x,y_m)$, along the top edge approximately $\mathbf{G}(x_m,y_0+\Delta y)$, and along the left $\mathbf{G}(x_0,y_m)$. The product of the tangential component of $\mathbf{G}$ on an edge times the length of the edge is the *circulation* of $\mathbf{G}$ along that edge. The circulation of $\mathbf{G}$ around the rectangle is

$$\mathbf{G}(x_m,y_0)\cdot(1,0)\Delta x+\mathbf{G}(x_0+\Delta x,y_m)\cdot(0,1)\Delta y$$

$$+\mathbf{G}(x_m,y_0+\Delta y)\cdot(-1,0)\Delta x+\mathbf{G}(x_0,y_m)\cdot(0,-1)\Delta y.$$

Applying the dot products the circulation of $\mathbf{G}$ around the rectangle can be rewritten

$$=g_1(x_m,y_0)\Delta x+g_2(x_0+\Delta x,y_m)\Delta y-g_1(x_m,y_0+\Delta y)\Delta x-g_2(x_0,y_m)\Delta y.$$

By the Mean Value Theorem this equals

$$(-g_{1y}(x_m,\overline{y})\Delta y)\Delta x+(g_{2x}(\overline{x},y_m)\Delta x)\Delta y.$$

As Δx and Δy tend to zero, $\overline{x}$ and x_m tend to x_0, and $\overline{y}$ and y_m tend to y_0. Since the partial derivatives are continuous, the circulation densities

$$\frac{(-g_{1y}(x_m,\overline{y})\Delta y)\Delta x+(g_{2x}(\overline{x},y_m)\Delta x)\Delta y}{\Delta x\Delta y}$$

tend to

$$-g_{1y}(x_0,y_0)+g_{2x}(x_0,y_0)=\operatorname{curl}\mathbf{G}(x_0,y_0).$$

Example 3.29. Let $\mathbf{G}(x,y) = (-y,x)$ be a velocity field. A brief calculation shows that $\operatorname{curl}\mathbf{G} = 2$ at every point. We observe that

$$\mathbf{G}(h,0) = (0,h)$$
$$\mathbf{G}(0,h) = (-h,0)$$
$$\mathbf{G}(-h,0) = (0,-h)$$
$$\mathbf{G}(0,-h) = (h,0).$$

If you sketch this vector field with $h > 0$ small, it suggests that near the origin $\mathbf{G}$ is the velocity of a counterclockwise rotation.   □

Example 3.30. Let $\mathbf{G}(x,y) = (y,0)$. Then

$$\operatorname{curl}\mathbf{G}(x,y) = \frac{\partial}{\partial x}(0) - \frac{\partial}{\partial y}(y) = -1.$$

As Figure 3.17 shows, small objects floating in this field rotate clockwise.   □

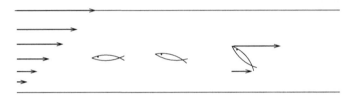

Fig. 3.17 The field in Example 3.30.

The next example shows that the overall rotation of a fluid may differ from the local rotation or curl.

Example 3.31. Let $\mathbf{F}$ be the vector field $\mathbf{F}(x,y) = \dfrac{(-y,x)}{(x^2 + y^2)^p}$, where $p > 0$. A sketch of $\mathbf{F}$ is illustrated in Figure 3.18. The field appears to rotate counterclockwise about the origin. The $\operatorname{curl}\mathbf{F} = \dfrac{2 - 2p}{(x^2 + y^2)^p}$ is negative, zero, or positive when the exponent p is greater than 1, equal to 1, or less than 1 as we ask you to verify in Problem 3.47.   □

Curl in $\mathbb{R}^3$. Take a small box with one vertex at (x_0, y_0, z_0) and edge lengths $\Delta x, \Delta y, \Delta z$. Denote the center point (x_m, y_m, z_m). At the center of each face form the product $(\mathbf{N} \times \mathbf{F})$(Area of the face), sum over all faces and divide by the volume of the box. We get

$$\Big((-1,0,0) \times \mathbf{F}(x_0, y_m, z_m)\Delta y \Delta z + (1,0,0) \times \mathbf{F}(x_0 + \Delta x, y_m, z_m)\Delta y \Delta z$$

$$+ (0,-1,0) \times \mathbf{F}(x_m, y_0, z_m)\Delta x \Delta z + (0,1,0) \times \mathbf{F}(x_m, y_0 + \Delta y, z_m)\Delta x \Delta z$$

$$+ (0,0,-1) \times \mathbf{F}(x_m, y_m, z_0)\Delta x \Delta y + (0,0,1) \times \mathbf{F}(x_m, y_m, z_0 + \Delta z)\Delta x \Delta y\Big)\frac{1}{\Delta x \Delta y \Delta z}.$$

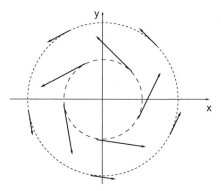

Fig. 3.18 A sketch typical of the vector fields **F** in Example 3.31. The curl of **F** may be positive or negative.

In Problem 3.52 we ask you to show that this tends to $\mathrm{curl}\,\mathbf{F}((x_0, y_0, z_0))$ as Δx, Δy, and Δz tend to zero.

Gradient. In Problem 3.53 we ask you to explore how the formula for the gradient of the pressure f at a point in a fluid arises by looking at the pressure forces on the surface of a small box. The pressure force on each face is approximately the product $(-f\mathbf{N}(\text{Area of the face}))$ where f is the pressure at the midpoint of the face. We ask you to show that the approximate total pressure force on the surface of a small box,

$$-f(x_0, y_m, z_m)(-1, 0, 0)\Delta y\Delta z - f(x_0 + \Delta x, y_m, z_m)(1, 0, 0)\Delta y\Delta z,$$

$$-f(x_m, y_0, z_m)(0, -1, 0)\Delta x\Delta z - f(x_m, y_0 + \Delta y, z_m)(0, 1, 0)\Delta x\Delta z$$

$$-f(x_m, y_m, z_0)(0, 0, -1)\Delta x\Delta y - f(x_m, y_m, z_0 + \Delta z)(0, 0, 1)\Delta x\Delta y$$

is equal to

$$-(f_x(\overline{x}, y_m, z_m), f_y(x_m, \overline{y}, z_m), f_z(x_m, y_m, \overline{z}))\Delta x\Delta y\Delta z$$

for some $\overline{x}$ between x_0 and $x_0 + \Delta x$, $\overline{y}$ between y_0 and $y_0 + \Delta y$, and $\overline{z}$ between z_0 and $z_0 + \Delta z$. Dividing by the volume we see that for continuous partial derivatives the force per volume

$$-(f_x(\overline{x}, y_m, z_m), f_y(x_m, \overline{y}, z_m), f_z(x_m, y_m, \overline{z}))$$

tends to

$$-\nabla f(x_0, y_0, z_0)$$

as Δx, Δy, and Δz tend to zero. Thus the gradient of the pressure at a point can be interpreted as a vector that arises from pressure differences on opposite sides of a small box and represents the total force per volume at that location.

Laplace operator. Another important operator defined for scalar functions is the composite of divergence and gradient

$$\text{div grad } f = \frac{\partial^2 f}{\partial x^2} + \frac{\partial^2 f}{\partial y^2} + \frac{\partial^2 f}{\partial z^2}.$$

It is denoted as Δf and called the *Laplace* operator.

The gradient and the Laplace operators are linear: that is for all once, respectively twice, differentiable functions f and g and all numbers a and b,

$$\nabla(af + bg) = a\nabla f + b\nabla g$$
$$\Delta(af + bg) = a\Delta f + b\Delta g$$

We define the Laplace operator Δ for a twice differentiable vector valued function componentwise. It is a linear operator, and so are the operators curl and div; that is, for all twice differentiable functions $\mathbf{F}$ and $\mathbf{G}$ and numbers a and b,

$$\Delta(a\mathbf{F} + b\mathbf{G}) = a\Delta\mathbf{F} + b\Delta\mathbf{G}$$
$$\text{curl}(a\mathbf{F} + b\mathbf{G}) = a\text{curl}\,\mathbf{F} + b\text{curl}\,\mathbf{G}$$
$$\text{div}(a\mathbf{F} + b\mathbf{G}) = a\text{div}\,\mathbf{F} + b\text{div}\mathbf{G}$$

Vector differential identities. Let f and g be scalar valued functions from $\mathbb{R}^3$ to $\mathbb{R}$ and $\mathbf{F}$ and $\mathbf{G}$ be vector fields from $\mathbb{R}^3$ to $\mathbb{R}^3$. Assuming that all required first, second, and sometimes third, partial derivatives exist and are continuous, we have lots of identities.

We ask you in Problems 3.56–3.59 to verify the following identities. In (e) and (f) we define
$$\mathbf{F} \cdot \nabla\mathbf{G} = f_1\mathbf{G}_x + f_2\mathbf{G}_y + f_3\mathbf{G}_z = (D\mathbf{G})\mathbf{F}.$$

(a) $\nabla(fg) = f\nabla g + g\nabla f$

(b) $\text{div}(f\mathbf{F}) = f\text{div}\,\mathbf{F} + \mathbf{F} \cdot \nabla f$

(c) $\text{div}(\mathbf{F} \times \mathbf{G}) = (\text{curl}\,\mathbf{F}) \cdot \mathbf{G} - \mathbf{F} \cdot \text{curl}\,\mathbf{G}$

(d) $\text{curl}(f\mathbf{F}) = f\text{curl}\,\mathbf{F} + (\nabla f) \times \mathbf{F}$

(e) $\text{curl}(\mathbf{F} \times \mathbf{G}) = (\text{div}\,\mathbf{G})\mathbf{F} + \mathbf{G} \cdot \nabla\mathbf{F} - ((\text{div}\,\mathbf{F})\mathbf{G} + \mathbf{F} \cdot \nabla\mathbf{G})$

(f) $\nabla(\mathbf{F} \cdot \mathbf{G}) = \mathbf{F} \times \text{curl}\,\mathbf{G} + \mathbf{G} \times \text{curl}\,\mathbf{F} + \mathbf{F} \cdot \nabla\mathbf{G} + \mathbf{G} \cdot \nabla\mathbf{F}$

(g) $\text{curl}\,\text{curl}\,\mathbf{F} = \nabla(\text{div}\,\mathbf{F}) - \Delta\mathbf{F}$ where $\Delta\mathbf{F} = (\Delta f_1, \Delta f_2, \Delta f_3)$

(h) $\Delta(fg) = f\Delta g + 2\nabla f \cdot \nabla g + g\Delta f$.

(i) $\text{div}(\nabla f \times \nabla g) = 0$

In the next two examples we prove the identities

- $\operatorname{div}\operatorname{curl}\mathbf{F} = 0$
- $\operatorname{curl}\nabla f = \mathbf{0}$

Example 3.32. Assume f has continuous second partial derivatives. We verify that $\operatorname{curl}(\nabla f) = \mathbf{0}$.

$$\operatorname{curl}(\nabla f) = \operatorname{curl}(f_x, f_y, f_z)$$

$$= \left(\frac{\partial}{\partial y}f_z - \frac{\partial}{\partial z}f_y, -\left(\frac{\partial}{\partial x}f_z - \frac{\partial}{\partial z}f_x\right), \frac{\partial}{\partial x}f_y - \frac{\partial}{\partial y}f_x\right).$$

Since the mixed second partial derivatives of a twice continuously differentiable function are equal, this is $(0,0,0)$. ☐

Example 3.33. Let $\mathbf{F} = (f_1, f_2, f_3)$ be a twice continuously differentiable vector function. We show that $\operatorname{div}\operatorname{curl}\mathbf{F} = 0$.

$$\operatorname{curl}\mathbf{F} = \nabla \times \mathbf{F} = \left(\frac{\partial}{\partial y}f_3 - \frac{\partial}{\partial z}f_2, -\left(\frac{\partial}{\partial x}f_3 - \frac{\partial}{\partial z}f_1\right), \frac{\partial}{\partial x}f_2 - \frac{\partial}{\partial y}f_1\right).$$

So

$$\operatorname{div}\operatorname{curl}\mathbf{F} = \nabla \cdot (\nabla \times \mathbf{F})$$

$$= \frac{\partial}{\partial x}\left(\frac{\partial}{\partial y}f_3 - \frac{\partial}{\partial z}f_2\right) + \frac{\partial}{\partial y}\left(\frac{\partial}{\partial z}f_1 - \frac{\partial}{\partial x}f_3\right) + \frac{\partial}{\partial z}\left(\frac{\partial}{\partial x}f_2 - \frac{\partial}{\partial y}f_1\right)$$

$$= \frac{\partial}{\partial x}\frac{\partial}{\partial y}f_3 - \frac{\partial}{\partial x}\frac{\partial}{\partial z}f_2 + \frac{\partial}{\partial y}\frac{\partial}{\partial z}f_1 - \frac{\partial}{\partial y}\frac{\partial}{\partial x}f_3 + \frac{\partial}{\partial z}\frac{\partial}{\partial x}f_2 - \frac{\partial}{\partial z}\frac{\partial}{\partial y}f_1 = 0.$$

☐

Problems

3.44. Compute the indicated curls and divergences.

(a) $\operatorname{curl}(x,0,0)$, $\operatorname{div}(x,0,0)$
(b) $\operatorname{curl}(0,x,0)$, $\operatorname{div}(0,x,0)$
(c) $\operatorname{curl}(0,0,x)$, $\operatorname{div}(0,0,x)$

3.45. Let $\mathbf{F}(x,y,z)$ and $\mathbf{G}(x,y,z)$ be differentiable vector fields that satisfy

$$\operatorname{curl}\mathbf{F}(x,y,z) = (5y + 7z, 3x, 0), \quad \operatorname{curl}\mathbf{G}(1,2,-3) = (5,7,9),$$

$$\operatorname{div}\mathbf{F}(1,2,-3) = 6, \quad \operatorname{div}\mathbf{G}(x,y,z) = x^2 - zy.$$

(a) Find $\operatorname{div}(3\mathbf{F} + 4\mathbf{G})$ at the point $(1,2,-3)$.
(b) Show that $\operatorname{div}\operatorname{curl}\mathbf{F}(x,y,z) = 0$.
(c) Find $\operatorname{curl}(3\mathbf{F} + 4\mathbf{G})$ at the point $(1,2,-3)$.

3.46. Let $\mathbf{F}(x,y,z) = (xy^2, yz^2, zx^2)$. Find curl $\mathbf{F}(1,2,3)$ and div $\mathbf{F}(1,2,3)$.

3.47. Consider the vector field $\mathbf{H}(x,y) = \dfrac{(-y,x)}{(x^2+y^2)^p}$ for points other than $(0,0)$. Draw the vectors $\mathbf{H}(1,0)$, $\mathbf{H}(0,1)$, $\mathbf{H}(-1,0)$, and $\mathbf{H}(0,-1)$. Show that

$$\text{curl}\,\mathbf{H} = \frac{(2-2p)}{(x^2+y^2)^p}$$

which is negative, zero, and positive when $p = 1.05$, 1, $.95$, respectively.

3.48. Verify that $\text{curl}\,(u(x,y), v(x,y), 0) = (0, 0, v_x - u_y)$.

3.49. Suppose (u,v,w) satisfies $\text{curl}\,(u,v,w) = \mathbf{0}$ in $\mathbb{R}^3$, i.e.,

$$u_y = v_x, \quad u_z = w_x, \quad v_z = w_y.$$

Prove that there is p such that $\nabla p = (u,v,w)$:

$$p_x = u, \quad p_y = v, \quad p_z = w,$$

by verifying the following steps.

(a) Define $p_1 = \displaystyle\int_0^x u(s,y,z)\,ds$, Then $p_{1x} = u$, and $p_{1y} = v + n$, where n does not depend on x.

(b) Define $p_2 = p_1 - c(y,z)$ where $c = \displaystyle\int_0^y n(t,z)\,dt$. Then $p_{2x} = u$ and $p_{2y} = v$.

(c) Write $p_{2z} = w + m$. Then m doesn't depend on x and y.

(d) Define $p_3 = p_2 - f(z)$ where $f_z = m$. Then $p_{3x} = u$, $p_{3y} = v$, and $p_{3z} = w$.

3.50. A vector field $\mathbf{F}$ has a *vector potential* $\mathbf{G}$ when it can be expressed $\mathbf{F} = \text{curl}\,\mathbf{G}$. Show that the inverse square field $-\dfrac{\mathbf{X}}{\|\mathbf{X}\|^3} = -\dfrac{(x_1,x_2,x_3)}{(x_1^2+x_2^2+x_3^2)^{3/2}}$ has a vector potential

$$\frac{x_3}{(x_1^2+x_2^2+x_3^2)^{1/2}} \frac{(-x_2,x_1,0)}{x_1^2+x_2^2}$$

at all points except along the x_3 axis.

3.51. Suppose a differentiable vector function $\mathbf{V}(\mathbf{X},t)$, that is, three components depending on the four variables of space and time, gives the velocity of fluid particles moving in $\mathbb{R}^3$, so that

$$\frac{d\mathbf{X}}{dt} = \mathbf{V}(\mathbf{X}(t),t)$$

for one of the particles whose position is given by $\mathbf{X}(t)$ at time t.

(a) Show that the acceleration of the particle is given by

$$\frac{d^2\mathbf{X}}{dt^2} = \mathbf{V}_t + (D\mathbf{V})\mathbf{V}.$$

(b) Verify that an example is $X(t) = C_1 + t^{-1}C_2$, $V(X,t) = t^{-1}(C_1 - X)$ where C_1 and C_2 are constant vectors.

3.52. Show that

$$\Big((-1,0,0)\times F(x_0,y_m,z_m)\varDelta y\varDelta z + (1,0,0)\times F(x_0+\varDelta x,y_m,z_m)\varDelta y\varDelta z$$

$$+ (0,-1,0)\times F(x_m,y_0,z_m)\varDelta x\varDelta z + (0,1,0)\times F(x_m,y_0+\varDelta y,z_m)\varDelta x\varDelta z$$

$$+ (0,0,-1)\times F(x_m,y_m,z_0)\varDelta x\varDelta y + (0,0,1)\times F(x_m,y_m,z_0+\varDelta z)\varDelta x\varDelta y\Big)\frac{1}{\varDelta x\varDelta y\varDelta z}.$$

tends to $\operatorname{curl} F((x_0,y_0,z_0)$ as $\varDelta x$, $\varDelta y$, and $\varDelta z$ tend to zero. We used this in our geometric description of $\operatorname{curl} F$.

3.53. Show that the sum of pressure forces over the faces of a small cube

$$-f(x_0,y_m,z_m)(-1,0,0)\varDelta y\varDelta z - f(x_0+\varDelta x,y_m,z_m)(1,0,0)\varDelta y\varDelta z$$
$$-f(x_m,y_0,z_m)(0,-1,0)\varDelta x\varDelta z - f(x_m,y_0+\varDelta y,z_m)(0,1,0)\varDelta x\varDelta z$$
$$-f(x_m,y_m,z_0)(0,0,-1)\varDelta x\varDelta y - f(x_m,y_m,z_0+\varDelta z)(0,0,1)\varDelta x\varDelta y$$

is equal to

$$-(f_x(\overline{x},y_m,z_m), f_y(x_m,\overline{y},z_m), f_z(x_m,y_m,\overline{z}))\varDelta x\varDelta y\varDelta z$$

for some $\overline{x}$ between x_0 and $x_0+\varDelta x$, $\overline{y}$ between y_0 and $y_0+\varDelta y$, $\overline{z}$ between z_0 and $z_0+\varDelta z$.

3.54. Show that $\operatorname{div}(v\nabla u) = v\varDelta u + \nabla u \cdot \nabla v$ for C^2 functions u, v from $\mathbb{R}^3$ to $\mathbb{R}$.

3.55. Show that $\varDelta(uv) = u\varDelta v + v\varDelta u + 2\nabla u \cdot \nabla v$ for C^2 functions u, v from $\mathbb{R}^3$ to $\mathbb{R}$.

3.56. Prove the vector differential identities (a)–(d) listed in the text.

3.57. Prove the vector differential identity (e) listed in the text.

3.58. Prove the vector differential identity (f) listed in the text.

3.59. Prove the vector differential identities (g)–(i) listed in the text.

3.60. Let $f(y,z)$ be a C^2 function from $\mathbb{R}^2$ to $\mathbb{R}$, and suppose there is a number c so that

$$f_{yy} + f_{zz} = -c^2 f.$$

(a) Show that the vector field $F = (cf, f_z, -f_y)$ has the property

$$\operatorname{curl} F = cF.$$

(b) Show that the function $f(y,z) = \sin(3y+4z)$ has this property, and find the corresponding vector field F.

Chapter 4
More about differentiation

Abstract This chapter describes applications of the derivative to methods for finding extreme values of functions of several variables, and to methods for approximating functions of several variables by polynomials.

4.1 Higher derivatives of functions of several variables

In Theorem 3.10 we showed that if all the second partial derivatives of a function $f(x,y)$ are continuous on an open set in $\mathbb{R}^2$, then

$$f_{xy} = f_{yx}.$$

Analogous results hold for differentiable functions of several variables, and for partial derivatives of all orders.

Definition 4.1. A function f from an open set in $\mathbb{R}^k$ to $\mathbb{R}$ is called *n times continuously differentiable* if it has all mixed partial derivatives up to order n, and if all n-th partial derivatives are continuous functions.

An n times continuously differentiable function is called a C^n function.

Theorem 4.1. *Let f be a C^n function from an open set in $\mathbb{R}^k$ to $\mathbb{R}$. Then two partial derivatives of f of order less than or equal to n, in which differentiation with respect to each variable occurs the same number of times, are equal, regardless of the order in which these partial differentiations are carried out.*

This result can be proved by repeated application of Theorem 3.10. For example, for a C^2 function $f(x,y,z)$ if we hold z fixed then the theorem gives $f_{xy} = f_{yx}$. Similarly $f_{yz} = f_{zy}$ and $f_{zx} = f_{xz}$. And, if f is C^3, then five applications of Theorem 3.10

© Springer International Publishing AG 2017
P. D. Lax and M. S. Terrell, *Multivariable Calculus with Applications*,
Undergraduate Texts in Mathematics, https://doi.org/10.1007/978-3-319-74073-7_4

give

$$f_{xyz} = f_{xzy} = f_{zxy} = f_{zyx} = f_{yzx} = f_{yxz}.$$

Example 4.1. Let

$$f(x,y,z) = x^2 y z^4.$$

Find f_{xzyy}. We can compute the partial derivatives in the given order.

$$f_x = 2xyz^4, \qquad f_{xz} = 8xyz^3, \qquad f_{xzy} = 8xz^3, \qquad f_{xzyy} = 0.$$

Alternatively by Theorem 4.1 we can write

$$f_{xzyy} = f_{yyxz}.$$

Since $f_{yy} = 0$, we get $f_{yyxz} = 0$. □

A *partial differential equation* or *pde* relates partial derivatives of a function. We say that a function is a solution of the pde if it satisfies the equation.

Example 4.2. Let $u(x,y,z,t) = e^{-kt}(\cos x + \cos y + \cos z)$. We show that u is a solution of the partial differential equation

$$u_t - k\Delta u = 0,$$

where Δ is the Laplace operator defined as $\Delta u = u_{xx} + u_{yy} + u_{zz}$. A brief calculation shows that

$$u_t = -k e^{-kt}(\cos x + \cos y + \cos z),$$
$$u_{xx} = e^{-kt}(-\cos x),$$
$$u_{yy} = e^{-kt}(-\cos y),$$
$$u_{zz} = e^{-kt}(-\cos z).$$

Therefore

$$k\Delta u = k e^{-kt}(-\cos x - \cos y - \cos z) = u_t$$

and u satisfies the equation $u_t - k\Delta u = 0$. □

Problems

4.1. Compute the partial derivatives.

(a) f_{xx} and f_{zz} for $(x,y,z) \neq (0,0,0)$ if $f(x,y,z) = (x^2 + y^2 + z^2)^{-1/2}$
(b) f_{xyz} if $f(x,y,z) = xy + yz + zx$
(c) $g_{xxy} - g_{xyx}$ if g is three times continuously differentiable from $\mathbb{R}^3$ to $\mathbb{R}$
(d) h_{x_j} and $h_{x_j x_k}$ if $h(\mathbf{X}) = \mathbf{A} \cdot \mathbf{X}$, where $\mathbf{A} = (a_1, \ldots, a_n)$ is constant and $\mathbf{X} = (x_1, \ldots, x_n)$.

4.2. Find the partial derivatives.

(a) $\dfrac{\partial}{\partial x_3}(x_1^2 + 2^2 x_2^2 + 3^3 x_3^2)$

(b) $\dfrac{\partial}{\partial x_k}\left(\displaystyle\sum_{j=1}^{n} j^2 x_j^2\right)$

(c) $(x^3 y + y^3 z + z^3 w)_{yyw}$

(d) $\dfrac{\partial}{\partial x_5}(\|\mathbf{X}\|^2)$ for $\mathbf{X}$ in $\mathbb{R}^n$, $n \geq 5$

(e) $\dfrac{\partial^2}{\partial x_5 \partial x_3}(\|\mathbf{X}\|^2)$ for $\mathbf{X}$ in $\mathbb{R}^n$, $n \geq 5$.

4.3. Let $f(x,y) = (x^2 + y^2)^{3/2}$. Find f_{xy}.

4.4. Let $f(\mathbf{X}) = \cos(\mathbf{A} \cdot \mathbf{X})$ where $\mathbf{X}$ is in $\mathbb{R}^n$ and $\mathbf{A}$ is a constant vector in $\mathbb{R}^n$. That is,

$$f(x_1, x_2, \ldots, x_n) = \cos(a_1 x_1 + a_2 x_2 + \cdots + a_n x_n).$$

(a) Show that $f_{x_1} = -a_1 \sin(\mathbf{A} \cdot \mathbf{X})$.
(b) Find the derivatives $f_{x_2 x_2}$ and $f_{x_3 x_2 x_2 x_4}$.
(c) Verify that $f_{x_1 x_1} + f_{x_2 x_2} + \cdots + f_{x_n x_n} = -\|\mathbf{A}\|^2 f$.

4.5. Define $g(x, y, z, w) = e^{ax+by+cz+dw}$, where a, b, c, and d are constants.

(a) Show that $g_x = ag$.
(b) Find the partial derivatives g_{xxww} and g_{yyzz}.
(c) Find a relation among the constants a, b, c, and d so that g satisfies the equation

$$g_{xxww} + g_{yyzz} - 2g_{xyzw} = 0.$$

(d) Find a relation among the constants a, b, c, and d so that g satisfies the equation

$$g_{xxww} + g_{yyzz} - 2g = 0.$$

4.6. For a linear combination of differentiable functions f and g from $\mathbb{R}^n$ to $\mathbb{R}$, with a and b constants, the partial derivatives satisfy

$$\frac{\partial}{\partial x_k}(af + bg) = a\frac{\partial f}{\partial x_k} + b\frac{\partial g}{\partial x_k}, \qquad k = 1, \ldots, n.$$

Show that the following formulas are true for C^4 functions f, g, and h.

(a) $\dfrac{\partial^2}{\partial x_k^2}(af + bg) = a\dfrac{\partial^2 f}{\partial x_k^2} + b\dfrac{\partial^2 g}{\partial x_k^2}$.

(b) $\dfrac{\partial^3}{\partial x_k^2 \partial x_\ell}(af + bg) = a\dfrac{\partial^3 f}{\partial x_k^2 \partial x_\ell} + b\dfrac{\partial^3 g}{\partial x_k^2 \partial x_\ell}$, $\quad \ell = 1, 2, \ldots, n, \ k = 1, 2, \ldots, n$.

(c) $(af + bg + ch)_{x_1} = af_{x_1} + g_{x_1} + ch_{x_1}$

(d) If f, g, h are C^4 functions from $\mathbb{R}^5$ to $\mathbb{R}$ then

$$(af + bg + ch)_{x_1 x_3 x_4 x_2} = af_{x_1 x_4 x_3 x_2} + bg_{x_2 x_3 x_4 x_1} + ch_{x_1 x_3 x_4 x_2}.$$

4.7. Show that the function $u(\mathbf{X}, t) = t^{-n/2} e^{-\|\mathbf{X}\|^2/4t}$, where $\mathbf{X}$ is in $\mathbb{R}^n$ and $t > 0$, satisfies the equation

$$u_t = u_{x_1 x_1} + u_{x_2 x_2} + \cdots + u_{x_n x_n}.$$

4.8. There are at most three distinct second partial derivatives of a C^2 function $f(x, y)$, namely f_{xx}, f_{xy}, and f_{yy}. How many distinct third partial derivatives could there be of a C^3 function $f(x_1, x_2, x_3, x_4)$?

4.9. For $\mathbf{X}$ in $\mathbb{R}^n$ $(n \geq 3)$ and for every twice differentiable function $u(\mathbf{X})$ we define $\Delta u = u_{x_1 x_1} + u_{x_2 x_2} + \cdots + u_{x_n x_n}$. Let $r = \|\mathbf{X}\|$. Show that

(a) For every constant p, $(r^p)_{x_j} = p r^{p-2} x_j$, $(r \neq 0)$.
(b) $\Delta(r^{2-n}) = 0$, $(r \neq 0)$.

4.10. Suppose u is a C^4 function from $\mathbb{R}^2$ to $\mathbb{R}$. Define $\Delta u = u_{xx} + u_{yy}$.

(a) Show that $\Delta \Delta u = u_{xxxx} + 2u_{xxyy} + u_{yyyy}$.
(b) Find examples of polynomial functions $u(x, y) = ax^4 + bx^2 y^2 + cy^4$, $a, b, c \neq 0$, for which $\Delta \Delta u$ is always positive, zero, negative.
(c) Show that for $(x, y) \neq (0, 0)$, $\Delta(x^2 + y^2)^{1/2} = (x^2 + y^2)^{-1/2}$.
(d) Suppose that $u(x, y) = v(r)$, where v is a C^2 function from $\mathbb{R}$ to $\mathbb{R}$ for $r > 0$ and $r = \sqrt{x^2 + y^2}$. Show that

$$\Delta u = v_{rr} + r^{-1} v_r.$$

(e) Show that for $(x, y) \neq (0, 0)$, $\Delta \Delta(x^2 + y^2)^{1/2} = (x^2 + y^2)^{-3/2}$.

4.11. Suppose that $u(x, y, z)$ is a C^2 function from $\mathbb{R}^3$ to $\mathbb{R}$. Define $\Delta u = u_{xx} + u_{yy} + u_{zz}$ and let $r = \sqrt{x^2 + y^2 + z^2}$.

(a) Show that for $(x, y, z) \neq (0, 0, 0)$, $\Delta(x^2 + y^2 + z^2)^{1/2} = 2(x^2 + y^2 + z^2)^{-1/2}$.
(b) Suppose $u(x, y, z) = v(r)$, where v is a C^2 function from $\mathbb{R}$ to $\mathbb{R}$. Show that for $r > 0$

$$\Delta u = v_{rr} + 2r^{-1} v_r.$$

(c) Show that for $(x, y, z) \neq (0, 0, 0)$, $\Delta \Delta(x^2 + y^2 + z^2)^{1/2} = 0$.

4.12. Let $\mathbf{F}(x, y, z, t) = (f_1(x, y, z, t), f_2(x, y, z, t), f_3(x, y, z, t))$ be a C^2 function on an open set in $\mathbb{R}^4$. Define

$$\operatorname{curl} \mathbf{F} = \left(\frac{\partial f_3}{\partial y} - \frac{\partial f_2}{\partial z}, \frac{\partial f_1}{\partial z} - \frac{\partial f_3}{\partial x}, \frac{\partial f_2}{\partial x} - \frac{\partial f_1}{\partial y} \right),$$

$$\operatorname{div} \mathbf{F} = \frac{\partial f_1}{\partial x} + \frac{\partial f_2}{\partial y} + \frac{\partial f_3}{\partial z},$$

$$\Delta \mathbf{F} = (\Delta f_1, \Delta f_2, \Delta f_3).$$

(a) Show that $\operatorname{curl}(\mathbf{F}_t) = (\operatorname{curl} \mathbf{F})_t$.
(b) Show that $\operatorname{curl} \operatorname{curl} \mathbf{F} = \nabla \operatorname{div} \mathbf{F} - \Delta \mathbf{F}$.

(c) Suppose that

$$\mathbf{E}(x,y,z,t) = (E_1(x,y,z,t), E_2(x,y,z,t), E_3(x,y,z,t)) \text{ and}$$
$$\mathbf{B}(x,y,z,t) = (B_1(x,y,z,t), B_2(x,y,z,t), B_3(x,y,z,t))$$

are C^2 functions on an open set in $\mathbb{R}^4$ that satisfy the following equations, called *Maxwell* equations,

$$\mathbf{E}_t = \operatorname{curl}\mathbf{B}, \quad \mathbf{B}_t = -\operatorname{curl}\mathbf{E}, \quad \operatorname{div}\mathbf{E} = 0, \quad \operatorname{div}\mathbf{B} = 0.$$

Show that $\mathbf{E}_{tt} = \varDelta\mathbf{E}$ and $\mathbf{B}_{tt} = \varDelta\mathbf{B}$.

(d) Verify that the functions

$$\mathbf{B}(\mathbf{X},t) = (\cos(y-t), 0, 0), \qquad \mathbf{E}(\mathbf{X},t) = (0,0,\cos(y-t))$$

satisfy the Maxwell equations. The fields are sketched in Figure 4.1 at one particular time t at the left, and at a slightly later time at the right.

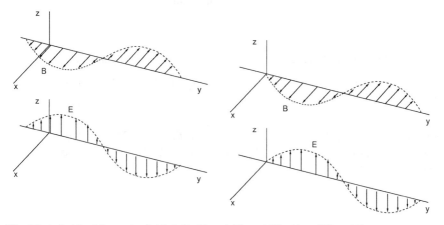

Fig. 4.1 A sketch of the vector fields in Problem 4.12, part (d), at two different times t.

4.2 Extrema of functions of two variables

Just as in single variable calculus a function f from $\mathbb{R}^n$ to $\mathbb{R}$ may have points in the domain at which it attains a local maximum or minimum.

Definition 4.2. Let $f : D \subset \mathbb{R}^n \to \mathbb{R}$. We say that f has a *local minimum* $f(\mathbf{A})$ at a point $\mathbf{A}$ in D if there is an open ball in $\mathbb{R}^n$ centered at $\mathbf{A}$ so that for every point $\mathbf{X}$ in D that is in the ball,

$$f(\mathbf{X}) \geq f(\mathbf{A}).$$

f has a *local maximum* $f(\mathbf{A})$ at a point $\mathbf{A}$ in D if there is an open ball in $\mathbb{R}^n$ centered at $\mathbf{A}$ so that for every point $\mathbf{X}$ in D that is in the ball,

$$f(\mathbf{X}) \leq f(\mathbf{A}).$$

A point that is either a local maximum or a local minimum is called a *local extremum* of f on D. We say a local maximum or local minimum is *strict* if the corresponding strict inequality $f(\mathbf{X}) > f(\mathbf{A})$ or $f(\mathbf{X}) < f(\mathbf{A})$ holds except for $\mathbf{X} = \mathbf{A}$.

Recall that a differentiable function f of a single variable that has a local extremum at an interior point a of an interval satisfies the condition

$$f'(a) = 0.$$

This result has the following extension to functions of several variables.

A first derivative test. We have the following test for local extrema at interior points.

Theorem 4.2. *Let f be a differentiable function from $D \subset \mathbb{R}^n$ to $\mathbb{R}$. Suppose f has a local extremum $f(\mathbf{A})$ at the interior point $\mathbf{A}$. Then the first partial derivatives of f are zero at $\mathbf{A}$, that is,*

$$\nabla f(\mathbf{A}) = \mathbf{0}.$$

Proof. This result is an immediate consequence of the result for functions of a single variable. Suppose $f(x_1, \ldots, x_n)$ has a local extremum at an interior point $\mathbf{A}$ of D. Hold all but the i-th coordinate fixed and let x_i vary. (See Figure 4.2) Since $\mathbf{A}$ is an interior point there is an interval on which the single variable function

$$f(a_1, \ldots, a_{i-1}, x_i, a_{i+1}, \ldots, a_n)$$

has a local extremum at $x_i = a_i$. At that point

$$\frac{\partial f}{\partial x_i}(\mathbf{A}) = 0.$$

Since this is true for each $i = 1, \ldots, n$, this completes the proof. $\square$

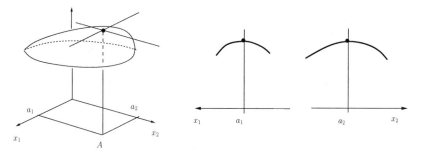

Fig. 4.2 *Left:* A graph of f, that has a local maximum at (a_1,a_2). *Center:* $f(x_1,a_2)$ has a local maximum at a_1. *Right:* $f(a_1,x_2)$ has a local maximum at a_2.

Example 4.3. Consider the graphs in Figure 4.3 of

$$f(x,y) = x^2 + y^2, \quad g(x,y) = -x^2 - y^2, \quad h(x,y) = y^3.$$

f has its smallest value at the point $(0,0)$, and its partial derivatives $f_x = 2x$ and $f_y = 2y$ are zero there. g has its largest value at $(0,0)$, and again the partial derivatives $g_x(0,0) = 0$ and $g_y(0,0) = 0$. It is also true that

$$\nabla h(0,0) = \mathbf{0}$$

since $\nabla h(x,y) = (0,3y^2)$. But h has neither a maximum nor a minimum at $(0,0)$:

$$h(0,y) = y^3$$

is positive for some points and negative for other points arbitrarily close to $(0,0)$.   □

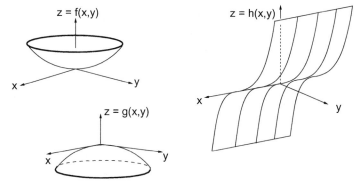

Fig. 4.3 Graphs of f, g, h in Example 4.3.

A second derivative test. As we saw in Example 4.3, a function f from $\mathbb{R}^n$ to $\mathbb{R}$ that has $\nabla f(\mathbf{A}) = \mathbf{0}$ may have a local maximum, a local minimum, or neither at $\mathbf{A}$.

We turn to the second partial derivatives to test for local extrema at interior points of the domain. Recall the case of functions of a single variable:

Taylor's Theorem for a twice continuously differentiable function gives

$$f(a+h) = f(a) + f'(a)h + \frac{f''(c)}{2}h^2$$

for some c between a and $a+h$. If $f'(a) = 0$ we get

$$f(a+h) - f(a) = \frac{f''(c)}{2}h^2.$$

Since f'' is continuous and $f''(a)$ is positive, then f'' is positive at all points close enough to a. By taking h small enough, c will be close to a, and $f''(c) > 0$. In that case

$$f(a+h) = f(a) + \frac{f''(c)}{2}h^2 > f(a).$$

Thus f has a strict local minimum at a. Similarly if $f'(a) = 0$ and $f''(a) < 0$, then $f''(c)$ is negative for h small enough and

$$f(a+h) = f(a) + \frac{f''(c)}{2}h^2 < f(a)$$

so f has a strict local maximum at a.

A function f from $\mathbb{R}^n$ to $\mathbb{R}$ has many second partial derivatives. We arrange them as a matrix:

Definition 4.3. Given a C^2 function from an open set in $\mathbb{R}^n$ to $\mathbb{R}$, the n by n matrix of second partial derivatives at $(x_1, \ldots, x_n)$

$$\mathcal{H}f(x_1, \ldots, x_n) = \begin{bmatrix} f_{x_1 x_1} & f_{x_1 x_2} & \cdots & f_{x_1 x_n} \\ f_{x_2 x_1} & f_{x_2 x_2} & \cdots & f_{x_2 x_n} \\ \vdots & & & \vdots \\ f_{x_n x_1} & f_{x_n x_2} & \cdots & f_{x_n x_n} \end{bmatrix}$$

is called the *Hessian* matrix of f at $(x_1, \ldots, x_n)$.

$\mathcal{H}$ is a *symmetric matrix* since $f_{x_i x_j} = f_{x_j x_i}$.

We introduce the notion of positive definite matrix to use information about the second partial derivatives of f at $\mathbf{X}$. We begin with the two-dimensional case.

Definition 4.4. A symmetric 2×2 matrix $\mathbf{S}$,

$$\mathbf{S} = \begin{bmatrix} p & q \\ q & r \end{bmatrix}$$

is called *positive definite* if the associated quadratic function

$$S(u,v) = \mathbf{U} \cdot \mathbf{S}\mathbf{U}$$

is positive for all $\mathbf{U} = (u,v) \neq \mathbf{0}$ in $\mathbb{R}^2$. The matrix $\mathbf{S}$ is *negative definite* if $-\mathbf{S}$ is positive definite, and $\mathbf{S}$ is *indefinite* if $\mathbf{U} \cdot \mathbf{S}\mathbf{U}$ has both positive and negative values.

The next result is an important property of positive definite matrices.

Theorem 4.3. *Let* $\mathbf{S} = \begin{bmatrix} p & q \\ q & r \end{bmatrix}$. $\mathbf{S}$ *is positive definite if and only if there exists a positive number* m *such that for all* (u,v) *in* $\mathbb{R}^2$,

$$S(u,v) = [u\ v]\begin{bmatrix} p & q \\ q & r \end{bmatrix}\begin{bmatrix} u \\ v \end{bmatrix} = pu^2 + 2quv + rv^2 \geq m(u^2 + v^2). \qquad (4.1)$$

Proof. If there is such a number m, then certainly $S(u,v) > 0$ for all $(u,v) \neq \mathbf{0}$, and $\mathbf{S}$ is positive definite.

Conversely, suppose $\mathbf{S}$ is a positive definite matrix. Since $S(u,v)$ is a polynomial in u and v, S is a continuous function in $\mathbb{R}^2$ and in particular on the unit circle $u^2 + v^2 = 1$. By the Extreme Value Theorem, S attains a minimum m at some point (c,d) on this circle. The minimum $m = S(c,d)$ is positive since $(c,d) \neq \mathbf{0}$ and $\mathbf{S}$ is positive definite. It follows that (4.1) holds for all (u,v) on the unit circle.

Next we show that (4.1) holds for every vector (u,v). Vector (u,v) can be written as a multiple of a unit vector (z,w):

$$(u,v) = a(z,w) = (az,aw), \qquad a = \sqrt{u^2 + v^2}, \qquad z^2 + w^2 = 1.$$

$S(u,v)$ is a^2 times $S(z,w)$ because

$$S(u,v) = S(az,aw) = pa^2z^2 + 2qazaw + ra^2w^2$$

$$= a^2(pz^2 + 2qzw + rz^2) = a^2 S(z,w).$$

For (z,w) on the unit circle we have shown that $S(z,w) \geq m = m(z^2 + w^2)$. Therefore

$$S(u,v) = a^2 S(z,w) \geq a^2 m(z^2 + w^2) = m(u^2 + v^2),$$

which proves the inequality (4.1). $\square$

The following result is a consequence of Theorem 4.3:

Theorem 4.4. *Suppose that a symmetric matrix* **S** *is positive definite, and for all* $\mathbf{U} = (u,v)$ *in* $\mathbb{R}^2$
$$S(u,v) = \mathbf{U} \cdot \mathbf{SU} \geq m(u^2 + v^2).$$
Then every symmetric matrix of the form $\mathbf{S} + \mathbf{T}$, *where the elements of the symmetric matrix* **T** *are small enough, is positive definite and*
$$(S+T)(u,v) = \mathbf{U} \cdot (\mathbf{S}+\mathbf{T})\mathbf{U} \geq \frac{m}{2}(u^2 + v^2).$$

Proof. First we show that if the entries of the symmetric matrix **T** are small enough, the associated quadratic function $T(u,v)$ satisfies

$$T(u,v) \geq -\frac{m}{2}(u^2 + v^2).$$

Denote
$$\mathbf{T} = \begin{bmatrix} a & b \\ b & c \end{bmatrix}, \quad T(u,v) = [u\ v]\begin{bmatrix} a & b \\ b & c \end{bmatrix}\begin{bmatrix} u \\ v \end{bmatrix} = au^2 + 2buv + cv^2.$$

If $b \geq 0$ we use the fact that $(u+v)^2 = u^2 + 2uv + v^2 \geq 0$ to get $2buv \geq b(-u^2 - v^2)$ and

$$au^2 + 2buv + cv^2 \geq au^2 + b(-u^2 - v^2) + cv^2$$
$$= (a-b)u^2 + (c-b)v^2.$$

If $b < 0$ we use the fact that $(u-v)^2 = u^2 - 2uv + v^2 \geq 0$ to get $b(u^2 + v^2) \leq 2buv$ and

$$au^2 + 2buv + cv^2 \geq au^2 + b(u^2 + v^2) + cv^2$$
$$= (a+b)u^2 + (c+b)v^2.$$

Take $|a|, |b|,$ and $|c|$ all less than $\frac{m}{4}$. Then $|a-b| \leq |a| + |b| < \frac{m}{2}$, so $a-b \geq -\frac{m}{2}$. Similarly for $c-b$, $a+b$, and $c+b$. Therefore when $|a|, |b|,$ and $|c|$ are all less than $\frac{m}{4}$

$$T(u,v) = au^2 + 2buv + cv^2$$
$$\geq -\frac{m}{2}(u^2 + v^2).$$

Now add the inequalities for S and T to get

$$(S+T)(u,v) = S(u,v) + T(u,v)$$
$$\geq m(u^2 + v^2) - \frac{m}{2}(u^2 + v^2) = \frac{m}{2}(u^2 + v^2)$$

for all (u,v). Since $\frac{m}{2} > 0$ the matrix $\mathbf{S} + \mathbf{T}$ is positive definite.   □

Theorem 4.5. *(**Second derivative test***) Suppose f is a C^2 function on the interior of $D \subset \mathbb{R}^2$ and that at some interior point (c,d) of D,*

$$\nabla f(c,d) = \mathbf{0}.$$

Let $\mathcal{H}f(c,d)$ be the Hessian matrix of second partial derivatives of f at (c,d),

$$\mathcal{H}f(c,d) = \begin{bmatrix} f_{xx}(c,d) & f_{xy}(c,d) \\ f_{yx}(c,d) & f_{yy}(c,d) \end{bmatrix}.$$

(a) If $\mathcal{H}f(c,d)$ is positive definite then f has a strict local minimum at (c,d).
(b) If $\mathcal{H}f(c,d)$ is negative definite then f has a strict local maximum at (c,d).
(c) If $\mathcal{H}f(c,d)$ is indefinite then f has neither a local maximum nor local minimum at (c,d).

Proof. (a) Suppose $\mathcal{H}f(c,d)$ is positive definite. By Theorem 4.3

$$[u\ v]\mathcal{H}f(c,d)\begin{bmatrix} u \\ v \end{bmatrix} \geq m(u^2 + v^2)$$

for some $m > 0$. In the proof of Theorem 4.4 we showed that if the difference $\mathbf{T}$ of the matrices $\mathcal{H}f(c,d)$ and $\mathcal{H}f(x,y)$ has entries that have absolute value less than $\frac{m}{4}$ then

$$\mathcal{H}f(x,y) = \mathcal{H}f(c,d) + \mathbf{T}$$

is positive definite. Since f has continuous second partial derivatives we can find a disk of radius r centered at (c,d) where all these differences are within $\frac{m}{4}$. Let (x,y) be a point of the disk and let (u,v) be the vector from (c,d) to (x,y), $(u,v) = (x-c, y-d)$. Define a function of the single variable t on an open interval containing 0 and 1 by

$$g(t) = f(c + tu, d + tv).$$

The function g evaluates f along the line segment from (c,d) to (x,y), with

$$g(0) = f(c,d), \qquad g(1) = f(x,y).$$

Differentiating with respect to t, the Chain Rule gives

$$g'(t) = u f_x(c + tu, d + tv) + v f_y(c + tu, d + tv)$$

and

$$g''(t) = u^2 f_{xx}(c + tu, d + tv) + 2uv f_{xy}(c + tu, d + tv) + v^2 f_{yy}(c + tu, d + tv).$$

By Taylor's Theorem for functions of a single variable we have

$$f(x,y) = g(1) = g(0) + g'(0)(1) + \frac{g''(\theta)}{2}(1)^2$$

for some θ between 0 and 1. Since $g'(0) = uf_x(c,d) + vf_y(c,d) = 0u + 0v = 0$ we have

$$f(x,y) = f(c,d) + \frac{g''(\theta)}{2}. \qquad (4.2)$$

$$g''(\theta) = u^2 f_{xx}(c + \theta u, d + \theta v) + 2uv f_{xy}(c + \theta u, d + \theta v) + v^2 f_{yy}(c + \theta u, d + \theta v)$$

$$= [u \ v] \mathcal{H} f(c + \theta u, d + \theta v) \begin{bmatrix} u \\ v \end{bmatrix}$$

where $(c + \theta u, d + \theta v)$ is on the segment between (c,d) and (x,y) and hence within distance r of (c,d). Since $\mathcal{H} f(c + \theta u, d + \theta v)$ is positive definite $g''(\theta)$ is positive. By equation (4.2) we get

$$f(x,y) > f(c,d).$$

(b) If $\mathcal{H} f(c,d)$ is negative definite apply part (a) to the function $-f$. The function f has a strict local maximum wherever $-f$ has a strict local minimum.

(c) If $\mathcal{H} f(c,d)$ is indefinite then

$$g''(0) = u^2 f_{xx}(c,d) + 2uv f_{xy}(c,d) + v^2 f_{yy}(c,d)$$

is positive for some values of (u,v) and negative for others. For r small enough, $g''(\theta)$ is close enough to $g''(0)$ so that $g''(\theta)$ will be positive for some values of (u,v) and negative for others. So for some points (x,y), $f(x,y) > f(c,d)$ and for others $f(x,y) < f(c,d)$. That is, at (c,d) the function f has neither a maximum nor minimum, but $\nabla f(c,d) = 0$. Such a point is called a *saddle point* for f, a name dating back to the days when people rode horses.

$\square$

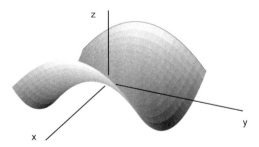

Fig. 4.4 The graph of a function with a saddle point.

The next theorem gives a useful criterion for a 2 by 2 matrix to be positive definite or negative definite.

Theorem 4.6. *The symmetric matrix*

$$S = \begin{bmatrix} p & q \\ q & r \end{bmatrix}$$

is positive definite if

$$p > 0 \quad \text{and} \quad pr - q^2 > 0.$$

S *is negative definite if* $p < 0$ *and* $pr - q^2 > 0$ *and* **S** *is indefinite if* $pr - q^2 < 0.$

Proof. The quadratic function associated to **S** is

$$S(u,v) = pu^2 + 2quv + rv^2.$$

For $v \neq 0$

$$S(u,v) = v^2 \left(p\left(\frac{u}{v}\right)^2 + 2q\frac{u}{v} + r \right).$$

Let $f(t) = pt^2 + 2qt + r$. Then

$$v^2 f\left(\frac{u}{v}\right) = S(u,v).$$

Rewrite $f(t)$ by "completing the square"

$$f(t) = p\left[t^2 + 2\frac{q}{p}t + \left(\frac{q}{p}\right)^2 \right] + r - \frac{q^2}{p} = p\left(t + \frac{q}{p} \right)^2 + \left(r - \frac{q^2}{p} \right).$$

Now $f(\frac{u}{v})$ is positive, and therefore $S(u,v)$ is positive, if

$$p > 0 \quad \text{and} \quad r - \frac{q^2}{p} > 0,$$

that is $p > 0$ and $pr - q^2 > 0$. And $f(\frac{u}{v})$ is negative, and therefore $S(u,v)$ is negative, if

$$p < 0 \quad \text{and} \quad r - \frac{q^2}{p} < 0,$$

that is $p < 0$ and $pr - q^2 > 0$. The same criteria apply when $v = 0$ because

$$S(u,0) = pu^2$$

is positive or negative with p.

Now consider the case $pr - q^2 < 0$. If $p > 0$ then $\frac{pr-q^2}{p} < 0$. Pick (u,v) so that the ratio $\frac{u}{v}$ is large enough that

$$p\left(\frac{u}{v} + \frac{q}{p} \right)^2 > -\frac{pr - q^2}{p}.$$

In this case $f(\frac{u}{v}) > 0$. Then $S(u,v) > 0$.

Alternatively, pick (u,v) so that the ratio $\frac{u}{v} = -\frac{q}{p}$ then

$$f(\tfrac{u}{v}) = 0 + r - \tfrac{q^2}{p}.$$

Since $p > 0$ and $pr - q^2 < 0$, $f(\frac{u}{v}) = r - \frac{q^2}{p} < 0$ and we get

$$S(u,v) = v^2 f(\tfrac{u}{v}) < 0.$$

Thus for $p > 0$ we have shown that $S(u,v)$ is positive for some (u,v) and negative for other (u,v). If $p < 0$ use a similar argument to show that again $S(u,v)$ is positive for some (u,v) and negative for other (u,v).   □

Example 4.4. Find the local extrema of $f(x,y) = x^2 + 2x + y^2 + 2$. The gradient

$$\nabla f(x,y) = (2x+2, 2y)$$

so $\nabla f(-1,0) = (0,0)$. The Hessian matrix of f at $(-1,0)$ is

$$\mathcal{H}f(-1,0) = \begin{bmatrix} f_{xx}(-1,0) & f_{xy}(-1,0) \\ f_{yx}(-1,0) & f_{yy}(-1,0) \end{bmatrix} = \begin{bmatrix} 2 & 0 \\ 0 & 2 \end{bmatrix}.$$

Since $f_{xx}(-1,0) = 2 > 0$ and $f_{xx}(-1,0)f_{yy}(-1,0) - f_{xy}^2(-1,0) = 4 > 0$, $\mathcal{H}f(-1,0)$ is positive definite and f has a strict local minimum at $(-1,0)$. Since $\nabla f(x,y)$ exists at all points (x,y) of $\mathbb{R}^2$ and is zero only at $(-1,0)$, there is only one local extremum.   □

Example 4.5. Find the local extrema of $f(x,y) = x^2 - y^2$. The gradient

$$\nabla f(x,y) = (2x, -2y)$$

so $\nabla f(0,0) = (0,0)$. The second partial derivatives of f are

$$f_{xx}(x,y) = 2, \quad f_{yy}(x,y) = -2, \quad f_{xy}(x,y) = f_{yx}(x,y) = 0.$$

The matrix

$$\mathcal{H}f(0,0) = \begin{bmatrix} 2 & 0 \\ 0 & -2 \end{bmatrix}$$

has $f_{xx}(0,0)f_{yy}(0,0) - f_{xy}^2(0,0) = (2)(-2) - 0^2 < 0$, so f has a saddle at $(0,0)$. See Figure 4.4 for a sketch of the graph of f.   □

Example 4.6. Find the local extrema of $f(x,y) = 2x^2 - xy + y^4$. The gradient

$$\nabla f(x,y) = (4x - y, -x + 4y^3)$$

is zero when $x = \frac{1}{4}y$ and $-x + 4y^3 = 0$. These imply $-\frac{1}{4}y + 4y^3 = 0$. The solutions are $y = 0, \pm\frac{1}{4}$, and since $x = \frac{1}{4}y$ the gradient is zero at

$$(0,0), \quad (\tfrac{1}{16},\tfrac{1}{4}), \quad (-\tfrac{1}{16},-\tfrac{1}{4}).$$

The Hessian $\mathcal{H}f(x,y) = \begin{bmatrix} 4 & -1 \\ -1 & 12y^2 \end{bmatrix}$ so

$$\mathcal{H}f(0,0) = \begin{bmatrix} 4 & -1 \\ -1 & 0 \end{bmatrix}, \quad \mathcal{H}f(\tfrac{1}{16},\tfrac{1}{4}) = \begin{bmatrix} 4 & -1 \\ -1 & \tfrac{12}{16} \end{bmatrix}, \quad \mathcal{H}f(-\tfrac{1}{16},-\tfrac{1}{4}) = \begin{bmatrix} 4 & -1 \\ -1 & \tfrac{12}{16} \end{bmatrix}.$$

Since $4(0) - (-1)^2 < 0$, f has a saddle at $(0,0)$. Since 4 and $4(\tfrac{12}{16}) - (-1)^2$ are positive, f has a local minimum at $(\tfrac{1}{16},\tfrac{1}{4})$. The last two Hessians are the same, therefore f also has a local minimum at $(-\tfrac{1}{16},-\tfrac{1}{4})$. □

Example 4.7. Consider the three functions

$$f(x,y) = x^2, \ x^3, \ -x^2.$$

Each has $\nabla f(0,0) = \mathbf{0}$ and Hessian

$$\mathcal{H}f(0,0) = \begin{bmatrix} 0 & 0 \\ 0 & 0 \end{bmatrix}.$$

$\mathcal{H}f(0,0)$ is not positive or negative definite and it is not indefinite. But for $f(x,y) = x^2$, $f(0,0)$ is a local minimum; for $f(x,y) = x^3$, $f(0,0)$ is neither a maximum or a minimum; and for $f(x,y) = -x^2$, $f(0,0)$ is a local maximum. In these examples the Hessian gives no information. □

Problems

4.13. Let $f(x,y) = (x-1)^2 + 2(y-2)^2 + (y-2)^3 = 1 - 2x + x^2 + 4y - 4y^2 + y^3$.

(a) Calculate ∇f and find the two points where $\nabla f = \mathbf{0}$.
(b) Calculate the Hessian matrix $\mathcal{H}f$ of f at the two points in part (a).
(c) Determine whether f has a local maximum, minimum, or saddle at the points you found.

4.14. A symmetric 2 by 2 matrix $\mathbf{A}$ has been computed numerically with small errors as a symmetric matrix $\mathbf{S}$, and each i, j entry of $\mathbf{S}$ is within 10^{-3} of the i, j entry of $\mathbf{A}$. If $S(u,v) \geq 3 \times 10^{-2}(u^2 + v^2)$ show that $\mathbf{A}$ is positive definite.

4.15. Write the following quadratic functions in the form $\mathbf{X} \cdot \mathbf{SX}$ for some symmetric matrix $\mathbf{S}$.

(a) $3x_1^2 + 4x_1 x_2 + x_2^2$
(b) $-x_1^2 + 5x_1 x_2 + 3x_2^2$

4.16. Let $f(x,y) = x^2 + 2xy + y^3$. Determine all local extrema.

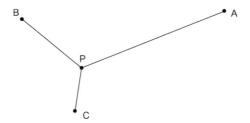

Fig. 4.5 The sum of distances from **P** to **A**, **B**, and **C** is minimized in Problem 4.18.

4.17. Let $f(x,y) = -x^3 + x^2 + xy + 3y^2$. Show that $f(0,0)$ is a local minimum.

4.18. Let $\mathbf{A} = (a_1, a_2)$, $\mathbf{B} = (b_1, b_2)$, and $\mathbf{C} = (c_1, c_2)$ be three points in $\mathbb{R}^2$, and let $f(x,y)$ be the sum of the distances from point (x,y) to **A**, **B**, and **C**.

(a) Show that if $\nabla f(\mathbf{P}) = \mathbf{0}$ then the sum of the unit vectors

$$\frac{\mathbf{P} - \mathbf{A}}{\|\mathbf{P} - \mathbf{A}\|} + \frac{\mathbf{P} - \mathbf{B}}{\|\mathbf{P} - \mathbf{B}\|} + \frac{\mathbf{P} - \mathbf{C}}{\|\mathbf{P} - \mathbf{C}\|} = \mathbf{0}.$$

(b) Show that the sum of three unit vectors is zero if and only if the angle between each pair of them is 120 degrees. See Figure 4.5.

4.19. Find a number m so that

$$x^2 + qxy + y^2$$

is positive definite when $|q| < m$.

4.20. Suppose a C^2 function $f(x,y)$ satisfies

$$f_{xx} + f_{yy} = 0.$$

(a) Suppose also that f_{xx} is not zero at any point. Show that f does not have any local maximum or minimum values.
(b) Suppose that at every point (x,y), at least one of the second partial derivatives f_{xx}, f_{xy}, or f_{yy} is not zero. Show that f does not have any local maximum or minimum values.

4.21. Let $f(x,y) = x^2 + 2y^2$ and $g(x,y) = x^4 + 2y^2$. Show that f and g attain their strict local minimums at the origin. Show that the matrix of second partial derivatives $\mathcal{H}f(0,0)$ is positive definite but $\mathcal{H}g(0,0)$ is not positive definite.

4.22. Find the point on the plane

$$z = x - 2y + 3$$

that is closest to the origin, by finding where the square of the distance between the origin and a point of the plane is a local minimum.

4.23. Confirm that the point found in Problem 4.22 is also a normal vector to the plane there, and make a sketch to show why this is the case.

4.3 Extrema of functions of several variables

Theorems developed in Section 4.2 for functions of two variables can be extended to functions of several variables.

Taylor's Theorem Recall that a function f from an open set in $\mathbb{R}^n$ to $\mathbb{R}$ is differentiable at $\mathbf{A}$ if $f(\mathbf{X})$ can be well approximated by $f(\mathbf{A}) + \nabla f(\mathbf{A}) \cdot (\mathbf{X} - \mathbf{A})$,

$$f(\mathbf{X}) = f(\mathbf{A}) + \nabla f(\mathbf{A}) \cdot (\mathbf{X} - \mathbf{A}) + R(\mathbf{X} - \mathbf{A})$$

where $\dfrac{R(\mathbf{X} - \mathbf{A})}{\|\mathbf{X} - \mathbf{A}\|}$ tends to zero as $\|\mathbf{X} - \mathbf{A}\|$ tends to zero. Let

$$p_1(\mathbf{X}) = f(\mathbf{A}) + \nabla f(\mathbf{A}) \cdot (\mathbf{X} - \mathbf{A}) = f(\mathbf{A}) + \sum_{i=1}^{n} f_{x_i}(\mathbf{A})(x_i - a_i).$$

p_1 is called the *first order Taylor approximation* of f at $\mathbf{A}$. Let $\mathbf{H} = \mathbf{X} - \mathbf{A}$, then

$$p_1(\mathbf{X}) = p_1(\mathbf{A} + \mathbf{H}) = f(\mathbf{A}) + \nabla f(\mathbf{A}) \cdot \mathbf{H} = f(\mathbf{A}) + \sum_{i=1}^{n} f_{x_i}(\mathbf{A})h_i.$$

Example 4.8. Write the first order Taylor approximation of

$$f(x,y,z) = \sin x + 2y + e^{yz}$$

at $(0,0,0)$, and use it to approximate $f(.1,.2,.01)$.
     The gradient of f is

$$\nabla f(x,y,z) = (\cos x, 2 + ze^{yz}, ye^{yz})$$

and $\nabla f(0,0,0) = (1,2,0)$. The first order Taylor approximation of f at $(0,0,0)$ is

$$p_1(\mathbf{0} + \mathbf{H}) = p_1(h_1, h_2, h_3) = f(0,0,0) + \nabla f(0,0,0) \cdot (h_1, h_2, h_3) = 1 + h_1 + 2h_2,$$

and $f(.1,.2,.01) \approx p_1(.1,.2,.01) = 1 + .1 + .4 = 1.5$.   □
     Suppose now f from $\mathbb{R}^n$ to $\mathbb{R}$ is C^3 on an open set that contains an open ball centered at $\mathbf{A}$, and let $\mathbf{X} = \mathbf{A} + \mathbf{H}$ be a point in the ball. Define the function

$$g(t) = f(\mathbf{A} + t\mathbf{H})$$

for t in an open interval that contains 0 and 1. For $0 \le t \le 1$ the points $\mathbf{A} + t\mathbf{H}$ are on a line segment from $\mathbf{A}$ to $\mathbf{A} + \mathbf{H}$, and $g(0) = f(\mathbf{A})$, $g(1) = f(\mathbf{A} + \mathbf{H})$. Since g is a C^3 function on an open interval that contains $[0,1]$, we can compute g', g'', and g''' and write the order 2 Taylor approximation of g, with the remainder. By the Chain Rule,

$$g'(t) = \nabla f(\mathbf{A} + t\mathbf{H}) \cdot \mathbf{H} = \sum_{i=1}^{n} f_{x_i}(\mathbf{A} + t\mathbf{H})h_i.$$

By the Chain Rule again

$$g''(t) = \frac{d}{dt}\left(\sum_{i=1}^{n} h_i f_{x_i}(\mathbf{A} + t\mathbf{H})\right) = \sum_{i=1}^{n} h_i \left(\sum_{j=1}^{n} f_{x_i x_j}(\mathbf{A} + t\mathbf{H})h_j\right) = \sum_{i,j=1}^{n} h_i f_{x_i x_j}(\mathbf{A} + t\mathbf{H})h_j.$$

At $t = 0$,

$$g'(0) = \sum_{i=1}^{n} f_{x_i}(\mathbf{A})h_i, \qquad g''(0) = \sum_{i,j=1}^{n} h_i f_{x_i x_j}(\mathbf{A})h_j = [h_1\ h_2\ \cdots\ h_n]\left[f_{x_i x_j}(\mathbf{A})\right]\begin{bmatrix} h_1 \\ h_2 \\ \vdots \\ h_n \end{bmatrix}$$

where $\left[f_{x_i x_j}(\mathbf{A})\right]$ is the Hessian matrix of f at $\mathbf{A}$. By the Chain Rule again

$$g'''(t) = \sum_{i,j,k=1}^{n} h_i h_j h_k f_{x_i x_j x_k}(\mathbf{A} + t\mathbf{H}).$$

According to Taylor's Theorem for functions of a single variable, there is a number θ between 0 and 1 so that

$$g(1) = g(0) + g'(0) + \tfrac{1}{2}g''(0) + \tfrac{1}{3!}g'''(\theta).$$

This expresses $f(\mathbf{A} + \mathbf{H})$ as the *order 2 Taylor approximation*, $p_2(\mathbf{A} + \mathbf{H})$, plus a remainder. Denote by $\mathbf{H}$ the column vector $\begin{bmatrix} h_1 \\ \vdots \\ h_n \end{bmatrix}$ and denote by $\mathbf{H}^T$ the row vector $(h_1, \ldots, h_n)$. Then we have

$$f(\mathbf{A} + \mathbf{H}) = f(\mathbf{A}) + \nabla f(\mathbf{A}) \cdot \mathbf{H} + \tfrac{1}{2}\mathbf{H}^T\left[f_{x_i x_j}(\mathbf{A})\right]\mathbf{H} + R_2(\mathbf{A}, \mathbf{H}) = p_2(\mathbf{A} + \mathbf{H}) + R_2(\mathbf{A}, \mathbf{H}).$$

By the triangle inequality

$$\left|R_2(\mathbf{A}, \mathbf{H})\right| = \left|\tfrac{1}{3!}\sum_{i,j,k=1}^{n} h_i h_j h_k f_{x_i x_j x_k}(\mathbf{A} + \theta\mathbf{H})\right| \le \tfrac{1}{3!}\sum_{i,j,k=1}^{n} \left|h_i h_j h_k f_{x_i x_j x_k}(\mathbf{A} + \theta\mathbf{H})\right|.$$

Each $|h_i| \le \|\mathbf{H}\|$, so we get $\left|R_2(\mathbf{A}, \mathbf{H})\right| \le \tfrac{1}{3!}\|\mathbf{H}\|^3 \sum_{i,j,k=1}^{n} \left|f_{x_i x_j x_k}(\mathbf{A} + \theta\mathbf{H})\right|.$

Since the third partial derivatives are continuous on the closed ball of radius $\|\mathbf{H}\|$ centered at $\mathbf{A}$, they are bounded. Let K be the sum of the bounds of the third partial derivatives. Then

$$\left|R_2(\mathbf{A}, \mathbf{H})\right| \le \tfrac{1}{3!}K\|\mathbf{H}\|^3 = k\|\mathbf{H}\|^3.$$

When $\mathbf{H} \neq \mathbf{0}$ then $0 \leq \dfrac{|R_2(\mathbf{A},\mathbf{H})|}{\|\mathbf{H}\|^2} \leq k\|\mathbf{H}\|$, so $\dfrac{|R_2(\mathbf{A},\mathbf{H})|}{\|\mathbf{H}\|^2}$ tends to zero as $\mathbf{H}$ tends to zero.

For C^m functions we define the *order m Taylor approximation* to f at $\mathbf{A}$ as

$$p_m(\mathbf{A}+\mathbf{H}) = f(\mathbf{A}) + \sum_{i_1=1}^{n} h_{i_1} f_{x_{i_1}}(\mathbf{A}) + \tfrac{1}{2} \sum_{i_1,i_2=1}^{n} h_{i_1} h_{i_2} f_{x_{i_1} x_{i_2}}(\mathbf{A})$$

$$+ \cdots + \tfrac{1}{m!} \sum_{i_1,\ldots,i_m=1}^{n} (h_{i_1} h_{i_2} \cdots h_{i_m}) f_{x_{i_1} x_{i_2} \cdots x_{i_m}}(\mathbf{A}).$$

An analogous argument to the one we used to prove Taylor's second order approximation can be made to prove the following theorem.

Theorem 4.7. (Taylor) *Suppose f from $\mathbb{R}^n$ to $\mathbb{R}$ is a C^{m+1} function on an open ball centered at $\mathbf{A}$. Then for $\mathbf{A}+\mathbf{H}$ in the ball,*

$$f(\mathbf{A}+\mathbf{H}) = f(\mathbf{A}) + \sum_{i_1=1}^{n} h_{i_1} f_{x_{i_1}}(\mathbf{A}) + \tfrac{1}{2} \sum_{i_1,i_2=1}^{n} h_{i_1} h_{i_2} f_{x_{i_1} x_{i_2}}(\mathbf{A})$$

$$+ \cdots + \tfrac{1}{m!} \sum_{i_1,\ldots,i_m=1}^{n} (h_{i_1} h_{i_2} \cdots h_{i_m}) f_{x_{i_1} x_{i_2} \cdots x_{i_m}}(\mathbf{A}) + R_m(\mathbf{A},\mathbf{H}),$$

where $\left|R_m(\mathbf{A},\mathbf{H})\right| \leq k\|\mathbf{H}\|^{m+1}$ for some constant k. The remainder goes to zero faster than $\|\mathbf{H}\|^m$ in the sense that

$$0 \leq \frac{\left|R_m(\mathbf{A},\mathbf{H})\right|}{\|\mathbf{H}\|^m} \leq k\|\mathbf{H}\|.$$

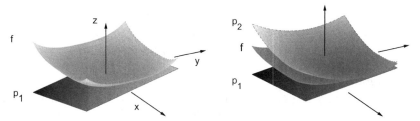

Fig. 4.6 Graphs for Example 4.9. *Left:* f and its Taylor polynomial p_1. *Right:* f, p_1, and p_2.

Example 4.9. Let $f(x,y) = x^2 + \tfrac{3}{2}y^2 + \tfrac{3}{4}(x^4 + y^4)$. Find the first and second order Taylor approximations to f at $(1,1)$.

$$f(1,1) = 4, \quad \nabla f(x,y) = (2x + 3x^3, 3y + 3y^3), \quad \nabla f(1,1) = (5,6),$$

$$\mathcal{H}f(x,y) = \begin{bmatrix} 2 + 9x^2 & 0 \\ 0 & 3 + 9y^2 \end{bmatrix}, \quad \mathcal{H}f(1,1) = \begin{bmatrix} 11 & 0 \\ 0 & 12 \end{bmatrix}.$$

So

$$p_1(1 + h_1, 1 + h_2) = f(1,1) + \nabla f(1,1) \cdot (h_1, h_2) = 4 + 5h_1 + 6h_2,$$

$$p_2(1 + h_1, 1 + h_2) = p_1(1 + h_1, 1 + h_2) + \tfrac{1}{2}[h_1 \ h_2]\mathcal{H}f(1,1)\begin{bmatrix} h_1 \\ h_2 \end{bmatrix}$$

$$= 4 + 5h_1 + 6h_2 + \tfrac{11}{2}h_1^2 + 6h_2^2.$$

We can also write p_1 and p_2 in the form

$$p_1(x,y) = 4 + 5(x-1) + 6(y-1),$$

$$p_2(x,y) = 4 + 5(x-1) + 6(y-1) + \tfrac{11}{2}(x-1)^2 + 6(y-1)^2.$$

Figure 4.6 shows graphs of f, p_1, and p_2.   □

Example 4.10. Find the second order Taylor approximation to

$$f(x,y,z) = \sin x + 2y + e^{yz}$$

at $\mathbf{A} = (0,0,0)$, and use it to estimate $f(.1,.2,.01)$. Recall we found

$$\nabla f(x,y,z) = (\cos x, 2 + ze^{yz}, ye^{yz})$$

and $\nabla f(0,0,0) = (1,2,0)$ in Example 4.8. The second partial derivatives are

$$\mathcal{H}f(x,y,z) = \begin{bmatrix} -\sin x & 0 & 0 \\ 0 & z^2 e^{yz} & (1 + yz)e^{yz} \\ 0 & (1 + yz)e^{yz} & y^2 e^{yz} \end{bmatrix}, \quad \mathcal{H}f(0,0,0) = \begin{bmatrix} 0 & 0 & 0 \\ 0 & 0 & 1 \\ 0 & 1 & 0 \end{bmatrix}.$$

The second order Taylor approximation to f at $(0,0,0)$ is

$$p_2(\mathbf{H}) = f(0) + \nabla f(0,0,0) \cdot \mathbf{H} + \tfrac{1}{2}\mathbf{H}^T \begin{bmatrix} 0 & 0 & 0 \\ 0 & 0 & 1 \\ 0 & 1 & 0 \end{bmatrix} \mathbf{H} = 1 + h_1 + 2h_2 + h_2 h_3,$$

and $p_2(.1,.2,.01) = 1 + .1 + 2(.2) + (.2)(.01) = 1.502$.   □

Extrema. Suppose f is a C^3 function from an open set in $\mathbb{R}^n$ to $\mathbb{R}$, and that the gradient of f is zero at a point $\mathbf{A}$. By Taylor's Theorem 4.7

$$f(\mathbf{A} + \mathbf{H}) = f(\mathbf{A}) + \tfrac{1}{2}\mathbf{H}^T \big[f_{x_i x_j}(\mathbf{A}) \big] \mathbf{H} + R_2(\mathbf{A}, \mathbf{H})$$

where $\big[f_{x_i x_j}(\mathbf{A}) \big]$ is the Hessian matrix of second partial derivatives of f at $\mathbf{A}$, and

$$\left| R_2(\mathbf{A}, \mathbf{H}) \right| \le k \|\mathbf{H}\|^3$$

for some number k. Since $\left[f_{x_i x_j}(\mathbf{A}) \right]$ is symmetric we can determine whether f has a strict local minimum or strict local maximum at $\mathbf{A}$ by considering the sign of $\mathbf{H}^T \left[f_{x_i x_j}(\mathbf{A}) \right] \mathbf{H}$. Suppose first that the matrix is positive definite, that is

$$\mathbf{H}^T \left[f_{x_i x_j}(\mathbf{A}) \right] \mathbf{H} \ge m \|\mathbf{H}\|^2$$

for some number $m > 0$. Then

$$f(\mathbf{A} + \mathbf{H}) - f(\mathbf{A}) = \tfrac{1}{2} \mathbf{H}^T \left[f_{x_i x_j}(\mathbf{A}) \right] \mathbf{H} + R_2(\mathbf{A}, \mathbf{H}) \ge \tfrac{1}{2} m \|\mathbf{H}\|^2 + R_2(\mathbf{A}, \mathbf{H}).$$

By Taylor's Theorem $\left| R_2(\mathbf{A}, \mathbf{H}) \right| \le k \|\mathbf{H}\|^3$, so

$$f(\mathbf{A} + \mathbf{H}) - f(\mathbf{A}) \ge \tfrac{1}{2} m \|\mathbf{H}\|^2 + R_2(\mathbf{A}, \mathbf{H}) > \tfrac{1}{2} m \|\mathbf{H}\|^2 - k \|\mathbf{H}\|^3 = m \|\mathbf{H}\|^2 (\tfrac{1}{2} - \tfrac{k}{m} \|\mathbf{H}\|).$$

Since $\tfrac{1}{2} - \tfrac{k}{m} \|\mathbf{H}\|$ is positive for $\|\mathbf{H}\|$ small enough, we see that f has a strict local minimum at $\mathbf{A}$. This proves the second derivative test, that we state as the following theorem.

Theorem 4.8. Second derivative test. *Let f be a C^3 function in an open set of $\mathbb{R}^n$ containing $\mathbf{A}$. If $\nabla f(\mathbf{A}) = \mathbf{0}$ and the Hessian matrix $\left[f_{x_i x_j}(\mathbf{A}) \right]$ is positive definite at $\mathbf{A}$, then f has a local minimum at $\mathbf{A}$.*

Looking further at our proof of Theorem 4.8, we can make an observation about the sign of the error in the approximation $f(\mathbf{A} + \mathbf{H}) \approx f(\mathbf{A}) + \nabla f(\mathbf{A}) \cdot \mathbf{H}$ in the general case where $\nabla f(\mathbf{A})$ might or might not be $\mathbf{0}$. We state this in the following theorem.

Theorem 4.9. *Let f be a C^3 function in an open set of $\mathbb{R}^n$ containing $\mathbf{A}$. If the Hessian matrix $\left[f_{x_i x_j}(\mathbf{A}) \right]$ is positive definite at $\mathbf{A}$, then the degree one Taylor approximation*

$$f(\mathbf{A} + \mathbf{H}) \approx p_1(\mathbf{A} + \mathbf{H}) = f(\mathbf{A}) + \nabla f(\mathbf{A}) \cdot \mathbf{H}$$

is an underestimate for all sufficiently small $\mathbf{H}$,

$$f(\mathbf{A} + \mathbf{H}) \ge p_1(\mathbf{A} + \mathbf{H}).$$

Proof. By Taylor's Theorem

$$f(\mathbf{A} + \mathbf{H}) - p_1(\mathbf{A} + \mathbf{H}) = \tfrac{1}{2} \mathbf{H}^T \left[f_{x_i x_j} \right] \mathbf{H} + R_2(\mathbf{A}, \mathbf{H})$$

where $|R_2(\mathbf{A}, \mathbf{H})| \le k \|\mathbf{H}\|^3$ for some $k \ge 0$. Since the Hessian is positive definite there is some $m > 0$ so that

182   4 More about differentiation

$$f(\mathbf{A} + \mathbf{H}) - p_1(\mathbf{A} + \mathbf{H}) \ge \tfrac{1}{2}m\|\mathbf{H}\|^2 - k\|\mathbf{H}\|^3 = m\|\mathbf{H}\|^2\left(\tfrac{1}{2} - \tfrac{k}{m}\|\mathbf{H}\|\right).$$

Since $\tfrac{1}{2} - \tfrac{k}{m}\|\mathbf{H}\|$ is positive for $\|\mathbf{H}\|$ small enough we see that for such $\mathbf{H}$,

$$f(\mathbf{A} + \mathbf{H}) \ge p_1(\mathbf{A} + \mathbf{H})$$

and p_1 is an underestimate for f.   □

In order to apply the second derivative test, Theorem 4.8, it is useful to have a criterion for determining whether a symmetric matrix is positive definite. We state the following result that generalizes Theorem 4.6.

Theorem 4.10. *Let* $\mathbf{M} = [m_{ij}]$ *be a symmetric matrix, $n \times n$. Suppose*

$$m_{11}, \quad \det\begin{bmatrix} m_{11} & m_{12} \\ m_{21} & m_{22} \end{bmatrix}, \quad \det\begin{bmatrix} m_{11} & m_{12} & m_{13} \\ m_{21} & m_{22} & m_{23} \\ m_{31} & m_{32} & m_{33} \end{bmatrix}, \quad \cdots, \quad \det\mathbf{M}$$

are all positive. Then $\mathbf{M}$ *is positive definite.*

For a proof of Theorem 4.10 please consult a text on matrix theory.

Example 4.11. Let $f(x,y,z) = x^2 + y^2 + z^2 + 2xyz$. At which of the points $(0,0,0), (1,1,1), (-1,-1,-1)$ does f have a strict local minimum?

$$\nabla f(x,y,z) = (2x + 2yz, 2y + 2xz, 2z + 2xy)$$

so $\nabla f(0,0,0) = (0,0,0)$, $\nabla f(1,1,1) = (4,4,4)$, $\nabla f(-1,-1,-1) = (0,0,0)$, Since $\nabla f(1,1,1) \ne \mathbf{0}$, $f(1,1,1)$ cannot be a local minimum. The matrices of second partial derivatives are

$$\mathcal{H}f(x,y,z) = \begin{bmatrix} 2 & 2z & 2y \\ 2z & 2 & 2x \\ 2y & 2x & 2 \end{bmatrix}, \quad \mathcal{H}f(0,0,0) = \begin{bmatrix} 2 & 0 & 0 \\ 0 & 2 & 0 \\ 0 & 0 & 2 \end{bmatrix}, \quad \mathcal{H}f(-1,-1.-1) = \begin{bmatrix} 2 & -2 & -2 \\ -2 & 2 & -2 \\ -2 & -2 & 2 \end{bmatrix}.$$

Since $\mathbf{U}^T\mathcal{H}f(0,0,0)\mathbf{U} = 2\|\mathbf{U}\|^2$ is positive except when $\mathbf{U} = \mathbf{0}$, $\mathcal{H}f(0,0,0)$ is positive definite and $f(0,0,0)$ is a strict local minimum. We could also use the determinant criterion to see that $\mathcal{H}f(0,0,0)$ is positive definite, since $2 > 0$, $\det\begin{bmatrix} 2 & 0 \\ 0 & 2 \end{bmatrix} = 4 > 0$, and $\det\mathcal{H}f(0,0,0) = 8 > 0$. At the point $(-1,-1,-1)$ we check for positive definiteness of $\mathcal{H}f(-1,-1,-1)$ by the determinant criterion:

$$2 > 0, \ \det\begin{bmatrix} 2 & -2 \\ 2 & 2 \end{bmatrix} = 4 - 4 = 0, \ \det\begin{bmatrix} 2 & -2 & -2 \\ -2 & 2 & -2 \\ -2 & -2 & 2 \end{bmatrix} = 2(0) + 2(-4-4) - 2(4+4) = -32.$$

Since the determinant of the 2 by 2 part is zero the theorem gives no information. If we experiment we find

$$[1\ 0\ 0]\mathcal{H}f(-1,-1,-1)\begin{bmatrix}1\\0\\0\end{bmatrix}=2, \quad [1\ 1\ 1]\mathcal{H}f(-1,-1,-1)\begin{bmatrix}1\\1\\1\end{bmatrix}=-8,$$

so the matrix is indefinite, and $f(-1,-1,-1)$ is a saddle.   □

Problems

4.24. Let $f(\mathbf{X})=\mathbf{C}\cdot\mathbf{X}$ be a linear function from $\mathbb{R}^n$ to $\mathbb{R}$, and let $\mathbf{A}$ be a point in $\mathbb{R}^n$. Show that the first and second order Taylor approximations to f at $\mathbf{A}$, p_1 and p_2, are equal to f.

4.25. Let $f(x,y,z)=e^y\log(1+x)+\sin z$.

(a) Show that f has no local extrema.
(b) Find the second order Taylor approximation $p_2(h_1,h_2,h_3)$ to f at $(0,0,0)$.

4.26. Let

$$f(x,y,z)=\frac{1}{1-xyz}.$$

Find the first order Taylor approximation to f at $\mathbf{A}=(\frac{1}{2},\frac{1}{2},\frac{1}{2})$.

4.27. Let $f(x,y,z)=x^2+xy+\frac{1}{2}y^2+2yz+z^3$.

(a) Show that ∇f is the zero vector at $\mathbf{0}$ and at one other point $\mathbf{A}$.
(b) Use Theorem 4.10 to show that f has a local minimum at $\mathbf{A}$.
(c) Find points arbitrarily near $(0,0,0)$ where f is positive and where f is negative, to show that f has a saddle at $(0,0,0)$.

4.28. Consider the set S of points in (x,y,z) space that satisfy

$$z^2=x^2+2y^2+1.$$

Let $(a,b,0)$ be a point in the $z=0$ plane. Find the local extrema of a function of (x,y) to determine the points of S that are closest to $(a,b,0)$.

4.29. Let $f(\mathbf{X})=\|\mathbf{X}\|^{-1}$ for $\mathbf{X}\neq\mathbf{0}$ in $\mathbb{R}^3$, and let $\mathbf{A}$ be a nonzero vector in $\mathbb{R}^3$.

(a) Find all first and second order partial derivatives of f.
(b) Find the second order Taylor approximation to f at $\mathbf{A}$.

4.30. Use Theorem 4.10 to show that the following matrices are positive definite.

(a) $\begin{bmatrix}2 & -1\\-1 & 1\end{bmatrix}$

(b) $\begin{bmatrix}2 & -1 & 0\\-1 & 1 & 0\\0 & 0 & 4\end{bmatrix}$

(c) $\begin{bmatrix} 2 & -1 & 0 \\ -1 & 1 & k \\ 0 & k & 6 \end{bmatrix}$, $k^2 < 3$

4.31. Let $f(x_1, x_2, x_3, x_4) = x_1 + x_2 + x_3 + x_4 + x_1 x_2 x_3 x_4$. Find the order 1 through 5 Taylor approximations p_1, p_2, p_3, p_4, p_5 for f at $\mathbf{A} = (0, 0, 0, 0)$.

4.4 Extrema on level sets

The Extreme Value Theorem 2.11 guarantees that a continuous function f from a closed and bounded set $D \subset \mathbb{R}^n$ to $\mathbb{R}$ has an absolute maximum and absolute minimum value. In Section 4.2 we saw that if a C^1 function f from $D \subset \mathbb{R}^n$ to $\mathbb{R}$ has a maximum or minimum at an *interior point* $\mathbf{A}$ then $\nabla f(\mathbf{A}) = \mathbf{0}$. This "first derivative test" gives us a way to find candidates for extrema that occur in the interior of D. In this section we develop a method for identifying candidates for extrema of a C^1 function f on a level set that has no interior points in $\mathbb{R}^n$. We state and prove the theorem for functions from $\mathbb{R}^3$ to $\mathbb{R}$ but an analogous result is true in $\mathbb{R}^2$ and in higher dimensions.

Recall that a value $f(\mathbf{P})$ is a local extremum of f on a set S, where $\mathbf{P}$ is in S, if $f(\mathbf{P}) \geq f(\mathbf{Q})$ for all $\mathbf{Q}$ in S near $\mathbf{P}$, or if $f(\mathbf{P}) \leq f(\mathbf{Q})$ for all $\mathbf{Q}$ in S near $\mathbf{P}$.

Theorem 4.11. Lagrange multiplier. *Let f and g be C^1 functions from an open set in $\mathbb{R}^3$ to $\mathbb{R}$, and denote by S a level set $g(x, y, z) = c$. Let $\mathbf{P}$ be a point of S such that*

(a) $\nabla g(\mathbf{P}) \neq \mathbf{0}$, and
(b) f has a local extremum on S at $\mathbf{P}$.

Then there is a number λ such that

$$\nabla f(\mathbf{P}) = \lambda \nabla g(\mathbf{P}).$$

Proof. By the Implicit Function Theorem it follows from (a) that there is a portion of S containing $\mathbf{P}$ where one of the variables (x, y, z) can be expressed as a function of the other two, say $g_z(\mathbf{P}) \neq 0$ and $z = \phi(x, y)$. Then

$$g(x, y, \phi(x, y)) = c.$$

Since f has a local extreme value at $\mathbf{P} = (p_1, p_2, p_3)$, the function

$$h(x, y) = f(x, y, \phi(x, y))$$

has a local extremum at the center of an open disk about (p_1, p_2). Therefore the derivatives of $h(x, y)$ with respect to x and y are zero at (p_1, p_2):

$$h_x = \frac{\partial f}{\partial x} + \frac{\partial f}{\partial z}\frac{\partial \phi}{\partial x} = 0, \qquad h_y = \frac{\partial f}{\partial y} + \frac{\partial f}{\partial z}\frac{\partial \phi}{\partial y} = 0. \tag{4.3}$$

We have by differentiating $g(x, y, \phi(x, y)) = c$ that

$$\frac{\partial g}{\partial x} + \frac{\partial g}{\partial z}\frac{\partial \phi}{\partial x} = 0, \qquad \frac{\partial g}{\partial y} + \frac{\partial g}{\partial z}\frac{\partial \phi}{\partial y} = 0. \tag{4.4}$$

Therefore, using $g_z \neq 0$,

$$\frac{\partial \phi}{\partial x} = -\frac{\frac{\partial g}{\partial x}}{\frac{\partial g}{\partial z}}, \qquad \frac{\partial \phi}{\partial y} = -\frac{\frac{\partial g}{\partial y}}{\frac{\partial g}{\partial z}}.$$

Substitute these formulas for the partial derivatives of ϕ at (p_1, p_2) into formula (4.3) for h_x and h_y. We get at **P**,

$$\frac{\partial f}{\partial x} + \frac{\partial f}{\partial z}\left(-\frac{\frac{\partial g}{\partial x}}{\frac{\partial g}{\partial z}}\right) = 0, \qquad \frac{\partial f}{\partial y} + \frac{\partial f}{\partial z}\left(-\frac{\frac{\partial g}{\partial y}}{\frac{\partial g}{\partial z}}\right) = 0.$$

Let $\lambda = \frac{\frac{\partial f}{\partial z}}{\frac{\partial g}{\partial z}}(\mathbf{P})$. Then $\nabla f(\mathbf{P}) = \lambda \nabla g(\mathbf{P})$ as asserted. $\qquad\qquad\square$

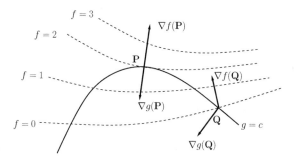

Fig. 4.7 The maximum value of f on $g = c$ is $f(\mathbf{P}) = 2$, where $\nabla f(\mathbf{P}) = \lambda \nabla g(\mathbf{P})$.

Example 4.12. Find the rectangular box of largest volume that has sides parallel to the x, y and a axes and that can be inscribed in the ellipsoid given by

$$5x^2 + 3y^2 + 7z^2 = 1.$$

We maximize the volume, $f(x, y, z) = (2x)(2y)(2z) = 8xyz$, for (x, y, z) in the level set $g(x, y, z) = 5x^2 + 3y^2 + 7z^2 - 1 = 0$, where (x, y, z) is in the first octant.

The equation $g = 0$ is often called a *constraint*. It is possible in this problem to solve the constraint $g = 0$ for z in terms of (x, y) and use the methods of Section 4.2. But let's use the Lagrange multiplier method. We solve the system of equations

$$5x^2 + 3y^2 + 7z^2 - 1 = 0, \quad \nabla f = (8yz, 8xz, 8xy) = \lambda\nabla g = \lambda(10x, 6y, 14z)$$

by considering cases. Either $\lambda = 0$ or not. If $\lambda = 0$ then at least two of the coordinates x, y, z are zero, so the volume f is zero. So assume $\lambda \neq 0$. We consider only points where x, y, z are all nonzero since we seek a maximum volume. Then we can form quotients; we find

$$\frac{8yz}{8xz} = \frac{y}{x} = \frac{\lambda(10x)}{\lambda(6y)} = \frac{5x}{3y}, \qquad \frac{8yz}{8xy} = \frac{z}{x} = \frac{\lambda(10x)}{\lambda(14z)} = \frac{5x}{7z},$$

so $y^2 = \frac{5}{3}x^2$ and $z^2 = \frac{5}{7}x^2$. The constraint $g = 0$ then gives

$$5x^2 + 3y^2 + 7z^2 - 1 = 5x^2 + 3(\tfrac{5}{3}x^2) + 7(\tfrac{5}{7}x^2) - 1 = 15x^2 - 1 = 0.$$

This gives as candidates for the maximum, the eight points with coordinates

$$x = \pm\frac{1}{\sqrt{3}\,\sqrt{5}}, \qquad y = \pm\frac{1}{\sqrt{3}\,\sqrt{3}}, \qquad z = \pm\frac{1}{\sqrt{3}\,\sqrt{7}}.$$

Since the domain of f is the first octant we take all the plus signs. The maximum volume of the box is then $f(\frac{1}{\sqrt{3}\sqrt{5}}, \frac{1}{\sqrt{3}\sqrt{3}}, \frac{1}{\sqrt{3}\sqrt{7}}) = \frac{8}{9\sqrt{35}}$ for a box that measures $\frac{2}{\sqrt{3}\sqrt{5}}$ by $\frac{2}{\sqrt{3}\sqrt{3}}$ by $\frac{2}{\sqrt{3}\sqrt{7}}$. $\qquad\square$

Example 4.13. Let $\mathbf{Q} = \begin{bmatrix} p & q \\ q & r \end{bmatrix}$ and consider the quadratic function

$$Q(\mathbf{X}) = \mathbf{X} \cdot \mathbf{QX} = [x\ y]\begin{bmatrix} p & q \\ q & r \end{bmatrix}\begin{bmatrix} x \\ y \end{bmatrix} = px^2 + 2qxy + ry^2$$

on the unit circle $\|\mathbf{X}\|^2 - 1 = x^2 + y^2 - 1 = 0$. Since Q is continuous and the circle is closed and bounded, Q has a maximum on the circle. If $\mathbf{X}$ is a point where Q is maximum then according to Theorem 4.1

$$\nabla Q(\mathbf{X}) = \lambda\nabla(\|\mathbf{X}\|^2).$$

or $(2px + 2qy, 2qx + 2ry) = \lambda(2x, 2y)$. After dividing by 2, this is neatly expressed as

$$\mathbf{QX} = \lambda\mathbf{X}.$$

Such a number λ is known as an *eigenvalue* of the matrix $\mathbf{Q}$, and vector $\mathbf{X}$ is a corresponding *eigenvector*. Dot the equation $\mathbf{QX} = \lambda\mathbf{X}$ with $\mathbf{X}$ to get

$$\mathbf{X} \cdot \mathbf{QX} = \lambda \|\mathbf{X}\|^2 = \lambda,$$

which shows that this eigenvalue is the maximum value of the quadratic function on the circle.

This example also shows that *every 2 by 2 symmetric matrix has an eigenvalue.* □

Example 4.14. Suppose there are three commodities A, B, C that have unit prices p, q, r, and are purchased in amounts x, y, z, respectively. The budget or "wealth constraint" is

$$px + qy + rz = w \tag{4.5}$$

where $w > 0$ is given. Let $U(x, y, z) = x^a y^b z^c$ be the utility function measuring consumer satisfaction from consuming x amount of A, y of B, and z of C, where a, b, and c are positive. By Theorem 4.11 the maximum of U given the wealth constraint satisfies

$$ax^{a-1} y^b z^c = \lambda p$$
$$bx^a y^{b-1} z^c = \lambda q$$
$$cx^a y^b z^{c-1} = \lambda r$$
$$w = px + qy + rz.$$

If $\lambda = 0$ then at least one of x, y or z is 0 and the utility is zero. If $\lambda \neq 0$ multiply the first three equations by $\dfrac{x}{a}, \dfrac{y}{b}, \dfrac{z}{c}$, respectively, to get

$$x^a y^b z^c = \lambda \frac{x}{a} p = \lambda \frac{y}{b} q = \lambda \frac{z}{c} r.$$

Divide by λ to get

$$\frac{x}{a} p = \frac{y}{b} q = \frac{z}{c} r. \tag{4.6}$$

So $yq = \frac{b}{a} xp$ and $zr = \frac{c}{a} xp$. Substituting this into the wealth constraint (4.5) we get

$$w = px\left(1 + \frac{b}{a} + \frac{c}{a}\right),$$

so that $px = w \dfrac{a}{a+b+c}$. The amounts the consumer spends on commodities A, B, C are according to (4.6)

$$px = \frac{aw}{a+b+c}, \quad qy = \frac{bw}{a+b+c}, \quad rz = \frac{cw}{a+b+c}$$

respectively. □

Example 4.15. Suppose there are N particles each having one of the energies

$$e_1, e_2, \ldots e_m.$$

Let x_i denote the number of particles with energy e_i. The number of ways this can be arranged is known to be

$$W = \frac{(x_1 + x_2 + \cdots + x_m)!}{x_1! x_2! \cdots x_m!}$$

since $x_1 + \cdots + x_m = N$ is the number of particles. In physics one wants to maximize W assuming constant total energy

$$g(x_1, \ldots, x_m) = e_1 x_1 + e_2 x_2 + \cdots + e_m x_m = E.$$

We maximize $\log W$ subject to the same constraint and use Stirling's approximation $\log(x!) \approx x \log x$ to approximate $\log W$ as

$$f(x_1, \ldots, x_m) = (x_1 + \cdots + x_m) \log(x_1 + \cdots + x_m) - x_1 \log x_1 - \cdots - x_m \log x_m.$$

By the n-dimensional version of Theorem 4.11 a local extreme value of f occurs at a point $\mathbf{X}$ that satisfies

$$\nabla f(\mathbf{X}) = \left(\log(x_1 + \cdots + x_m) - \log x_1, \ldots, \log(x_1 + \cdots + x_m) - \log x_m \right)$$

$$= \lambda \nabla g(\mathbf{X}) = \lambda(e_1, \ldots, e_m).$$

This gives $\log(x_1 + \cdots + x_m) - \log x_i = \lambda e_i$, or $\frac{x_i}{x_1 + \cdots + x_m} = e^{-\lambda e_i}$. In the case where $(x_1 + \cdots + x_m) = N$ we get

$$x_i = N e^{-\lambda e_i}.$$

In applications to statistical mechanics, it turns out that the multiplier λ is the reciprocal temperature!   □

Problems

4.32. In calculus we learn that the farmer with 400 meters of fencing material can maximize the area of his rectangular field by making it square, 100 meters on each side. Use Lagrange multipliers to verify that result.

4.33. Show using Theorem 4.11 that the maximum value of the function

$$Q(x, y) = 3x^2 + 2xy + 3y^2$$

on the unit circle $x^2 + y^2 = 1$ is 4.

4.34. Use the Lagrange multiplier method to find the point of the line $y = mx + b$ that is closest to $(0, 0)$.

4.35. Use the Lagrange multiplier method to find the point of the hyperplane

$$\mathbf{C} \cdot \mathbf{X} = 0$$

that is closest to a given point **A** in $\mathbb{R}^n$.

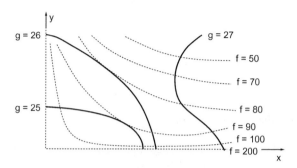

Fig. 4.8 Level sets in Problem 4.36.

4.36. Differentiable functions f and g in the first quadrant have some level sets as shown in Figure 4.8. Determine from the figure:

(a) The minimum of f on the curve $g = 25$.
(b) The sign of the Lagrange multiplier λ in (a).
(c) The minimum of g on the curve $f = 80$.
(d) The maximum of f on the curve $g = 27$. Does the equation $\nabla f = \lambda \nabla g$ apply here?

4.37. Let $\mathbf{A} = \begin{bmatrix} 1 & 2 \\ 2 & -2 \end{bmatrix}$. Express the quadratic function and constraint

$$\mathbf{X} \cdot \mathbf{A}\mathbf{X}, \qquad \mathbf{X} \cdot \mathbf{X} = 1$$

at point $\mathbf{X} = (x, y)$, in terms of x and y. Solve a Lagrange multiplier problem to find the constrained maximum of the quadratic function and express the results as an eigenvalue equation

$$\mathbf{A}\mathbf{X} = \lambda \mathbf{X}.$$

4.38. Use the Lagrange multiplier method to find the points of the curve

$$x^3 + y^2 = 1$$

that are closest to the point $(-1, 0)$. See Figure 4.9.

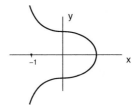

Fig. 4.9 The curve in Problem 4.38.

4.39. Use the Lagrange multiplier method to find points of the level set

$$x^3 + y^2 + z^2 = 1$$

that are closest to the point $(-1, 0, 0)$.

4.40. Let $f(x, y, z) = x^2 y^3 z^4$. Use the Lagrange multiplier method to find the maximum value of f on the plane $12x + 12y + 24z = 1$.

Chapter 5
Applications to motion

Abstract The concepts and techniques of calculus are indispensable for the description and study of dynamics, the science of motion in space under the action of forces. Both were created by Isaac Newton in the late seventeenth century and they revolutionized both mathematics and physics. In this chapter we describe the basic concepts and laws of the dynamics of point masses and deduce some of their mathematical consequences.

5.1 Motion in space

There are of course no point masses in nature—each body has a nonzero size—but in many situations a small body can be very well approximated by a point. Small here is a relative term; for instance in the study of the motion of planets around the sun the earth is regarded as a point mass.

The position of a point in three-dimensional space is described by its three Cartesian coordinates x, y, and z. These coordinates form a vector $\mathbf{X}$ in $\mathbb{R}^3$. When the point is moving, its position is described as a vector valued function of time, denoted $\mathbf{X}(t) = (x(t), y(t), z(t))$. The function $\mathbf{X}$ parametrizes the curve on which the particle moves. Informally we may speak of the motion, or curve, $\mathbf{X}(t)$.

As we saw in Section 3.1 the derivative with respect to time of the position function is the velocity of the particle at time t; it is denoted

$$\mathbf{V}(t) = \frac{d\mathbf{X}(t)}{dt} = \mathbf{X}'(t).$$

The norm of the velocity, $\|\mathbf{V}(t)\|$, is the speed of the particle.

Example 5.1. Let $\mathbf{X}(t) = (\cos t, \sin t, t)$, $-\frac{\pi}{2} \leq t \leq 2\pi$ be the position of a particle at time t. Find the position, velocity, and speed of the particle at $t = 0$ and at $t = \frac{\pi}{4}$.

$$\mathbf{V}(t) = \mathbf{X}'(t) = (-\sin t, \cos t, 1).$$

© Springer International Publishing AG 2017
P. D. Lax and M. S. Terrell, *Multivariable Calculus with Applications*,
Undergraduate Texts in Mathematics, https://doi.org/10.1007/978-3-319-74073-7_5

At $t = 0$ the particle is at $\mathbf{X}(0) = (1,0,0)$ and its velocity is $\mathbf{V}(0) = (0,1,1)$. At $t = \frac{\pi}{4}$ the particle is at $\mathbf{X}(\frac{\pi}{4}) = (\frac{1}{\sqrt{2}}, \frac{1}{\sqrt{2}}, \frac{\pi}{4})$ and $\mathbf{V}(\frac{\pi}{4}) = (-\frac{1}{\sqrt{2}}, \frac{1}{\sqrt{2}}, 1)$. The speed at time t is

$$\|\mathbf{V}(t)\| = \sqrt{(-\sin t)^2 + (\cos t)^2 + 1^2} = \sqrt{2}.$$

The particle moves with constant speed along the curve shown in Figure 5.1.
□

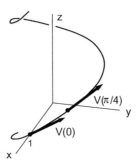

Fig. 5.1 The curve in Example 5.1 showing position and velocity at times $t = 0$ and $t = \frac{\pi}{4}$.

The derivative with respect to t of the velocity function is the *acceleration* of the particle at time t; it is denoted

$$\frac{d\mathbf{V}(t)}{dt} = \frac{d^2\mathbf{X}(t)}{dt^2} = \mathbf{X}''(t).$$

Example 5.2. Suppose a particle moves along a C^2 curve $\mathbf{X}(t)$ so that the speed is constant. We show that the velocity and acceleration are orthogonal to each other at each time t. Since the speed is constant we have $\|\mathbf{X}'(t)\| = c$, Squaring both sides we get $\|\mathbf{X}'(t)\|^2 = \mathbf{X}'(t) \cdot \mathbf{X}'(t) = c^2$. By the product rule

$$\frac{d}{dt}(\mathbf{X}'(t) \cdot \mathbf{X}'(t)) = \mathbf{X}'(t) \cdot \mathbf{X}''(t) + \mathbf{X}''(t) \cdot \mathbf{X}'(t) = 2\mathbf{X}'(t) \cdot \mathbf{X}''(t) = \frac{d(c^2)}{dt} = 0.$$

so $\mathbf{X}'(t)$ and $\mathbf{X}''(t)$ are orthogonal when the speed is constant.   □

Newtonian mechanics. The basic notions of mechanics are velocity, acceleration, mass, and force. Force is a vector quantity, denoted as $\mathbf{F}$. A force $\mathbf{F}$, acting on a particle with mass m, accelerates it according to Newton's law:

Force equals mass times acceleration.

In symbols

$$\mathbf{F} = m\frac{d^2\mathbf{X}}{dt^2}. \tag{5.1}$$

Equation (5.1) is called the equation of motion, *Newton's Law of Motion.*
Next we prove the following important result.

Theorem 5.1. *Suppose a particle moves in* $\mathbb{R}^3$ *with position* **X** *in a field of force* **F(X)** *that is parallel to* **X**, *i.e., of the form*

$$\mathbf{F(X)} = k(\mathbf{X})\mathbf{X}$$

where $k(\mathbf{X})$ *is a scalar valued function. Then the particle travels in a plane.*

Proof. First we show that for a particle moving in a force field that is parallel to **X**, the cross product of $\mathbf{X}(t)$ and $\mathbf{X}'(t)$ is a constant vector. Consider $(\mathbf{X} \times \mathbf{X}')'$. By the product rule for cross products

$$(\mathbf{X} \times \mathbf{X}')' = \mathbf{X}' \times \mathbf{X}' + \mathbf{X} \times \mathbf{X}''. \tag{5.2}$$

Since $\mathbf{F(X)} = k(\mathbf{X})\mathbf{X}$ we get from Newton's Law of Motion

$$\mathbf{X}'' = \frac{k(\mathbf{X})}{m}\mathbf{X}.$$

Set this expression into equation (5.2). Since the cross product of a vector with itself is zero we get

$$(\mathbf{X} \times \mathbf{X}')' = \mathbf{X} \times \left(\frac{k(\mathbf{X})}{m}\mathbf{X}\right) = \frac{k(\mathbf{X})}{m}\mathbf{X} \times \mathbf{X} = \mathbf{0}.$$

Therefore $\mathbf{X} \times \mathbf{X}'$ is a constant vector **C**. Since the cross product is orthogonal to each factor,

$$\mathbf{C} \cdot \mathbf{X}(t) = 0$$

for all times t. If $\mathbf{C} \neq \mathbf{0}$ the equation $\mathbf{C} \cdot \mathbf{X} = 0$ is the equation of a plane through the origin, so the motion takes place in this plane. If $\mathbf{C} = \mathbf{0}$ then $\mathbf{X}(t)$ and $\mathbf{X}'(t)$ are parallel to each other at each t and the particle moves along a line. Or $\mathbf{X}'(t) = \mathbf{0}$ and the particle does not move. In any case the particle remains in a plane. □

Problems

5.1. Find the velocity, the acceleration, and the speed of particles with the following positions.

(a) $\mathbf{X}(t) = \mathbf{C}$ constant,
(b) $\mathbf{X}(t) = (t, t, t)$,
(c) $\mathbf{X}(t) = (1 - t, 2 - t, 3 + t)$,
(d) $\mathbf{X}(t) = (t, 2t, 3t)$,
(e) $\mathbf{X}(t) = (t, t^2, t^3)$.

5.2. A particle moves on a line. Let $\mathbf{X}(t)$ be its position at time t.

(a) Suppose $\mathbf{X}(t) = \mathbf{A} + \mathbf{B}t$ for some constant vectors $\mathbf{A}$ and $\mathbf{B} \neq \mathbf{0}$. What is the location of the particle at time $t = 0$?
(b) Find the velocity and acceleration at time t of the particle in part (a).
(c) Suppose instead $\mathbf{X}(t) = \mathbf{A} + (t^3 - t)\mathbf{B}$. Find the velocity and acceleration at time t. When is the speed of the particle zero?
(d) At what times is the particle of part (c) located at $\mathbf{X}(0)$?

5.3. A particle moves on a circle of radius r in the plane according to

$$\mathbf{X}(t) = (r\cos(\omega t), r\sin(\omega t))$$

where r and ω are constant.

(a) Find the velocity and show that it is orthogonal to $\mathbf{X}(t)$, hence tangent to the circle.
(b) Show that the speed is $r\omega$.
(c) Find the acceleration and show that it is parallel to $\mathbf{X}(t)$ but directed toward the origin.
(d) Show that the magnitude of acceleration is $r\omega^2$.

5.4. A particle of unit mass moves in a gradient force field, so that

$$\mathbf{X}'' = -\nabla p(\mathbf{X})$$

for some function p. Show that the *energy*

$$\tfrac{1}{2}\|\mathbf{X}'(t)\|^2 + p(\mathbf{X}(t))$$

does not change with time.

5.5. A particle moves with nonzero velocity along a differentiable curve $\mathbf{X}(t)$ on a sphere centered at the origin, so that $\|\mathbf{X}(t)\|$ is constant. Show that its velocity is tangent to the sphere.

5.6. Give an example of a particle motion $\mathbf{X}(t) = (x(t), y(t))$, $a < t < b$, on the circle $x^2 + y^2 = 4$, such that the acceleration is not toward the origin.

5.7. Let $\mathbf{A}$, $\mathbf{B}$, $\mathbf{K}$, and $\mathbf{F}$ be constant vectors in $\mathbb{R}^3$ and let m be a number. Verify that the function

$$\mathbf{X}(t) = \mathbf{A} + \mathbf{B}t + \mathbf{K}t^2$$

satisfies the equation of motion $\mathbf{F} = m\mathbf{X}''$ when $\mathbf{K} = \frac{1}{2m}\mathbf{F}$. Find the relation of $\mathbf{A}$ and $\mathbf{B}$ to the initial position and velocity $\mathbf{X}(0)$, $\mathbf{X}'(0)$.

5.8. Verify that the following functions satisfy the indicated equations of motion in $\mathbb{R}^3$. $\mathbf{A}$ and $\mathbf{B}$ denote constant vectors.

(a) $\mathbf{X}(t) = \mathbf{A}\cos t$ is a solution of $\mathbf{X}'' = -\mathbf{X}$,
(b) $\mathbf{X}(t) = \mathbf{A}\sin(2t)$ is a solution of $\mathbf{X}'' = -4\mathbf{X}$,
(c) $\mathbf{X}(t) = \mathbf{A}\cos(3t) + \mathbf{B}\sin(3t)$ is a solution of $\frac{1}{9}\mathbf{X}'' = -\mathbf{X}$.

5.9. For each motion $\mathbf{X}(t)$ given in Problem 5.8, describe a plane (See Theorem 5.1) in which the motion occurs.

5.10. Consider the equation of motion

$$\mathbf{X}'' = \mathbf{MX},$$

where $\mathbf{M}$ is a constant matrix. Suppose there is a constant vector $\mathbf{U}$ and number ω such that

$$\mathbf{MU} = -\omega^2 \mathbf{U}.$$

Show that $\mathbf{X}(t) = \cos(\omega t)\mathbf{U}$ is one solution of the equation of motion.

Such a number $-\omega^2$ is known as an *eigenvalue* of the matrix $\mathbf{M}$ and vector $\mathbf{U}$ is a corresponding *eigenvector*.

5.11. Verify that $\mathbf{X}(t) = \mathbf{C} + \dfrac{1 - e^{-kt}}{k}\mathbf{D}$, where k is a positive constant and $\mathbf{C}, \mathbf{D}$ are constant vectors, satisfies the equation of motion

$$\mathbf{X}'' = -k\mathbf{X}'.$$

Find the relation of $\mathbf{C}$ and $\mathbf{D}$ to the initial position and velocity $\mathbf{X}(0), \mathbf{X}'(0)$. Find the total displacement of the particle

$$\lim_{t \to \infty} \mathbf{X}(t) - \mathbf{X}(0)$$

in terms of the initial velocity.

5.12. An ancient society once believed that the moon moved on a circle about the earth, being pulled across the sky by horses. In view of Newton's law, and of Problem 5.3 part (c), in which direction is the moon really being pulled?

5.13. A particle moves in $\mathbb{R}^3$ with velocity $\mathbf{V} = (v_1, v_2, v_3)$, and acceleration

$$\mathbf{V}' = \mathbf{V} \times \mathbf{B},$$

where $\mathbf{B}$ is some vector field. For example, positively charged particles follow such a rule when $\mathbf{B}$ is a magnetic field. Suppose $\mathbf{B}$ is a constant field, $\mathbf{B} = (0, 0, b)$. See Figure 5.2.

(a) Show that the components of acceleration are given by

$$\begin{aligned}
v_1' &= bv_2 \\
v_2' &= -bv_1 \\
v_3' &= 0
\end{aligned} \tag{5.3}$$

(b) Show that the position function $\mathbf{X}(t) = (a\sin(\omega t), a\cos(\omega t), bt)$ satisfies equations (5.3) and that $\mathbf{X}''(t) = \mathbf{X}'(t) \times \mathbf{B}$.

(c) Describe the direction of $\mathbf{V}'(t)$ at points along the helical curve $\mathbf{X}(t)$ given in part (b).

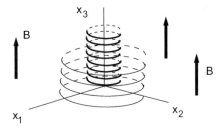

Fig. 5.2 Particle motions in Problem 5.13.

5.14. Suppose a particle of mass m with position $\mathbf{X}(t)$ at time t is joined by a spring to another particle of mass m with position $\mathbf{Y}(t)$. Then the equations of motion are

$$m\mathbf{X}'' = k(\mathbf{Y} - \mathbf{X}),$$
$$m\mathbf{Y}'' = -k(\mathbf{Y} - \mathbf{X})$$

where k is a constant depending on the strength of the spring. Denote by $\mathbf{W}$ the vector $\mathbf{W} = (\mathbf{X}, \mathbf{Y}) = (x_1, x_2, x_3, y_1, y_2, y_3)$ with 6 components. Define the function p on $\mathbb{R}^6$ by

$$p(\mathbf{W}) = p(\mathbf{X}, \mathbf{Y}) = \tfrac{1}{2}k\|\mathbf{X} - \mathbf{Y}\|^2.$$

(a) Show that the equations of motion can be expressed in the form

$$m\mathbf{W}''(t) = -\nabla p(\mathbf{W}(t))$$

where $\nabla p(\mathbf{W})$ is the vector of partial derivatives of p with respect to the 6 variables of $\mathbf{W}$.

(b) Show that the quantity

$$\tfrac{1}{2}m\|\mathbf{W}'(t)\|^2 + p(\mathbf{W}(t))$$

does not change with time.

5.2 Planetary motion

There is a large variety of force fields in nature; gravitational fields are an important case. Take in particular the motion of a particle of mass m in the gravitational field of a unit mass located at the origin.

According to the law of gravitation, the force $\mathbf{F}$ exerted by a particle of unit mass at the origin on a particle of mass m located at the point $\mathbf{X}$ is directed from $\mathbf{X}$ to the origin, and its strength is m times the reciprocal of the square of the distance of $\mathbf{X}$ from the origin times some constant. To simplify the model we take the gravitational constant to be 1. That is,

$$\mathbf{F}(\mathbf{X}) = -m \frac{1}{\|\mathbf{X}\|^2} \frac{\mathbf{X}}{\|\mathbf{X}\|} = -m \frac{\mathbf{X}}{\|\mathbf{X}\|^3}$$

For this choice of force Newton's law of motion reads

$$m\mathbf{X}'' = -m \frac{\mathbf{X}}{\|\mathbf{X}\|^3}. \tag{5.4}$$

Notice that mass m occurs on both sides of (5.4) as a factor; dividing both sides by m gives an equation of motion

$$\mathbf{X}'' = -\frac{\mathbf{X}}{\|\mathbf{X}\|^3}. \tag{5.5}$$

That the motion under the force of gravity does not depend on the mass of the moving object is intuitively clear, for if we imagine the moving particle cut into two parts, the trajectory of each part is the same, although the two parts may have different masses. So from now on we take the mass m of the moving particle to be 1.

Next we observe that the term on the right side of equation (5.5) is a gradient:

$$-\frac{\mathbf{X}}{\|\mathbf{X}\|^3} = \nabla p, \quad \text{where} \quad p(\mathbf{X}) = \frac{1}{\|\mathbf{X}\|}. \tag{5.6}$$

So we can rewrite equation (5.5) as

$$\mathbf{X}'' = \nabla p(\mathbf{X}). \tag{5.7}$$

Conservation of energy. We turn now to studying the motion of a particle of unit mass in a force field that is the gradient of a differentiable function p. We have seen that the gravitational field of a point mass is such a field. Since the sum of gradients is a gradient, it follows that the gravitational field of any distribution of point masses is a gradient field.

We derive now an important property of solutions of equations of the form (5.7) where p is a differentiable function. Take the dot product of both sides with $\mathbf{X}'$; we get

$$\mathbf{X}'' \cdot \mathbf{X}' = \nabla p(\mathbf{X}) \cdot \mathbf{X}'.$$

The left side is the derivative with respect to t of $\frac{1}{2}\mathbf{X}' \cdot \mathbf{X}'$. The right side is, by the Chain Rule, the t derivative of $p(\mathbf{X}(t))$. Since their derivatives are equal, $\frac{1}{2}\mathbf{X}' \cdot \mathbf{X}'$ and $p(\mathbf{X})$ differ by a constant E independent of t:

$$\frac{1}{2}\mathbf{X}' \cdot \mathbf{X}' - p(\mathbf{X}) = E. \tag{5.8}$$

The first term $\frac{1}{2}\mathbf{X}' \cdot \mathbf{X}'$ in (5.8) is called the *kinetic energy* of the moving particle; the second term $-p(\mathbf{X})$ is called the *potential energy;* their sum is the total energy of the moving particle. Equation (5.8) says that the total energy of the particle remains the same during motion. This is the Law of *Conservation of Energy*.

Thus we have shown that for the motion of a point mass in a gradient force field, conservation of energy is a consequence of Newton's law of motion.

Kepler's Laws. We return to equation (5.5), governing the motion of a particle under the gravitational force of a unit mass at the origin.

Since by Theorem 5.1 the motion occurs in a plane through the origin, it simplifies our calculations to choose coordinates in which that plane is the x, y plane. Therefore we set $z = 0$ in equations (5.5), giving

$$x'' + \frac{x}{r^3} = 0, \quad y'' + \frac{y}{r^3} = 0, \qquad r = \sqrt{x^2 + y^2}. \tag{5.9}$$

The energy equation (5.8) can be written as

$$\tfrac{1}{2}\left((x')^2 + (y')^2\right) - \frac{1}{\sqrt{x^2 + y^2}} = E. \tag{5.10}$$

where E is constant. We obtain another relation between x, y, x', and y' by multiplying the first equation in (5.9) by y, the second equation by x and taking the difference of the two; we get

$$xy'' - yx'' = 0.$$

Since $xy'' - yx''$ is the t derivative of $xy' - yx'$, we conclude that $xy' - yx'$ is a constant, call it A:

$$xy' - yx' = A. \tag{5.11}$$

We introduce polar coordinates:

$$x = r\cos\phi, \quad y = r\sin\phi. \tag{5.12}$$

Differentiating these equations with respect to t gives

$$x' = r'\cos\phi - r\phi'\sin\phi, \quad y' = r'\sin\phi + r\phi'\cos\phi. \tag{5.13}$$

A brief calculation using formulas (5.13) shows that

$$(x')^2 + (y')^2 = (r')^2 + r^2(\phi')^2.$$

Setting formulas (5.12), (5.13) into equation (5.11) gives

$$xy' - yx' = r^2\phi' = A.$$

So we can express ϕ' as

$$\phi' = \frac{A}{r^2}. \tag{5.14}$$

When A is zero, ϕ is a constant, so the motion takes place along a straight line. Such a one-dimensional motion is uninteresting, so we take the case that A is nonzero. In fact we may assume $A > 0$ (see Problem 5.19). The energy equation (5.10) becomes

$$(r')^2 + r^2(\phi')^2 - \frac{2}{r} = 2E.$$

Using (5.14) for ϕ' in the equation above we get

$$(r')^2 + \frac{A^2}{r^2} - \frac{2}{r} = 2E. \tag{5.15}$$

Both r and ϕ are functions of t. It follows from (5.14) that ϕ' is of one sign; it follows that ϕ is a monotonic function of t. Therefore we can express r as a function of ϕ. By the Chain Rule

$$r' = \frac{dr}{dt} = \frac{dr}{d\phi}\phi'.$$

Using (5.14) to express ϕ' as $\frac{A}{r^2}$ we get

$$r' = \frac{dr}{d\phi}\frac{A}{r^2}.$$

Set this expression of r' into the left side of equation (5.15); we obtain

$$\left(\frac{dr}{d\phi}\right)^2 \frac{A^2}{r^4} + \frac{A^2}{r^2} - \frac{2}{r} = 2E. \tag{5.16}$$

Since A is not zero, we can multiply this equation by $\frac{r^4}{A^2}$. We get

$$\left(\frac{dr}{d\phi}\right)^2 + r^2 - \frac{2r^3}{A^2} = 2E\frac{r^4}{A^2}. \tag{5.17}$$

We introduce the abbreviations

$$a = \frac{2E}{A^2}, \qquad b = \frac{1}{A^2} \tag{5.18}$$

and rewrite the equation (5.17) as

$$\left(\frac{dr}{d\phi}\right)^2 = ar^4 + 2br^3 - r^2. \tag{5.19}$$

The derivatives $\frac{dr}{d\phi}$ and $\frac{d\phi}{dr}$ are reciprocals of each other. (See Problem 5.20.) Taking the reciprocal of both sides of (5.19) we get

$$\left(\frac{d\phi}{dr}\right)^2 = \frac{1}{ar^4 + 2br^3 - r^2}. \tag{5.20}$$

We introduce $u = \dfrac{1}{r}$ as new variable. By the Chain Rule

$$\frac{d\phi}{dr} = \frac{d\phi}{du}\frac{du}{dr} = \frac{d\phi}{du}\left(-\frac{1}{r^2}\right).$$

Use this to express the left side of (5.20), and multiply both sides by r^4; we get

$$\left(\frac{d\phi}{du}\right)^2 = \frac{r^4}{ar^4 + 2br^3 - r^2} = \frac{1}{a + 2bu - u^2}.$$

Taking the square root we get

$$\frac{d\phi}{du} = \frac{1}{\sqrt{a + 2bu - u^2}}. \tag{5.21}$$

To determine ϕ as a function of u we have to find the integral of the function of u on the right in (5.21). We recall that the derivative of the inverse sine function is

$$\frac{d}{dy}\sin^{-1}y = \frac{1}{\sqrt{1 - y^2}}.$$

Set $y = p + qu$, p and q constants, $q > 0$. We get

$$\frac{d}{du}(\sin^{-1}(p + qu)) = \frac{q}{\sqrt{1 - (p + qu)^2}} = \frac{1}{\sqrt{a + 2bu - u^2}},$$

where a and b are

$$a = \frac{1 - p^2}{q^2}, \qquad b = -\frac{p}{q}. \tag{5.22}$$

This shows that we may take $\phi(u)$ in (5.21) to be

$$\phi(u) = \sin^{-1}(p + qu),$$

where p and q are related to a and b by (5.22). Therefore

$$\sin\phi = p + qu.$$

We omit a brief calculation with (5.22) that confirms q is positive except perhaps for circular orbits. We guide you in Problem 5.15 to discover all circular orbits.

We recall that u is $\dfrac{1}{r}$; multiplying the equation above by r we get

$$r\sin\phi = pr + q. \tag{5.23}$$

We claim that this is the equation of a conic in polar coordinates. The equation is easily rewritten in Cartesian coordinates x and y:

$$y = p \sqrt{x^2 + y^2} + q. \tag{5.24}$$

Hence

$$p^2(x^2 + y^2) = (y - q)^2. \tag{5.25}$$

It follows from equation (5.18) that b is positive. By equation (5.22), $b = -\dfrac{p}{q}$, therefore p is negative, in particular p is nonzero. This shows that equation (5.25) is quadratic in x and y, hence its zero set is a conic.

In Problems 5.21 and 5.22 we ask you to discover that the conic is an ellipse when $p < -1$, parabola when $p = -1$, and hyperbola when $-1 < p < 0$. Figure 7.20 is a sketch of the elliptical orbit of the moon around the earth.

Kepler based his laws on observations of the planets. His first law states that the orbits of planets in the solar system are conics with one of the foci at the sun. The calculation above shows that Kepler's first law is a consequence of the inverse square law of gravitational force. Several of the Problems discuss Kepler's other laws.

Newton showed that all of Kepler's laws are consequences of the inverse square law of gravitational force. This striking result led to the universal acceptance of the inverse square law of gravitation.

Problems

5.15. Consider the circular orbit defined by

$$\mathbf{X}(t) = (x(t), y(t)) = (a \cos \omega t, a \sin \omega t)$$

where a and ω are constants.

(a) Show that the radius function $r = \sqrt{x^2 + y^2}$ is the constant a.
(b) Show that if $\omega^2 a^3 = 1$ then $\mathbf{X}$ is a solution of equations (5.9):

$$x'' + \frac{x}{(x^2 + y^2)^{3/2}} = 0, \quad y'' + \frac{y}{(x^2 + y^2)^{3/2}} = 0.$$

(c) Show that the constant A defined in equation (5.11) is equal to ωa^2.

Remark. The constancy of $\omega^2 a^3$ is one case of Kepler's laws shown in this problem for circular orbits, and generally in Problem 5.21. An example is the prediction of the orbit radius of Saturn:

$$\frac{(1 \text{ Earth orbit radius})^3}{(1 \text{ Earth year})^2} = \frac{(a \text{ Earth orbit radii})^3}{(29.5 \text{ Earth year})^2}$$

that gives $a_{\text{Saturn}} = 9.5$, so that Saturn is 9.5 times as far from the sun as Earth is.

5.16. We ask you to verify that the intersection of a circular cone with a plane is an ellipse. There are two spheres tangent to the cone and the plane, see Figure 5.3. Points A, B, D, and F are colinear. The points of tangency C and E of the spheres with the plane are the foci of the ellipse. To verify this show the following propositions.

(a) Segments BD and DC have the same length.
(b) Segments FD and DE have the same length.
(c) The sum of the lengths of BD and FD is the same for every point D of the intersection.
(d) The sum of the lengths of DC and DE is the same for every point D of the intersection.

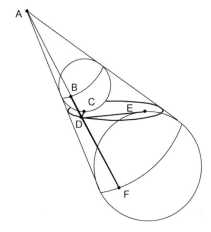

Fig. 5.3 A cone, two spheres, and an ellipse in Problem 5.16.

5.17. One of Kepler's laws is that a planet "sweeps" equal areas in equal times, that is, the rate of change of area suggested in Figure 5.4 is independent of the time. Explain the following items that prove this.

(a) If the planet is located at $\mathbf{U} = (x(t), y(t))$ at time t, then its location at time $t + h$ is approximately $\mathbf{W} = (x(t) + x'(t)h, y(t) + y'(t)h)$.
(b) The signed area of the ordered triangle $\mathbf{0UW}$ is $\frac{h}{2}(xy' - yx')$.
(c) The rate of change of area is $\frac{1}{2}A$, where A is the constant in equation (5.11).
(d) If the orbit is a closed loop, deduce from part (c) that the area enclosed is $\frac{1}{2}A$ times the period T of the orbit.

5.18. Newton's law (5.4), $m\mathbf{X}'' = -m\dfrac{\mathbf{X}}{\|\mathbf{X}\|^3}$, is denoted in physics texts as

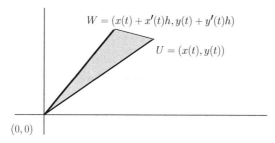

$W = (x(t) + x'(t)h, y(t) + y'(t)h)$

$U = (x(t), y(t))$

$(0,0)$

Fig. 5.4 Approximate area swept by a planet in time h, for Kepler's law in Problem 5.17.

$$mr'' = -mMG\frac{\mathbf{r}}{|\mathbf{r}|^3}.$$

Define $\mathbf{X}(t) = \mathbf{r}(kt)$, where the constant k is to be chosen. Use the Chain Rule to find k so that the equation becomes (5.4), that is, we can rescale time to eliminate MG.

5.19. Suppose $(x,y) = (f(t), g(t))$ is a solution of the equations of motion (5.9) with $xy' - yx'$ negative. Show that

$$(x,y) = (-f(t), g(t))$$

defines another solution with $xy' - yx'$ positive.

5.20. Use equation (5.16) to show that there are at most two values of r at which $\dfrac{dr}{d\phi}$ could be zero. Deduce that away from those two values we can solve for ϕ as a function of r and there $\dfrac{dr}{d\phi}$ and $\dfrac{d\phi}{dr}$ are reciprocals.

5.21. Writing the orbit curve (5.24) in the form $-pr = q - y$ and taking the case $p < -1$, prove the following statements.

(a) The motion occurs in the region where $y < q$.
(b) There are two points of the orbit located on the y axis. Deduce that this is an elliptical orbit.
(c) Show that the semimajor axis is $\dfrac{-pq}{p^2-1}$ and the semiminor is $\dfrac{q}{\sqrt{p^2-1}}$.
(d) Deduce that the semiminor axis is A times the square root of the semimajor axis.
(e) The area of an ellipse is π times the product of the semimajor and semiminor axes. Deduce from part (d) and from part (d) of Problem 5.17 that

$$\tfrac{1}{2}AT = \pi A(\text{semimajor axis})^{3/2}.$$

Hence deduce Kepler's *third law*: the square of the period of the orbit is proportional to the cube of the semimajor axis.

5.22. Writing the orbit curve (5.24) in the form $-pr = q - y$ prove the following statements.

(a) The motion occurs in the region where $y < q$.
(b) If $p = -1$ the orbit is a parabola.
(c) If $-1 < p < 0$ the orbit is half a hyperbola.

5.23. We've seen in equation (5.14) that motion toward a gravitating mass is possible along a line, and in fact that is the only trajectory on which a collision can occur (with a *point* mass). Consider now an inverse fifth power law,

$$\mathbf{X}'' = -\|\mathbf{X}\|^{-6}\mathbf{X}.$$

(a) Show that a potential function is $\frac{1}{4}\|\mathbf{X}\|^{-4}$, so that $\frac{1}{2}\|\mathbf{X}'\|^2 - \frac{1}{4}\|\mathbf{X}\|^{-4}$ is conserved.
(b) Let a be a positive number. Suppose a semicircular trajectory is expressed as

$$\mathbf{X}(t) = (a + a\cos\theta(t), a\sin\theta(t)).$$

Show that $\dfrac{1}{2}a^2(\theta')^2 - \dfrac{1}{16a^4(1+\cos\theta)^2}$ is constant.

(c) Take the case of zero constant in part (b), and θ' positive. Find the constant k such that

$$\theta'(t) = \frac{k}{1 + \cos\theta(t)}.$$

(d) Multiply the equation above by $1 + \cos\theta$ and integrate with respect to t. Assuming $\theta(0) = 0$, deduce that $\theta(t) + \sin\theta(t) = kt$. Sketch a graph of $f(\theta) = \theta + \sin\theta$ as a function of θ to show that there is a well-defined function $\theta(t)$, $0 \le t \le \frac{\pi}{k}$ with this property.
(e) Show that there are trajectories that are not straight lines that tend to the origin in finite time.

5.24. Suppose N masses $m_1, \ldots, m_N$ are attracted to each other by gravity and have position functions $\mathbf{X}_1(t), \ldots, \mathbf{X}_N(t)$. Newton's laws give

$$\mathbf{X}_k'' = -\sum_{j \ne k,\; j=1}^{N} m_j G \frac{\mathbf{X}_k - \mathbf{X}_j}{\|\mathbf{X}_k - \mathbf{X}_j\|^3}.$$

The center of mass of the system is $\mathbf{C}(t) = \dfrac{\sum_{k=1}^{N} m_k \mathbf{X}_k(t)}{\sum_{j=1}^{N} m_j}$. Show that the acceleration of the center of mass is zero: $\mathbf{C}''(t) = \mathbf{0}$.

5.25. Suppose $\mathbf{X}(t)$ is a solution of Newton's law $\mathbf{X}'' = -\|\mathbf{X}\|^{-3}\mathbf{X}$. Set $\mathbf{Y}(t) = a\mathbf{X}(bt)$, where a and b are nonzero numbers. Show that $\mathbf{Y}$ is also a solution of the same equation, that is, $\mathbf{Y}'' = -\|\mathbf{Y}\|^{-3}\mathbf{Y}$, provided that $a^3b^2 = 1$. (This includes the case $(a,b) = (1,-1)$, where the solution is run backward, $\mathbf{X}(-t)$.)

5.26. Let $\mathbf{X}_j$ be solutions of the system of Newton's laws in Problem 5.24, and set $\mathbf{Y}_j(t) = a\mathbf{X}_j(bt)$, where a and b are nonzero numbers. Show that the $\mathbf{Y}_j$ are also solutions of the same system provided that $a^3b^2 = 1$.

Chapter 6
Integration

Abstract In this chapter we introduce the concept of the multiple integral—the precise mathematical expression for finding the total amount of a quantity in a region in the plane or in space. Examples include area, volume, the total mass of a body, the total electrical charge in a region, or total population of a country.

6.1 Introduction to area, volume, and integral

Examples of integrals We introduce the concept of the integral of a function of two or more variables through two problems.

Mass. Let D be a set in space. Let $f(x,y,z)$ [mass/volume] be the density at the point (x,y,z) of some material distributed in D. How can we find the total mass $M(f,D)$ contained in D?

If the density is between lower and upper bounds ℓ and u then the total mass is between the bounds

$$\ell \operatorname{Vol}(D) \leq M(f,D) \leq u \operatorname{Vol}(D)$$

where $\operatorname{Vol}(D)$ denotes the volume of D. This estimate for the mass in D is a good start. But perhaps we can do better. Split D into two subsets D_1 and D_2. On each one f has a lower and upper bound,

$$\ell_1 \leq f(x,y,z) \leq u_1 \quad \text{in } D_1,$$

$$\ell_2 \leq f(x,y,z) \leq u_2 \quad \text{in } D_2.$$

Thus we have

$$\ell_1 \operatorname{Vol}(D_1) \leq M(f,D_1) \leq u_1 \operatorname{Vol}(D_1),$$
$$\ell_2 \operatorname{Vol}(D_2) \leq M(f,D_2) \leq u_2 \operatorname{Vol}(D_2).$$

P. D. Lax and M. S. Terrell, *Multivariable Calculus with Applications*,
Undergraduate Texts in Mathematics, https://doi.org/10.1007/978-3-319-74073-7_6

Adding these inequalities we get that

$$\ell_1 \operatorname{Vol}(D_1) + \ell_2 \operatorname{Vol}(D_2)$$
$$\leq M(f,D_1) + M(f,D_2)$$
$$\leq u_1 \operatorname{Vol}(D_1) + u_2 \operatorname{Vol}(D_2).$$

The sum of the masses in D_1 and D_2 is the mass in D, $M(f,D)$. The upper bounds for the density on D_1 and D_2 are less than or equal to u:

$$u_1 \leq u, \qquad u_2 \leq u.$$

The lower bounds are greater than or equal to ℓ:

$$\ell \leq \ell_1, \qquad \ell \leq \ell_2.$$

Putting this all together we get that the total mass $M(f,D)$ satisfies

$$\ell \operatorname{Vol}(D) \leq \ell_1 \operatorname{Vol}(D_1) + \ell_2 \operatorname{Vol}(D_2)$$
$$\leq M(f,D)$$
$$\leq u_1 \operatorname{Vol}(D_1) + u_2 \operatorname{Vol}(D_2) \leq u \operatorname{Vol}(D).$$

By subdividing D into n nonoverlapping subsets $D_1, \dots, D_n$ we can repeat this process and get a sequence of inequalities that get us closer to the value of the total mass in D,

$$\sum_{j=1}^{n} \ell_j \operatorname{Vol}(D_j) \leq M(f,D) \leq \sum_{j=1}^{n} u_j \operatorname{Vol}(D_j).$$

Population. In a similar manner let D be a set in the plane, say the map of a country. Suppose the population density at (x,y) is $f(x,y)$ [population/area] and that on D, the population density is between ℓ and u,

$$\ell \leq f(x,y) \leq u.$$

If we know the area of D we can estimate the total population $P(f,D)$ in D as

$$\ell \operatorname{Area}(D) \leq P(f,D) \leq u \operatorname{Area}(D).$$

Using an approach like we did for mass, we refine the estimate by splitting D into nonoverlapping subsets D_1 and D_2. Let ℓ_1 and ℓ_2 be lower bounds, and u_1 and u_2 upper bounds, for the population density in D_1 and D_2, respectively. Then

$$\ell_1 \operatorname{Area}(D_1) + \ell_2 \operatorname{Area}(D_2) \leq P(f,D) \leq u_1 \operatorname{Area}(D_1) + u_2 \operatorname{Area}(D_2).$$

Splitting D into n nonoverlapping subsets we get that the total population in D satisfies

$$\sum_{j=1}^{n} \ell_j \operatorname{Area}(D_j) \leq P(f,D) \leq \sum_{j=1}^{n} u_j \operatorname{Area}(D_j).$$

These two examples raise two important questions.

- How do we find the area of a subset of the plane or the volume of a subset in space?
- What properties of f and D assure that the process of taking upper and lower estimates squeezes in on a single number?

That number, if it exists, is what we call the integral of f over D.

First we look at the question of area and volume.

Area Our discussion of area is based on three basic properties of area in the plane. We shall explain later that what is said here about area applies, with appropriate changes, to volume in $\mathbb{R}^3$, as well as n-dimensional volume in $\mathbb{R}^n$.

(a) If two sets C and D in the plane have area, and C is contained in D, then the area of C is less than or equal to the area of D.
(b) If two sets C and D in the plane have area and have only boundary points in common then the union of C and D has area that is the sum of the areas of C and D.
(c) The area of a rectangle whose edges are parallel to the x and y axes is the product of the lengths of the edges of the rectangle. It doesn't matter whether all, some, or none of the boundary points are included in the rectangle.

To find out whether a subset D of the plane has area we introduce the notions of *lower area* and *upper area* of D. Take $h > 0$. Divide the whole plane into squares of edge length h by the lines $x = kh$ and $y = mh$, where k and m are integers. The boundary of each h-square is included in the h-square.

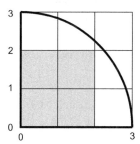 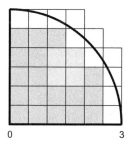

Fig. 6.1 D is the quarter disk including its boundary. There are four 1-squares in D, and twenty-two $\frac{1}{2}$-squares filling more area.

For each bounded set D, let $N_L(D,h)$ be the number of h-squares that are contained in D. Let $A_L(D,h)$ be the area of all those squares,

$$A_L(D,h) = h^2 N_L(D,h).$$

The squares of edge length $\frac{1}{2}h$ that are contained in D fill out all the squares of edge length h that are contained in D and possibly more. See Figure 6.1. Therefore their total area is greater than or equal to the total area of squares of edge length h:

$$A_L(D, \tfrac{1}{2}h) \geq A_L(D, h). \tag{6.1}$$

Setting $h = \frac{1}{2^n}$ in inequality (6.1) shows that the sequence

$$A_L(D, \tfrac{1}{2^n}) = \left(\tfrac{1}{2^n}\right)^2 N_L(D, \tfrac{1}{2^n}) \qquad n = 1, 2, 3, \dots \tag{6.2}$$

is an increasing sequence. Since D is bounded, the total area of h-squares contained in D is less than the area of a large square containing the set D. Therefore (6.2) is a bounded sequence. We remind you of the Monotone Convergence Theorem, that states that a bounded increasing sequence has a limit. Call this limit the lower area of D and denote it as $A_L(D)$:

$$A_L(D) = \lim_{n \to \infty} A_L(D, \tfrac{1}{2^n}).$$

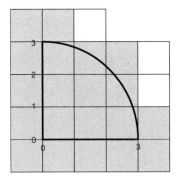 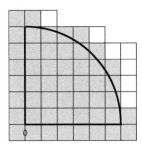

Fig. 6.2 D is the quarter disk including its boundary. Twenty 1-squares intersect D, and fifty $\frac{1}{2}$-squares intersect D, covering a smaller area than the 1-squares.

We next define the upper area. Denote by $N_U(D, h)$ the number of all h-squares that intersect D. Denote by $A_U(D, h)$ the area of the union of these squares:

$$A_U(D, h) = h^2 N_U(D, h).$$

If a square of edge length $\frac{1}{2}h$ intersects D, so does the square of edge length h that contains it. Therefore

$$A_U(D, \tfrac{1}{2}h) \leq A_U(D, h). \tag{6.3}$$

This shows that the sequence

$$A_U(D, \tfrac{1}{2^n}) = \left(\tfrac{1}{2^n}\right)^2 N_U(D, \tfrac{1}{2^n}) \qquad n = 1, 2, 3, \dots$$

is decreasing. Again by the Monotone Convergence Theorem a decreasing sequence of nonnegative numbers has a limit; call this limit the *upper area* of D, and denote it as $A_U(D)$:

$$A_U(D) = \lim_{n \to \infty} A_U(D, \tfrac{1}{2^n}). \tag{6.4}$$

Since fewer h-squares are contained in D than h-squares that intersect D, $A_L(D,h)$ is less than or equal to $A_U(D,h)$. See Figures 6.1 and 6.2. It follows that so are their limits:

$$A_L(D) \le A_U(D). \tag{6.5}$$

In words: the lower area is less than or equal to the upper area. We define now the concept of area.

Definition 6.1. If the lower area of a set D is equal to its upper area, this common value is called the Area(D), and we say that D *has area*.

Next we verify that area so defined has the three properties we listed at the start of this section.

(a) *If two sets C and D in the plane have area, and C is contained in D, then the area of C is less than, or equal, to the area of D.*
   Since C is contained in D, for each h, $A_L(C,h) \le A_L(D,h)$. So the lower area $A_L(C) \le A_L(D)$. Since C and D have area,

$$A_L(C) = \text{Area}(C), \qquad A_L(D) = \text{Area}(D),$$

   so Area$(C) \le$ Area(D).

(b) *If two sets C and D in the plane have area and have only boundary points in common then the union of C and D has area that is the sum of the areas of C and D.*
   Since C and D have only boundary points in common every h-square in C or D will be an h-square in the union $C \cup D$. However, $C \cup D$ may contain more h-squares if C and D share part of their boundaries. So

$$A_L(C,h) + A_L(D,h) \le A_L(C \cup D,h).$$

Similarly the number of h-squares that intersect C plus the number that intersect D is greater than or equal to the number of h-squares that intersect $C \cup D$ since some h-squares intersect both. Therefore

$$A_U(C \cup D,h) \le A_U(C,h) + A_U(D,h).$$

Let $h = \tfrac{1}{2^n}$ tend to zero in the inequalities

$$A_L(C,h) + A_L(D,h) \le A_L(C \cup D,h) \le A_U(C \cup D,h) \le A_U(C,h) + A_U(D,h).$$

The right side tends to Area(C) + Area(D), as does the left. Therefore the two center terms tend to a common limit, that is by definition Area$(C \cup D)$.

(c) *The area of a rectangle whose edges are parallel to the x and y axes is the product of the lengths of the edges of the rectangle. It doesn't matter if all, some, or none of the boundary points are included in the rectangle.*
Let R be the rectangle $[a,b] \times [c,d]$. The number of h-squares that meet the boundary of R does not exceed twice the perimeter divided by h (when h is smaller than the sides of R). Therefore

$$0 \le (b-a)(d-c) - A_L(R,h) \le \frac{4((b-a)+(d-c))}{h}h^2$$

and

$$0 \le A_U(R,h) - (b-a)(d-c) \le \frac{4((b-a)+(d-c))}{h}h^2.$$

Letting h tend to zero we see that the lower and upper areas of R are both equal to $(b-a)(d-c)$. A similar argument works whether the rectangle R contains some or none of its boundary.

Smoothly bounded sets. We show now that all sets that we intuitively think of having area—smoothly bounded geometric figures—have area in this sense.

Definition 6.2. A *smoothly bounded* set D in the plane is a closed bounded set whose boundary is the union of a finite number of curves each of which is the graph of a continuously differentiable function, either

$$y = f(x), \qquad x \text{ in some closed interval,}$$

or

$$x = f(y), \qquad y \text{ in some closed interval.}$$

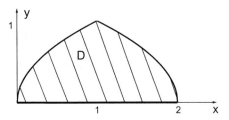

Fig. 6.3 The smoothly bounded set D in Example 6.1.

Example 6.1. The set D shown in Figure 6.3 is a smoothly bounded set. The boundary, included in D, is the union of three curves that are graphs of continuously differentiable functions:

$$x = f_1(y) = y^2 \qquad (0 \le y \le 1),$$

$$x = f_2(y) = 2 - y^2 \qquad (0 \le y \le 1),$$

and

$$y = f_3(x) = 0 \qquad (0 \le x \le 2).$$

$\square$

Example 6.2. The boundary of the set D shown in Figure 6.4 is the union of curves that are graphs of continuously differentiable functions. D is a smoothly bounded set. $\square$

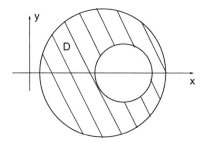

Fig. 6.4 The smoothly bounded set D in Example 6.2.

Theorem 6.1. *If D is a smoothly bounded set then its upper and lower areas are equal, so D has area.*

The key result needed to prove Theorem 6.1 is the following result.

Theorem 6.2. *Let C be the boundary of a smoothly bounded set D in $\mathbb{R}^2$. Denote by $C(h)$ the number of squares of edge length h that intersect C. Then*

$$C(h) \le \frac{c}{h} \qquad (6.6)$$

where c is some number that depends on C.

Proof. D is smoothly bounded so its boundary, C, is the union of a finite number of graphs of continuously differentiable functions. To prove the theorem we prove inequality (6.6) for the graph of a single continuously differentiable function

$$y = g(x) \qquad a \le x \le b$$

and add the inequalities. Denote by m an upper bound for the derivative of g on $[a, b]$,

$$|g'(x)| \le m.$$

The number of h-squares that intersect the graph of g in the strip

$$nh \leq x \leq (n+1)h$$

is at most $m+2$. Therefore the total number of h-squares that intersect the graph of g on the interval $[a,b]$ is at most $m+2$ times the number of h-intervals in $[a,b]$. The number of h-intervals in $[a,b]$ is less than or equal to $\frac{b-a}{h}$. See Figure 6.5. Therefore

$$C(h) \leq (m+2)\frac{b-a}{h} = \frac{(m+2)(b-a)}{h}.$$

This proves that we can take $c = (m+2)(b-a)$ in inequality (6.6) for the part of the boundary that consists of just one smooth graph. If the boundary of D consists of a finite number of smooth graphs, the estimate (6.6) follows by adding the coefficients c for each part. This completes the proof of Theorem 6.2.   □

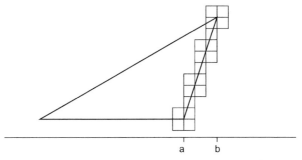

Fig. 6.5 On $[a,b]$ the graph of $y = 3x$ is covered by no more than $(3+2)(\frac{b-a}{h})$ h-squares.

We now show that the upper and lower area of a smoothly bounded set are equal.

Proof. (of Theorem 6.1) Denote by $C(h)$ the number of h-squares that intersect the boundary of D. Then

$$C(h) \geq N_U(D,h) - N_L(D,h).$$

It follows from Theorem 6.2 that

$$0 \leq h^2 N_U(D,h) - h^2 N_L(D,h) \leq h^2 C(h) < ch \qquad (6.7)$$

for some number c. Let h tend to zero through the sequence $h = \frac{1}{2^n}$. We saw before that the sequences $h^2 N_U(D,h)$ and $h^2 N_L(D,h)$ tend to the upper and lower area of D. It follows from inequality (6.7) that the limits

$$A_U(D) = \lim_{h \to 0} h^2 N_U(D,h) = \lim_{h \to 0} h^2 N_L(D,h) = A_L(D) \qquad (6.8)$$

exist and are equal. This common limit was defined as the area. Therefore the area exists.   □

Remark. In the special case where B is the boundary of a smoothly bounded set, $A_L(B,h) = 0$ for all h and $A_L(B) = 0$. By equation (6.8), $A_U(B) = A_L(B)$ and the area of B is zero. By properties (a) and (b) of area we see that the area of a smoothly bounded set D with boundary B is the same whether any boundary points are included in D or not.

Volume Divide 3-space into cubes of edge length h by the planes $x = kh$, $y = mh$, $z = nh$, where k, m, and n are integers. See Figure 6.6.

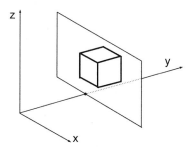

Fig. 6.6 An h-cube in 3-space, bounded by planes. A plane $y = mh$ is indicated.

We define in an entirely analogous fashion to area, using h-cubes in the place of h-squares, the notion of lower volume $V_L(D)$ and upper volume $V_U(D)$ of a bounded set D in 3-space.

Definition 6.3. If the lower and upper volumes of D are equal, this common value is called the *volume* of D, $\mathrm{Vol}(D)$, and we say that D *has volume.*

Next we define a class of sets in space that we will show have volume.

Definition 6.4. A *smoothly bounded* set in $\mathbb{R}^3$ is a closed bounded set whose boundary is the union of a finite number of graphs of continuously differentiable functions

$$z = f(x,y), \quad \text{or} \quad y = f(x,z), \quad \text{or} \quad x = f(y,z),$$

defined on smoothly bounded sets in the coordinate planes.

A solid spherical ball and a solid cube are examples of smoothly bounded sets in $\mathbb{R}^3$.

The next result is an extension to three dimensions of Theorem 6.2.

Theorem 6.3. *Let S be the boundary of a smoothly bounded set in* $\mathbb{R}^3$ *and denote by $S(h)$ the number of h-cubes that intersect S. Then*

$$S(h) \leq \frac{s}{h^2},$$

where s is a number that depends on S.

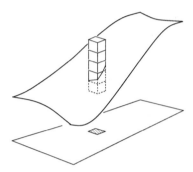

Fig. 6.7 The number of h-cubes in a stack that may intersect the graph of f is determined by the derivative ∇f.

See Figure 6.7. The proof of this theorem is modeled on the proof of Theorem 6.2.

Proof. Since S is the union of the graphs of a finite number of continuously differentiable functions, it suffices to prove the inequality for the graph of a single differentiable function, say $z = f(x,y)$, defined in a smoothly bounded set D in the x,y plane. Consider an h-square contained D. For points $\mathbf{P}_1$ and $\mathbf{P}_2$ in this h-square there is a point $\mathbf{P}$ so that

$$|f(\mathbf{P}_1) - f(\mathbf{P}_2)| = |\nabla f(\mathbf{P}) \cdot (\mathbf{P}_1 - \mathbf{P}_2)| \leq \max_{\mathbf{Q} \text{ in } D} \|\nabla f(\mathbf{Q})\| \|\mathbf{P}_1 - \mathbf{P}_2\|$$

$$\leq \left(\sqrt{2} \max_{\mathbf{Q} \text{ in } D} \left\{ \left|\frac{\partial f}{\partial x}(\mathbf{Q})\right|, \left|\frac{\partial f}{\partial y}(\mathbf{Q})\right| \right\} \right) \sqrt{2} h.$$

This bound helps us see that the number of h-cubes that intersect the graph of f over a single h-square that intersects D is less than $2m$, where m is an upper bound for the magnitudes of the first partial derivatives of f on D. Since the number of h-squares that intersect D is less than $\frac{a}{h^2}$, where a is the area of a rectangle containing D, the number of h-cubes that intersect the graph of f is less than $\frac{2am}{h^2}$, as claimed. □

It follows from Theorem 6.3 that the total volume of the h-cubes that intersect S is less than $\frac{s}{h^2}h^3 = sh$, which tends to zero as h tends to zero.

It also follows from Theorem 6.3 that the lower and upper volume of a smoothly bounded set D in 3-space are equal, so we can state the following theorem.

Theorem 6.4. *Smoothly bounded sets in $\mathbb{R}^3$ have volume.*

Volume in $\mathbb{R}^n$. Volume in n-dimensional space can be defined entirely analogously, using n-dimensional boxes. An n-dimensional h-box consists of all points $\mathbf{X} = (x_1, x_2, \dots x_n)$ for which

$$n_j h \le x_j \le (n_j + 1)h, \qquad (j = 1, \dots, n). \qquad (6.9)$$

where the n_j are integers. The n-dimensional volume of the h-box is defined to be h^n.

The concepts of a smoothly bounded set in n-dimensional space $\mathbb{R}^n$ and its volume are a direct generalization of the three-dimensional case. The statement and proof of Theorems 6.3 and 6.4 can be readily extended to n dimensions. In Problem 6.6 we ask you to determine the volume of several h-boxes.

We have shown that smoothly bounded regions in $\mathbb{R}^2$ have area and similarly smoothly bounded regions in $\mathbb{R}^3$ have volume. Next we look at two examples of bounded sets, one in the plane whose area is not defined and the other in space whose volume is not defined.

Example 6.3. Define D to be the set of points (x, y) in the unit square $0 \le x \le 1$, $0 \le y \le 1$ where both x and y are rational numbers. The h-squares of the upper area will have total area greater than 1. As h tends to zero we see $A_U(D) = 1$. Since every h-square no matter how small contains points that have rational and irrational coordinates, the interior of D is empty and we get $A_L(D) = 0$. Even though D is contained in a square with area we cannot say what the area of D is, since $A_L(D) \ne A_U(D)$. See Figure 6.8.   □

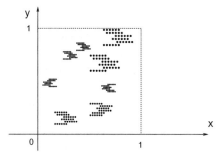

Fig. 6.8 A set in $\mathbb{R}^2$ with undefined area in Example 6.3.

Example 6.4. Define D to be the set of points (x, y, z) in the unit cube $0 \le x \le 1$, $0 \le y \le 1$, $0 \le z \le 1$ where x is irrational. That is, remove from the solid cube

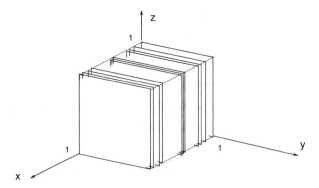

Fig. 6.9 A set in $\mathbb{R}^3$ with undefined volume in Example 6.4.

all points that lie on planes satisfying $x = r$ rational. The h-cubes that intersect
D have a total volume greater than 1 and as h tends to zero, we get $V_U(D) = 1$.
Since D has no interior points, $V_L(D) = 0$. So $V_L(D) \neq V_U(D)$. D does not
have volume. See Figure 6.9.   □

Examples and properties of multiple integrals Before making our formal definition of the integral we give you two examples of the integral of a function f over a
set D, and investigate some of the properties common to both examples.

Example 6.5. Let D be a smoothly bounded set in $\mathbb{R}^2$, and $z = f(x,y) \geq 0$ the
height of the graph of f above D. The volume $V(f,D)$ of the region R in $\mathbb{R}^3$
defined by the inequality

$$0 \leq z \leq f(x,y), \quad (x,y) \text{ in } D,$$

is an example of the integral of f over D with respect to area,

$$V(f,D) = \int_D f(x,y)\,\mathrm{d}A.$$

See Figure 6.10. Another notation for this integral is

$$V(f,D) = \int_D f(x,y)\,\mathrm{d}x\,\mathrm{d}y.$$

The significance of the $\mathrm{d}x\,\mathrm{d}y$ notation will be explained in Section 6.3.   □

Example 6.6. Let D be a smoothly bounded set in $\mathbb{R}^3$. Denote by $f(x,y,z)$ the
density at the point (x,y,z) of D of some material distributed in D. The total
mass $M(f,D)$ contained in D is an example of the integral of f over D with
respect to volume, denoted as

$$M(f,D) = \int_D f(x,y,z)\,\mathrm{d}V.$$

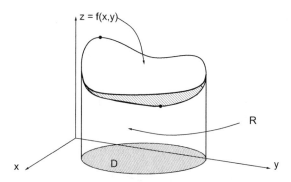

Fig. 6.10 The region R lies under the graph of f and above D.

Again as we will see in Section 6.3 another notation for this integral is

$$M(f,D) = \int_D f(x,y,z)\,dx\,dy\,dz.$$

□

We use these two examples to illustrate key properties that we will use to define the integral, namely dependence of the integral on the function f and on the set D.

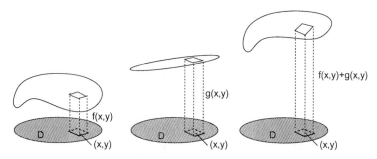

Fig. 6.11 Adding the heights of thin columns.

(a) In Example 6.5 consider three regions in space with the same base D and with heights f, g, and $f + g$. See Figure 6.11. The volume of a thin column with height $(f + g)(x,y)$, where (x,y) is in the small base B of the column, is the sum of the volumes of two thin columns with base B, one with height $f(x,y)$ and the other with height $g(x,y)$. Therefore the volume of the region defined by

$$0 \le z \le f(x,y) + g(x,y), \quad (x,y) \text{ in } D$$

is the sum

$$V(f + g, D) = V(f,D) + V(g,D).$$

Denoting the volume as the integral of height with respect to area we write this as

$$\int_D (f+g)\,dA = \int_D f\,dA + \int_D g\,dA.$$

In Example 6.6 take the case that there are two materials distributed in D, with densities f and g. Their total density is $f+g$, and the total mass in the region is the sum of the total mass of each material:

$$M(f+g,D) = M(f,D) + M(g,D).$$

Denoting the total mass as the integral of density with respect to volume we write this as

$$\int_D (f+g)\,dV = \int_D f\,dV + \int_D g\,dV.$$

(b) In Example 6.5 change the height f by a factor c. Volume is changed by the same factor:

$$V(cf,D) = cV(f,D).$$

Denoting volume as an integral we write this as

$$\int_D cf\,dA = c\int_D f\,dA.$$

In Example 6.6 introduce a different unit for mass. Then both density and the total mass are changed by the same factor:

$$M(cf,D) = cM(f,D).$$

Denoting mass as the integral of density f we write this as

$$\int_D cf\,dV = c\int_D f\,dV.$$

Properties (a) and (b) can be summarized by saying that the integral of f over D depends *linearly* on f.

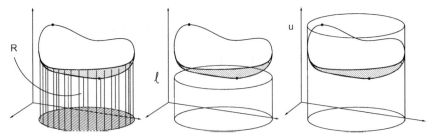

Fig. 6.12 *Left:* The region R under the graph of f. *Center:* The cylinder of height ℓ is contained in R. *Right:* The cylinder of height u contains R.

(c) In Example 6.5 let ℓ and u be lower and upper bounds, $\ell \leq f(x.y) \leq u$, for the height at all (x,y) in D. Then the volume over D lies between the bounds

$$\ell \operatorname{Area}(D) \leq V(f,D) \leq u \operatorname{Area}(D).$$

See Figure 6.12. We can write

$$\ell \operatorname{Area}(D) \leq \int_D f \, dA \leq u \operatorname{Area}(D).$$

Similarly in Example 6.6 if the density at all points of D lies between two bounds $\ell \leq f(x,y,z) \leq u$, then the total mass contained in D lies between the bounds

$$\ell \operatorname{Vol}(D) \leq M(f,D) \leq u \operatorname{Vol}(D).$$

This is called the *lower and upper bound property*. We can write this as

$$\ell \operatorname{Vol}(D) \leq \int_D f \, dV \leq u \operatorname{Vol}(D).$$

(d) In Example 6.5 if D is the union of two disjoint sets C and E, the total volume is the sum of the volume over C and the volume over E, $V(f,C \cup E) = V(f,C) + V(f,E)$. We can write it as

$$\int_{C \cup E} f \, dA = \int_C f \, dA + \int_E f \, dA, \qquad C \text{ and } E \text{ disjoint interiors.}$$

Similarly in Example 6.6 the total mass contained in the union of two disjoint sets C and E is the sum of the masses contained in each set, $M(f,C \cup E) = M(f,C) + M(f,E)$. See Figure 6.13. This property is called the *additivity property*. We can write it as

$$\int_{C \cup E} f \, dV = \int_C f \, dV + \int_E f \, dV.$$

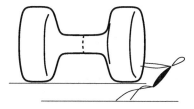

Fig. 6.13 Additivity of mass when objects are joined at their boundaries.

Problems

6.1. Let D be the set of points (x,y) in $\mathbb{R}^2$ such that $0 \leq y \leq 1 - x^2$.

(a) How many $\frac{1}{4}$-squares are in D?
(b) Does the $\frac{1}{8}$-square $[\frac{1}{2}, \frac{5}{8}] \times [\frac{5}{8}, \frac{3}{4}]$ intersect D?

6.2. Let D be the closed quarter disk in the first quadrant, with the disk center at the origin and radius 3. See Figure 6.1.

(a) Find $A_U(D, \frac{1}{2})$.
(b) Find $A_L(D, \frac{1}{2})$.

6.3. Make a sketch of the rectangle D given by $0 \leq x \leq \pi$, $0 \leq y \leq \pi$. Find

(a) the number $N_L(D, \frac{1}{2})$ of $\frac{1}{2}$-squares contained in D, and their total area;
(b) the number $N_U(D, \frac{1}{2})$ of $\frac{1}{2}$-squares that intersect D, and their total area;
(c) the number $C(\frac{1}{2})$ of $\frac{1}{2}$-squares that intersect the boundary of D, and their total area;
(d) Verify that $C(\frac{1}{2}) \geq N_U(D, \frac{1}{2}) - N_L(D, \frac{1}{2})$

6.4. Show that the ball $|\mathbf{X}| \leq 1$ in $\mathbb{R}^3$ is a smoothly bounded set.

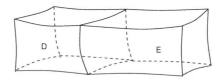

Fig. 6.14 Sets D and E in Problem 6.5.

6.5. The intersection of two smoothly bounded sets D and E in $\mathbb{R}^3$ is along a smoothly bounded common surface in Figure 6.14. Show that

$$\text{vol}(D \cup E) = \text{vol}\,D + \text{vol}\,E.$$

6.6. Find the n-dimensional volume of each set.

(a) A union of 57 h-boxes in $\mathbb{R}^n$,
(b) The box in $\mathbb{R}^3$ where $0 \leq x_j \leq 10$, $j = 1,2,3$, $n = 3$,
(c) The box in $\mathbb{R}^6$ where $0 \leq x_j \leq 10$, $j = 1,2,3,4,5,6$, $n = 6$.
(d) The box in $\mathbb{R}^3$ where $0 \leq x_j \leq \frac{1}{10}$, $j = 1,2,3$, $n = 3$,
(e) The box in $\mathbb{R}^6$ where $0 \leq x_j \leq \frac{1}{10}$, $j = 1,2,3,4,5,6$, $n = 6$.

6.7. A plate shown in Figure 6.15 has density $f(x,y)$.

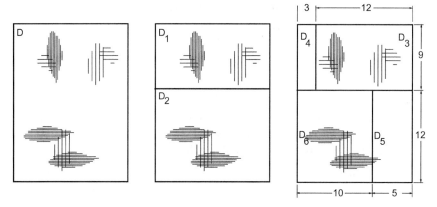

Fig. 6.15 A plate with variable density in Problem 6.7. Three subdivisions are shown.

(a) If $2 \leq f(x,y) \leq 7$ in D show that

$$630 \leq \text{mass}(D) \leq 2205.$$

(b) If $2 \leq f(x,y) \leq 4$ in D_1, and $4 \leq f(x,y) \leq 7$ in D_2 use additivity to show that $990 \leq \text{mass}(D) \leq 1800$.

(c) Further information becomes available, that $2 \leq f(x,y) \leq 4$ in D_3, $f = 2$ in D_4, $4 \leq f(x,y) \leq 6$ in D_5, and $6 \leq f(x,y) \leq 7$ in D_6. Use the additive property and the lower and upper bound properties to further narrow the mass estimate to

$$1230 \leq \text{mass}(D) \leq 1686.$$

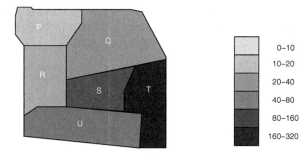

Fig. 6.16 A map of population density, for Problem 6.8.

6.8. A region is subdivided into counties, with the range of population density (people/square mile) f indicated on the map in Figure 6.16. The map key gives lower and upper bounds in each region:

$$P: \quad 0 \le f(x,y) \le 10$$
$$Q: \quad 20 \le f(x,y) \le 40$$
$$R: \quad 10 \le f(x,y) \le 20$$
$$S: \quad 80 \le f(x,y) \le 160$$
$$T: \quad 160 \le f(x,y) \le 320$$
$$U: \quad 40 \le f(x,y) \le 80$$

(a) The area of county Q is 250 square miles. Give bounds for the population of Q.

(b) The Western region consists of counties P, R, and U, that have areas 80, 100, 120 square miles, respectively. Give bounds for the population of the Western region.

(c) Counties S and T have about the same area. Is it possible for $\int_S f(x,y)\,dA$ to be larger than $\int_T f(x,y)\,dA$?

(d) Show that $-4600 \le \int_Q f\,dA - \int_U f\,dA \le 5200$.

Fig. 6.17 The liquid volumes in Problem 6.9.

6.9. A liquid of variable density ρ fills the region D in the large beaker in Figure 6.17, and another liquid of constant density $\delta = 200$ [kg/m^3] fills region C. Each of C and D has volume .05 [m^3]. The total mass of liquids is

$$\int_C \delta\,dV + \int_D \rho\,dV = 30.$$

(a) Find the value of each integral.

(b) Show that the minimum and maximum of ρ must satisfy

$$\rho_{min} \le 400 \le \rho_{max}.$$

6.10. Let D denote the closed unit disk $x^2 + y^2 \le 1$ in the plane and let U denote the open disk $x^2 + y^2 < 1$.

(a) Describe the boundary of D.

(b) Describe the boundary of U.

(c) What is Area(D)? Area(U)?

6.2 The integral of a continuous function of two variables

We focus in this section on the definition of the integral in the case where f is a continuous function on a smoothly bounded set D in the plane.

We shall first define the upper and lower integrals, $I_U(f,D)$ and $I_L(f,D)$, for continuous functions f that are nonnegative and show that they are equal; their common value is defined as the integral of f over D and denoted $I(f,D)$. We then verify that $I(f,D)$ has the four basic properties of the integral introduced in Section 6.1.

(a) $I(f+g,D) = I(f,D) + I(g,D)$
(b) $I(cf,D) = cI(f,D)$ for every number c
(c) If $\ell \le f(x,y) \le u$ for all (x,y) in D, then

$$\ell\,\text{Area}(D) \le I(f,D) \le u\,\text{Area}(D).$$

(d) If C and D are disjoint or have only boundary points in common,

$$I(f,C \cup D) = I(f,C) + I(f,D).$$

We then extend the definition of the integral to functions that may have negative values, and verify the four properties listed above.

In Theorem 6.10 we will show that all properties of the integral follow from these four properties.

Upper and lower integrals of a continuous function f over D. According to the Extreme Value Theorem 2.11, f is bounded on D. Divide the plane into squares of side h as described in Section 6.1, and take all h-squares that intersect D. In each of these h-squares B_j we denote by u_j and ℓ_j an upper and a lower bound for f:

$$0 \le \ell_j \le f(x,y) \le u_j \le M, \quad (x,y) \text{ in } B_j. \tag{6.10}$$

Call the sum

$$\sum_j h^2 u_j \tag{6.11}$$

over all h-squares that intersect D an *upper h-sum*. Call the sum

$$\sum_j h^2 \ell_j \tag{6.12}$$

over all h-squares that are contained in D a *lower h-sum*.

We claim: For a nonnegative function f every upper h-sum is greater than or equal to every lower h-sum.

To see this we observe that for an h-square that is contained in D, the term $h^2 u_j$ in (6.11) is greater than or equal to the corresponding term $h^2 \ell_j$ in (6.12), since an upper bound u_j is greater than or equal to the lower bound ℓ_j. The terms in the sum (6.11) corresponding to h-squares that intersect D but are not contained in D do not appear in (6.12). Since the function f is nonnegative, so are the upper bounds

u_j and the terms $u_j h^2$. It follows that the upper sum is greater than or equal to the lower sum for the same value of h.

Next we show: For a nonnegative function f, given an upper h-sum, there is an upper $h/2$-sum that is less than or equal to the upper h-sum.

To prove this we observe that if an $h/2$-square intersects D, so does the h-square that contains it. An upper bound u_j for f in the h-square is an upper bound for f in each $h/2$-square contained in it. If all four $h/2$-squares contained in the h-square intersect D, their contribution to the $h/2$ upper sum (using the same u_j) equals the contribution of the h-square to the upper h-sum. If some of the $h/2$-squares don't intersect D, their contribution to the sum of $h/2$-terms using the same u_j is less than or equal to the contribution of the h-square to the h-upper sum. This shows that the upper $h/2$-sum can be chosen to be less than or equal to the upper h-sum.

Since $f \geq 0$ all upper h-sums are nonnegative, i.e., zero is a lower bound for the set of upper h-sums. The Greatest Lower Bound Theorem states that if a set of numbers has a lower bound then the greatest lower bound of that set exists. Call it $U(f,D,h)$. We have shown that for $f \geq 0$ every upper h-sum has a corresponding upper $h/2$-sum that is less than or equal to the upper h-sum; it follows that $U(f,D,h/2)$ is less than or equal to $U(f,D,h)$. This shows that

$$U(f,D,h), \quad h = \tfrac{1}{2^n}, \quad n = 1,2,\ldots \tag{6.13}$$

is a decreasing sequence of numbers that are greater than or equal to zero. Therefore by the Monotone Convergence Theorem it has a limit as n tends to infinity. This limit is called the *upper integral* of f. We denote it as

$$I_U(f,D) = \lim_{n \to \infty} U(f,D,2^{-n}).$$

There is an analogous result for lower sums: For $f \geq 0$, given a lower h-sum, there is a lower $h/2$-sum that is greater than or equal to the lower h-sum.

To prove this we note that the union of all $h/2$-squares contained in D includes all h-squares contained in D. A lower bound ℓ_j for f in an h-square is a lower bound for f in all $h/2$-squares contained in it. An $h/2$-square contained in D but that is not part of an h-square that is contained in D gives a nonnegative contribution to the $h/2$-sum, provided we choose a nonnegative lower bound for f. The resulting lower $h/2$-sum is greater than or equal to the corresponding lower h-sum.

Each lower h-sum is less than uA, where u is an upper bound for f on D and A is the area of a square that contains D. The Least Upper Bound Theorem states that if a set of numbers has an upper bound then it has a least upper bound. Denote by $L(f,D,h)$ the least upper bound of the lower h-sums. We showed that for $f \geq 0$, for each lower h-sum there is a greater or equal lower $h/2$-sum, so it follows that $L(f,D,h/2)$ is greater than or equal to $L(f,D,h)$. Therefore

$$L(f,D,h), \quad h = \tfrac{1}{2^n}, \quad n = 1,2,\ldots \tag{6.14}$$

is an increasing sequence less than or equal uA. By the Monotone Convergence Theorem this bounded, increasing sequence has a limit, called the *lower integral* of

f over D; we denote it as

$$I_L(f,D) = \lim_{n\to\infty} L(f,D,2^{-n}).$$

Theorem 6.5. *The upper and lower integrals of a continuous nonnegative function f over a smoothly bounded set D in $\mathbb{R}^2$ are equal,*

$$I_L(f,D) = I_U(f,D).$$

Proof. To show that the upper integral and lower integral of a continuous function are equal we show that for every positive tolerance ϵ, no matter how small, the upper h-sum and lower h-sum differ by less than ϵ when h is sufficiently small.

We have shown that for $f \geq 0$ every lower h-sum is less than or at most equal to every upper h-sum.

Differences due to the boundary. The difference between the number of terms in an upper h-sum (6.11) and the number of terms in a lower h-sum (6.12) is less than or equal to the number $C(h)$ of h-squares that intersect the boundary of D. We have shown in Theorem 6.2 that $C(h)$ is less than or equal to c/h for some constant c determined by the boundary of D. Therefore the total contribution of these terms to the upper h-sum is less than or equal to

$$Mh^2 C(h) \leq Mh^2 \frac{c}{h} = Mch,$$

where M is an upper bound for the u_j in that set. This tends to zero as h tends to zero. This shows that the sum of the terms in the upper sum that are not in the lower sum tends to zero as h tends to zero, so is less than $\frac{\epsilon}{2}$ if h is sufficiently small.

Differences due to h-squares in D. We show that the difference of the sum of the terms that are in both the upper sum and the lower sum tends to zero as h tends to zero. The difference of the upper and lower sums over h-squares in D is

$$\sum_j h^2(u_j - \ell_j). \tag{6.15}$$

The function f is continuous on a closed bounded set D; therefore it is uniformly continuous: for every tolerance t, we can choose a precision p so small that if the distance between two points in D is less than p, the values of the function f at these two points differ less than t. Take h so small that the distance between two points in an h-square is less than p. If u_i is the maximum of f and ℓ_i is the minimum of f on the h-square then

$$u_j - \ell_j < t.$$

Using this estimate for the terms in the sum (6.15) shows that the difference between the sums over interior squares is less than $Nh^2 t$, where N is the number of terms in the sum. N is the number of h-squares contained in D, which is less than A/h^2, where

A is the area of a rectangle containing D. This shows that the difference (6.15) is less than At. Given the tolerance ϵ, take $t = \frac{\epsilon}{2A}$, then the difference (6.15) is less than $\frac{\epsilon}{2}$.

Adding our two estimates then we have shown that there are upper and lower h-sums with difference less than ϵ if h is sufficiently small. This shows that the upper and lower integrals are equal, and thereby completes the proof of Theorem 6.5. ☐

Upper and lower integrals of bounded functions. Although we focused on the basic case where f is continuous and D includes its boundary, the definitions we made for $I_U(f,D)$ and $I_L(f,D)$ apply to bounded functions f, not necessarily continuous, on a smoothly bounded set D. In that case the lower and upper integrals $I_L(f,D)$ and $I_U(f,D)$ both exist but they might not be equal.

Definition 6.5. Let f be a bounded function on a smoothly bounded set D. If $I_U(f,D) = I_L(f,D)$ we say that f is *integrable* on D and that the integral of f over D exists. We write the integral as

$$I_L(f,D) = I_U(f,D) = \int_D f \, dA.$$

The function f is called the *integrand*. For Cartesian coordinates x,y in the plane the integral of f over D is also denoted

$$\int_D f(x,y)\,dx\,dy.$$

and is called a *double integral*.

Remark. By essentially the same argument as for Theorem 6.5 it can be shown that a bounded function that is continuous on the interior of a smoothly bounded set is integrable, i.e., that $I_L(f,D) = I_U(f,D)$.

Example 6.7. Let D denote the rectangle $[0,1) \times (0,2)$, and

$$f(x,y) = y \sin \frac{1}{1-x^2}.$$

Then $0 \le f(x,y) \le y \le 2$ and f is continuous in the interior of D. Therefore by the remark above, the integral

$$\int_D y \sin \frac{1}{1-x^2} \, dA$$

exists. ☐

Approximate integrals. Next we show that the integral of f over D can be estimated with arbitrary accuracy without calculating upper or lower sums.

Definition 6.6. Let f be defined on a smoothly bounded set D in the plane. Divide the plane as before into h-squares and denote by $\mathbf{C}_j = (x_j, y_j)$ some point of the j-th h-square contained in D. Define an *approximate integral*, or *Riemann sum*,

$$S(f, D, h) = \sum_j f(\mathbf{C}_j) h^2, \qquad (6.16)$$

where the sum is taken over all h-squares contained in D.

Theorem 6.6. *Let $f \geq 0$ be an integrable function on a smoothly bounded set D. As $h = 2^{-n}$ tends to zero, every sequence of approximate integrals*

$$S(f, D, h) = \sum_j f(\mathbf{C}_j) h^2$$

converges to

$$\int_D f \, dA,$$

independent of the choice of the points $\mathbf{C}_j$.

Proof. Since $f(\mathbf{C}_j)$ lies between every lower and upper bound for f on the j-th h-square, $S(f, D, h)$ is greater than or equal to every lower h-sum. It is also less than or equal to every upper h-sum since the terms in the upper h-sum corresponding to h-squares that are not contained in D are greater than or equal to zero, since $f \geq 0$. Therefore $S(f, D, h)$ is contained between the least upper bound of all lower h-sums and the greatest lower bound of all upper h-sums:

$$L(f, D, h) \leq S(f, D, h) \leq U(f, D, h). \qquad (6.17)$$

Because f is integrable on D, $L(f, D, h)$ and $U(f, D, h)$ tend to the same limit, that we called the integral of f over D. It follows from inequality (6.17) that $S(f, D, h)$ also tends to the same limit:

$$\lim_{h \to 0} S(f, D, h) = \int_D f \, dA,$$

as asserted.   □

Example 6.8. Let $f \geq 0$ be a continuously differentiable function on a smoothly bounded set D in the plane, and let R be the region above D and under the graph of f. See Figure 6.18. R is a smoothly bounded set in $\mathbb{R}^3$. We show that $\int_D f \, dA$ is the volume of R. By Theorem 6.6 the Riemann sums $\sum_i f(x_i, y_i) h^2$

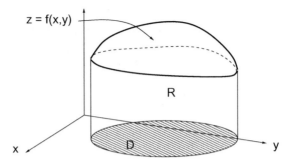

Fig. 6.18 The sets D and R in Example 6.8.

converge to $\int_D f\, dA$ as h tends to zero. Every Riemann sum using h-squares is greater than or equal to the h-cube lower volume of R, and is less than or equal to the upper h-cube volume of R,

$$V_L(R,h) \le \sum_i f(x_i, y_i)h^2 \le V_U(R,h).$$

Since R is smoothly bounded the upper and lower volumes have the same limit as h tends to zero, that is the limit of the Riemann sums as well. Therefore

$$\int_D f\, dA = \mathrm{Vol}(R).$$

□

Example 6.9. Let $f(x,y) = 1$ on a smoothly bounded set D, and let R be the region under the graph, over D. By Example 6.8

$$\mathrm{Vol}(R) = \int_D 1\, dA.$$

From our definition of volume we see that

$$\mathrm{Vol}(R) = \mathrm{Area}(D)(1).$$

Therefore

$$\int_D 1\, dA = \mathrm{Area}(D).$$

□

Example 6.10. Evaluate

$$\int_D y\, dA$$

where D is the rectangle $0 \le x \le 5$, $0 \le y \le 3$. The graph of $f(x,y) = y$ over D is shown in Figure 6.19. By Example 6.8 the integral is the volume of the set

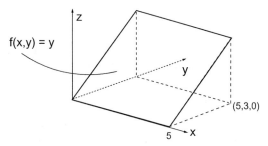

Fig. 6.19 The graph of $f(x,y) = y$ where $0 \le x \le 5$ and $0 \le y \le 3$ in Example 6.10.

R of points (x,y,z) with $0 \le z \le f(x,y)$, $0 \le x \le 5$, $0 \le y \le 3$.

$$\int_D y\,dA = \mathrm{Vol}(R) = \tfrac{1}{2}(3)(3)(5) = 22.5.$$

□

We next show that our definition of $\int_D f\,dA$ for nonnegative continuous functions over smoothly bounded sets satisfies the four properties we stated at the start of this section.

Theorem 6.7. *Let D denote a smoothly bounded set, f and g nonnegative continuous functions in D, and c a nonnegative number. Then*

(a) $\displaystyle \int_D (f+g)\,dA = \int_D f\,dA + \int_D g\,dA,$

(b) $\displaystyle \int_D cf\,dA = c\int_D f\,dA,$

(c) If $\ell \le f(x,y) \le u$ on D then $\ell\,\mathrm{Area}(D) \le \displaystyle\int_D f\,dA \le u\,\mathrm{Area}(D),$

(d) If C and D are smoothly bounded sets that are disjoint or have only boundary points in common, and if $C \cup D$ issmoothly bounded, then

$$\int_{C \cup D} f\,dA = \int_C f\,dA + \int_D f\,dA.$$

Proof. **(a)** It follows from the definition of approximate integrals that

$$S(f+g,D,h) = S(f,D,h) + S(g,D,h)$$

provided we use the same points $\mathbf{C}_j$ in forming all three approximate integrals. Letting h tend to zero gives part (a).

(b) Similarly,

$$S(cf,D,h) = cS(f,D,h).$$

Letting h tend to zero gives part (b).

(c) To get an upper bound for the integral of f over D we replace each $f(\mathbf{C}_i)$ in the Riemann sum (6.16) by the upper bound u for the values of the function f. This gives an upper bound for the Riemann sum. The limit of this upper bound as h tends to zero is $u\,\mathrm{Area}(D)$; it is an upper bound for the integral, as asserted in part (c). By a similar argument $\ell\,\mathrm{Area}(D) \le \int_D f\,dA$.

(d) If C and D are disjoint, then for h small enough an h-square may belong to C or D, but not both. In this case a Riemann sum for f over $C \cup D$ is the sum of a Riemann sum for f over C and a Riemann sum over D. If C and D have a boundary curve in common, there are h-squares that belong to $C \cup D$ but belong neither to C nor to D. These h-squares intersect the common boundary of C and D. Since the boundaries of C and D are smooth curves, the number of h-squares they intersect is less than some constant multiple of $1/h$. Therefore their total area tends to zero as h tends to zero. Therefore the Riemann sum over these squares tends to zero; this proves part (d). □

Example 6.11. Let $f(x,y) = 4 + y$ and let $D = [0,5] \times [0,3]$ be the rectangular region

$$0 \le x \le 5, \qquad 0 \le y \le 3.$$

From Example 6.10 we know that $\int_D y\,dA = 22.5$ and from Example 6.9 we know that $4\int_D 1\,dA = 4\mathrm{Area}(D) = 4(15)$. So by Theorem 6.7

$$\int_D (4+y)\,dA = 4\int_D 1\,dA + \int_D y\,dA = 4(15) + 22.5 = 82.5.$$

□

We now extend the definition of the integral to functions that may have negative values.

Definition 6.7. Let f be a continuous function on a smoothly bounded set D in $\mathbb{R}^2$. Write f as the difference of two continuous functions $g \ge 0$ and $k \ge 0$:

$$f = g - k.$$

We define the integral of f over D as the difference of the integrals of g and k over D,

$$\int_D f\,dA = \int_D g\,dA - \int_D k\,dA.$$

There are many ways of writing f as such a difference. We show that all such differences give the same value for the integral. Take two decompositions of f

$$f = g - k, \quad f = m - n. \tag{6.18}$$

where g, k, m, and n are nonnegative and continuous on D. We show that calculating the integral of f using either of these decompositions of f gives the same value, i.e., we want to show that

$$\int_D g\,dA - \int_D k\,dA = \int_D m\,dA - \int_D n\,dA.$$

To prove this relation we rewrite it as

$$\int_D g\,dA + \int_D n\,dA = \int_D m\,dA + \int_D k\,dA.$$

By part (a) of Theorem 6.7, since $g + n \geq 0$ and $m + k \geq 0$, we can rewrite the two sides as

$$\int_D (g+n)\,dA = \int_D (m+k)\,dA$$

Since it follows from relations (6.18) that $g + n$ and $m + k$ are equal, so are their integrals. This proves that the definition of the integral of f does not depend on how we write f as a difference of two nonnegative functions.

If now $f \geq 0$, use the difference

$$-f = 0 - f.$$

We get that

$$\int_D (-f)\,dA = \int_D 0\,dA - \int_D f\,dA.$$

Since $\int_D 0\,dA = 0$ this gives

$$\int_D (-f)\,dA = -\int_D f\,dA.$$

Next we claim that Theorem 6.6—asserting that the integral of f over D is the limit of the Riemann sums $S(f,D,h)$ as h tends to zero—is valid for all continuous functions f, not just for nonnegative ones. To see this represent f as the difference $g - k$ of two nonnegative functions g and k, and use the same points $\mathbf{C}_i$ for each function. Then

$$\sum f(\mathbf{C}_i)h^2 = S(f,D,h)$$

$$= S(g-k,D,h) = \sum (g(\mathbf{C}_i) - k(\mathbf{C}_i))h^2$$

$$= \sum g(\mathbf{C}_i)h^2 - \sum k(\mathbf{C}_i)h^2 = S(g,D,h) - S(k,D,h).$$

Then apply Theorem 6.6 to the functions g and k. As h tends to zero, the right side tends to

$$\int_D g \, dA - \int_D k \, dA,$$

therefore so does the left side. Since $g - k = f$, this is by Definition 6.7 equal to $\int_D f \, dA$.

The same reasoning, representing the integrand as the difference of nonnegative functions, shows that Theorem 6.7 on the linearity, boundedness, and additivity of the integral holds for continuous functions and numbers that may be negative. We formulate this as follows.

Theorem 6.8. *For continuous functions f and g on smoothly bounded sets C and D, and all numbers c,*

(a) $\displaystyle\int_D (f+g)\,dA = \int_D f\,dA + \int_D g\,dA,$

(b) $\displaystyle\int_D cf\,dA = c\int_D f\,dA,$

(c) If $\ell \le f(x,y) \le u$ then $\ell\,\mathrm{Area}(D) \le \displaystyle\int_D f\,dA \le u\,\mathrm{Area}(D)$,

(d) If C and D are disjoint or have only boundary points in common and $C \cup D$ is smoothly bounded then

$$\int_{C \cup D} f\,dA = \int_C f\,dA + \int_D f\,dA.$$

Example 6.12. Evaluate

$$\int_D x^3 y^2 \, dA$$

where D is the closed unit disk centered at the origin. Observe that the integrand $f(x,y) = x^3 y^2$ has a symmetry:

$$f(-x,y) = -x^3 y^2 = -f(x,y).$$

Consider forming a Riemann sum, where for each point $\mathbf{C}_j = (x_j, y_j)$ that we choose in an h-square with $x_j > 0$ we choose the corresponding point $(-x_j, y_j)$ for the h-square on the other side of the y-axis. Then the Riemann sum is exactly zero. Since this can be done for every h, and since the integral is the limit of Riemann sums, we get

$$\int_D x^3 y^2 \, dA = 0.$$

$\square$

In Problem 6.23 we ask you to use the properties above to prove the next theorem.

Theorem 6.9. *For continuous functions f and g defined on smoothly bounded sets C and D in* $\mathbb{R}^2$,

(a) If $g(x,y) \le f(x,y)$ *for all* (x,y) *in D then*

$$\int_D g \, dA \le \int_D f \, dA,$$

(b) If C is a subset of D and $f \ge 0$ *on D then*

$$\int_C f \, dA \le \int_D f \, dA.$$

Example 6.13. Let $f \ge 0$ be a continuous population density function on D and denote by $a > 0$ a number in the range of f. Suppose $f \ge a$ on a smoothly bounded set $C \subset D$. Using Theorem 6.9 we estimate the population of D as

$$\int_D f \, dA \ge \int_C f \, dA \ge \int_C a \, dA = a \operatorname{Area}(C).$$

See Figure 6.20. □

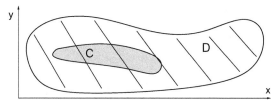

Fig. 6.20 $f \ge 0$ and $f \ge a$ in C. So $\displaystyle\int_D f \, dA \ge \int_C f \, dA \ge a \operatorname{Area}(C)$. See Example 6.13.

As we noted earlier in the chapter, the case of continuous functions on smoothly bounded sets D is basic, and we focused our attention on them. In fact analogues to Definition 6.7 and Theorems 6.7, 6.8, and 6.9 hold for bounded functions that are integrable on the interior of smoothly bounded sets.

Next we show that if a number $\mathscr{I}(f,D)$ is defined for all continuous functions f on smoothly bounded sets D, that satisfies the four properties listed in Theorem 6.8 then $\mathscr{I}(f,D)$ *is* the integral of f over D.

Theorem 6.10. *Suppose $\mathscr{I}(f,D)$ is defined for every continuous function f on a smoothly bounded set D and that it has the following properties.*

(a) $\mathscr{I}(f+g,D) = \mathscr{I}(f,D) + \mathscr{I}(g,D)$
(b) $\mathscr{I}(cf,D) = c\mathscr{I}(f,D)$ for every number c
(c) If $\ell \leq f(x,y) \leq u$ for all (x,y) in D, then

$$\ell \operatorname{Area}(D) \leq \mathscr{I}(f,D) \leq u \operatorname{Area}(D),$$

(d) For all pairs of smoothly bounded sets C and D that are disjoint or have only boundary points in common,

$$\mathscr{I}(f, C \cup D) = \mathscr{I}(f,C) + \mathscr{I}(f,D),$$

Then

$$\mathscr{I}(f,D) = \int_D f \, dA.$$

Proof. We shall deduce from these four properties that for $f \geq 0$, $\mathscr{I}(f,D)$ is less than or equal every upper sum and greater than or equal to every lower sum. To see this, take an upper h-sum and lower h-sum and partition D into subsets D_i that are the intersections of D with the h-squares. Denote by ℓ_i and u_i lower and upper bounds for f on D_i. By properties (c) and (d) we get

$$\sum_{h\text{-squares in } D} \ell_i h^2 \leq \sum \ell_i \operatorname{Area}(D_i)$$

$$\leq \mathscr{I}(f,D) \leq \sum u_i \operatorname{Area}(D_i) \leq \sum_{h\text{-squares intersecting } D} u_i h^2,$$

therefore

$$L(f,D,h) \leq \mathscr{I}(f,D) \leq U(f,D,h).$$

As h tends to zero we get

$$I_L(f,D) \leq \mathscr{I}(f,D) \leq I_U(f,D).$$

Since $I_U(f,D) = I_L(f,D) = \int_D f \, dA$, their common value is $\mathscr{I}(f,D)$.

Now given a continuous function f not necessarily nonnegative, choose a positive function p so that $f + p$ is positive as well. According to property (a)

$$\mathscr{I}(f+p,D) = \mathscr{I}(f,D) + \mathscr{I}(p,D).$$

So

$$\int_D (f+p)\, dA = \mathscr{I}(f,D) + \int_D p \, dA.$$

Subtract to get

$$\mathscr{I}(f,D) = \int_D (f+p)\,dA - \int_D p\,dA = \int_D f\,dA.$$

$\square$

Problems

6.11. Use an area or a volume interpretation to evaluate the integrals.

(a) $\displaystyle\int_D 1\,dA$ where D is the region $1 \le x \le 2,\ 0 \le y \le \log x$.

(b) $\displaystyle\int_D x\,dA$ where D is the rectangle $0 \le x \le 3,\ -1 \le y \le 1$.

(c) $\displaystyle\int_U \sqrt{1-x^2-y^2}\,dA$ where U is the unit disk centered at the origin.

(d) $\displaystyle\int_H 1\,dV$ where H is the half ball $x^2+y^2+z^2 \le 1,\ z \ge 0$.

6.12. Use a volume interpretation to evaluate $\displaystyle\int_D f\,dA$ for each function f.

(a) $f(x,y) = 3$ and D is the disk of radius 5 centered at $(0,0)$.
(b) $f(x,y) = \frac{1}{2}y$ and D is the rectangle where $-2 \le x \le 3$ and $0 \le y \le 4$.
(c) $f(x,y) = \sqrt{x^2+y^2}$ and D is the unit disk centered at the origin.

6.13. Sketch rectangles $D = [-a,0]\times[0,b]$ and $E = [0,1]\times[-c,c]$ in the plane, where a,b,c are positive. Determine without calculation which of these integrals are positive.

(a) $\displaystyle\int_D x\,dA$

(b) $\displaystyle\int_E x\,dA$

(c) $\displaystyle\int_E (1-x)\,dA$

(d) $\displaystyle\int_E y^2\,dA$

6.14. Let $f(x,y)$ be a continuous function on a smoothly bounded set D that is symmetric about the origin, that is, D contains the negative of each of its points, and assume $f(-x,-y) = -f(x,y)$ for all points of D.

(a) Find $f(0,0)$ if $(0,0)$ is in D.
(b) Show that there are approximate integrals of f over D that are exactly zero.
(c) Show that $\displaystyle\int_D f\,dA = 0$.

(d) Evaluate $\int_D xy\,dA$ where D is the unit disk centered at the origin.

(e) Which of the following functions satisfy $f(-x,-y) = -f(x,y)$?

$$x, \quad y^2, \quad x\cos y, \quad xy^2, \quad x-y.$$

6.15. Use properties of the integral, symmetry, and a volume interpretation to evaluate

$$\int_D (y^3 + 3xy + 2)\,dA$$

where D is the unit disk centered at the origin.

6.16. Suppose f is a bounded function and D a smoothly bounded set in $\mathbb{R}^2$.

(a) Show that if $\mathrm{Area}(D) = 0$ then the interior of D is empty.

(b) Show that if the interior of D is empty then $I_L(f,D) = 0$.

(c) Conclude that if f is an integrable function and $\mathrm{Area}(D) = 0$ then $\int_D f\,dA = 0$.

6.17. Justify the following items which prove:

If f is continuous on $\mathbb{R}^2$ and $\int_R f\,dA = 0$ for all smoothly bounded sets R, then f is zero at all points of $\mathbb{R}^2$.

(a) If $f(a,b) = p > 0$ then there is a disc D of radius $r > 0$ centered at (a,b) in which $f(x,y) > \frac{1}{2}p$.

(b) If f is continuous and $f(x,y) \geq p_1 > 0$ on a disk R then

$$\int_R f\,dA \geq p_1(\mathrm{Area}(R)) > 0.$$

(c) If f is continuous and $\int_R f\,dA = 0$ for all smoothly bounded regions R, then f cannot be positive at any point.

(d) f is not negative at any point either.

6.18. Write a Riemann sum for

$$\int_{[0,1]\times[0,1]} x^2 y^3\,dA$$

using points (x_j,y_j) at the upper right corner of each $\frac{1}{4}$-square.

6.19. The sum over integers i and j,

$$\sum_{i^2+j^2\leq 10h^{-2}} ((ih)^2 + (jh)^2)^2 h^2$$

approximates which integral,

$$\int_{x^2+y^2\le10^2}(x^2+y^2)^2\,dA, \quad\text{or}\quad \int_{x^2+y^2\le10}(x^2+y^2)^2\,dA?$$

6.20. Let D be a rectangle $[a,b]\times[c,d]$ and consider integrals of the form

$$\int_D f(x)g(y)\,dA$$

where f and g are continuous functions of one variable. Use the notion of Riemann sum to prove that the integral of the product is the product of the integrals:

$$\int_D f(x)g(y)\,dA = \int_a^b f(x)\,dx \int_c^d g(y)\,dy.$$

6.21. The integral

$$J = \int_{[0,1]\times[0,1]} \sin\!\left(\frac{1}{(1-x^2)(1-y^2)}\right)dA$$

exists because the integrand is bounded and is continuous in the interior of the square.

(a) Find upper and lower bounds on the integrand.
(b) Calculation using Riemann sums indicates that

$$\int_{[0,.999]\times[0,.999]} \sin\!\left(\frac{1}{(1-x^2)(1-y^2)}\right)dA = .423$$

approximately. Assuming that is correct, find bounds on J.

6.22. Let $f(x,y) = (4-x^2-y^2)^{-1/2}$ in the disk D of radius r centered at the origin, where $r<2$. Is f bounded? integrable on D?

6.23. Prove Theorem 6.9 by justifying the following steps.

(a) For part (a):

(i) $0 \le f(x,y) - g(x,y)$.

(ii) $0 \le \int_D (f-g)\,dA.$

(iii) $0 \le \int_D (f-g)\,dA = \int_D f\,dA - \int_D g\,dA.$

(iv) $\int_D g\,dA \le \int_D f\,dA.$

(b) For part (b):

(i) Every h-square in C is in D.

(ii) For every Riemann sum for $\int_C f\,dA$ there is a Riemann sum for $\int_D f\,dA$ that is equal or larger.

(iii) Every Riemann sum for $\int_C f\,\mathrm{d}A$ does not exceed $\int_D f\,\mathrm{d}A$.

(iv) $\int_C f\,\mathrm{d}A \leq \int_D f\,\mathrm{d}A$.

6.3 Double integrals as iterated single integrals

Now that we have defined the integral

$$\int_D f\,\mathrm{d}A$$

for a continuous function f over a smoothly bounded set D in $\mathbb{R}^2$ we show how to use definite integrals of a single variable to compute it. We show in this section that such integrals can be evaluated by performing integration with respect to x, followed by integration with respect to y.

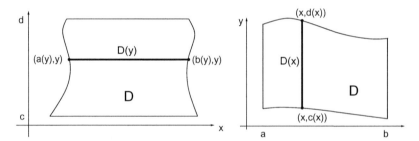

Fig. 6.21 *Left: D is x simple. Right: D is y simple.*

We say that a bounded set D is *x simple* if it has the following property: The set of all points in D whose second coordinate is y is an interval $D(y)$ parallel to the x axis, whose endpoints $a(y)$ and $b(y)$ are continuous functions of y:

$$a(y) \leq x \leq b(y), \qquad c \leq y \leq d.$$

See Figure 6.21. The integral of a continuous function $f(x,y)$ with respect to x over the interval $D(y)$

$$\int_{D(y)} f(x,y)\,\mathrm{d}x = \int_{x=a(y)}^{x=b(y)} f(x,y)\,\mathrm{d}x$$

is a continuous function of y. We integrate this function with respect to y between c and d, getting

$$\int_c^d \left(\int_{x=a(y)}^{x=b(y)} f(x,y)\,\mathrm{d}x \right) \mathrm{d}y.$$

We call this an *iterated integral*.

We can make similar calculations for functions over *y simple* sets: The set of all points in D whose first coordinate is x is an interval $D(x)$ that is parallel to the y axis and whose endpoints $c(x)$ and $d(x)$ are continuous functions of x:

$$c(x) \le y \le d(x), \qquad a \le x \le b.$$

See Figure 6.21. We can then calculate an iterated integral

$$\int_a^b \left(\int_{D(x)} f(x,y)\,dy \right) dx = \int_a^b \left(\int_{y=c(x)}^{y=d(x)} f(x,y)\,dy \right) dx.$$

Let's look at some examples.

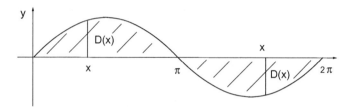

Fig. 6.22 The set D in Example 6.14.

Example 6.14. Let D be the region bounded by the graphs of $y = \sin x$ and $y = 0$ for $0 \le x \le 2\pi$. Find the iterated integral of the constant function $f(x,y) = 7$ over D. We sketch the region D in Figure 6.22. D is the union of two y simple sets. For $0 \le x \le \pi$ we have $0 \le y \le \sin x$, and for $\pi \le x \le 2\pi$ we have $\sin x \le y \le 0$. So we set up two iterated integrals

$$\int_{x=0}^{x=\pi} \left(\int_{y=0}^{y=\sin x} 7\,dy \right) dx + \int_{x=\pi}^{x=2\pi} \left(\int_{y=\sin x}^{y=0} 7\,dy \right) dx$$

$$= \int_{x=0}^{x=\pi} 7\sin x\,dx + \int_{x=\pi}^{x=2\pi} -7\sin x\,dx = \left[-7\cos x \right]_{x=0}^{x=\pi} + \left[7\cos x \right]_{x=\pi}^{x=2\pi}$$

$$= -7(-1-1) + 7(1-(-1)) = 28.$$

$\square$

Example 6.15. Let D be the region bounded by the graphs of $x = y^2$ and $x = y$, and let $f(x,y) = 2xy^3$. See Figure 6.23. Region D is both x simple and y simple. Every point (x,y) in D has $0 \le y \le 1$, and for each y between 0 and 1, $D(y)$ is the interval of points with x values between y^2 and y. The iterated integral

$$\int_0^1 \left(\int_{D(y)} 2xy^3\,dx \right) dy = \int_0^1 \left(\int_{x=y^2}^{x=y} 2xy^3\,dx \right) dy = \int_0^1 \left[x^2 y^3 \right]_{x=y^2}^{x=y} dy$$

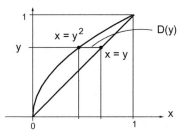

Fig. 6.23 The set D in Example 6.15.

$$= \int_0^1 (y^2 y^3 - y^4 y^3)\,dy = \int_0^1 (y^5 - y^7)\,dy = \left[\tfrac{1}{6}y^6 - \tfrac{1}{8}y^8\right]_0^1 = \tfrac{1}{6} - \tfrac{1}{8} = \tfrac{1}{24}.$$

Viewing D as y simple we see that every point in D has $0 \le x \le 1$. The set $D(x)$ is between x and $\sqrt{x}$. The iterated integral

$$\int_0^1 \left(\int_{D(x)} 2xy^3\,dy \right) dx = \int_0^1 \left(\int_{y=x}^{y=\sqrt{x}} 2xy^3\,dy \right) dx$$

$$= \int_0^1 \left[\tfrac{1}{2}xy^4 \right]_{y=x}^{y=\sqrt{x}} dx = \int_0^1 (\tfrac{1}{2}x^3 - \tfrac{1}{5}x^5)\,dx = \tfrac{1}{8} - \tfrac{1}{12} = \tfrac{1}{24}.$$

□

We can generalize the notion of an iterated integral over an x simple set to deal with more complicated sets. Denote by $D(y)$ the set of all points of a set D whose second coordinate is y. For each value of y, suppose $D(y)$ consists of a finite number of intervals parallel to the x axis whose endpoints are piecewise continuous functions of y. Since D is bounded, $D(y)$ is empty for all y outside some interval (c,d). From single variable calculus we know that for such a set D, the integral of a continuous function $f(x,y)$ with respect to x over $D(y)$ is a piecewise continuous function of y. Integrate this function with respect to y between c and d. Denote the resulting number as

$$II(f,D) = \int_c^d \left(\int_{D(y)} f(x,y)\,dx \right) dy.$$

We call this the iterated integral of f over D. We show now the following.

Theorem 6.11. *Let f be a continuous function on a smoothly bounded set D. Then the iterated integral $II(f,D)$ is equal to the integral $\int_D f\,dA$.*

Proof. To see this we show that the iterated integral has all four properties of the integral listed in Theorem 6.10.

Property (a),

$$II(f + g, D) = II(f, D) + II(g, D),$$

follows from properties of integrals for functions of a single variable, since for each y, the integral of $f + g$ with respect to x over $D(y)$ is the sum of the integral of f with respect to x and the integral of g with respect to x over $D(y)$. The integral of this sum with respect to y is the sum of the integrals with respect to y of the two terms. A similar argument shows property (b):

$$II(cf, D) = cII(f, D).$$

(c) To derive the lower and upper bound property of the iterated integral we apply first the lower and upper bound property of integrals with respect to x. Denote by $L(y)$ the sum of the lengths of the intervals constituting $D(y)$. Suppose the values $f(x, y)$ are between ℓ and u. The integral of $f(x, y)$ with respect to x over $D(y)$ lies between some bounds $\ell L(y)$ and $u L(y)$:

$$\ell L(y) \le \int_{D(y)} f(x, y) \, dx \le u L(y)$$

for all y. It follows from properties of integrals of a function of a single variable that

$$\ell \int_c^d L(y) \, dy \le \int_c^d \left(\int_{D(y)} f(x, y) \, dx \right) dy \le u \int_c^d L(y) \, dy.$$

From single variable calculus, or by an argument similar to that in Example 6.8, we know that the integral of $L(y)$ with respect to y is the area of D. So the inequality above can be rewritten as

$$\ell \operatorname{Area}(D) \le II(f, D) \le u \operatorname{Area}(D),$$

which is property (c) for iterated integrals.

To show property (d), additivity with respect to the domain of integration, we note that if C and D are disjoint or have only boundary points in common, then $C(y)$ and $D(y)$ are disjoint or have only boundary points in common. Therefore for each y

$$\int_{(C \cup D)(y)} f(x, y) \, dx = \int_{C(y)} f(x, y) \, dx + \int_{D(y)} f(x, y) \, dx.$$

Integrating both sides with respect to y we get

$$II(f, C \cup D) = \int_c^d \left(\int_{(C \cup D)(y)} f(x, y) \, dx \right) dy$$

$$= \int_c^d \left(\int_{C(y)} f(x, y) \, dx \right) dy + \int_c^d \left(\int_{D(y)} f(x, y) \, dx \right) dy = II(f, C) + II(f, D).$$

This completes the proof of additivity for the iterated integral.

Now that we have shown that the iterated integral has all four properties of the integral listed at the beginning of Section 6.2, we appeal to Theorem 6.10, that these four properties imply $II(f,D)$ is equal to the integral of f over D.

Reversing the roles of x and y we have the analogous result for iterated integrals,

$$\int_a^b \left(\int_{D(x)} f(x,y)\,dy \right) dx.$$

$\square$

In the special case where f is continuous on a rectangle $D = [a,b] \times [c,d]$ then by Theorem 6.11

$$\int_D f\,dA = \int_{y=c}^{y=d} \left(\int_{x=a}^{x=b} f(x,y)\,dx \right) dy,$$

and

$$\int_D f\,dA = \int_{x=a}^{x=b} \left(\int_{y=c}^{y=d} f(x,y)\,dy \right) dx.$$

Therefore

$$\int_a^b \left(\int_c^d f\,dy \right) dx = \int_c^d \left(\int_a^b f\,dx \right) dy.$$

Example 6.16. Let D be the rectangle given by

$$2 \le x \le 3, \qquad 0 \le y \le 1.$$

Then

$$\int_D xy^2\,dA = \int_2^3 \left(\int_0^1 xy^2\,dy \right) dx = \int_2^3 \left[\tfrac{1}{3}xy^3 \right]_{y=0}^1 dx = \int_2^3 \tfrac{1}{3}x\,dx = \tfrac{5}{6}.$$

Integrating first with respect to x and then with respect to y we get

$$\int_D xy^2\,dA = \int_0^1 \left(\int_2^3 xy^2\,dx \right) dy = \int_0^1 \tfrac{5}{2}y^2\,dy = \tfrac{5}{6}.$$

$\square$

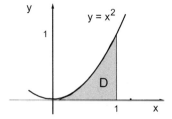

Fig. 6.24 The region in Example 6.17.

Example 6.17. We can compute the double integral of $f(x,y) = xy^3$ over the domain D shown in Figure 6.24. D is bounded by the lines $y = 0$, $y = 1$, and the graphs of the functions (x in terms of y) $x = 1$ and $x = \sqrt{y}$. Then

$$\int_D xy^3\, dA = \int_0^1 \left(\int_{\sqrt{y}}^1 xy^3\, dx \right) dy = \int_0^1 \tfrac{1}{2}(y^3 - y^4)\, dy = \tfrac{1}{2}(\tfrac{1}{4} - \tfrac{1}{5}) = \tfrac{1}{40}.$$

This double integral may also be evaluated by

$$\int_D xy^3\, dA = \int_0^1 \left(\int_{y=0}^{y=x^2} xy^3\, dy \right) dx$$

$$= \int_0^1 x\left(\tfrac{1}{4}(x^2)^4\right) - \tfrac{1}{4}(0)^2\,dx = \int_0^1 \tfrac{1}{4}x^9\, dx = \tfrac{1}{40}.$$

$\square$

Problems

6.24. Compute

$$\int_D f(x,y)\, dA$$

where D is the set in the plane bounded by the graphs of $y = 1$, $x = y$, and $x = 4 - y$, and $f(x,y) = e^{x+y}$.

6.25. Evaluate

$$\int_D y\, dA$$

where D is the half disk where $x^2 + y^2 \leq 1$ and $y \geq 0$.

6.26. Evaluate the integrals.

(a) $\displaystyle\int_D (x^2 - y^2)\, dA$ where $D = [-1,1] \times [0,2]$.

(b) $\displaystyle\int_D x^2 y^2 (x+y)\, dx dy$ where $D = [0,1] \times [0,1]$

6.27. Consider the integrals

$$\int_D \sqrt{y}\, dA, \quad \int_D x\, dA, \quad \int_D (\sqrt{y} + x)\, dA,$$

where D is the triangular region in Figure 6.25.

(a) List the three integrals in order from least to greatest, using inequalities among the functions $\sqrt{y}$, x and $\sqrt{y} + x$ on D.

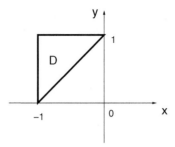

Fig. 6.25 The set D in Problem 6.27.

(b) Evaluate the integrals as iterated integrals.

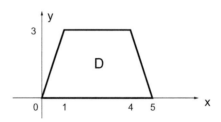

Fig. 6.26 County D in Problem 6.28.

6.28. The population density [people/area] of County D, Figure 6.26, where every-one wants to live near the Southwest corner, is

$$p(x,y) = \frac{c}{(1+3x+y)^2}.$$

The total population is 10^5 and c is a constant.

(a) Find the average population density.

(b) Express the total population as an integral over D, and set it up as an iterated integral.

(c) Find c.

6.29. Evaluate

$$\int_R \sin y \, dA$$

where R is the region $0 \le x \le 1$, $0 \le y \le \sqrt{x}$.

6.30. Evaluate

$$\int_R e^y x \, dA$$

where R is the region bounded by $y = x + 2$ and $y = x^2$.

6.31. Supply the missing numbers.

$$\int_{[0,2]\times[-1,(?)]} (3xy^2 + 5x^4y^3)\,dA = (?)\int_0^{(?)} x\,dx \int_{-1}^1 y^2\,dy + (?)\int_0^2 x^4\,dx \int_{-1}^{(?)} y^3\,dy.$$

6.4 Change of variables in a double integral

A change of variables relates two pairs of variables (x,y) and (u,v) by a function $\mathbf{F}$ from $\mathbb{R}^2$ to $\mathbb{R}^2$.

Definition 6.8. A continuously differentiable function

$$\mathbf{F}(u,v) = (x(u,v), y(u,v))$$

from an open set U in the plane to $\mathbb{R}^2$ is called a *smooth change of variables* if $\mathbf{F}$ is one to one and its derivative matrix is invertible at each point of U.

If $\mathbf{F}$ is a smooth change of variables then the Jacobian

$$J\mathbf{F}(u,v) = \det D\mathbf{F}(u,v) = \det \begin{bmatrix} x_u & x_v \\ y_u & y_v \end{bmatrix}$$

is nonzero and its absolute value can be interpreted as the local magnification factor at (u,v) of area under the mapping. To see this take the triangle S whose vertices are

$$(u,v), \quad (u+h,v), \quad (u,v+h).$$

These points are mapped into the points

$$(x(u,v), y(u,v)), \quad (x(u+h,v), y(u+h,v)), \quad (x(u,v+h), y(u,v+h))$$

in the x,y plane and are vertices of a triangle T. See Figure 6.27. We approximate $(x(u+h,v), y(u+h,v))$ by $(x+hx_u, y+hy_u)$ and approximate $(x(u,v+h), y(u,v+h))$ by $(x+hx_v, y+hy_v)$, where the functions x, y and their partial derivatives are evaluated at (u,v). Denote by T' the triangle with vertices

$$(x,y), \quad (x+hx_u, y+hy_u), \quad (x+hx_v, y+hy_v).$$

For h small the shaded area of $\mathbf{F}(S)$ in Figure 6.27 is close to the area of the triangular region T. The error in the linear approximation of sides of T by the sides of T' is less than a multiple of h^2. Two of the side lengths of T' are $\|(hx_u, hy_u)\|$ and $\|(hx_v, hy_v)\|$, multiples of h, so changing the length of a side of T' by a multiple of h^2 changes the area by less than kh^3, for some number k. (See Problem 6.33.) So the area of T differs from the area of T' by an amount that is less than kh^3. Therefore

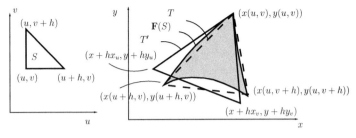

Fig. 6.27 Area(T) − Area(T') is less than a constant multiple of h^3.

$$0 \le \frac{\text{Area}(T) - \text{Area}(T')}{\text{Area}(S)} \le \frac{kh^3}{\frac{1}{2}h^2}$$

and the limit of the quotient as h tends to zero is zero. Therefore as h tends to zero, $\dfrac{\text{Area}(T')}{\text{Area}(S)}$ and $\dfrac{\text{Area}(T)}{\text{Area}(S)}$ have the same limit. The area of the triangle T' whose vertices are given above is

$$\text{Area}(T') = \left| \tfrac{1}{2}(hx_u hy_v - hx_v hy_u) \right| = \tfrac{1}{2}|x_u y_v - x_v y_u| h^2.$$

The area of S is $\frac{1}{2}h^2$. The ratio of these two areas is

$$|J\mathbf{F}(u,v)| = |x_u y_v - x_v y_u|, \tag{6.19}$$

the absolute value of the Jacobian at (u,v). If the mapping $\mathbf{F}$ *preserves orientation.* That is if it maps small positively oriented triangles in the u,v plane to positively oriented approximate triangles in the x,y plane then the ratio of the areas is simply the Jacobian J of the mapping.

Here is the change of variables theorem for integrals.

Theorem 6.12. *Let $F(u,v) = (x(u,v), y(u,v))$ denote a smooth change of variables that maps a smoothly bounded set C onto a smoothly bounded set D, so that the boundary of C is mapped to the boundary of D. Denote by f a continuous function on D. Then*

$$\int_D f(x,y)\,dx\,dy = \int_C f(x(u,v), y(u,v))|J\mathbf{F}(u,v)|\,du\,dv \tag{6.20}$$

where $J\mathbf{F}$ is the Jacobian of the mapping.

Proof. Divide the u,v plane into h-squares and approximate

$$\int_C f(x(u,v), y(u,v))|J\mathbf{F}(u,v)|\,du\,dv$$

by a finite sum

$$S(f,C,h) = \sum_j f(\mathbf{F}(u_j,v_j))|J\mathbf{F}(u_j,v_j)|h^2,$$

where (u_j,v_j) is some point in the j-th h-square, and the sum taken over all h-squares contained in C. The mapping $\mathbf{F}$ carries the h-squares in the u,v plane into smoothly bounded sets in the x,y plane that we denote as D_j. The area of D_j is approximately $|J\mathbf{F}(u_j,v_j)|h^2$, with an error less than some multiple of h^3.

Let D' be the union of the sets D_j. The integral of f over D' is the sum of the integrals of f over D_j:

$$\int_{D'} f\,dx\,dy = \sum_j \int_{D_j} f\,dx\,dy.$$

We approximate the j-th integral above as $f(u_j,v_j)\,\mathrm{Area}(D_j)$, and we approximate the $\mathrm{Area}(D_j)$ by $|J\mathbf{F}(u_j,v_j)|h^2$. The error thus committed in each term in the sum is less than ch^3, for some constant c. The total error committed is less than or equal to the sum of the individual errors. Since each individual error is less than ch^3, and since the number of terms in the sum is less $\dfrac{\mathrm{Area}(D)}{h^2}$, the total error is less than

$$\frac{\mathrm{Area}(D)}{h^2}(ch^3) = \mathrm{Area}(D)ch.$$

This shows that

$$\left| S(f,C,h) - \int_{D'} f(x,y)\,dA \right| \le Ch.$$

So $\displaystyle\int_{D'} f(x,y)\,dA$ and $\displaystyle\int_{C} f(x(u,v),y(u,v))|J\mathbf{F}(u,v)|\,du\,dv$ differ less than some constant times h. For h small enough the area of the set of points that are in D but not in D' does not exceed Mch where M is an upper bound on $|J\mathbf{F}(u,v)|$ on the closure of C and c is a number that depends on the boundary of C. Hence the difference between $\displaystyle\int_{D'} f(x,y)\,dA$ and $\displaystyle\int_{D} f(x,y)\,dA$ tends to zero as h tends to zero. $\square$

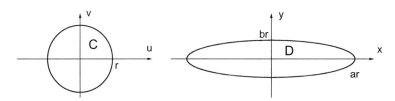

Fig. 6.28 The smooth change of variables in Example 6.18 maps the circle to the ellipse.

Example 6.18. Define sets

$$C = \{(u,v) : u^2 + v^2 \le r^2\}, \qquad D = \left\{(x,y) : \left(\frac{x}{a}\right)^2 + \left(\frac{y}{b}\right)^2 \le r^2\right\}.$$

The boundary of C is a circle and the boundary of D is an ellipse. See Figure 6.28. The mapping $\mathbf{F}(u,v) = (au, bv)$, $x = au$, $y = bv$, $a > 0$, $b > 0$ sends points (u,v) in the set C one to one onto the points of D. Find the area of D. We have

$$J\mathbf{F}(u,v) = \det D\mathbf{F} = \det \begin{bmatrix} a & 0 \\ 0 & b \end{bmatrix} = ab > 0,$$

so $\mathbf{F}$ is orientation preserving. By the change of variables theorem

$$\text{Area}(D) = \int_D 1 \, dx \, dy = \int_C 1(ab) \, du \, dv = ab\pi r^2.$$

$\square$

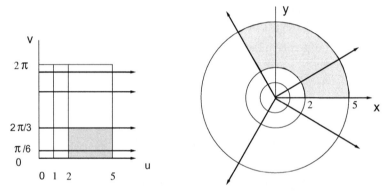

Fig. 6.29 The mapping $\mathbf{F}$ in Examples 6.19 and 6.20.

Example 6.19. Let D be the shaded set on the right in Figure 6.29. Find $\int_D e^{x^2+y^2} \, dA$. The iterated integrals we would obtain in x,y coordinates will not help us evaluate this integral because we cannot find a simple antiderivative for $e^{x^2+y^2}$ with respect to x or y. So we try a change of variables $\mathbf{F}$ that is suggested by the polar coordinate transformation

$$x = u\cos v, \quad y = u\sin v.$$

Note that $\mathbf{F}$ is one to one on an open set that contains the rectangle C given by $2 \le u \le 5$, $0 \le v \le 2\pi/3$. The Jacobian is

$$J\mathbf{F}(u,v) = \det D\mathbf{F}(u,v) = \det \begin{bmatrix} \cos v & -u\sin v \\ \sin v & u\cos v \end{bmatrix} = u\cos^2 v + u\sin^2 v = u.$$

The Jacobian is positive in C. By the change of variables theorem,

$$\int_D e^{x^2+y^2}\,dx\,dy = \int_C e^{u^2\cos^2 v + u^2\sin^2 v}\,u\,du\,dv = \int_{v=0}^{v=2\pi/3}\left(\int_{u=2}^{u=5} e^{u^2}u\,du\right)dv$$

$$= \int_{v=0}^{v=2\pi/3}\left[\tfrac{1}{2}e^{u^2}\right]_{u=2}^{u=5}dv = \int_{v=0}^{v=2\pi/3}\tfrac{1}{2}(e^{25}-e^4)\,dv = \tfrac{1}{2}(e^{25}-e^4)\tfrac{2\pi}{3}.$$

□

Remark: It is customary to write (r,θ) for polar coordinates instead of (u,v). Then the change of variables formula is

$$\int_D f(x,y)\,dx\,dy = \int_C f(r\cos\theta, r\sin\theta)\,r\,dr\,d\theta.$$

Example 6.20. Let $D = \{(x,y) : x^2 + y^2 \le 25\}$ and let C denote the rectangle $0 \le u \le 5,\ 0 \le v \le 2\pi$. See Figure 6.29. We evaluate

$$\int_D e^{x^2+y^2}\,dA.$$

The polar coordinate mapping $\mathbf{F}(u,v) = (u\cos v, u\sin v)$ is not a smooth change of variables on a set that contains C: $\det D\mathbf{F}(0,0) = 0$ and also $\mathbf{F}$ is not one to one, $\mathbf{F}(u,0) = \mathbf{F}(u,2\pi)$. For $0 < \epsilon$ let

$$C_\epsilon = \{(u,v) : \epsilon \le u \le 5,\ 0 \le v \le 2\pi - \epsilon\}$$

and set $D_\epsilon = \mathbf{F}(C_\epsilon)$. Then the change of variables theorem applies. The integrand $e^{x^2+y^2}$ is bounded on D and the area of the region $C - C_\epsilon$ tends to zero as ϵ tends to zero, as does the area of the region $D - D_\epsilon$. Therefore as ϵ tends to zero, the relation

$$\int_{D_\epsilon} e^{x^2+y^2}\,dx\,dy = \int_{C_\epsilon} e^{u^2}u\,du\,dv$$

tends to

$$\int_D e^{x^2+y^2}\,dx\,dy = \int_C e^{u^2}u\,du\,dv.$$

We evaluate the last integral as an iterated integral

$$= \int_{v=0}^{v=2\pi}\left(\int_{u=0}^{u=5} e^{u^2}u\,du\right)dv = \int_{v=0}^{v=2\pi}\left[\tfrac{1}{2}e^{u^2}\right]_{u=0}^{u=5}dv = \tfrac{1}{2}(e^{25}-1)2\pi.$$

□

Example 6.21. Let D be the parallelogram in $\mathbb{R}^2$ bounded by the lines

$$y = -x + 5,\quad y = -x + 2,\quad y = 2x - 1,\quad y = 2x - 4.$$

and let R be the unit square $[0,1] \times [0,1]$. See Figure 6.30. Compute

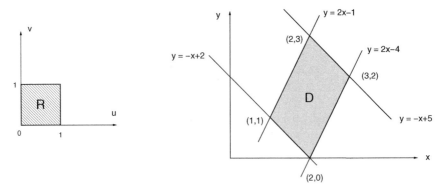

Fig. 6.30 The region in Example 6.21.

$$\int_D e^x \, dA.$$

The mapping $\mathbf{F}(u,v) = (u + v + 1, -u + 2v + 1)$ takes R one to one onto the parallelogram D.

$$D\mathbf{F}(u,v) = \begin{bmatrix} 1 & 1 \\ -1 & 2 \end{bmatrix}, \quad J\mathbf{F}(u,v) = \det(D\mathbf{F}(u,v)) = 3.$$

So

$$\int_D e^x \, dx\,dy = \int_R e^{u+v+1} |3| \, du\,dv = 3e \int_0^1 \int_0^1 e^u e^v \, du\,dv$$

$$= 3e \int_0^1 e^u \, du \int_0^1 e^v \, dv = 3e(e-1)^2.$$

$\square$

Problems

6.32. Consider the mapping

$$\begin{bmatrix} x \\ y \end{bmatrix} = \begin{bmatrix} 1 & 1 \\ 0 & 1 \end{bmatrix} \begin{bmatrix} u \\ v \end{bmatrix}.$$

(a) Show that the Jacobian is 1.
(b) Sketch the triangle with vertices $(1,0)$, $(1,1)$, and $(0,1)$ and its image under the mapping. Show that its image has the same area.

6.33. Figure 6.31 shows a triangle whose sides are multiples of h. The size of the triangle has been changed by moving one side so that one edge is changed by kh^2 for some k. Use the steps below to show that the change in area is less than or equal to some constant times h^3.

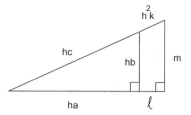

Fig. 6.31 The triangles in Problem 6.33.

(a) Show that $\ell = k_1 h^2$ for some k_1.
(b) Show that $m = k_2 h + k_3 h^2$ for some k_2, k_3.
(c) Show that the change in area is less than some multiple of h^3 if h is sufficiently small.

6.34. For a mapping of the form $(x,y) = (f(u), g(v))$, verify that the Jacobian is given by

$$\det \begin{bmatrix} x_u & x_v \\ y_u & y_v \end{bmatrix} = f'(u)g'(v).$$

6.35. Consider the mapping defined by the complex square

$$x + iy = (u + iv)^2,$$

that is, $x = u^2 - v^2$, $y = 2uv$. It maps the quarter annulus shown in Figure 6.32 to the half annulus, for example. Find the Jacobian at the point $(1,0)$, and sketch the approximate image of the small indicated triangle, that has vertices at $(1,0)$, $(1.1,0)$, and $(1,.1)$.

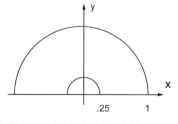
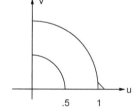

Fig. 6.32 The mapping in Problem 6.35.

6.36. Use the change of variables $(x,y) = (u\cos v, u\sin v)$ to evaluate the integrals, where U is the unit disk centered at the origin.

(a) $\displaystyle\int_U \sqrt{x^2 + y^2}\, dA,$

(b) $\displaystyle\int_U (3 + x^4 + 2x^2 y^2 + y^4)\, dA,$

(c) $\displaystyle\int_U y\,dA.$

6.37. Use the change of variables $(x,y) = (r\cos\theta, r\sin\theta)$ to polar coordinates to evaluate the integral

$$\int_D e^{-\|X\|}\,dA$$

where D is the set of points (x,y) with $a^2 \le x^2 + y^2 \le b^2$. a,b are positive constants.

6.38. Use symmetries or the change of variables $(x,y) = (r\cos\theta, r\sin\theta)$ to evaluate the integrals, where D is the annulus

$$1 \le \sqrt{x^2 + y^2} \le 8.$$

(a) $\displaystyle\int_D (x^2 + y^2)^p\,dA,\ \ p$ any number

(b) $\displaystyle\int_D (3 + y - x^2)\,dA,$

(c) $\displaystyle\int_D \log(x^2 + y^2)\,dA.$

6.39. Verify that each of the following formulas defines a mapping of the unit square C to the rectangle D shown in Figure 6.33. In each case, make a sketch that shows your initials written in C, and their image in D.

(a) $(x,y) = (5u, 3v)$
(b) $(x,y) = (5v, 3u)$
(c) $(x,y) = (5v, 3 - 3u)$

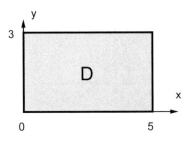

Fig. 6.33 The sets in Problem 6.39.

6.40. For each mapping in Problem 6.39,

(a) calculate the Jacobian $\det \begin{bmatrix} x_u & x_v \\ y_u & y_v \end{bmatrix}$, and

(b) state whether the mapping is orientation preserving or not.

6.41. Substitutions are made in the integral $\int_D (1+xy)\,dxdy$ using the mappings in Problem 6.39 and their Jacobians from Problem 6.40. Verify the formulas:

(a) $\displaystyle\int_D (1+xy)\,dxdy = \int_C (1+15uv)15\,dudv.$

(b) $\displaystyle\int_D (1+xy)\,dxdy \neq \int_C (1+15vu)(-15)\,dudv.$

(c) $\displaystyle\int_D (1+xy)\,dxdy = \int_C (1+15v(1-u))15\,dudv.$

6.5 Integration over unbounded sets

To integrate a function f over an unbounded set we will require that the unbounded set D is the union of an increasing sequence

$$D_1 \subset D_2 \subset D_3 \subset \cdots$$

of smoothly bounded sets.

We start by giving some examples.

Example 6.22. (a) The first quadrant D, consisting of all points (x,y) with x and y nonnegative, is unbounded. Let D_n be the square $0 \le x \le n,\ 0 \le y \le n$. See Figure 6.34. Quadrant D is the union of the sets D_n.

(b) Let D again be the first quadrant and let D_n be the points where $x \ge 0, y \ge 0$, and $x+y \le n$. Again D is the union of the D_n. See Figure 6.34.

(c) D is the half plane consisting of all points (x,y) with $x \ge 0$, and D_n is the rectangle $0 \le x \le n,\ -n \le y \le n$.

(d) D is the whole plane and D_n is the square $-n \le x \le n,\ -n \le y \le n$.

(e) D is the whole plane and D_n is the disc of radius n centered at the origin, $x^2 + y^2 \le n^2$,

□

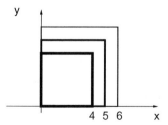 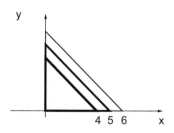

Fig. 6.34 The first quadrant is the union of increasing squares or triangles.

Integration of a continuous function f over an unbounded set D is defined by considering the limit of a sequence of integrals of f over smoothly bounded sets.

Definition 6.9. Let D be an unbounded set in $\mathbb{R}^2$. Denote by $D(n)$ the set of points in D whose distance from the origin is less than or equal to n. See Figure 6.35. A nonnegative continuous function f defined on D is *integrable* over D if the sequence of numbers

$$\int_{D(n)} f \, dA, \qquad n = 1, 2, 3, \ldots$$

converges. The limit of this sequence is called the *integral* of f over D. We say that the integral *exists*, and write

$$\int_D f \, dA = \lim_{n \to \infty} \int_{D(n)} f \, dA.$$

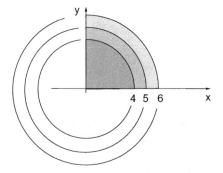

Fig. 6.35 The sets $D(n)$ engulf the first quadrant D. The sets $D(4)$, $D(5)$, and $D(6)$ are indicated.

We show now that we may replace the sequence of subsets $D(n)$ by any increasing sequences of bounded subsets D_n whose union is D:

Theorem 6.13. *Let D be an unbounded set, and*

$$D_1 \subset D_2 \subset D_3 \subset \cdots$$

an increasing sequence of smoothly bounded subsets of D whose union is D, and let f be a nonnegative continuous function on D. Then

$$\lim_{m \to \infty} \int_{D_m} f \, dA = \lim_{n \to \infty} \int_{D(n)} f \, dA$$

in the sense that if one limit exists, so does the other, and the limits are equal.

Proof. Suppose first that $\lim\limits_{m\to\infty}\int_{D_m} f\,dA$ exists. For a nonnegative continuous function f both sequences of integrals above are increasing. We observe that for each $D(n)$, all subsets D_m for m large enough contain $D(n)$; it follows that

$$\lim_{m\to\infty}\int_{D_m} f\,dA \geq \int_{D(n)} f\,dA.$$

Therefore by the Monotone Convergence Theorem $\lim\limits_{n\to\infty}\int_{D(n)} f\,dA$ exists and

$$\lim_{m\to\infty}\int_{D_m} f\,dA \geq \lim_{n\to\infty}\int_{D(n)} f\,dA.$$

Now since the sets D_m are bounded, all subsets $D(n)$ for n large enough contain D_m. From this the opposite inequality follows. Therefore the two limits are equal. The same argument applies if we reverse the roles of D_n and $D(n)$.   □

Here is an example of an integral over the whole plane:

$$\int_{\mathbb{R}^2} e^{-x^2-y^2}\,dA. \tag{6.21}$$

We evaluate this integral over the plane in two different ways.

(i) Choose as the sequence D_n the squares $-n \leq x \leq n,\ -n \leq y \leq n$. According to Theorem 6.11

$$\int_{D_n} e^{-x^2-y^2}\,dA = \int_{-n}^{n}\left(\int_{-n}^{n} e^{-x^2} e^{-y^2}\,dx\right) dy = \int_{-n}^{n} e^{-y^2}\,dy\int_{-n}^{n} e^{-x^2}\,dx.$$

The two integrals on the right, one with respect to x, the other with respect to y, are equal. Therefore the right side is equal to

$$\left(\int_{-n}^{n} e^{-x^2}\,dx\right)^2.$$

Does this sequence of numbers converge? For $x \geq 1$ we have $0 \leq e^{-x^2} \leq e^{-x}$ and since $\int_{0}^{\infty} e^{-x}\,dx$ exists, $\int_{0}^{\infty} e^{-x^2}\,dx$ exists as well. By symmetry

$$\int_{-\infty}^{0} e^{-x^2}\,dx = \int_{0}^{\infty} e^{-x^2}\,dx.$$

So $\int_{-n}^{n} e^{-x^2}\,dx$ converges to $\int_{-\infty}^{\infty} e^{-x^2}\,dx$, and

$$\int_{\mathbb{R}^2} e^{-x^2-y^2}\,dA = \left(\int_{-\infty}^{\infty} e^{-x^2}\,dx\right)^2 \tag{6.22}$$

exists. Next we find its numerical value.

(ii) Choose as the sequence D_n the discs $x^2 + y^2 \le n^2$. In the integral over D_n, make a change to polar coordinates $x = r\cos\theta$, $y = r\sin\theta$. As we saw in Example 6.20, when we make the change of variables we get

$$\int_{D_n} e^{-x^2-y^2}\, dA = \int_{P_n} e^{-r^2} r\, dr\, d\theta$$

where P_n is the rectangle $0 \le r \le n$, $0 \le \theta < 2\pi$.

This integral can be carried out explicitly; it is equal to

$$\int_0^n \left(\int_0^{2\pi} d\theta \right) e^{-r^2} r\, dr = \pi(1 - e^{-n^2}).$$

The limit is π as n tends to infinity. This shows that $\int_{\mathbb{R}^2} e^{-x^2-y^2}\, dA = \pi$. By equation (6.22) we conclude that

$$\int_{-\infty}^{\infty} e^{-x^2}\, dx = \sqrt{\pi}. \tag{6.23}$$

It is curious that we have found the value of the integral (6.23) with respect to a single variable by finding the value of the integral (6.21) with respect to two variables.

We remark that the evaluation of the integral (6.23) is basic in the theory of probability. The graph of the function

$$f(x) = \frac{1}{\sqrt{\pi}} e^{-x^2}$$

is the classic bell curve, of the normal distribution. Probability is an important application of integration over unbounded sets.

Not all functions describing probabilities are continuous on all of $\mathbb{R}^2$. So we extend the concept of integration over an unbounded set D to functions that are not continuous on D.

Definition 6.10. Let D be an unbounded set in the plane. Let f be a bounded nonnegative function on D that is integrable on $D(m)$ for some m (as described in Definition 6.9) and is continuous on $D - D(m)$. We say that f is *integrable* on D if the sequence of integrals

$$\int_{D(m)} f\, dA, \int_{D(m+1)} f\, dA, \int_{D(m+2)} f\, dA, \dots$$

converges. We denote this limit

$$\lim_{n\to\infty} \int_{D(n)} f\, dA = \int_D f\, dA.$$

Definition 6.11. We say that a function p is a *probability density* function if

(a) $p(x,y) \geq 0$ for all (x,y) in $\mathbb{R}^2$, and

(b) $\int_{\mathbb{R}^2} p(x,y)\, dA = 1$.

If p is integrable on a set D then the *probability* that (x,y) is in D is

$$\int_D p(x,y)\, dA.$$

Example 6.23. Show that

$$p(x,y) = \frac{1}{\pi} e^{-(x^2+y^2)}$$

is a probability density function, and find the probability that (x,y) is in the first quadrant. We showed that $\int_{\mathbb{R}^2} e^{-x^2-y^2}\, dA = \pi$, so

$$\int_{\mathbb{R}^2} \frac{1}{\pi} e^{-(x^2+y^2)}\, dA = 1.$$

Since p is nonnegative, p is a probability density function. The probability that (x,y) is in the first quadrant is

$$\int_{x\geq 0,\, y\geq 0} \frac{1}{\pi} e^{-(x^2+y^2)}\, dA = \lim_{n\to\infty} \int_0^{\pi/2} \frac{1}{\pi} \int_0^n e^{-r^2} r\, dr\, d\theta = \frac{\pi}{2}\frac{1}{\pi} \lim_{n\to\infty} \left[\frac{-1}{2} e^{-r^2} \right]_0^n = \frac{1}{4}.$$

We might also have reasoned that by symmetry the integral of p on each of the four quadrants is equal; therefore the probability is $\frac{1}{4}$. □

In the next example p is not continuous on $\mathbb{R}^2$ but it is integrable on $\mathbb{R}^2$.

Example 6.24. Let

$$p(x,y) = \begin{cases} \frac{2x+c-y}{4} & 0 \leq x \leq 1 \text{ and } 0 \leq y \leq 2 \\ 0 & \text{otherwise} \end{cases}$$

Find c so that p is a probability density function. p is not continuous on $\mathbb{R}^2$ but by Definition 6.10 p is integrable over $\mathbb{R}^2$. The integral of p over $\mathbb{R}^2$ is

$$\int_{\mathbb{R}^2} p\, dA = \int_{D=[0,1]\times[0,2]} p\, dA + \int_{\mathbb{R}^2-D} p\, dA$$

$$= \int_0^2 \int_0^1 \frac{2x+c-y}{4}\, dx\, dy + 0 = \int_0^2 \frac{1+c-y}{4}\, dy = \frac{2+2c-2}{4} = \frac{c}{2}.$$

So $c = 2$. We check that p is nonnegative: in the rectangle $0 \leq x \leq 1$, $0 \leq y \leq 2$ we have $x \geq 0$ and $2 - y \geq 0$, so

$$p(x,y) = \frac{2x+(2-y)}{4} \geq 0.$$

□

We end this section by extending the notion of integrability in an unbounded domain to continuous functions that take on both positive and negative values.

Theorem 6.14. *Let D be an unbounded set, and f a continuous function on D whose absolute value is integrable over D,*

$$\int_D |f|\,dA \quad \text{exists.}$$

Let D_n be an increasing sequence of smoothly bounded subsets of D whose union is D,

$$D_1 \subset D_2 \subset \cdots \subset D_n \subset \cdots, \qquad \cup_n D_n = D.$$

Then the limit of the integrals of f over the sequence D_n,

$$\lim_{n\to\infty} \int_{D_n} f\,dA \quad \text{exists,}$$

and this limit is the same for all such sequences D_n.

Proof. Decompose f as the difference of its positive and negative parts, $f = f_+ - f_-$, where

$$f_+(x,y) = \max(f(x,y),0), \quad f_-(x,y) = \max(-f(x,y),0).$$

The absolute value of f is the function

$$|f| = f_+ + f_-.$$

This shows that the nonnegative functions f_+ and f_- are less than or equal to $|f|$. Since $|f|$ is integrable over D, so are f_+ and f_- (see Problem 6.50).

According to Theorem 6.13, the integral of f_+ and f_- over D is the limit of the integral of f_+ and f_- over any increasing sequence of smoothly bounded subsets D_n of D whose union is D. It follows that the limit of the integral of their difference $f_+ - f_-$ over D_n is the difference of the limits of the integral f_+ and of f_- over the sequence of subsets D_n. Since $f_+ - f_- = f$, this completes the proof of Theorem 6.14.

□

Based on Theorem 6.14 we make the following definition.

Definition 6.12. A continuous function f that takes on both positive and negative values in an unbounded set D is called *integrable* over D if its absolute value $|f|$ is integrable over D.

Problems

6.42. Describe the sets in which the functions below are positive.

(a) $f(x,y) = 1 - x^2$
(b) $f(x,y) = xy$
(c) $f(x,y) = \cos(2\pi \sqrt{x^2 + y^2})$

6.43. Let D be the region in $\mathbb{R}^2$ where $0 \le x$ and $0 \le y \le 1$.

(a) Evaluate $\displaystyle\int_D e^{-x}\,dxdy$

(b) Evaluate $\displaystyle\int_D e^{-x\sqrt{y}}\,dxdy$

(c) Show that e^{-xy} is not integrable over D.

6.44. Let $a > 0, b > 0$. Use the change of variables theorem to show that

$$\int_{x^2+y^2\le n^2} e^{-(ax^2+by^2)}\,dxdy = \int_{\frac{u^2}{a}+\frac{v^2}{b}\le n^2} e^{-(u^2+v^2)}\,\frac{1}{\sqrt{ab}}\,dudv,$$

and evaluate

$$\int_{\mathbb{R}^2} e^{-(ax^2+by^2)}\,dA.$$

6.45. Let U be the open unit disk in $\mathbb{R}^2$ centered at the origin, $\sqrt{x^2 + y^2} = r < 1$. Which of these functions are integrable over $\mathbb{R}^2 - U$?

(a) $\log r$

(b) $\dfrac{1}{r^2}$

(c) $\dfrac{1}{r^{2.1}}$

(d) r^{-3}

6.46. Verify the following identities for the positive and negative parts of numbers x and functions f.

(a) $|x| = x_+ + x_-$
(b) $x = x_+ - x_-$
(c) $|f| = f_+ + f_-$
(d) $f = f_+ - f_-$

6.47. Evaluate $\displaystyle\int_{-\infty}^{\infty} e^{-(x^2/4t)}\,dx$ by a change of variables.

6.48. Evaluate $\displaystyle\int_0^{\infty} e^{-x^2}\,dx$ by a symmetry argument.

6.49. Show that if f is continuous then the positive part of f, f_+, is continuous.

6.50. Justify the following steps to prove that if f is integrable on $\mathbb{R}^2$ and g is a continuous function with $0 \le g \le f$ then g is integrable on $\mathbb{R}^2$.

(a) $\displaystyle\int_{D(n)} g\,dA$ exists.

(b) $\displaystyle 0 \le \int_{D(n)} g\,dA \le \int_{D(n)} f\,dA.$

(c) The numbers $\displaystyle\int_{D(n)} g\,dA$ are an increasing sequence bounded above.

(d) $\displaystyle\lim_{n\to\infty}\int_{D(n)} g\,dA$ exists.

6.51. Let p be the probability density function in Example 6.24.

(a) Sketch the set in $\mathbb{R}^2$ where $x+y \le 2$.
(b) Find the probability that $x+y \le 2$.

6.52. Let $p(x,y) = 2x$ when $0 \le x \le 1$ and $0 \le y \le 1$, and $p(x,y) = 0$ otherwise.

(a) Show that p is a probability density function.
(b) Sketch the set in $\mathbb{R}^2$ where $x \ge y$.
(c) Find the probability that $x \ge y$.

6.53. Let p be a probability density function in $\mathbb{R}^2$. Each point of $\mathbb{R}^2$ is either in a set D or not in D. What does the equation

$$\int_D p\,dA + \int_{\mathbb{R}^2-D} p\,dA = \int_{\mathbb{R}^2} p\,dA$$

tell you about the probability that (x,y) is not in D?

6.54. Suppose f and g are continuous functions on $\mathbb{R}^2$ so that f^2 and g^2 are integrable on $\mathbb{R}^2$. Use the inequality

$$2ab \le a^2 + b^2$$

to show that the following functions are also integrable:

(a) $|fg|$
(b) fg
(c) $(f+g)^2$

6.55. Show that f^2 is integrable on $\mathbb{R}^2$ for each function f:

(a) $\dfrac{r}{1+r^4}$, where $r = \sqrt{x^2+y^2}$
(b) e^{-r}
(c) $\dfrac{|x|}{1+r^4}$
(d) ye^{-r}

6.56. Justify steps (a)–(d) to prove that if a continuous function f from $\mathbb{R}^2$ to $\mathbb{R}$ is integrable on an unbounded set D then

$$\left| \int_D f \, dA \right| \le \int_D |f| \, dA.$$

(a) $\displaystyle \int_D f \, dA = \int_D f_+ \, dA - \int_D f_- \, dA \le \int_D f_+ \, dA + \int_D f_- \, dA = \int_D |f| \, dA$

(b) $\displaystyle \int_D (-f) \, dA \le \int_D |f| \, dA$

(c) $\displaystyle -\int_D f \, dA \le \int_D |f| \, dA$

(d) $\displaystyle \left| \int_D f \, dA \right| \le \int_D |f| \, dA.$

6.6 Triple and higher integrals

We outline key definitions, state theorems, and look at many examples of integrals of functions of three or more variables.

The notion of a smoothly bounded set in $\mathbb{R}^n$ is defined inductively on the dimension n.

Definition 6.13. A closed set D in $\mathbb{R}^n$ is *smoothly bounded* when its boundary is the union of finitely many graphs of C^1 functions of the form

$$x_k = g(x_1, \ldots, x_{k-1}, x_{k+1}, \ldots, x_n) \quad \text{some} \quad k = 1, \ldots, n$$

defined on smoothly bounded sets in the hyperplane where $x_k = 0$.

The notions needed to define the volume of a smoothly bounded set in $\mathbb{R}^n$ are analogous to those we used to define the concept of area of a smoothly bounded set in $\mathbb{R}^2$. In a similar manner we define the *lower* and *upper volumes* and say that a set *has volume* $\mathrm{Vol}(D)$ if its upper and lower volumes are equal. The following theorem is analogous to Theorem 6.4.

Theorem 6.15. *Smoothly bounded sets in $\mathbb{R}^n$ have volume.*

The definition of the integral of a function of n variables mirrors the definition of the integral of a function of two variables.

Definition 6.14. Let f from $\mathbb{R}^n$ to $\mathbb{R}$ be a bounded function on a smoothly bounded set D in $\mathbb{R}^n$. We define the upper and lower integrals

$$I_U(f,D), \quad I_L(f,D)$$

for bounded nonnegative functions as we did in $\mathbb{R}^2$. If the upper integral equals the lower integral we say f is *integrable* over D, and we write the integral

$$I_U(f,D) = I_L(f,D) = \int_D f \, d^n\mathbf{X}.$$

Just as we saw for functions from $\mathbb{R}^2$ to $\mathbb{R}$, we can show that continuous functions are integrable over smoothly bounded sets.

Theorem 6.16. *For a continuous function f on a smoothly bounded set D in $\mathbb{R}^n$, the upper integral of f over D is equal to the lower integral, and the integral of f exists:*

$$I_U(f,D) = I_L(f,D) = \int_D f \, d^n\mathbf{X}.$$

We have the four properties of linearity, boundedness, and additivity:

Theorem 6.17. *For all continuous functions f and g from $\mathbb{R}^n$ to $\mathbb{R}$ on smoothly bounded sets C and D, and all numbers c,*

(a) $\displaystyle\int_D (f+g) \, d^n\mathbf{X} = \int_D f \, d^n\mathbf{X} + \int_D g \, d^n\mathbf{X},$

(b) $\displaystyle\int_D cf \, d^n\mathbf{X} = c \int_D f \, d^n\mathbf{X},$

(c) If $\ell \leq f(\mathbf{X}) \leq u$ then $\ell \operatorname{Vol}(D) \leq \displaystyle\int_D f \, d^n\mathbf{X} \leq u \operatorname{Vol}(D)$.

(d) If C and D are disjoint or have only boundary points in common and $C \cup D$ is smoothly bounded then $\displaystyle\int_{C \cup D} f \, d^n\mathbf{X} = \int_C f \, d^n\mathbf{X} + \int_D f \, d^n\mathbf{X}$.

Example 6.25. The constant function $f(x_1, \ldots, x_n) = 1$ is continuous on $\mathbb{R}^n$ and we have $1 \leq f(\mathbf{X}) \leq 1$. It is integrable over every smoothly bounded set D. By the boundedness property, Theorem 6.17 part (c),

$$\operatorname{Vol}(D) \leq \int_D 1 \, d^n\mathbf{X} \leq \operatorname{Vol}(D).$$

Therefore $\mathrm{Vol}(D) = \displaystyle\int_D 1\,\mathrm{d}^n\mathbf{X}.$ $\qquad\qquad\square$

The integral is characterized by the four properties in Theorem 6.17; we have the following theorem.

Theorem 6.18. *Suppose a number $\mathscr{I}(f,D)$ is defined for every continuous function f on a smoothly bounded set D in $\mathbb{R}^n$ and that it has the following properties.*

(a) $\mathscr{I}(f+g,D) = \mathscr{I}(f,D) + \mathscr{I}(g,D)$
(b) $\mathscr{I}(cf,D) = c\mathscr{I}(f,D)$ for every number c
(c) If $\ell \le f(x_1,x_2,\dots,x_n) \le u$ for all $(x_1,x_2,\dots,x_n)$ in D, then

$$\ell\,\mathrm{Vol}(D) \le \mathscr{I}(f,D) \le u\,\mathrm{Vol}(D),$$

(d) For all pairs of smoothly bounded sets C and D that are disjoint or have only boundary points in common,

$$\mathscr{I}(f,C \cup D) = \mathscr{I}(f,C) + \mathscr{I}(f,D),$$

Then

$$\mathscr{I}(f,D) = \int_D f\,\mathrm{d}^n\mathbf{X}.$$

The proof is similar to the proof of Theorem 6.10 for double integrals.

As with double integrals we often calculate integrals using approximate integrals (Riemann sums).

Theorem 6.19. *The integral $\displaystyle\int_D f\,\mathrm{d}^n\mathbf{X}$ of a continuous function f on a smoothly bounded set D in $\mathbb{R}^n$ is equal to the limit as h tends to zero of approximate integrals*

$$\sum_j f(\mathbf{P}_j)h^n$$

that are sums over h-boxes in D, with any chosen points $\mathbf{P}_j$ in the j-th box.

Average of a function. The integral allows a definition of the average of a function.

Definition 6.15. The *average* of an integrable function over a smoothly bounded set D is

$$\frac{1}{\mathrm{Vol}(D)}\int_D f\,\mathrm{d}^n\mathbf{X} = \frac{\int_D f\,\mathrm{d}^n\mathbf{X}}{\int_D 1\,\mathrm{d}^n\mathbf{X}}.$$

The next theorem gives conditions under which the average of f is a value of f. We say that a set D in $\mathbb{R}^n$ is *connected* if for every two points $\mathbf{A}$ and $\mathbf{B}$ of D, there is a continuous function $\mathbf{X}$ from some interval $[a,b]$ to D so that $\mathbf{X}(a) = \mathbf{A}$ and $\mathbf{X}(b) = \mathbf{B}$.

Theorem 6.20. Mean Value Theorem for Integrals. *Suppose f from $\mathbb{R}^n$ to $\mathbb{R}$ is a continuous function on a smoothly bounded connected set D in $\mathbb{R}^n$, with $\mathrm{Vol}(D) \neq 0$. Then there is a point $\mathbf{P}$ in D at which*

$$f(\mathbf{P}) = \frac{\int_D f \, \mathrm{d}^n \mathbf{X}}{\mathrm{Vol}(D)}.$$

Proof. By the Extreme Value Theorem there are points $\mathbf{A}$ and $\mathbf{B}$ in D where f has its minimum, m, and maximum M, on D,

$$m = f(\mathbf{A}) \leq f(\mathbf{X}) \leq f(\mathbf{B}) = M.$$

By the boundedness property of integrals

$$f(\mathbf{A}) \leq \frac{\int_D f(\mathbf{X}) \, \mathrm{d}^n \mathbf{X}}{\mathrm{Vol}(D)} \leq f(\mathbf{B}).$$

Since D is connected there is a continuous function $\mathbf{X}$ from some interval $[a,b]$ to D with $\mathbf{X}(a) = \mathbf{A}$ and $\mathbf{X}(b) = \mathbf{B}$. Thus

$$f \circ \mathbf{X}(a) \leq \frac{\int_D f(\mathbf{X}) \, \mathrm{d}^n \mathbf{X}}{\mathrm{Vol}(D)} \leq f \circ \mathbf{X}(b).$$

Since $f \circ \mathbf{X}$ is continuous, it follows from the Intermediate Value Theorem that there is a number c between a and b where

$$f \circ \mathbf{X}(c) = \frac{\int_D f(\mathbf{X}) \, \mathrm{d}^n \mathbf{X}}{\mathrm{Vol}(D)}.$$

Take $\mathbf{P} = \mathbf{X}(c)$. $\square$

We focus in the rest of this section on integrals of functions of three variables.

Triple integrals. The integral of a continuous function f on a smoothly bounded set D in $\mathbb{R}^3$ is called a *triple integral* and is denoted

$$\int_D f \, \mathrm{d}V.$$

Example 6.26. Let $f(x,y,z) = x^5$ and let D be the set $1 \leq x^2 + y^2 + z^2 \leq 4$. Find

$$\int_D x^5 \, \mathrm{d}V.$$

We observe that $f(-x,y,z) = -f(x,y,z)$ and D is symmetric about the y,z plane. We form approximate integrals $\sum_j f(\mathbf{P}_j)h^3$ by choosing symmetric points $(-x_i,y_i,z_i)$ and (x_i,y_i,z_i) in the h-boxes of D on opposite sides of the y,z plane. If we number the h-boxes on the $x > 0$ side of the y,z plane from $i = 1$ to N, these approximate integrals cancel

$$\sum_{i=1}^{N}((x_i)^5h^3 + (-x_i)^5h^3) = 0$$

for all h. Therefore $\displaystyle\int_D x^5\, dV = 0$.   □

Example 6.27. Let D be the set $1 \le x^2 + y^2 + z^2 \le 4$ and let

$$f(x,y,z) = \sin(z^3) + x^5 + 200.$$

We saw in Example 6.26 that $\displaystyle\int_D x^5\, dV = 0$. Similarly since $\sin(z^3)$ is odd and D is symmetric about the x,y plane we have that $\displaystyle\int_D \sin(z^3)\, dV = 0$. Since the integral of the constant function 200 over D is $200\,\mathrm{Vol}(D)$,

$$\int_{1 \le x^2+y^2+z^2 \le 4}(\sin(z^3) + x^5 + 200)\, dV = 0 + 0 + 200\,\mathrm{Vol}(D).$$

The volume of D is the volume inside the sphere $x^2 + y^2 + z^2 = 4$ minus the volume inside the smaller sphere $x^2 + y^2 + z^2 = 1$, so $\mathrm{Vol}(D) = \frac{4}{3}\pi(2^3 - 1^3)$, and

$$\int_{1 \le x^2+y^2+z^2 \le 4}(200 + x^5 + \sin(z^3))\, dV = \tfrac{5600}{3}\pi.$$

□

Example 6.28. Let D_1 be the set of points (x,y,z) in $\mathbb{R}^3$ where $x^2 + y^2 + z^2 \le 4$, let D_2 be the set where $4 < x^2 + y^2 + z^2 \le 9$, and let

$$f(x,y,z) = \begin{cases} 2 & (x,y,z) \text{ in } D_1 \\ 1 & (x,y,z) \text{ in } D_2. \end{cases}$$

Then $D = D_1 \cup D_2$ is the solid ball of radius 3 centered at the origin. The average value of f on D is

$$\frac{\int_D f\, dV}{\mathrm{Vol}(D)} = \frac{\int_{D_1} 2\, dV + \int_{D_2} 1\, dV}{\mathrm{Vol}(D)} = \frac{2\,\mathrm{Vol}(D_1) + \mathrm{Vol}(D_2)}{\mathrm{Vol}(D)} = \tfrac{35}{27}.$$

□

Similar to the case of double integrals, triple integrals can sometimes be computed by computing three iterated integrals. Suppose that D is the set of points in $\mathbb{R}^3$ that satisfy

$$g_1(x,y) \le z \le g_2(x,y), \qquad (x,y) \text{ in } D_{xy}$$

where D_{xy} is a smoothly bounded set in the x,y plane and g_1 and g_2 are differentiable. See Figure 6.36. Then D is a smoothly bounded set in $\mathbb{R}^3$. If f is continuous on D, $\int_D f \, dV$ exists. For each (x,y) in D_{xy} let $D(x,y)$ be the interval of points (x,y,z) with

$$g_1(x,y) \le z \le g_2(x,y).$$

See Figure 6.36. Integrate f with respect to z from $g_1(x,y)$ to $g_2(x,y)$,

$$\int_{D(x,y)} f(x,y,z)\,dz = \int_{z=g_1(x,y)}^{z=g_2(x,y)} f(x,y,z)\,dz.$$

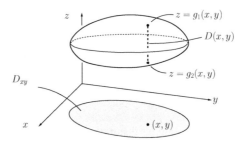

Fig. 6.36 A set D is between graphs, $g_1(x,y) \le z \le g_2(x,y)$.

The result is a continuous function of (x,y) that we can integrate over D_{xy}. By an argument similar to the argument for double integrals we can show that iterated integrals satisfy the four basic properties of linearity, boundedness, and additivity. Therefore by Theorem 6.18 the iterated integral equals the triple integral,

$$\int_{D_{xy}} \left(\int_{z=g_1(x,y)}^{z=g_2(x,y)} f(x,y,z)\,dz \right) dA = \int_D f \, dV.$$

If D_{xy} is y simple, with $a \le x \le b$ and $c(x) \le y \le d(x)$, where c and d are C^1 functions, see Figure 6.37, then we can compute the triple integral by

$$\int_D f \, dV = \int_{x=a}^{x=b} \left(\int_{y=c(x)}^{y=d(x)} \left(\int_{z=g_1(x,y)}^{z=g_2(x,y)} f(x,y,z)\,dz \right) dy \right) dx.$$

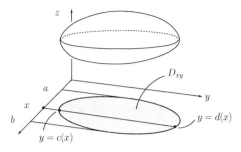

Fig. 6.37 D_{xy} is y simple.

Example 6.29. Let D be the set in the first octant that is bounded by the surface $z = x^2 + y^2$ and the plane $z = 4$, and define $f(x,y,z) = x$. Find

$$\int_D f \, dV.$$

Figure 6.38 shows a sketch of D. D_{xy} is the quarter disk in the x,y plane, bounded by the x and y axes and the graph of $y = \sqrt{4 - x^2}$.

$$\int_D x \, dV = \int_{D_{xy}} \left(\int_{z=x^2+y^2}^{z=4} x \, dz \right) dA.$$

Since D_{xy} consists of the points (x,y) with $0 \leq x \leq 2$ and $0 \leq y \leq \sqrt{4 - x^2}$ we have

$$\int_D x \, dV = \int_{D_{xy}} \left[xz \right]_{z=x^2+y^2}^{z=4} dA = \int_{D_{xy}} x(4 - x^2 - y^2) \, dA$$

$$= \int_{x=0}^{x=2} \left(\int_{y=0}^{y=\sqrt{4-x^2}} x(4 - x^2 - y^2) \, dy \right) dx = \int_{x=0}^{x=2} \left[x((4 - x^2)y - \tfrac{1}{3}y^3 \right]_{y=0}^{y=\sqrt{4-x^2}} dx$$

$$= \int_{x=0}^{x=2} \tfrac{2}{3} x(4 - x^2)^{3/2} \, dx = -\tfrac{2}{15}(4 - x^2)^{5/2} \Big|_{x=0}^{x=2} = \tfrac{64}{15}.$$

□

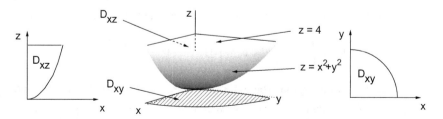

Fig. 6.38 The set D in Examples 6.29 and 6.30.

Example 6.30. Let D and $f(x,y,z) = x$ be as in Example 6.29. Now view D as a smoothly bounded set between the graphs of the functions of x and z given by $y = 0$ and $y = \sqrt{z - x^2}$ over the smoothly bounded set D_{xz} indicated in Figure 6.38.

$$\int_D f \, dV = \int_{D_{xz}} \left(\int_{y=0}^{y=\sqrt{z-x^2}} x \, dy \right) dA$$

$$= \int_{z=0}^4 \left(\int_{x=0}^{x=\sqrt{z}} \left(\int_{y=0}^{y=\sqrt{z-x^2}} x \, dy \right) dx \right) dz = \int_{z=0}^4 \left(\int_{x=0}^{x=\sqrt{z}} x(z-x^2)^{1/2} \, dx \right) dz$$

$$= \int_{z=0}^4 \left[-\tfrac{1}{3}(z-x^2)^{3/2} \right]_{x=0}^{x=\sqrt{z}} dz = \int_{z=0}^4 \tfrac{1}{3} z^{3/2} \, dz = \tfrac{64}{15}.$$

□

Change of variables. The change of variables theorem for double integrals, Theorem 6.12, has an analogue for triple integrals and integrals of functions of n variables.

A *smooth change of variables* $\mathbf{F}$ is a continuously differentiable mapping from an open set U in $\mathbb{R}^n$ to $\mathbb{R}^n$ that is one to one on U and $D\mathbf{F}(\mathbf{P})$ is invertible for each point $\mathbf{P}$ in U. We recall Definition 3.4 of the Jacobian of a mapping at point $\mathbf{P}$, $J\mathbf{F}(\mathbf{P}) = \det(D\mathbf{F}(\mathbf{P}))$.

Theorem 6.21. *Let $\mathbf{F}$ be a smooth change of variables that maps a smoothly bounded set C onto a smoothly bounded set D, so that the boundary of C is mapped to the boundary of D. Let f be a continuous function on D. Then*

$$\int_D f(\mathbf{X}) \, d^n\mathbf{X} = \int_C f(\mathbf{F}(\mathbf{U})) |J\mathbf{F}(\mathbf{U})| \, d^n\mathbf{U}.$$

In some of the following examples of the change of variables we make tacit use of limiting arguments similar to the extension of polar coordinates in Example 6.20.

Example 6.31. Find the integral of $f(x,y,z) = z\sqrt{x^2+y^2}$ over the region D that is bounded by the graphs of $z = 6$, $z = 0$, $x^2 + y^2 = 4$, and $x^2 + y^2 = 1$, and where $x \geq 0$. See Figure 6.39. The iterated integral in rectangular coordinates is

$$\int_{D_{xy}} \left(\int_{z=0}^{z=6} z\sqrt{x^2+y^2} \, dz \right) dx \, dy = \int_{D_{xy}} 18\sqrt{x^2+y^2} \, dx \, dy$$

where D_{xy} is the region where $1 \leq x^2 + y^2 \leq 4$ and $x \geq 0$. The double integral involves integrating $\sqrt{x^2+y^2}$ with respect to x or y. Using a change of variables to cylindrical coordinates, $x = r\cos\theta$, $y = r\sin\theta$, $z = z$ the Jacobian is

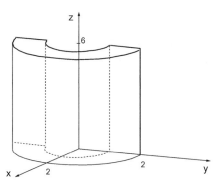

Fig. 6.39 The region D in Example 6.31.

$$\det \begin{bmatrix} \cos\theta & -r\sin\theta & 0 \\ \sin\theta & r\cos\theta & 0 \\ 0 & 0 & 1 \end{bmatrix} = r(\cos^2\theta + \sin^2\theta) = r.$$

The map $\mathbf{F}(r,\theta,z) = (x(r,\theta,z), y(r,\theta,z), z(r,\theta,z))$ takes the rectangular box C given by $1 \le r \le 2$, $0 \le \theta \le \pi$, $0 \le z \le 6$ one to one onto D. By the change of variables theorem

$$\int_D f(x,y,z)\,dxdydz = \int_C f(\mathbf{F}(r,\theta,z))\,rdrd\theta dz.$$

Since $x^2 + y^2 = r^2$, $f(\mathbf{F}(r,\theta,z)) = zr$ and the integral is

$$\int_D f\,dxdydz = \int_{z=0}^{z=6} \left(\int_{\theta=0}^{\theta=\pi} \left(\int_{r=1}^{r=2} zr^2\,dr \right) d\theta \right) dz$$

$$= \int_{z=0}^{z=6} z\,dz \int_{\theta=0}^{\theta=\pi} d\theta \int_{r=1}^{r=2} r^2\,dr = (18)(\pi)(\tfrac{8-1}{3}) = 42\pi.$$

$\square$

Here is an example why we need the mapping $\mathbf{F}$ to be one to one.

Example 6.32. The mapping $\mathbf{F}(r,\theta,z) = (r\cos\theta, r\sin\theta, z)$ maps the points in the rectangular box C given by

$$1 \le r \le 2, \quad 0 \le \theta \le 4\pi, \quad 0 \le z \le 6$$

onto the region D between the two cylinders shown in Figure 6.40. The volume of D is $\pi(2^2 - 1^2)6 = 18\pi$. If we were to mistakenly use the change of variables formula to compute the volume of D we would get

$$\mathrm{Vol}(D) = \int_D 1\,dxdydz =_? \int_C |J\mathbf{F}(r,\theta,z)|\,drd\theta dz.$$

The right hand side is

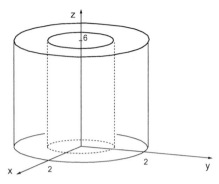

Fig. 6.40 The region in Example 6.32.

$$= \int_C r \, dr d\theta dz = \int_1^2 r \, dr \int_0^{4\pi} d\theta \int_0^6 dz$$

$$= \tfrac{1}{2}(2^2 - 1^2)(4\pi)6 = 36\pi.$$

What went wrong? Even though $J\mathbf{F}(r,\theta,z) = r \neq 0$ at each point, the mapping is not one to one on a set that has positive volume. The points with $2\pi \leq \theta \leq 4\pi$ are sent to the same points in the range as those with $0 \leq \theta \leq 2\pi$.   □

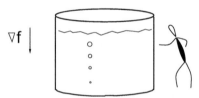

Fig. 6.41 The energy density f in Example 6.33 increases toward the bottom of the tank.

Example 6.33. A chemical reaction occurs in a tank D given by

$$x^2 + y^2 \leq 4, \quad 0 \leq z \leq 3.$$

Suppose the energy density [energy/vol] is $f(x,y,z) = 100 - 5z$ at each point (x,y,z) in D. Find the total energy and the average energy density in the tank. The total energy is

$$\int_D f \, dV = \int_{\theta=0}^{\theta=2\pi} \int_{r=0}^{r=2} \left(\int_{z=0}^{z=3} (100 - 5z) \, dz \right) r \, dr d\theta$$

$$= \int_{\theta=0}^{\theta=2\pi} d\theta \int_{r=0}^{r=2} r \, dr \int_{z=0}^{z=3} (100 - 5z) \, dz = (2\pi)\tfrac{4}{2}\left[-\tfrac{1}{2(5)}(100 - 5z)^2 \right]_{z=0}^{z=3} = \tfrac{4\pi}{10}(100^2 - 85^2).$$

The volume of the cylindrical tank is $\pi(2^2)(3) = 12\pi$, so the average energy density in D is

$$\frac{1}{\text{Vol}(D)} \int_D f \, dV = \frac{1}{30}(100^2 - 85^2).$$

□

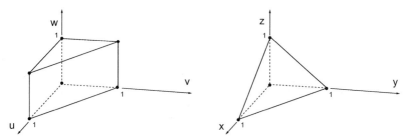

Fig. 6.42 *Left:* The prism C, *Right:* the tetrahedron D in Example 6.34.

Example 6.34. Let C be the prism in $\mathbb{R}^3$ given by

$$0 \le w \le 1, \quad u + v \le 1, \quad u \ge 0, \quad v \ge 0.$$

Let D be the tetrahedron given by

$$x + y + z \le 1, \quad x \ge 0, \quad y \ge 0, \quad z \ge 0.$$

See Figure 6.42. The function $(x,y,z) = \mathbf{F}(u,v,w) = ((1-w)u, (1-w)v, w)$ maps the interior of C one to one onto the interior of D. The Jacobian is

$$\det D\mathbf{F}(u,v,w) = \det \begin{bmatrix} 1-w & 0 & -u \\ 0 & 1-w & v \\ 0 & 0 & 1 \end{bmatrix} = (1-w)^2.$$

By the change of variables theorem,

$$\text{Vol}(D) = \int_D 1 \, dV = \int_C (1-w)^2 \, du \, dv \, dw.$$

As an iterated integral this is equal to

$$\int_{C_{uv}} \left(\int_{w=0}^{w=1} (1-w)^2 \, dw \right) dA = \int_{C_{uv}} \left[-\tfrac{1}{3}(1-w)^3 \right]_{w=0}^{w=1} dA = \tfrac{1}{3} \int_{C_{uv}} dA = \tfrac{1}{3} \tfrac{1}{2} = \tfrac{1}{6}.$$

□

Spherical coordinates. Let

$$(x,y,z) = \mathbf{F}(\rho.\phi,\theta) = (\rho \sin \phi \cos \theta, \rho \sin \phi \sin \theta, \rho \cos \phi)$$

be the mapping between rectangular and spherical coordinates, $\rho \geq 0$, $0 \leq \phi \leq \pi$, $0 \leq \theta \leq 2\pi$. $\mathbf{F}$ maps regions in (ρ, ϕ, θ) space to (x, y, z) space.

$$DF(\rho.\phi,\theta) = \begin{bmatrix} \sin\phi\cos\theta & \rho\cos\phi\cos\theta & -\rho\sin\phi\sin\theta \\ \sin\phi\sin\theta & \rho\cos\phi\sin\theta & \rho\sin\phi\cos\theta \\ \cos\phi & -\rho\sin\phi & 0 \end{bmatrix}.$$

We ask you in Problem 6.62 to verify that

$$J\mathbf{F}(\rho,\phi,\theta) = \det DF(\rho,\phi,\theta) = \rho^2 \sin\phi.$$

Note that for $\rho = 0$, $\phi = 0$ and $\phi = \pi$ the matrix derivative is not invertible and $\mathbf{F}$ is not one to one when $\theta = 0$ and 2π. The change of variable formula can be proved by taking the limit as ρ tends to zero, ϕ tends to zero or π, and θ tends to 0 or 2π. See Example 6.20 for an analogous argument for polar coordinates.

Example 6.35. Take D to be the region between the spheres of radius a and b centered at the origin. Find the integral of $f(x,y,z) = \sqrt{x^2 + y^2 + z^2}$ over D. Use the spherical change of coordinate mapping $\mathbf{F}(\rho,\phi,\theta)$ that maps the rectangular box C given by $a \leq \rho \leq b$, $0 \leq \phi \leq \pi$, $0 \leq \theta \leq 2\pi$ onto D. By the change of variables theorem,

$$\int_D \sqrt{x^2 + y^2 + z^2}\,dxdydz = \int_C \sqrt{\rho^2}\rho^2 \sin\phi\,d\rho d\phi d\theta$$

$$= \int_{\rho=a}^{\rho=b} \rho^3\,d\rho \int_{\phi=0}^{\phi=\pi} \sin\phi\,d\phi \int_{\theta=0}^{\theta=2\pi} d\theta = \tfrac{1}{4}(b^4 - a^4)(2)(2\pi) = \pi(b^4 - a^4).$$

□

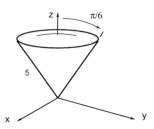

Fig. 6.43 The cone region D in Example 6.36.

Example 6.36. Let D be the cone-shaped region in (x,y,z) space shown in Figure 6.43. D is described in spherical coordinates by $0 \leq \rho \leq 5$, $0 \leq \phi \leq \frac{\pi}{6}$, $0 \leq \theta \leq 2\pi$. The volume of D is

$$\text{Vol}(D) = \int_D dx\,dy\,dz = \int_{\theta=0}^{2\pi}\int_{\phi=0}^{\pi/6}\int_{\rho=0}^{5}\rho^2\sin\phi\,d\rho d\phi d\theta$$

$$= \int_{\theta=0}^{2\pi}\left(\int_{\phi=0}^{\pi/6}\left(\int_{\rho=0}^{5}\rho^2\,d\rho\right)\sin\phi\,d\phi\right)d\theta = \int_{\theta=0}^{2\pi}d\theta\int_{\phi=0}^{\pi/6}\sin\phi\,d\phi\int_{\rho=0}^{5}\rho^2\,d\rho$$

$$= (2\pi)\left[-\cos\phi\right]_{\phi=0}^{\pi/6}(\tfrac{1}{3}5^3) = (2\pi)\left(1 - \tfrac{\sqrt{3}}{2}\right)\tfrac{1}{3}(125).$$

□

Example 6.37. A function f gives the electric charge density [charge/volume] inside the unit sphere $\rho = 1$ as

$$f(\rho,\phi,\theta) = \rho^2\sin^2\theta.$$

Find the total charge inside the sphere.

$$\int_{\rho\le1} f\,dV = \int_{\theta=0}^{2\pi}\int_{\phi=0}^{\pi}\int_{\rho=0}^{1}\rho^2\sin^2\theta\rho^2\sin\phi\,d\rho d\phi d\theta$$

$$= \int_{\theta=0}^{2\pi}\sin^2\theta\,d\theta\int_{\phi=0}^{\pi}\sin\phi\,d\phi\int_{\rho=0}^{1}\rho^4\,d\rho = \pi(2)(\tfrac{1}{5}) = \tfrac{2\pi}{5}.$$

□

Variation with a parameter. Often there is a fourth variable in a triple integral. If $f(x,y,z,t)$ is the energy density at (x,y,z) at time t, then the total energy in D at time t is

$$\int_D f(x,y,z,t)\,dx\,dy\,dz,$$

a function of t. What is the rate of change of the total energy in D with respect to time?

Theorem 6.22. *Suppose f is a continuous function on $D\times[a,b]$ where D is a smoothly bounded set in $\mathbb{R}^3$.*

(a) Then $\int_D f(\mathbf{X},t)\,dV$ is a continuous function of t.
(b) If f is continuously differentiable in t then

$$\int_D f(\mathbf{X},t)\,dV$$

is a continuously differentiable function of t and

$$\frac{d}{dt}\int_D f(\mathbf{X},t)\,dV = \int_D f_t(\mathbf{X},t)\,dV$$

Proof. (a)

$$\int_D f(\mathbf{X},t)\,dV - \int_D f(\mathbf{X},s)\,dV = \int_D (f(\mathbf{X},t) - f(\mathbf{X},s))\,dV.$$

According to Theorem 2.12 a continuous function on a closed bounded set is uniformly continuous; so for every positive ϵ there is a δ such that

$$|f(\mathbf{X},t) - f(\mathbf{X},s)| < \epsilon$$

for $|t - s| < \delta$ and all $\mathbf{X}$ in D. Then for $|t - s| < \delta$ the integral of $f(\mathbf{X},t)$ and the integral of $f(\mathbf{X},s)$ over D differ by less than $\epsilon \operatorname{Vol}(D)$.

(b) Continuously differentiable in t implies that for every $\epsilon > 0$ there is a δ such that

$$\left| \frac{f(\mathbf{X},t+h) - f(\mathbf{X},t)}{h} - f_t(\mathbf{X},t) \right| < \epsilon \qquad (6.24)$$

for $|h| < \delta$ for all $\mathbf{X}$ in D. Here is why: f_t is uniformly continuous, so given ϵ there is δ so that if $|t - s| < \epsilon$ then $|f_t(\mathbf{X},t) - f_t(\mathbf{X},s)| < \epsilon$ for all $\mathbf{X}$. Take $|h| < \delta$. By the Mean Value Theorem

$$\left| \frac{f(\mathbf{X},t+h) - f(\mathbf{X},t)}{h} - f_t(\mathbf{X},t) \right| = \left| f_t(\mathbf{X},c) - f_t(\mathbf{X},t) \right| < \epsilon$$

where $|c - t| < |h| < \delta$.

It follows from estimate (6.24) that as h tends to zero

$$\frac{\int_D f(\mathbf{X},t+h)\,dV - \int_D f(\mathbf{X},t)\,dV}{h} = \int_D \frac{f(\mathbf{X},t+h) - f(\mathbf{X},t)}{h}\,dV$$

converges as h tends to zero to $\int_D f_t(\mathbf{X},t)\,dV$. This derivative is continuous by part (a).

$\square$

Example 6.38. Suppose now that the chemical reaction of Example 6.33 has energy density $f(x,y,z,t) = e^{-t^2}(100 - 5z)$ that depends on time. Find the rate of change of the total energy in the tank. The total energy is

$$\int_D f\,dV = \int_D e^{-t^2}(100 - 5z)\,dV.$$

We showed that $\int_D (100 - 5z)\,dV = \frac{4\pi}{10}(100^2 - 85^2)$, so the total energy is

$$\int_D e^{-t^2}(100 - 5z)\,dV = e^{-t^2}\frac{4\pi}{10}(100^2 - 85^2).$$

The time rate of change is then

$$\frac{\mathrm{d}}{\mathrm{d}t}\int_D e^{-t^2}(100-5z)\,\mathrm{d}V = \int_D (-2t)e^{-t^2}(100-5z)\,\mathrm{d}V = (-2t)e^{-t^2}\tfrac{4\pi}{10}(100^2-85^2).$$

$\square$

Example 6.39. Let $p(x,y,t)$ be a C^1 function that represents the population density [people/area] in a region D of $\mathbb{R}^2$ at each time t. Express the population of D at time t and its rate of change with respect to time [people/area/time] using integrals. The population at time t is

$$P(t) = \int_D p(x,y,t)\,\mathrm{d}A.$$

Its rate of change is $P'(t) = \dfrac{\mathrm{d}}{\mathrm{d}t}\displaystyle\int_D p(x,y,t)\,\mathrm{d}A = \int_D p_t(x,y,t)\,\mathrm{d}A.$ $\square$

Probability density functions. The definition of integrable functions over unbounded sets in $\mathbb{R}^3$ can be made analogously to the definitions in Section 6.5. We also make the following definition of probability density function p from $\mathbb{R}^3$ to $\mathbb{R}$ similar to the case in $\mathbb{R}^2$.

Definition 6.16. A *probability density function* in $\mathbb{R}^3$ is a nonnegative function that is integrable over $\mathbb{R}^3$ and whose integral over $\mathbb{R}^3$ is equal to 1.

Example 6.40. We have evaluated

$$\int_{-\infty}^{\infty} e^{-x^2}\,\mathrm{d}x = \pi^{1/2}.$$

Set $p(x,y,z) = \pi^{-3/2}e^{-x^2-y^2-z^2} > 0$. Then p is a probability density function:

$$\int_{\mathbb{R}^3} \pi^{-3/2}e^{-x^2-y^2-z^2}\,\mathrm{d}V = \pi^{-3/2}\lim_{n\to\infty}\int_{-n}^{n}\int_{-n}^{n}\int_{-n}^{n} e^{-z^2}e^{-y^2}e^{-x^2}\,\mathrm{d}x\,\mathrm{d}y\,\mathrm{d}z$$

$$= \pi^{-3/2}\int_{-\infty}^{\infty} e^{-z^2}\,\mathrm{d}z \int_{-\infty}^{\infty} e^{-y^2}\,\mathrm{d}y \int_{-\infty}^{\infty} e^{-x^2}\,\mathrm{d}x = \pi^{-3/2}\pi^{1/2}\pi^{1/2}\pi^{1/2} = 1.$$

$\square$

Example 6.41. A function of the form

$$f(\rho,\phi,\theta) = \rho e^{-\rho}\cos\phi$$

is used in chemistry to describe a "2p" orbital of an electron. (See also Problem 9.48.) In the application, the probability density function for finding the electron at point (ρ,ϕ,θ) is

$$p(\rho,\phi,\theta) = c(f(\rho,\phi,\theta))^2$$

where c is chosen to make $\displaystyle\int_{\mathbb{R}^3} p\,dV = 1$. To find c, we calculate the integral

$$\int_{\mathbb{R}^3} p\,dV = \int_{\theta=0}^{2\pi}\left(\int_{\phi=0}^{\pi}\left(\int_{\rho=0}^{\infty} c\rho^2 e^{-2\rho}\cos^2\phi\rho^2\sin\phi\,d\rho\right)d\phi\right)d\theta$$

$$= c(2\pi)\left[-\tfrac{1}{3}\cos^3\phi\right]_{\phi=0}^{\pi}\int_{\rho=0}^{\infty}\rho^4 e^{-2\rho}\,d\rho = c(2\pi)\tfrac{2}{3}\int_{\rho=0}^{\infty}\rho^4 e^{-2\rho}\,d\rho.$$

In Problem 6.63 we ask you to show that the last integral is equal to $\tfrac{3}{4}$, thus $c = \tfrac{1}{\pi}$. $\qquad\qquad\qquad\qquad\qquad\qquad\qquad\qquad\qquad\qquad\qquad\qquad\qquad\qquad\qquad\qquad\qquad$ $\square$

Problems

6.57. Let $D = [0,a]^5$ be the box in $\mathbb{R}^5$ consisting of all points $\mathbf{X} = (x_1,x_2,x_3,x_4,x_5)$ where $0 \le x_j \le a$ for $j = 1,2,3,4,5$. Evaluate the integrals.

(a) $\mathrm{Vol}(D)$

(b) $\displaystyle\int_D x_1^2\,d^5\mathbf{X}$

(c) $\displaystyle\int_D (x_1^2 - x_4^2 + 7x_5 x_3)\,d^5\mathbf{X}$.

6.58. Find the average value of the function $f(x,y,z) = x^2 + y^2 - z^2$ on the sets

(a) $[-1,1]^3$

(b) $x^2 + y^2 + z^2 \le 1$.

6.59. Evaluate the integral $\displaystyle\int_D xz^2\,dV$ for the two regions.

(a) the rectangular region $D = [1,2]\times[3,5]\times[-1,10]$

(b) the subset of $D = [1,2]\times[3,5]\times[-1,10]$ where $z > x + y$.

6.60. Let D be the cube where x,y,z each vary from -2 to 2.

(a) Explain by a symmetry argument why $\displaystyle\int_D xz^2\,dV = 0$.

(b) Evaluate $\displaystyle\int_D (8x + 2xz^2 - 4y^2z + 10)\,dV$.

6.61. A planet is represented by a ball of radius R centered at the origin. Its energy density [energy/volume] at (x,y,z) is $f(x,y,z) = e^{-\sqrt{x^2+y^2+z^2}}$. Find the total energy of the planet.

6.62. Verify that the Jacobian for the change from rectangular to spherical coordinates is

$$J\mathbf{F}(\rho,\phi,\theta) = \rho^2\sin\phi.$$

6.63. In this problem we evaluate the integral

$$\int_0^\infty \rho^4 e^{-2\rho}\, d\rho = \tfrac{3}{4},$$

that was used in Example 6.41.

(a) Verify that $\int_0^\infty e^{-2\rho}\, d\rho = \tfrac{1}{2}$.

(b) Let k and n be positive. Show that

$$\int_0^n \rho^k e^{-2\rho}\, d\rho = n^k \frac{-e^{-2n}}{2} - \int_0^n k\rho^{k-1}\frac{e^{-2\rho}}{-2}\, d\rho.$$

(c) Let $i_k = \int_0^\infty \rho^k e^{-2\rho}\, d\rho$. Show that for $k \ge 1$, $i_k = \dfrac{k}{2} i_{k-1}$.

(d) Deduce that $i_4 = \tfrac{3}{4}$.

6.64. An integral over the unbounded set $\mathbb{R}^n$ is the limit of integrals over an increasing sequence of smoothly bounded sets $D_1 \subset D_2 \subset \cdots D_n \subset \cdots$ whose union is $\mathbb{R}^n$.
Show by a change of variables that

$$\int_{\mathbb{R}^n} \frac{1}{(4\pi t)^{n/2}} e^{-\|\mathbf{X}\|^2/(4t)}\, d^n\mathbf{X} = 1.$$

6.65. Let p be a positive number and define a function f from $\mathbb{R}^3$ to $\mathbb{R}$ by $f(\mathbf{X}) = 0$ when $\|\mathbf{X}\| < p$, and $f(\mathbf{X}) = e^{-\|\mathbf{X}\|}$ when $\|\mathbf{X}\| \ge p$. Evaluate the integral

$$\int_{\mathbb{R}^3} f\, dV$$

using spherical coordinates.

6.66. Let $a_1, \ldots, a_n$ be positive constants. Show by a change of variables that

$$\int_{\mathbb{R}^n} e^{-(a_1 x_1^2 + \cdots + a_n x_n^2)}\, d^n\mathbf{X} = \frac{1}{(4\pi)^{n/2}\, \sqrt{a_1 a_2 \cdots a_n}}.$$

6.67. Consider the integral over the n-dimensional cube

$$\int_{[0,1]^n} \|\mathbf{X}\|^2\, d^n\mathbf{X} = \int_{[0,1]^n} x_1^2\, d^n\mathbf{X} + \cdots + \int_{[0,1]^n} x_n^2\, d^n\mathbf{X}.$$

(a) Evaluate $\displaystyle\int_{[0,1]^n} x_1^2\, d^n\mathbf{X}$ as an iterated integral. By symmetry each $\displaystyle\int_{[0,1]^n} x_j^2\, d^n\mathbf{X}$ has the same value.

(b) Evaluate $\displaystyle\int_{[0,1]^n} \|\mathbf{X}\|^2\, d^n\mathbf{X}$.

(c) Find the average value of $\|\mathbf{X}\|^2$ on $[0,1]^n$.
(d) Find the average value of $\|\mathbf{X}\|^2$ on $[0,2]^n$.

6.68. Let $0 < a < b$ and consider the integral

$$\int_D (x^2 + y^2 + z^2)^q \, dV$$

where D is the region of $\mathbb{R}^3$ between the spheres of radius a and b centered at the origin. For parts (c) and (d), see Definition 6.9 and use the corresponding concept of integration over unbounded sets in $\mathbb{R}^3$.

(a) Evaluate the integral for the case $q = -3/2$.
(b) Evaluate the integral for other values of q.
(c) Verify that $\displaystyle\int_{x^2+y^2+z^2>1} (x^2 + y^2 + z^2)^{-2} \, dV$ exists and find its value.
(d) Verify that $\displaystyle\int_{x^2+y^2+z^2>1} (x^2 + y^2 + z^2)^{-3/2} \, dV$ does not exist.

Chapter 7
Line and surface integrals

Abstract We use integrals to find the total amount of some quantity on a curve or surface in space. Examples include total mass of a wire, work along a curve, total charge on a surface, and flux across a surface.

7.1 Line integrals

In single variable calculus we saw that the total mass of a straight wire that lies on an interval between $x = a$ and $x = b$ and that has density $f(x)$ [mass/length] is $\int_a^b f(x)\,dx$. We also saw that if a force $f(x)$ varies continuously with position x and is in the direction of the motion then the total work done in moving from a to b on the x axis is $\int_a^b f(x)\,dx$. In this section we introduce *line integrals* to find the total mass of a curved wire and the total work along a curve.

Definition 7.1. Let $\mathbf{X}$ be a C^1 function from $[a,b]$ to $\mathbb{R}^3$, $\mathbf{X}(t) = (x(t), y(t), z(t))$, with $\mathbf{X}'(t) \neq \mathbf{0}$ on $[a,b]$. The range of $\mathbf{X}$, C, is called a *smooth curve* from $\mathbf{X}(a)$ to $\mathbf{X}(b)$. $\mathbf{X}$ is called a *smooth parametrization* of C. The *length* of the curve, or arclength is

$$\text{Length}(C) = \int_C ds = \int_a^b \|\mathbf{X}'(t)\|\,dt.$$

© Springer International Publishing AG 2017
P. D. Lax and M. S. Terrell, *Multivariable Calculus with Applications*,
Undergraduate Texts in Mathematics, https://doi.org/10.1007/978-3-319-74073-7_7

A *piecewise smooth curve* is a finite union of smooth curves $C_1, \ldots, C_m$ each joined continuously, but perhaps not differentiably, to the next. The length of a piecewise smooth curve is the sum of the lengths of the smooth parts.

Next we compute some arc lengths.

Example 7.1. The curve C_1 given by

$$\mathbf{X}_1(t) = (t, 2t, t+1), \qquad 0 \le t \le 1$$

is a line segment from $(0,0,1)$ to $(1,2,2)$. The length of C_1 is the distance between the endpoints,

$$\sqrt{(1-0)^2 + (2-0)^2 + (2-1)^2} = \sqrt{6}.$$

Using the definition for the length of C_1 we find

$$x'(t) = (t)' = 1, \qquad y'(t) = (2t)' = 2, \qquad z'(t) = (t+1)' = 1,$$

and we get $\displaystyle \int_{C_1} ds = \int_0^1 \sqrt{1^2 + 2^2 + 1^2} \, dt = \int_0^1 \sqrt{6} \, dt = \sqrt{6}.$ $\qquad\qquad$ □

Example 7.2. Find the length of the curve C_2 given by

$$\mathbf{X}_2(u) = (e^u - 1, 2e^u - 2, e^u), \qquad 0 \le u \le \log 2.$$

Since $\mathbf{X}_2'(u) = (e^u, 2e^u, e^u)$ we get

$$\int_{C_2} ds = \int_0^{\log 2} \sqrt{(e^u)^2 + (2e^u)^2 + (e^u)^2} \, du = \int_0^{\log 2} \sqrt{6} \, e^u \, du = \sqrt{6}(e^{\log 2} - e^0) = \sqrt{6}.$$

$\qquad\qquad$ □

Note that C_1 in Example 7.1 and C_2 in Example 7.2 are the same curve, parametrized differently. Let

$$0 \le u \le \log 2, \quad 0 \le t \le 1, \quad t = e^u - 1.$$

Then $\mathbf{X}_1$ and $\mathbf{X}_2$ are related by $\mathbf{X}_2(u) = \mathbf{X}_1(t(u))$. By the Chain Rule

$$\mathbf{X}_2'(u) = \mathbf{X}_1'(t(u))t'(u),$$

and by the change of variables theorem for integrals

$$\int_{t=0}^{t=1} \|\mathbf{X}_1'(t)\| \, dt = \int_{u=0}^{u=\log 2} \|\mathbf{X}_1'(t(u))\| \, |t'(u)| \, du = \int_{u=0}^{u=\log 2} \|\mathbf{X}_2'(u)\| \, du.$$

We show by the same argument that the definition of the length of a curve is independent of the parametrization of the curve. For if $t = t(u)$, $\alpha \le u \le \beta$ is another parameter with $\dfrac{dt}{du}$ positive, then

$$\frac{d\mathbf{X}(t(u))}{du} = \frac{d\mathbf{X}}{dt}\frac{dt}{du}$$

and so by the change of variables formula

$$\text{Length}(C) = \int_a^b \left\|\frac{d\mathbf{X}}{dt}\right\| dt = \int_\alpha^\beta \left\|\frac{d\mathbf{X}}{dt}\right\|\frac{dt}{du} du = \int_\alpha^\beta \left\|\frac{d\mathbf{X}}{dt}\frac{dt}{du}\right\| du = \int_\alpha^\beta \left\|\frac{d\mathbf{X}(t(u))}{du}\right\| du.$$

Two different interpretations of the integral of $\|\mathbf{X}'(t)\|$. We saw in Chapter 5 that if

$$\mathbf{X}(t) = (x(t), y(t), z(t))$$

is the position of a particle at time t then $\mathbf{X}'(t)$ is its velocity and $\|\mathbf{X}'(t)\|$ is its speed. Then the integral of the speed over the interval from $t = a$ to $t = b$ is the total distance or length of the path traveled from $t = a$ to $t = b$.

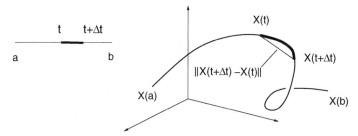

Fig. 7.1 A parametrization stretches or shrinks subintervals of the domain.

Alternatively, consider the interval $[a, b]$ on the number line and imagine that it is stretched like a rubber band onto a curve in space. See Figure 7.1. The part of the band that was originally between t and $t + \Delta t$ is now on the curve between $\mathbf{X}(t)$ and $\mathbf{X}(t + \Delta t)$. Originally the length of this part of the band was Δt. Now that part has length roughly the distance between $\mathbf{X}(t)$ and $\mathbf{X}(t + \Delta t)$, or

$$\|\mathbf{X}(t + \Delta t) - \mathbf{X}(t)\|.$$

By the Mean Value Theorem that is approximately

$$\|\mathbf{X}'(t)\|\,|\Delta t|.$$

That is, $\|\mathbf{X}'(t)\|$ is the factor by which the interval from t to $t + \Delta t$ was stretched when it was mapped to the part of the curve between $\mathbf{X}(t)$ and $\mathbf{X}(t + \Delta t)$. The sum of the lengths $\|\mathbf{X}'(t)\|\Delta t$ of the pieces tends to the length of the curve.

Arc length parametrization. Denote by C a smooth curve of length L. There are many smooth parametrizations for C. There is one called the *arc length parametrization* that is natural in the sense that it maps the interval $[0, L]$ one to one onto C without stretching or shrinking the interval at any point in the process. That is, the derivative $\dfrac{d\mathbf{X}(s)}{ds}$ at each s in $[0,L]$ is a unit vector, so that the stretching factor, $\|\mathbf{X}'(t)\|$, is 1. In Problem 7.3 we show you an example of such a parametrization. To see that there is such a parametrization of a smooth curve, let $\mathbf{X}$ be a smooth parametrization of C with domain $[a,b]$ and define $s(t)$ to be the length of the curve from $\mathbf{X}(a)$ to $\mathbf{X}(t)$:

$$s(t) = \int_a^t \|\mathbf{X}'(\tau)\| \, d\tau.$$

By the Fundamental Theorem of Calculus $s'(t) = \|\mathbf{X}'(t)\|$. Since $\mathbf{X}$ is smooth, $\|\mathbf{X}'(t)\|$ is not zero, so $s'(t)$ is positive. Therefore s is increasing and invertible, and we can write t as a function of s, $t = t(s)$. By the Inverse Function Theorem for single variable this inverse function is differentiable and

$$\frac{dt}{ds} = \frac{1}{\|\mathbf{X}'(t)\|}.$$

We define the *arc length parametrization* to be $\mathbf{X}(t(s))$. By the Chain Rule,

$$\frac{d(\mathbf{X}(t(s)))}{ds} = \frac{d\mathbf{X}}{dt}\frac{dt}{ds} = \frac{\mathbf{X}'(t)}{\|\mathbf{X}'(t)\|},$$

which is a unit vector. In terms of speed, the arc length parametrization has speed 1 at each point of the curve.

We have shown that every smooth curve C has an arc length parametrization. It is useful to know that it exists. The next example shows an unusual case where there is a simple formula for it.

Example 7.3. Let C be the helix given by the parametrization

$$\mathbf{X}(t) = (a\cos t, a\sin t, bt), \qquad 0 \le t \le 2\pi, \quad a^2 + b^2 \ne 0.$$

Then $\|\mathbf{X}'(t)\| = \sqrt{a^2 + b^2}$ for each t. If $a^2 + b^2 = 1$ then $\mathbf{X}$ is the arc length parametrization of the helix. If not, then $s = \displaystyle\int_0^t \|\mathbf{X}'(\tau)\| \, d\tau = \sqrt{a^2 + b^2}\, t$, and

$$\mathbf{X}(t(s)) = \left(a\cos\left(\tfrac{s}{\sqrt{a^2+b^2}}\right), a\sin\left(\tfrac{s}{\sqrt{a^2+b^2}}\right), b\tfrac{s}{\sqrt{a^2+b^2}}\right), \qquad 0 \le s \le 2\pi\sqrt{a^2+b^2},$$

is the arc length parametrization. $\qquad\qquad\qquad\qquad\qquad\qquad\qquad\qquad\qquad\qquad\qquad\qquad\square$

Line integral To motivate the definition of the line integral consider a wire with continuous density $f(x,y,z)$ [mass/length] and assume the wire lies along a smooth curve C parametrized by

$$\mathbf{X}(t) = (x(t), y(t), z(t)), \qquad a \le t \le b.$$

Partition the curve into short segments C_i. By the continuity of f, if the segments are short the density doesn't vary much in each segment. Estimate the mass m_i of the i-th segment by using the density $f(\mathbf{X}(t_i))$ at one end of the segment times the length of the segment

$$m_i \approx f(\mathbf{X}(t_i))\text{Length}(C_i).$$

The sum of these estimates is an estimate for the total mass of the wire. (See Figure 7.2) The length of C_i is close to the length of the secant, $\|\mathbf{X}(t_i) - \mathbf{X}(t_{i-1})\|$. By the Mean Value Theorem and continuity of $\mathbf{X}'$, $\|\mathbf{X}(t_i) - \mathbf{X}(t_{i-1})\|$ is close to $\|\mathbf{X}'(t_i)\|\,|t_i - t_{i-1}|$. So the mass of the wire is approximately the sum

$$\sum_i m_i \approx \sum_i f(\mathbf{X}(t_i))\|\mathbf{X}'(t_i)\|\,|t_i - t_{i-1}|.$$

The sum on the right tends to the integral

$$\int_a^b f(\mathbf{X}(t))\|\mathbf{X}'(t)\|\,dt$$

as the subinterval lengths $t_i - t_{i-1}$ tend to zero. This motivates the definition of the line integral of f over C with respect to arc length.

Definition 7.2. Let f be a continuous function on a smooth curve C parametrized by $\mathbf{X}(t) = (x(t), y(t), z(t))$, $a \le t \le b$. The *integral of f over C with respect to arc length* is

$$\int_C f\,ds = \int_a^b f(\mathbf{X}(t))\|\mathbf{X}'(t)\|\,dt.$$

For a piecewise smooth curve C, we define the line integral to be the sum of line integrals over the smooth pieces.

Just as we saw with the length of C, the integral of a continuous function f over a smooth curve is independent of the parametrization.

Example 7.4. A wire lies along a helical curve C given by

$$\mathbf{X}(t) = (\cos t, \sin t, t), \qquad 0 \le t \le 4\pi$$

and its density is $f(x, y, z) = z$ [mass/length]. Find the mass of the wire. We integrate f along C.

$$\int_C f\,ds = \int_a^b \underbrace{f(\mathbf{X}(t))}_{\text{mass/length}} \underbrace{\|\mathbf{X}'(t)\|\,dt}_{\text{length}}.$$

$\mathbf{X}'(t) = (-\sin t, \cos t, 1)$ and $f(\mathbf{X}(t)) = t$. The total mass is

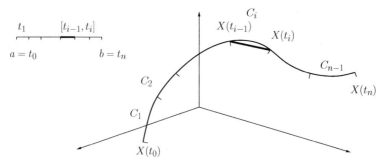

Fig. 7.2 The mass of a wire is approximated using short straight segments.

$$\int_C f\,ds = \int_0^{4\pi} t\sqrt{(-\sin t)^2 + (\cos t)^2 + 1^2}\,dt = \int_0^{4\pi} t\sqrt{2}\,dt = \tfrac{\sqrt{2}}{2}t^2\Big|_0^{4\pi} = 8\pi^2\sqrt{2}.$$

□

Example 7.5. The curve C given in Example 7.4 can also be parametrized by

$$\mathbf{X}_1(\tau) = (\cos(4\tau), \sin(4\tau), 4\tau), \qquad 0 \le \tau \le \pi.$$

Using this parametrization for C we get that the integral of $f(x,y,z) = z$ over C is

$$\int_C f\,ds = \int_0^{\pi} f(\mathbf{X}_1(\tau))\|\mathbf{X}_1'(\tau)\|\,d\tau = \int_0^{\pi} (4\tau)(4\sqrt{2})\,d\tau = 8\sqrt{2}\pi^2.$$

□

Definition 7.3. Let f be a continuous function on a smooth curve C parametrized by $\mathbf{X}(t) = (x(t), y(t), z(t))$, $a \le t \le b$. The *average* of f over C is

$$\frac{\int_C f\,ds}{\int_C ds} = \frac{\int_a^b f(\mathbf{X}(t))\|\mathbf{X}'(t)\|\,dt}{\int_a^b \|\mathbf{X}'(t)\|\,dt}.$$

Example 7.6. In Example 7.4 we saw that the density $f(x,y,z)$ varied at points along C. To find the average density on C we divide the mass of the wire, $\int_C f\,ds$, by its length, $\int_C ds$. We compute

$$\text{Length}(C) = \int_C ds = \int_0^{4\pi} \|\mathbf{X}'(t)\|\,dt = \int_0^{4\pi} \sqrt{2}\,dt = 4\pi\sqrt{2}.$$

The average density is $\dfrac{\int_C f\,ds}{\int_C ds} = \dfrac{8\pi^2\sqrt{2}}{4\pi\sqrt{2}} = 2\pi$ [mass/length]. $\qquad\square$

Work along a curve The work done by a constant force $\mathbf{F}$ in moving an object along a straight line from point $\mathbf{A}$ to $\mathbf{B}$ is

$$\mathbf{F}\cdot(\mathbf{B}-\mathbf{A}).$$

The work done by a variable force $\mathbf{F}$ in moving an object along a smooth curve C from $\mathbf{A}$ to $\mathbf{B}$ motivates finding the integral of the tangential component of $\mathbf{F}$ over C, called the integral of $\mathbf{F}$ along C.

Definition 7.4. Let $\mathbf{F}$ be a continuous vector field on a smooth curve C given by $\mathbf{X}(t) = (x(t), y(t), z(t))$, $a \le t \le b$. The *unit tangent vector* to C in the direction of increasing t is

$$\mathbf{T} = \frac{\mathbf{X}'(t)}{\|\mathbf{X}(t)\|},$$

and $\mathbf{T}$ orients C from $\mathbf{A} = \mathbf{X}(a)$ to $\mathbf{B} = \mathbf{X}(b)$. See Figure 7.3. The *integral of* $\mathbf{F}$ *along* C (in the tangential direction) from $\mathbf{A} = \mathbf{X}(a)$ to $\mathbf{B} = \mathbf{X}(b)$ is

$$\int_C \mathbf{F}\cdot\mathbf{T}\,ds = \int_a^b \mathbf{F}(\mathbf{X}(t))\cdot\mathbf{T}(\mathbf{X}(t))\|\mathbf{X}'(t)\|\,dt = \int_a^b \mathbf{F}(\mathbf{X}(t))\cdot\mathbf{X}'(t)\,dt.$$

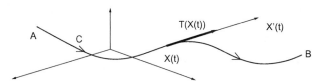

Fig. 7.3 The unit tangent vector $\mathbf{T}$ at the point $\mathbf{X}(t)$.

If C is piecewise smooth we define the integral of $\mathbf{F}$ along C to be the sum of the integrals of $\mathbf{F}$ along the smooth pieces.

In Figure 7.4 a curve C is traversed from $\mathbf{A}$ to $\mathbf{B}$. If C is traversed in the opposite direction from $\mathbf{B}$ to $\mathbf{A}$, we denote that curve by $-C$. Denote the unit tangents as $\mathbf{T}_1$ on C and as $\mathbf{T}_2$ on $-C$. The directions of $\mathbf{T}_1$ and $\mathbf{T}_2$ at each point are opposite, so $\mathbf{F}\cdot\mathbf{T}_1 = -\mathbf{F}\cdot\mathbf{T}_2$, and

$$\int_{-C} \mathbf{F}\cdot\mathbf{T}_2\,ds = -\int_C \mathbf{F}\cdot\mathbf{T}_1\,ds.$$

Example 7.7. Find the work done by $\mathbf{F}(x,y,z) = (y,-x,z)$ in moving from $(1,0,0)$ to $(1,0,2\pi)$ along the helical curve C given by $\mathbf{X}(t) = (\cos t, \sin t, t)$, $0 \le t \le 2\pi$. At the point $\mathbf{X}(t)$ on C,

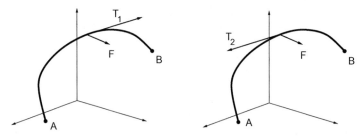

Fig. 7.4 Reversing the direction in which the curve is traversed reverses the direction of the unit tangents, and $\mathbf{F} \cdot \mathbf{T}_1 = -\mathbf{F} \cdot \mathbf{T}_2$.

$$\mathbf{F}(\mathbf{X}(t)) = (\sin t, -\cos t, t), \quad \mathbf{X}'(t) = (-\sin t, \cos t, 1), \quad \|\mathbf{X}'(t)\| = \sqrt{2}$$

and $\mathbf{T}(\mathbf{X}(t)) = \dfrac{(-\sin t, \cos t, 1)}{\sqrt{2}}$. The work is

$$\int_C \mathbf{F} \cdot \mathbf{T} ds = \int_0^{2\pi} (\sin t, -\cos t, t) \cdot \frac{(-\sin t, \cos t, 1)}{\sqrt{2}} \sqrt{2} dt$$

$$= \int_0^{2\pi} (t-1) dt = \left[\tfrac{1}{2} t^2 - t \right]_0^{2\pi} = 2\pi^2 - 2\pi.$$

□

Example 7.8. Find the work done by $\mathbf{F}(x,y,z) = (y, -x, z)$ in moving from $(1, 0, 2\pi)$ to $(1, 0, 0)$ along a straight curve C_2. We parametrize C_2 by

$$\mathbf{X}(t) = (1, 0, 2\pi) + t(0, 0, -2\pi) = (1, 0, -2\pi t + 2\pi), \quad 0 \le t \le 1.$$

$$\mathbf{F}(\mathbf{X}(t)) = (0, -1, -2\pi t + 2\pi), \quad \mathbf{X}'(t) = (0, 0, -2\pi),$$

$$\|\mathbf{X}'(t)\| = 2\pi, \quad \mathbf{T}(\mathbf{X}(t)) = (0, 0, -1).$$

The work done is

$$\int_{C_2} \mathbf{F} \cdot \mathbf{T} ds = \int_0^1 \mathbf{F}(\mathbf{X}(t)) \cdot \mathbf{T}(\mathbf{X}(t)) \|\mathbf{X}'(t)\| dt = \int_0^1 \mathbf{F}(\mathbf{X}(t)) \cdot \mathbf{X}'(t) dt$$

$$= \int_0^1 (0, -1, -2\pi t + 2\pi) \cdot (0, 0, -2\pi) dt = \int_0^1 4\pi^2 (t-1) dt = 2\pi^2 (t-1)^2 \Big|_0^1 = -2\pi^2.$$

□

There are many ways to parametrize the segment in Example 7.8. In Problem 7.12 we ask you to create two different parametrizations and verify that you get the same answer for the work.

Example 7.9. Find the work done by $\mathbf{F}(x,y,z) = (y, -x, z)$ in moving an object from $(1, 0, 0)$ to $(1, 0, 0)$ along the curve $C = C_1 \cup C_2$, where C_1 is the helical

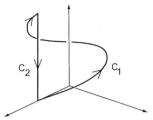

Fig. 7.5 Work done by $F(x,y,z) = (y,-x,z)$ along the loop $C_1 \cup C_2$ in Example 7.9 is not zero.

curve from $(1,0,0)$ to $(1,0,2\pi)$ in Example 7.7 and C_2 is the straight curve from $(1,0,2\pi)$ to $(1,0,0)$ in Example 7.8. The work done is the sum of the work along the smooth curves,

$$\int_C \mathbf{F} \cdot \mathbf{T} ds = \int_{C_1 \cup C_2} \mathbf{F} \cdot \mathbf{T} ds$$

$$= \int_{C_1} \mathbf{F} \cdot \mathbf{T} ds + \int_{C_2} \mathbf{F} \cdot \mathbf{T} ds = 2\pi^2 - 2\pi - 2\pi^2 = -2\pi.$$

Note that the work by the force $\mathbf{F}(x,y,z) = (y,-x,z)$ in moving an object around the loop $C_1 \cup C_2$ is not zero. $\square$

Example 7.10. Let $\mathbf{F}(x,y) = \left(\dfrac{-y}{x^2+y^2}, \dfrac{x}{x^2+y^2}\right)$ and let C_R be a circle of radius R centered at the origin traversed in the counterclockwise direction. One way to find

$$\int_{C_R} \mathbf{F} \cdot \mathbf{T} ds$$

is to parametrize C_R. Another way is to use the special geometric relationship between this vector field at $\mathbf{P}$, $\mathbf{F}(\mathbf{P})$, and the tangent to C_R at $\mathbf{P}$, $\mathbf{T}(\mathbf{P})$. See Figure 7.6. We observe that

$$\mathbf{F}(x,y) \cdot (x,y) = 0 \quad \text{and} \quad \mathbf{T}(x,y) \cdot (x,y) = 0$$

and that $\mathbf{F}$ is a multiple of $\mathbf{T}$ at each point on C_R. So the angle between $\mathbf{F}$ and $\mathbf{T}$ is zero and the tangential component of $\mathbf{F}$, $\mathbf{F} \cdot \mathbf{T} = \|\mathbf{F}\|\|\mathbf{T}\|\cos\theta$, is equal to the magnitude of $\mathbf{F}$. The magnitude of $\mathbf{F}(x,y)$ is $\|\mathbf{F}(x,y)\| = \dfrac{1}{\sqrt{x^2+y^2}}$, the reciprocal of the distance to the origin, which on C_R is $1/R$. Therefore

$$\int_{C_R} \mathbf{F} \cdot \mathbf{T} ds = \int_{C_R} \frac{1}{R} ds = \frac{1}{R}\text{Length}(C_R) = \frac{2\pi R}{R} = 2\pi.$$

$\square$

In Problem 7.6 we ask you to redo Example 7.10 by parametrizing C_R.

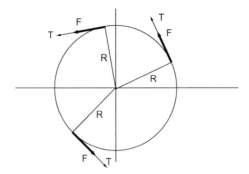

Fig. 7.6 In Example 7.10, $\mathbf{T}(\mathbf{P})$ and $\mathbf{F}(\mathbf{P})$ have the same direction at each point $\mathbf{P}$ of the circle.

Evaluation of line integrals For a piecewise smooth curve C parametrized by

$$\mathbf{X}(t) = (x(t), y(t), z(t)), \qquad a \le t \le b,$$

the integral of the tangential component of $\mathbf{F} = (f_1, f_2, f_3)$ over C is

$$\int_C \mathbf{F} \cdot \mathbf{T} \, ds = \int_a^b \mathbf{F}(\mathbf{X}(t)) \cdot \mathbf{X}'(t) \, dt.$$

If we denote

$$\mathbf{T} \, ds = (dx, dy, dz)$$

then

$$\int_C \mathbf{F} \cdot \mathbf{T} \, ds = \int_C f_1(x,y,z) \, dx + f_2(x,y,z) \, dy + f_3(x,y,z) \, dz$$

$$= \int_C f_1(x,y,z) \, dx + \int_C f_2(x,y,z) \, dy + \int_C f_3(x,y,z) \, dz.$$

In terms of the parametrization $\mathbf{X}$ we then have

$$\int_C f_1(x,y,z) \, dx = \int_a^b f_1(\mathbf{X}(t)) x'(t) \, dt,$$

$$\int_C f_2(x,y,z) \, dy = \int_a^b f_2(\mathbf{X}(t)) y'(t) \, dt,$$

$$\int_C f_3(x,y,z) \, dz = \int_a^b f_3(\mathbf{X}(t)) z'(t) \, dt.$$

The benefit of this approach is that we may use different parametrizations of C for evaluating the integrals of different components of $\mathbf{F}$. Let's look at an example.

Example 7.11. Find the integral of

$$\mathbf{F}(x, y) = (x^2 + 3, 2)$$

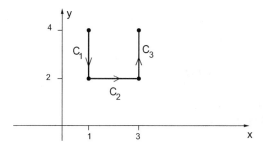

Fig. 7.7 The curve in Example 7.11.

along the curve C shown in Figure 7.7, from $(1,4)$ to $(3,4)$. One way to approach this problem is to parametrize the three segments C_1, C_2, C_3 and find the sum of the line integrals. Let's look at the integrals of the component functions over C and see whether it helps simplify the problem.

$$\int_C \mathbf{F} \cdot \mathbf{T} ds = \int_C (x^2 + 3) dx + 2 dy = \int_C (x^2 + 3) dx + \int_C 2 dy$$

Consider the integral of the first component

$$\int_C (x^2 + 3) dx = \int_{C_1} (x^2 + 3) dx + \int_{C_2} (x^2 + 3) dx + \int_{C_3} (x^2 + 3) dx.$$

On C_1, $x'(t) = 0$ because there is no change in x. So $\displaystyle\int_{C_1} (x^2 + 3) dx = 0$ and $\displaystyle\int_{C_3} (x^2 + 3) dx = 0$ for the same reason. The curve C_2 is parallel to the x axis so we can take $\mathbf{X}(t) = (t, 2)$, $1 \le t \le 3$. Then $x = t$, $dx = dt$ and

$$\int_{C_2} (x^2 + 3) dx = \int_{t=1}^3 (t^2 + 3) dt = \int_{x=1}^3 (x^2 + 3) dx = \left[\tfrac{1}{3} x^3 + 3x\right]_1^3 = \tfrac{26}{3} + 6 = \tfrac{44}{3}.$$

Similarly considering the second component function

$$\int_C 2 dy = 2 \left(\int_{C_1} dy + \int_{C_2} dy + \int_{C_3} dy \right) = 2(2 - 4) + 0 + 2(4 - 2) = 0,$$

so $\displaystyle\int_C \mathbf{F} \cdot \mathbf{T} ds = \tfrac{44}{3}$.   □

Application of line integrals to circulation The velocity of a fluid flow in the plane, at a moment in time, may be represented as a vector field

$$\mathbf{U}(x,y) = (u(x,y), v(x,y)).$$

The functions u and v are the x and y components of the velocity at (x,y).

Example 7.12. (i) The velocity field $\mathbf{U}(x,y) = (y,0)$ represents a flow parallel to the x axis, that depends only on y. The region $0 \leq y \leq 1$ models the flow of a river of depth 1. See Figure 7.8.

(ii) The velocity field $\mathbf{U}(x,y) = (-y,x)$ represents a flow that rotates counterclockwise around the origin, with speed $\|\mathbf{U}(x,y)\| = \sqrt{x^2 + y^2}$ equal to the distance to the origin.   □

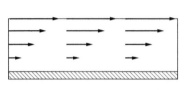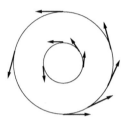

Fig. 7.8 Sketches of vector fields from Example 7.12. *Left:* $\mathbf{U}(x,y) = (y,0)$. *Right:* $\mathbf{U}(x,y) = (-y,x)$.

In contexts where $\mathbf{U}$ is the velocity field of a fluid flow, the line integral

$$\int_C \mathbf{U} \cdot \mathbf{T} \, ds = \int_C u\,dx + v\,dy$$

is called the *circulation* of $\mathbf{U}$ along C. Of particular interest is circulation along a curve that is a loop. Let's look at some examples.

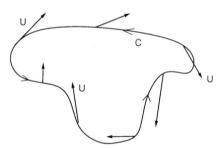

Fig. 7.9 The circulation of $\mathbf{U}$ along C is negative.

Example 7.13. The information about the vector field $\mathbf{U}$ and the curve C in Figure 7.9 indicates that $\mathbf{U} \cdot \mathbf{T} < 0$ at every point of C. So the circulation of $\mathbf{U}$ along C is negative,

$$\int_C \mathbf{U} \cdot \mathbf{T} \, ds < 0.$$

If we traverse C in the opposite direction, then the circulation of $\mathbf{U}$ along the resulting curve $-C$ is positive,

$$\int_{-C} \mathbf{U} \cdot \mathbf{T} ds > 0.$$

□

Example 7.14. Let C be a circle oriented clockwise as shown in Figure 7.10. Unit tangent vectors $\mathbf{T}$ are sketched lightly. The velocity vector field

$$\mathbf{U}(x,y) = (y,0)$$

is sketched using darker vectors. Is the circulation of $\mathbf{U}$ along C,

$$\int_C \mathbf{U} \cdot \mathbf{T} ds$$

positive, negative, or zero? Let $\mathbf{P}$ and $\mathbf{Q}$ be two diametrically opposite points on C, with $\mathbf{P}$ on the upper half of C, C_1, and $\mathbf{Q}$ on the lower half C_2. Unit tangent vectors $\mathbf{T}(\mathbf{P})$ and $\mathbf{T}(\mathbf{Q})$ point in opposite directions, and velocity vectors $\mathbf{U}(\mathbf{P})$ and $\mathbf{U}(\mathbf{Q})$ only differ in length, with $\mathbf{U}(\mathbf{P})$ longer than $\mathbf{U}(\mathbf{Q})$. We see that

$$\mathbf{U}(\mathbf{P}) \cdot \mathbf{T}(\mathbf{P}) > 0, \quad \mathbf{U}(\mathbf{Q}) \cdot \mathbf{T}(\mathbf{Q}) < 0,$$

and $\mathbf{U}(\mathbf{P}) \cdot \mathbf{T}(\mathbf{P}) > \mathbf{U}(\mathbf{Q}) \cdot \mathbf{T}(\mathbf{P}) = \mathbf{U}(\mathbf{Q}) \cdot (-\mathbf{T}(\mathbf{Q}))$. Therefore

$$\int_{C_1} \mathbf{U} \cdot \mathbf{T} ds > -\int_{C_2} \mathbf{U} \cdot \mathbf{T} ds$$

and the total circulation along C is positive:

$$\int_C \mathbf{U} \cdot \mathbf{T} ds = \int_{C_1} \mathbf{U} \cdot \mathbf{T} ds + \int_{C_2} \mathbf{U} \cdot \mathbf{T} ds > 0.$$

□

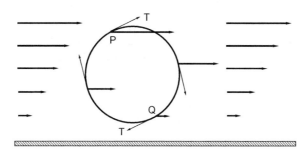

Fig. 7.10 The circulation of $\mathbf{U}$ along C is positive, in Example 7.14.

In Chapter 8 we will see that the circulation of a vector field around a small closed loop is related to the local rotation, called the *curl* of the velocity field, curl $\mathbf{U}(\mathbf{P})$, that we defined in Section 3.5.

Example 7.15. Let $\mathbf{U}(x,y) = \dfrac{(-y,x)}{(x^2+y^2)^a}$, a constant. Find the circulation of $\mathbf{U}$ along a circle C of radius r centered at the origin, and oriented counterclockwise.

At each point (x,y) of C, the vector $(-y,x)$ is orthogonal to (x,y), and its direction agrees with the unit tangent to C. Its norm is $\|(-y,x)\| = r$. Therefore the unit tangent vector is $\mathbf{T} = \dfrac{1}{r}(-y,x)$. So $\mathbf{U} \cdot \mathbf{T} = \dfrac{y^2+x^2}{r^{2a+1}} = \dfrac{1}{r^{2a-1}}$. The circulation is

$$\int_C \mathbf{U} \cdot \mathbf{T} ds = \frac{1}{r^{2a-1}} \int_C ds = \frac{1}{r^{2a-1}} 2\pi r = \frac{2\pi}{r^{2a-2}}.$$

$\square$

Example 7.16. Note the special case $a = 1$ in Example 7.15. The circulation of

$$\mathbf{U}(x,y) = \frac{(-y,x)}{x^2+y^2}$$

along circles $x^2 + y^2 = r^2$ is 2π, independent of the radius r. As we will see in Section 8.1, along circles C that do not enclose the origin, the circulation of $\mathbf{U}$ is *zero*. In Problem 7.17 we outline a geometric reason for this fact. $\square$

Flux across a curve in $\mathbb{R}^2$. The integral of the tangential component $\mathbf{F} \cdot \mathbf{T}$ of a vector field $\mathbf{F}$ along an oriented curve C can be interpreted as work or as circulation along the curve. The integral of the component of $\mathbf{F}$ normal to C is called the *flux* of $\mathbf{F}$ across C.

Let C be a smooth curve in the plane, parametrized by $\mathbf{X}(t) = (x(t),y(t))$ with $\mathbf{X}'(t) = (x'(t),y'(t)) \neq \mathbf{0}$. There are two unit normal vectors to C at $(x(t),y(t))$,

$$\mathbf{N} = \frac{(-y'(t),x'(t))}{\|\mathbf{X}'(t)\|}, \qquad -\mathbf{N} = \frac{(y'(t),-x'(t))}{\|\mathbf{X}'(t)\|}.$$

They are orthogonal to $\mathbf{X}'(t)$ and point in opposite directions.

Definition 7.5. Let C be a smooth curve parametrized by $\mathbf{X}(t) = (x(t),y(t))$, $a \leq t \leq b$, and let $\mathbf{F} = (f_1,f_2)$ be a continuous vector field on C. The *flux of* $\mathbf{F}$ *across* C in the direction of the normal

$$\mathbf{N} = \mathbf{N}(x(t),y(t)) = \frac{(-y'(t),x'(t))}{\|\mathbf{X}'(t)\|}$$

is

$$\int_C \mathbf{F} \cdot \mathbf{N} ds = \int_a^b \mathbf{F}(\mathbf{X}(t)) \cdot \frac{(-y'(t),x'(t))}{\|\mathbf{X}'(t)\|} \|\mathbf{X}'(t)\| dt = \int_C -f_1 \, dy + f_2 \, dx.$$

Example 7.17. Find the flux of $\mathbf{F}(x,y) = (x,y)$ outward across the circle C of radius 3 centered at the origin. C can be parametrized by

$$\mathbf{X}(t) = (3\cos t, 3\sin t), \qquad 0 \le t \le 2\pi.$$

Then $\mathbf{X}'(t) = (-3\sin t, 3\cos t)$ and $\mathbf{T}(t) = (-\sin t, \cos t)$. The two unit normals are

$$\mathbf{N}(t) = (\cos t, \sin t), \qquad -\mathbf{N}(t) = -(\cos t, \sin t).$$

$\mathbf{N}$ points outward at each point on C. The flux outward across C is

$$\int_C \mathbf{F} \cdot \mathbf{N} ds = \int_0^{2\pi} (3\cos t, 3\sin t) \cdot (\cos t, \sin t) \|\mathbf{X}'(t)\| dt$$

$$= \int_0^{2\pi} 3(\cos^2 t + \sin^2 t) 3 \, dt = 9(2\pi) = 18\pi.$$

An alternative way to compute the integral is to observe that at each point on C, $\mathbf{F}$ and $\mathbf{N}$ are parallel. So $\mathbf{F} \cdot \mathbf{N} = \|\mathbf{F}\| \|\mathbf{N}\| \cos 0 = \|\mathbf{F}\|$. This is equal to $\sqrt{x^2 + y^2} = 3$ since (x, y) is on C. Therefore

$$\int_C \mathbf{F} \cdot \mathbf{N} ds = \int_C 3 ds = 3 \mathrm{Length}(C) = 3(2\pi(3)) = 18\pi.$$

$\square$

Line integrals in $\mathbb{R}^n$. Analogous to Definition 7.1 we define a *smooth curve C* in $\mathbb{R}^n$ to be the range of a C^1 function $\mathbf{X}(t) = (x_1(t), \dots, x_n(t))$, $a \le t \le b$, with $\mathbf{X}'(t) \ne \mathbf{0}$. $\mathbf{X}$ is called a *smooth parametrization* of C. The *length* of C is defined as

$$\mathrm{Length}(C) = \int_a^b \|\mathbf{X}'(t)\| dt.$$

We note that the length of a curve as defined above is independent of the parametrization of the curve. For if $t = t(u)$, $\alpha \le u \le \beta$ is another parameter, with $\frac{dt}{du}$ positive, then

$$\frac{d\mathbf{X}(t(u))}{du} = \frac{d\mathbf{X}}{dt} \frac{dt}{du}$$

and so by the change of variables formula

$$\mathrm{Length}(C) = \int_a^b \left\| \frac{d\mathbf{X}}{dt} \right\| dt = \int_\alpha^\beta \left\| \frac{d\mathbf{X}}{dt} \right\| \frac{dt}{du} du$$

$$= \int_\alpha^\beta \left\| \frac{d\mathbf{X}}{dt} \frac{dt}{du} \right\| du = \int_\alpha^\beta \left\| \frac{d\mathbf{X}(t(u))}{du} \right\| du.$$

Given a function f from $\mathbb{R}^n$ to $\mathbb{R}$ that is continuous on C, the *line integral of f over C* is

$$\int_C f ds = \int_a^b f(\mathbf{X}(t)) \|\mathbf{X}'(t)\| dt.$$

In particular taking $f(\mathbf{X}) = 1$ we get

$$\text{Length}(C) = \int_C ds = \int_a^b \|\mathbf{X}'(t)\| \, dt.$$

We define the average of f on C as

$$\frac{1}{\text{Length}(C)} \int_C f ds.$$

Example 7.18. Show that the average of the first coordinates of the points on the line segment from $\mathbf{A} = (a_1, \ldots, a_n)$ to $\mathbf{B} = (b_1, \ldots, b_n)$ in $\mathbb{R}^n$ is

$$\tfrac{1}{2}(a_1 + b_1).$$

Let $\mathbf{X}(t) = \mathbf{A} + t(\mathbf{B} - \mathbf{A})$, $0 \le t \le 1$ parametrize the segment from $\mathbf{A}$ to $\mathbf{B}$ and let $f(x_1, \ldots, x_n) = x_1$. The average of f over the segment is

$$\frac{\int_0^1 f(\mathbf{X}(t)) \|\mathbf{X}'(t)\| \, dt}{\int_0^1 \|\mathbf{X}'(t)\| \, dt} = \frac{\int_0^1 (a_1 + t(b_1 - a_1)) \|\mathbf{B} - \mathbf{A}\| \, dt}{\int_0^1 \|\mathbf{B} - \mathbf{A}\| \, dt}$$

$$= \frac{\|\mathbf{B} - \mathbf{A}\| \int_0^1 (a_1 + t(b_1 - a_1)) \, dt}{\|\mathbf{B} - \mathbf{A}\|} = \tfrac{1}{2}(a_1 + b_1).$$

In Problem 7.5 we ask you to show this is true for each coordinate. $\square$

Let $\mathbf{F}$ be a vector field

$$\mathbf{F}(x_1, \ldots, x_n) = (f_1(x_1, \ldots, x_n), \ldots, f_n(x_1, \ldots, x_n))$$

that is continuous on a smooth curve C parametrized by $\mathbf{X}$,

$$\mathbf{X}(t) = (x_1(t), \ldots, x_n(t)), \qquad a \le t \le b.$$

We denote the *unit tangent* in the direction of increasing t as $\mathbf{T}(\mathbf{X}(t)) = \dfrac{\mathbf{X}'(t)}{\|\mathbf{X}'(t)\|}$. As in $\mathbb{R}^3$ the line integral of the tangential component of $\mathbf{F}$ along C is

$$\int_C \mathbf{F} \cdot \mathbf{T} ds = \int_a^b \mathbf{F}(\mathbf{X}(t)) \cdot \frac{\mathbf{X}'(t)}{\|\mathbf{X}'(t)\|} \|\mathbf{X}'(t)\| \, dt = \int_a^b \mathbf{F}(\mathbf{X}(t)) \cdot \mathbf{X}'(t) \, dt.$$

Example 7.19. Define $\mathbf{F}(\mathbf{X}) = \mathbf{X}$ in $\mathbb{R}^n$ and let C be a smooth curve from a point $\mathbf{A} = \mathbf{X}(a)$, to $\mathbf{B} = \mathbf{X}(b)$. Then

$$\int_C \mathbf{F}(\mathbf{X}) \cdot \mathbf{T} ds = \int_a^b \mathbf{X}(t) \cdot \mathbf{X}'(t) \, dt$$

$$= \int_a^b (x_1(t) x_1'(t) + x_2(t) x_2'(t) + \cdots + x_n(t) x_n'(t)) \, dt$$

$$= \int_a^b \frac{d}{dt}(\tfrac{1}{2}(x_1^2 + \cdots + x_n^2))\,dt = \left[\tfrac{1}{2}\|\mathbf{X}(t)\|^2\right]_a^b = \tfrac{1}{2}(\|\mathbf{B}\|^2 - \|\mathbf{A}\|^2).$$

$\square$

Problems

7.1. Let C be a line segment in the x,y plane with endpoints $\mathbf{A}$ and $\mathbf{B}$. Consider two parametrizations of C: first $\mathbf{X}(t) = \mathbf{A} + t(\mathbf{B} - \mathbf{A})$ for $0 \le t \le 1$, and second $\mathbf{Y}(u) = \mathbf{B} + \tfrac{1}{2}u(\mathbf{A} - \mathbf{B})$ for $0 \le u \le 2$. Calculate the following integrals using both parametrizations.

(a) $\displaystyle\int_C ds$,

(b) $\displaystyle\int_C y\,ds$,

(c) The average of y on C.

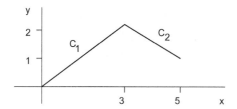

Fig. 7.11 The curves C_1 and C_2 in Problem 7.2 are line segments.

7.2. Let C_1 and C_2 be the segments in Figure 7.11, and denote by $C_1 \cup C_2$ the combined curve consisting of C_1 followed by C_2. Find the value of the line integrals by using their geometric meaning.

(a) $\displaystyle\int_{C_1} ds$

(b) $\displaystyle\int_{C_2} ds$

(c) $\displaystyle\int_{C_1 \cup C_2} ds$

7.3. Let $\mathbf{X}(t) = \mathbf{A} + t(\mathbf{B} - \mathbf{A})$, $0 \le t \le 1$, parametrize the line segment C from $\mathbf{A}$ to $\mathbf{B}$. Find a number k so that the substitution $t = ks$ gives a new parametrization $\mathbf{Y}(s) = \mathbf{X}(ks)$, with

$$\mathbf{Y}(0) = \mathbf{A}, \quad 0 \le s \le \text{Length}(C), \quad \mathbf{Y}(\text{Length}(C)) = \mathbf{B}, \quad \text{and} \quad \|\mathbf{Y}'(s)\| = 1.$$

$\mathbf{Y}(s)$ is the arc length parametrization of C.

7.4. Let C be the smooth curve in $\mathbb{R}^4$ given by $\mathbf{X} : [0, 2\pi] \to \mathbb{R}^4$,

$$\mathbf{X}(t) = (\cos t, \sin t, \cos kt, \sin kt)$$

and k is a constant.

(a) Find $\mathbf{X}'(t)$ and the unit tangent vector $\mathbf{T}$ at $\mathbf{X}(t)$.
(b) Find the length of C.
(c) Verify that for each number k the curve C lies on the sphere

$$x_1^2 + x_2^2 + x_3^2 + x_4^2 = 2.$$

(d) Let $f(x_1, x_2, x_3, x_4) = x_1$ and $g(x_1, x_2, x_3, x_4) = x_4$. Show that the average of f on C is 0, but the average of g is not zero unless k is an integer.

7.5. Show that the average of the i-th component x_i of the points on a line segment from $\mathbf{A} = (a_1, \ldots, a_n)$ to $\mathbf{B} = (b_1, \ldots, b_n)$, in any number of dimensions, is

$$\tfrac{1}{2}(a_i + b_i).$$

7.6. Let $\mathbf{F}(x, y) = \dfrac{(-y, x)}{x^2 + y^2}$, and let C_R be the circle of radius R centered at the origin and traversed in the counterclockwise direction. Use the parametrization $\mathbf{X}(t) = (R \cos t, R \sin t)$, $0 \le t \le 2\pi$ to calculate

$$\int_{C_R} \mathbf{F} \cdot \mathbf{T} ds.$$

7.7. Let C be the triangle with vertices $(0,0)$, $(1,0)$, and $(1,1)$, traversed counterclockwise. Evaluate the line integral

$$\int_C y^2 \, dx + x \, dy.$$

7.8. Let $\mathbf{F} = (p, q)$ be a constant vector field, and C a line segment in the plane. Choose one of the two unit normal vectors $\mathbf{N}$ for C. Show that the flux of $\mathbf{F}$ across C is given by

$$\int_C \mathbf{F} \cdot \mathbf{N} ds = \mathbf{F} \cdot \mathbf{N} \operatorname{Length}(C).$$

7.9. A smooth curve C has unit tangent vectors $\mathbf{T} = (t_1, t_2)$. Take the unit normal to C to be $\mathbf{N} = (t_2, -t_1)$ at each point. Let $\mathbf{F} = (f_1, f_2)$ be a vector field, and define a vector field $\mathbf{G} = (f_2, -f_1)$ at each point. Show that

$$\int_C \mathbf{F} \cdot \mathbf{T} ds = \int_C \mathbf{G} \cdot \mathbf{N} ds.$$

7.10. Let C be the line segment from $(0,0)$ to $(4,3)$, parametrized by

$$(x(t), y(t)) = (\tfrac{4}{5}t, \tfrac{3}{5}t), \quad (0 \le t \le 5)$$

and let C_a be the circle parametrized by

$$\mathbf{X}(t) = (a\cos t, a\sin t), \qquad 0 \le t \le 2\pi, \quad a > 0.$$

Let $\mathbf{F} = (8,0)$, a constant vector field.

(a) Find the normal vector $\mathbf{N}(x(t), y(t))$ to C, that is obtained by rotating the unit tangent vector $\mathbf{T}$ ninety degrees clockwise.
(b) Find the flux of $\mathbf{F}$ across C.
(c) Find the outward pointing unit normal vector $\mathbf{N}_a$ at each point of C_a.
(d) Find the flux of $\mathbf{F}$ outward across C_a.

7.11. Let a smooth curve C be the graph of $y = f(x)$ on $[a,b]$. Use the parametrization

$$\mathbf{X}(t) = (t, f(t)), \qquad a \le t \le b$$

to show that the length of C is given by

$$\int_a^b \sqrt{1 + (f'(t))^2}\, dt.$$

Verify that it gives the right answer for $y = 3x$, $0 \le x \le 1$.

7.12. Find the work done by $\mathbf{F}(x, y, z) = (y, -x, z)$ in moving from $(1, 0, 2\pi)$ to $(1, 0, 0)$ along a straight segment C. Create two smooth parametrizations for C and verify that you get the same total work with each one.

7.13. Let C_1 be the unit circle centered at the origin and C_2 be the boundary of the 2 by 2 square centered at the origin of $\mathbb{R}^2$. Use symmetry arguments and properties of integrals to show without calculation:

(a) $\displaystyle\int_{C_1} x^2 ds = \frac{1}{2}\int_{C_1}(x^2 + y^2)ds$

(b) $\displaystyle\int_{C_2} x^2 ds = \frac{1}{2}\int_{C_2}(x^2 + y^2)ds$

(c) $\displaystyle\int_{C_1}(x^2 + y^2)^{10}ds = \text{Length}(C_1)$

7.14. Let C be the piecewise smooth closed curve shown in Figure 7.12 consisting of graphs, C_1 and C_2, of two functions g_1 and g_2, traversed clockwise. Let $\mathbf{F}(x,y) = (y, 0)$. The work done by $\mathbf{F}$ is

$$\int_{C_1 \cup C_2} \mathbf{F} \cdot \mathbf{T} ds = \int_{C_1} \mathbf{F} \cdot \mathbf{T} ds + \int_{C_2} \mathbf{F} \cdot \mathbf{T} ds = \int_C y\, dx.$$

(a) Write a parametrization for C_2 in terms of g_2, and show that

$$\int_{C_2} \mathbf{F} \cdot \mathbf{T} ds = \int_a^b g_2(x)\, dx.$$

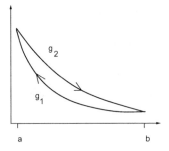

Fig. 7.12 The curves in Problem 7.14.

(b) Write a parametrization for C_1 in terms of g_1, and show that

$$\int_{C_1} \mathbf{F} \cdot \mathbf{T} ds = - \int_a^b g_1(x) \, dx.$$

(c) Show that the area between the graphs of g_1 and g_2 is $\int_C y \, dx$.

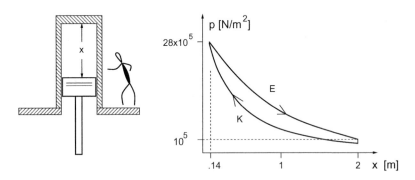

Fig. 7.13 The curves in Problem 7.15.

7.15. A container ship is driven by a large diesel engine. The engineer has measured the pressure p in the cylinders at each piston position x. See Figure 7.13. The curve marked K is while the piston moves upward to compress the air, and the curve marked E is while the burning fuel expands to force the piston down. The work [Joules] done in one cycle is the line integral

$$\int_{K \cup E} p \, dx.$$

(a) Do $\int_K p \, dx$ and $\int_E p \, dx$ have the same, or opposite signs? Why?

(b) Use the results from Problem 7.14 to show that the work done in one cycle is the area between the graphs.

7.16. Find the circulation of the vector fields

(a) $\mathbf{U}(x,y) = (-3y, 2x)$
(b) $\mathbf{V}(x,y) = -5\mathbf{U}(x,y) = (15y, -10x)$

counterclockwise around the circle of radius R, $x^2 + y^2 = R^2$.

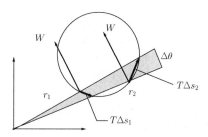

Fig. 7.14 Figure for Problem 7.17.

7.17. Justify the following items to show that the circulation of

$$\mathbf{U}(x,y) = \frac{(-y, x)}{x^2 + y^2}$$

around any circle that does not enclose the origin is zero. See Figure 7.14, where $\mathbf{W}$ denotes unit vectors orthogonal to each radial line.

(a) The integral

$$\int_C \mathbf{U} \cdot \mathbf{T} ds$$

can be well approximated by a sum over pairs of segments Δs_1, Δs_2 using small angles $\Delta\theta$ as shown, where r_1 and r_2 are distances from the origin to the segments.
(b) $\mathbf{U} = \frac{1}{r}\mathbf{W}$
(c) $\Delta\theta \approx -\frac{1}{r_1}\mathbf{W} \cdot (\mathbf{T}_1 \Delta s_1) \approx \frac{1}{r_2}\mathbf{W} \cdot (\mathbf{T}_2 \Delta s_2)$
(d) $\int_C \mathbf{U} \cdot \mathbf{T} ds \approx \sum \left(\frac{1}{r_2}\mathbf{W} \cdot (\mathbf{T}_2 \Delta s_2) + \frac{1}{r_1}\mathbf{W} \cdot (\mathbf{T}_1 \Delta s_1) \right)$ is approximately zero.

7.18. Let $\mathbf{F}(x,y) = \nabla \log r$, where $r = \sqrt{x^2 + y^2}$, and let C be the circle $x^2 + y^2 = a^2$. Find the flux of $\mathbf{F}$ outward across C, and show that it is independent of the radius a.

7.19. Evaluate the following line integrals along C, the line segment from $\mathbf{A} = (0,0)$ to $\mathbf{B} = (3,4)$. Which ones are equal to the expressions on the right hand sides?

(a) $\displaystyle\int_C x\,dx =_? \left[\tfrac{1}{2}x^2\right]_A^B$

(b) $\displaystyle\int_C x\,dy =_? \left[xy\right]_A^B$

(c) $\displaystyle\int_C dx + 5\,dy =_? \left[x + 5y\right]_A^B$

7.2 Conservative vector fields

Some vector fields have the property that if you integrate along any piecewise smooth *closed curve*, $\mathbf{X}(t)$, $a \le t \le b$ where $\mathbf{X}(a) = \mathbf{X}(b)$, then the integral is zero. We saw in Example 7.9 that not all vector fields have this property. In this section we develop some criteria for determining which fields do have it.

Definition 7.6. We say that a vector field $\mathbf{F}$ is *conservative* if it is the gradient of a continuously differentiable function g,

$$\mathbf{F}(\mathbf{P}) = \nabla g(\mathbf{P})$$

for all $\mathbf{P}$ in the domain of $\mathbf{F}$. The function g is called a *potential function* of $\mathbf{F}$.

Example 7.20. Let $g(x,y,z) = -(x^2 + y^2 + z^2)^{-1/2} = -\dfrac{1}{\|\mathbf{X}\|}$. We have seen that

$$\nabla g(x,y,z) = \frac{(x,y,z)}{(x^2+y^2+z^2)^{3/2}} = \frac{\mathbf{X}}{\|\mathbf{X}\|^3} = \frac{1}{\|\mathbf{X}\|^2}\frac{\mathbf{X}}{\|\mathbf{X}\|}$$

is the inverse square vector field. This shows that the inverse square vector field is conservative and that $-\frac{1}{\|\mathbf{X}\|}$ is a potential function.   □

Theorem 7.1. *Suppose $\mathbf{F}$ is a continuous vector field from an open connected set D in $\mathbb{R}^n$ to $\mathbb{R}^n$, $n \ge 2$. Then the following three statements are equivalent, that is, each implies the others.*

(a) $\mathbf{F}$ is conservative,

(b) For every piecewise smooth closed curve C in D, $\displaystyle\int_C \mathbf{F}\cdot\mathbf{T}\,ds = 0$.

(c) For any two points $\mathbf{A}$ and $\mathbf{B}$ in D, and for any two piecewise smooth curves C_1 and C_2 in D that start at $\mathbf{A}$ and end at $\mathbf{B}$

$$\int_{C_1} \mathbf{F}\cdot\mathbf{T}\,ds = \int_{C_2} \mathbf{F}\cdot\mathbf{T}\,ds$$

Proof. We show that (a) implies (b), (b) implies (c), and (c) implies (a). From that it follows that each statement implies both of the others, and so if one is false the others must be false too. We write out the case for a vector field from $\mathbb{R}^3$ to $\mathbb{R}^3$.

Suppose (a) holds. Then there is a function g such that

$$\mathbf{F}(x,y,z) = (g_x(x,y,z), g_y(x,y,z), g_z(x,y,z)).$$

For a smooth curve C in D given by $\mathbf{X}(t)$, $a \leq t \leq b$,

$$\int_C \mathbf{F} \cdot \mathbf{T} ds = \int_a^b \mathbf{F}(\mathbf{X}(t)) \cdot \mathbf{X}'(t) dt = \int_a^b \nabla g(\mathbf{X}(t)) \cdot \mathbf{X}'(t) dt.$$

By the Chain Rule for curves, $\nabla g(\mathbf{X}(t)) \cdot \mathbf{X}'(t) = \dfrac{d}{dt} g(\mathbf{X}(t))$. Therefore

$$\int_C \mathbf{F} \cdot \mathbf{T} ds = \int_a^b \nabla g(\mathbf{X}(t)) \cdot \mathbf{X}'(t) dt = \int_a^b \frac{d}{dt} g(\mathbf{X}(t)) dt = g(\mathbf{X}(b)) - g(\mathbf{X}(a)). \quad (7.1)$$

If C is closed then $\mathbf{X}(b) = \mathbf{X}(a)$ and $g(\mathbf{X}(b)) - g(\mathbf{X}(a))$ is zero. If C is piecewise smooth, say one smooth piece from $\mathbf{X}(a)$ to $\mathbf{X}(c)$ and a second smooth piece from $\mathbf{X}(c)$ to $\mathbf{X}(b) = \mathbf{X}(a)$, then we evaluate the integral over the smooth pieces, getting

$$g(\mathbf{X}(c)) - g(\mathbf{X}(a)) + g(\mathbf{X}(b)) - g(\mathbf{X}(c)) = -g(\mathbf{X}(a)) + g(\mathbf{X}(b)) = 0.$$

Similarly for any number of smooth pieces. This proves that (a) implies (b).

Suppose (b) holds. Let $\mathbf{A}$ and $\mathbf{B}$ be two points in D. Let C_1 and C_2 be any two piecewise smooth curves in D that start at $\mathbf{A}$ and end at $\mathbf{B}$, and let $C = C_1 \cup (-C_2)$. Then C is a piecewise smooth closed curve, and by (b)

$$\int_C \mathbf{F} \cdot \mathbf{T} ds = 0.$$

Using properties of line integrals we have

$$\int_{C_1 \cup (-C_2)} \mathbf{F} \cdot \mathbf{T} ds = \int_{C_1} \mathbf{F} \cdot \mathbf{T} ds + \int_{-C_2} \mathbf{F} \cdot \mathbf{T} ds = \int_{C_1} \mathbf{F} \cdot \mathbf{T} ds - \int_{C_2} \mathbf{F} \cdot \mathbf{T} ds = 0.$$

Therefore

$$\int_{C_1} \mathbf{F} \cdot \mathbf{T} ds = \int_{C_2} \mathbf{F} \cdot \mathbf{T} ds.$$

This proves that (b) implies (c).

Suppose (c) is true. We state without proof that open connected sets in $\mathbb{R}^3$ have the property, that every two points in the set can be joined by a piecewise smooth curve in the set. Now let $\mathbf{A}$ be a point of D. Because D is open and connected, for each point (x,y,z) in D there is a piecewise smooth curve C in D from $\mathbf{A}$ to (x,y,z). Define a function g by

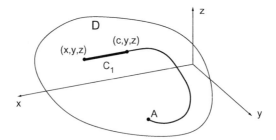

Fig. 7.15 A curve from **A** to (x,y,z) that ends parallel to the x axis.

$$g(x,y,z) = \int_C \mathbf{F} \cdot \mathbf{T} ds, \quad (x,y,z) \text{ in } D, \quad C \text{ any curve in } D \text{ from } \mathbf{A} \text{ to } (x,y,z).$$

We will show that $\nabla g = \mathbf{F}$. Since (c) is true, any curve we choose from **A** to (x,y,z) will result in the same number $g(x,y,z)$. Since D is open and connected there is a point (c,y,z) in D with $c < x$, so that the straight line C_1 from (c,y,z) to (x,y,z) is in D and can be combined with a curve C_2 in D from **A** to (c,y,z). See Figure 7.15. Using these curves,

$$g(x,y,z) = \int_{C=C_1 \cup C_2} \mathbf{F} \cdot \mathbf{T} ds = \int_{C_1} \mathbf{F} \cdot \mathbf{T} ds + \int_{C_2} \mathbf{F} \cdot \mathbf{T} ds.$$

Parametrize C_1 by $\mathbf{X}_1(t) = (t,y,z)$, $c \leq t \leq x$. Writing components of $\mathbf{F} = (f_1, f_2, f_3)$ we get

$$\int_{C_1} \mathbf{F} \cdot \mathbf{T} ds = \int_c^x \mathbf{F}(t,y,z) \cdot (1,0,0) dt = \int_c^x f_1(t,y,z) dt.$$

The derivative of g with respect to x is

$$g_x(x,y,z) = \frac{\partial}{\partial x} \left(\int_c^x f_1(t,y,z) dt + \int_{C_2} \mathbf{F} \cdot \mathbf{T} ds \right).$$

The second integral does not depend on x so its derivative with respect to x is zero. By the Fundamental Theorem of Calculus,

$$g_x(x,y,z) = \frac{\partial}{\partial x} \int_c^x f_1(t,y,z) dt = f_1(x,y,z).$$

Similar arguments show that $g_y = f_2$ and $g_z = f_3$. So $\nabla g = \mathbf{F}$, and $\mathbf{F}$ is conservative. This concludes the proof that (c) implies (a), and is the last implication in our chain.

□

   In the proof of Theorem 7.1 we showed that if $\mathbf{F} = \nabla g$ then the integral of $\mathbf{F}$ along every piecewise smooth curve from **A** to **B** is simply $g(\mathbf{B}) - g(\mathbf{A})$. (See equation (7.1).) This is often called the Fundamental Theorem of Line Integrals.

Theorem 7.2. Fundamental Theorem of Line Integrals. *If C is a piecewise smooth curve from $\mathbf{A}$ to $\mathbf{B}$ in the domain of a continuously differentiable function g, then*

$$\int_C \nabla g \cdot \mathbf{T} \, ds = g(\mathbf{B}) - g(\mathbf{A}).$$

We say the integral of a conservative vector field is *independent of path* since $g(\mathbf{B}) - g(\mathbf{A})$ depends only on the values of the potential function at the endpoints of the curve.

Knowing that a vector field is conservative is helpful in evaluating line integrals.

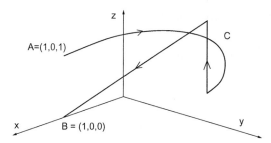

Fig. 7.16 The curve in Example 7.21.

Example 7.21. Find the integral of $\mathbf{F}(x,y,z) = (x,y,z)$ along the curve C shown in Figure 7.16. By inspection we see that $\mathbf{F}$ is conservative because

$$\mathbf{F}(x,y,z) = \nabla g(x,y,z) = \nabla(\tfrac{1}{2}(x^2 + y^2 + z^2)).$$

By the Fundamental Theorem of Line Integrals,

$$\int_C \mathbf{F} \cdot \mathbf{T} \, ds = g(1,0,0) - g(1,0,1) = \tfrac{1}{2} - 1 = -\tfrac{1}{2}.$$

□

Example 7.22. Find the integral of $\mathbf{F}(x,y,z) = (x^2 + y, \sin y, z)$ along the curve C shown in Figure 7.17. We find by inspection that

$$\nabla\left(\tfrac{1}{3}x^3 - \cos y + \tfrac{1}{2}z^2\right) = (x^2, \sin y, z).$$

We can write the integral as a sum,

$$\int_C \mathbf{F} \cdot \mathbf{T} \, ds = \int_C \left((x^2 + y)\,dx + \sin y \, dy + z \, dz\right)$$

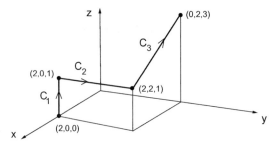

Fig. 7.17 The curve in Example 7.22.

$$= \int_C \left(x^2\,dx + \sin y\,dy + z\,dz \right) + \int_C y\,dx.$$

The first integral can be evaluated by the Fundamental Theorem of Line Integrals:

$$\int_C x^2\,dx + \sin y\,dy + z\,dz = \int_C \nabla\left(\tfrac{1}{3}x^3 - \cos y + \tfrac{1}{2}z^2 \right) \cdot \mathbf{T}\,ds$$

$$= \left[\tfrac{1}{3}x^3 - \cos y + \tfrac{1}{2}z^2 \right]_{(2,0,0)}^{(0,2,3)} = \left(-\cos(2) + \tfrac{1}{3}3^2 \right) - \left(\tfrac{1}{3}2^3 - 1 + 0 \right).$$

For the second integral we see there is no change in x along C_1 or C_2. Therefore

$$\int_C y\,dx = \int_{C_3} y\,dx = \int_{C_3} 2\,dx.$$

Again by the Fundamental Theorem

$$\int_{C_3} 2\,dx = \int_{C_3} \nabla(2x) \cdot \mathbf{T}\,ds = \left[2x \right]_{(2,2,1)}^{(0,2,3)} = -4.$$

Thus $\displaystyle \int_C \mathbf{F} \cdot \mathbf{T}\,ds = -\cos(2) + \tfrac{9}{2} - \tfrac{8}{3} + 1 - 4.$ □

In Example 7.21 we saw that $\mathbf{F}(x,y,z) = (x,y,z)$ is conservative by producing a potential function g. In Example 7.9 we saw, as a result of calculating some integrals, that $\mathbf{F}(x,y,z) = (y,-x,z)$ does not satisfy the independence of path property, so by Theorem 7.1 $(y,-x,z)$ cannot be a conservative vector field. The next theorem gives us a useful criterion for a vector field in $\mathbb{R}^3$ or $\mathbb{R}^2$ to be conservative.

Theorem 7.3.(a) *If* $\mathbf{F}$ *is a conservative* C^1 *vector field from* $\mathbb{R}^3$ *to* $\mathbb{R}^3$, *then*
$\operatorname{curl} \mathbf{F} = \mathbf{0}$.
(b) *If* $\mathbf{F}$ *is a conservative* C^1 *vector field from* $\mathbb{R}^2$ *to* $\mathbb{R}^2$, *then* $\operatorname{curl} \mathbf{F} = 0$.

Proof. We prove part (a). If $\mathbf{F} = \nabla g$ then $\mathbf{F} = (f_1, f_2, f_3) = (g_x, g_y, g_z)$ and

$$\operatorname{curl} \mathbf{F} = (f_{3y} - f_{2z}, -f_{3x} + f_{1z}, f_{2x} - f_{1y}) = (g_{zy} - g_{yz}, -(g_{zx} - g_{xz}), g_{yx} - g_{xy}).$$

Since $\mathbf{F} = \nabla g$ is continuously differentiable, g is twice continuously differentiable, so its second mixed partial derivatives do not depend on the order of differentiation. This shows that curl $\mathbf{F} = \mathbf{0}$.   □

Example 7.23. Let $\mathbf{F}(x,y,z) = (y, -x, z)$. Then

$$\operatorname{curl}\mathbf{F} = (0-0, -(0-0), -1-1) = (0,0,-2).$$

According to Theorem 7.3, since curl $\mathbf{F} \neq \mathbf{0}$, $\mathbf{F}$ is not conservative.   □

The next example shows that while curl $\mathbf{F} = \mathbf{0}$ is a necessary condition, it is not sufficient to imply that $\mathbf{F}$ is conservative.

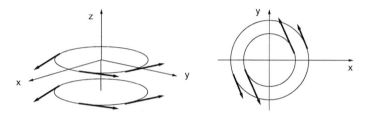

Fig. 7.18 Sketches of the vector field in Example 7.24. $\mathbf{F}$ is not conservative.

Example 7.24. Let $\mathbf{F}(x,y,z) = \left(\dfrac{-y}{x^2+y^2}, \dfrac{x}{x^2+y^2}, 0 \right)$, defined for $(x,y) \neq (0,0)$.
Then

$$\operatorname{curl}\mathbf{F} = \left(0-0, -(0-0), \frac{y^2-x^2}{(x^2+y^2)^2} - \frac{y^2-x^2}{(x^2+y^2)^2} \right) = \mathbf{0}$$

Let C_R be the circle of radius R in the x,y plane, centered at the origin traversed counterclockwise. See Figure 7.18. Notice that at each point along the circle C_R, the vector $\mathbf{F}(x,y,0)$ has magnitude R^{-1} and points in the same direction as the unit tangent $\mathbf{T}(x,y,0)$. Hence $\mathbf{F} \cdot \mathbf{T} = R^{-1}$ and

$$\int_{C_R} \mathbf{F} \cdot \mathbf{T} ds = R^{-1} \operatorname{Length}(C_R) = R^{-1} 2\pi R = 2\pi.$$

Even though curl $\mathbf{F} = \mathbf{0}$, $\mathbf{F}$ is not conservative since the integral along the closed curve C_R is not zero.   □

Example 7.25. Figure 7.19 shows some level sets for a C^1 function f in $\mathbb{R}^2$, and some piecewise smooth curves C_1 and C_2 in $\mathbb{R}^2$.
If $\mathbf{F} = \nabla f$, find $\displaystyle\int_{C_1} \mathbf{F} \cdot \mathbf{T} ds$ and $\displaystyle\int_{C_2} \mathbf{F} \cdot \mathbf{T} ds$.

$$\int_{C_1} \mathbf{F} \cdot \mathbf{T} ds = \int_{C_1} \nabla f \cdot \mathbf{T} ds = f(\mathbf{Q}) - f(\mathbf{P}) = 30 - 40 = -10.$$

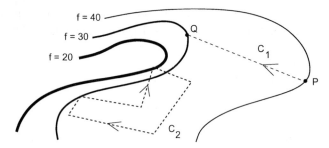

Fig. 7.19 Level sets and curves in Example 7.25.

The line integral

$$\int_{C_2} \mathbf{F} \cdot \mathbf{T} ds = 0,$$

because $\mathbf{F}$ is conservative and C_2 is a closed curve.   □

Example 7.26. Show that $\mathbf{F}(x,y) = (x+y, y^2+x)$ is conservative by finding a potential function g so that $\mathbf{F} = \nabla g$. That is,

$$g_x = x+y, \qquad g_y = y^2+x.$$

An antiderivative for g_x with respect to x is

$$g(x,y) = \tfrac{1}{2}x^2 + xy + h(y)$$

where $h(y)$ is some function of y. The partial derivative with respect to y is then

$$g_y = x + h'(y).$$

Take $h(y) = \tfrac{1}{3}y^3$ so that $g_y = y^2 + x$. We have found

$$\nabla(\tfrac{1}{2}x^2 + xy + \tfrac{1}{3}y^3) = (x+y, y^2+x).$$

□

Example 7.27. Is $\mathbf{F}(x,y) = (x+y, y^2)$ conservative? We compute

$$\operatorname{curl} \mathbf{F}(x,y) = f_{2x} - f_{1y} = 0 - 1 \neq 0.$$

By Theorem 7.3, $\mathbf{F}$ is not conservative.   □

Problems

7.20. Find a potential function for each of these conservative vector fields.

(a) $(x,0,0)$
(b) $(0,z,y)$
(c) a constant vector field (a,b,c)
(d) $\mathbf{F}+\mathbf{G}$, if $\mathbf{F}$ and $\mathbf{G}$ are each conservative.

7.21. Let

$$G(x,y) = \left(\frac{x}{x^2+y^2}, \frac{y}{x^2+y^2}\right), \qquad F(x,y) = \left(\frac{-y}{x^2+y^2}, \frac{x}{x^2+y^2}\right).$$

(a) Find a potential function $g(x,y)$ so that $\nabla g = \mathbf{G}$. What is the domain of g?
(b) Find $\nabla \tan^{-1}\left(\frac{y}{x}\right)$. Why is $\tan^{-1}\left(\frac{y}{x}\right)$ *not* a potential function for $\mathbf{F}$?

7.22. Show that the vector fields in the integrands below are conservative by finding potential functions, and evaluate the integrals on any smooth curve C from $(0,0,0)$ to (a,b,c) using the Fundamental Theorem of Line Integrals.

(a) $\displaystyle\int_C z^2\,dz$

(b) $\displaystyle\int_C \nabla(xy)\cdot \mathbf{T}\,ds$

(c) $\displaystyle\int_C z\,dx + y\,dy + x\,dz$

7.23. Let C be a smooth curve from $(0,0,0)$ to (a,b,c). Which of the integrals below can be evaluated by the Fundamental Theorem of Line Integrals? Find the values of those that can.

(a) $\displaystyle\int_C x^2\,dy$

(b) $\displaystyle\int_C (\nabla(xy) - 3\nabla(z^2\cos y))\cdot \mathbf{T}\,ds$

(c) $\displaystyle\int_C dx + dy$

7.24. (a) Use the fact that

$$\frac{2(x,y)}{(x^2+y^2)^2} = \nabla\left(\frac{-1}{x^2+y^2}\right)$$

to evaluate

$$\int_C \frac{x}{(x^2+y^2)^2}\,dx + \frac{y}{(x^2+y^2)^2}\,dy$$

where C is any smooth curve from $(1,2)$ to $(2,2)$ not passing through the origin.
(b) Why do we restrict C to curves that do not pass through the origin?

7.25. (a) Use the fact that $\nabla(\|\mathbf{X}\|^{-1}) = -\|\mathbf{X}\|^{-3}\mathbf{X}$ to evaluate

$$\int_C \frac{-x_1\,dx_1 - x_2\,dx_2 - x_3\,dx_3}{(x_1^2 + x_2^2 + x_3^2)^{3/2}}$$

where C is any smooth curve from $(1,1,2)$ to $(2,2,1)$ not passing through the origin.

(b) Why do we restrict C to curves that do not pass through the origin?

7.26. Each of the vector fields below is some variant of the inverse square field

$$\|\mathbf{X}\|^{-3}\mathbf{X} = \nabla(-\|\mathbf{X}\|^{-1}).$$

Find a potential function for each of them.

(a) $\dfrac{2(-x,-y,-z)}{(x^2+y^2+z^2)^{3/2}}$

(b) $\dfrac{3}{(x^2+y^2+z^2)^{3/2}}(x,y,z)$

(c) $\dfrac{(x,y-5,z)}{(x^2+(y-5)^2+z^2)^{3/2}}$

(d) $\dfrac{3(x+1,y-5,z)}{((x+1)^2+(y-5)^2+z^2)^{3/2}} + \dfrac{(x,y,z)}{(x^2+y^2+z^2)^{3/2}}$

7.27. The inverse square vector field in Problem 7.26 finds uses in modeling gravity forces of point masses, and of electrostatic forces of point charges. Let $\mathbf{P}_1,\dots,\mathbf{P}_k$ be k different points of $\mathbb{R}^3$ where the masses or charges are located. Let $c_1,\dots,c_k$ be numbers that represent the masses or charges, all of one sign in the case of gravity. Then the vector field

$$\mathbf{F}(\mathbf{X}) = \sum_{j=1}^{k} c_j\|\mathbf{X}-\mathbf{P}_j\|^{-3}(\mathbf{X}-\mathbf{P}_j)$$

can represent the force on another mass or charge due to the k given particles. Show that $\mathbf{F}$ is conservative by finding a function $g(\mathbf{X})$ so that $\mathbf{F} = \nabla g$.

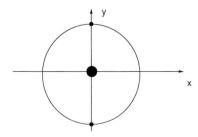

Fig. 7.20 The orbit of the moon around the earth, for Problem 7.28.

7.28. The elliptical orbit of the moon around the earth can be expressed in the form discussed in Chapter 5,

$$y = -19\sqrt{x^2 + y^2} + 7.2 \times 10^9. \qquad \text{[meters]}$$

See Figure 7.20, where the ellipse is to scale, and almost circular. (Moon and Earth are drawn too large so you can see them better.)

(a) Verify from this orbit equation that the Earth-Moon distance at apogee, that is farthest from Earth, at the bottom of the figure, is 400×10^6 meters, and at perigee, closest to Earth at top of figure, 360×10^6 meters.

(b) Over a six month interval, the work done on the moon to pull it from apogee down to perigee is

$$\int_{\text{half orbit}} \mathbf{F} \cdot \mathbf{T} \, ds$$

where the force is

$$\mathbf{F} = -mMG \, \nabla\left(\frac{1}{\sqrt{x^2 + y^2}} \right),$$

m and M are Moon and Earth masses, and G the gravity constant. Use the Fundamental Theorem 7.2 to show that the work is equal to

$$mMG\left(\frac{1}{\sqrt{\text{perigee}}} - \frac{1}{\sqrt{\text{apogee}}} \right).$$

7.29. Let the conservative field

$$\mathbf{F}(\mathbf{X}) = \|\mathbf{X}\|^{-3}\mathbf{X} = \nabla(-\|\mathbf{X}\|^{-1})$$

represent the electric field due to a positive charge at the origin.

(a) For each small $h > 0$ the field $\dfrac{\mathbf{F}(\mathbf{X} + (h,0,0)) - \mathbf{F}(\mathbf{X})}{h}$ can represent the electric field due to a strong positive charge at $(-h,0,0)$ and a strong negative charge at the origin. Show that it is conservative.

(b) Show that $\lim\limits_{h \to 0} \dfrac{\mathbf{F}(\mathbf{X} + (h,0,0)) - \mathbf{F}(\mathbf{X})}{h}$ is conservative.

(c) Denote by g a function that is three times continuously differentiable, and set $\mathbf{G} = \nabla g$. Show that each of the partial derivatives $\mathbf{G}_x$, $\mathbf{G}_y$, and $\mathbf{G}_z$ is a conservative field.

7.30. We have seen in Problem 7.27 that a finite number of point charges produce an electric field that is a linear combination of inverse square fields. Suppose there is a continuous distribution of charge density $c(\mathbf{X})$ [charge/volume] at each point $\mathbf{X}$ in a bounded set D in $\mathbb{R}^3$. The potential at any point $\mathbf{P}$ outside of D is

$$g(\mathbf{P}) = \int_D \frac{c(\mathbf{X})}{\|\mathbf{X} - \mathbf{P}\|} \, dV.$$

(a) Suppose $\mathbf{P}$ is at least one unit distance from every point of D. Show that the integrand

$$\frac{c(\mathbf{X})}{\|\mathbf{X}-\mathbf{P}\|}$$

is a bounded continuous function of $\mathbf{X}$ in D.

(b) Integrate the Taylor approximation (See Problem 4.29)

$$\|\mathbf{X}-\mathbf{P}\|^{-1} \approx \|P\|^{-1} + \|\mathbf{P}\|^{-3}\mathbf{P}\cdot\mathbf{X} + \tfrac{1}{2}\|\mathbf{P}\|^{-5}\big(3(\mathbf{P}\cdot\mathbf{X})^2 - \|\mathbf{P}\|^2\|\mathbf{X}\|^2\big)$$

to get an approximation for the potential:

$$g(\mathbf{P}) \approx \left(\int_D c(\mathbf{X})\,dV\right)\|\mathbf{P}\|^{-1} + \sum_{j=1}^{3}\left(\int_D c(\mathbf{X})x_j\,dV\right)p_j\|\mathbf{P}\|^{-3}$$

$$+ \tfrac{1}{2}\left(\sum_{j,k=1}^{3}\left(\int_D c(\mathbf{X})3x_kx_j\,dV\right)p_jp_k\|\mathbf{P}\|^{-5} - \int_D c(\mathbf{X})\|\mathbf{X}\|^2\,dV\|\mathbf{P}\|^{-3}\right).$$

(c) Show that the integrands $c(\mathbf{X})x_j$, $c(\mathbf{X})x_kx_j$, and $c(\mathbf{X})\|\mathbf{X}\|^2$ are bounded continuous functions in D.

(d) Show that as $\mathbf{P}$ tends to infinity the three terms of the approximation are bounded by $a_1\|\mathbf{P}\|^{-1}$, $a_2\|\mathbf{P}\|^{-2}$, $a_3\|\mathbf{P}\|^{-3}$, for various constants a_i.

Remark. The gradients of the three terms of the approximation are known in Physics as the Coulomb, dipole, and quadrupole parts of the field.

7.3 Surfaces and surface integrals

We introduce smooth surfaces in $\mathbb{R}^3$, using parametrizations.

Definition 7.7. Let $\mathbf{X}$ be a C^1 function from a smoothly bounded set D in $\mathbb{R}^2$ to $\mathbb{R}^3$, denoted $\mathbf{X}(u,v) = (x(u,v), y(u,v), z(u,v))$. Suppose $\mathbf{X}$ satisfies the following conditions on the interior of D.

(a) $\mathbf{X}$ is one to one.
(b) The partial derivatives of the component functions of $\mathbf{X}$ are bounded.
(c) The partial derivatives

$$\mathbf{X}_u(u,v) = \big(x_u(u,v), y_u(u,v), z_u(u,v)\big),$$
$$\mathbf{X}_v(u,v) = \big(x_v(u,v), y_v(u,v), z_v(u,v)\big)$$

are linearly independent so that $\mathbf{X}_u(u,v) \times \mathbf{X}_v(u,v) \neq \mathbf{0}$.

The range S of $\mathbf{X}$ is called a *smooth surface*, parametrized by $\mathbf{X}$. The plane that contains the point $\mathbf{X}(u,v)$ and has normal vector $\mathbf{X}_u(u,v) \times \mathbf{X}_v(u,v)$ is called the plane tangent to S at $\mathbf{X}(u,v)$, and $\mathbf{X}$ is called a *parametrization* of S.

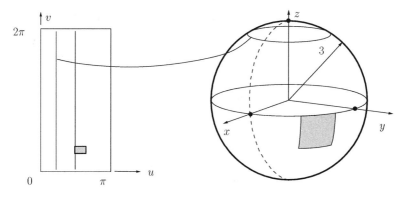

Fig. 7.21 The surface in Example 7.28.

Example 7.28. Let

$$\mathbf{X}(u,v) = (3\cos v \sin u, 3\sin v \sin u, 3\cos u)$$

with $0 \le u \le \pi$ and $0 \le v \le 2\pi$. Since

$$\|\mathbf{X}(u,v)\|^2 = 3^2(\cos^2 v + \sin^2 v)\sin^2 u + 3^2\cos^2 u = 3^2,$$

the points $\mathbf{X}(u,v)$ lie on a sphere of radius 3 centered at the origin. See Figure 7.21. $\mathbf{X}$ is one to one on the interior of the domain, where $0 < u < \pi$, $0 < v < 2\pi$. The partial derivatives

$$\mathbf{X}_u(u,v) = (3\cos v \cos u, 3\sin v \cos u, -3\sin u)$$
$$\mathbf{X}_v(u,v) = (-3\sin v \sin u, 3\cos v \sin u, 0)$$

are linearly independent where $\sin u \neq 0$. We compute a normal vector to the tangent plane,

$$\mathbf{X}_u(u,v) \times \mathbf{X}_v(u,v) = 9\sin u(\cos v \sin u, \sin v \sin u, \cos u),$$

and its norm

$$\|\mathbf{X}_u(u,v) \times \mathbf{X}_v(u,v)\| = 9\sin u.$$

$\square$

One way to parametrize the graph of a C^1 function $f(x,y) = z$ is to define a new function $\mathbf{X}(x,y) = (x,y,f(x,y))$ from the x,y plane to $\mathbb{R}^3$.

Example 7.29. Let $f(x,y) = \sqrt{R^2 - x^2 - y^2}$, $R > 0$. Define

$$\mathbf{X}(x,y) = \left(x, y, \sqrt{R^2 - x^2 - y^2}\right)$$

for $x^2 + y^2 \leq a^2 < R^2$. See Figure 7.22. The partial derivatives

$$X_x = \left(1, 0, -\frac{x}{\sqrt{R^2 - x^2 - y^2}}\right), \qquad X_y = \left(0, 1, -\frac{y}{\sqrt{R^2 - x^2 - y^2}}\right),$$

are linearly independent because the pattern of zeros and ones prevents either vector being a multiple of the other. Since $0 \leq x^2 + y^2 \leq a^2 < R^2$ the derivatives are bounded:

$$\left| -\frac{x}{\sqrt{R^2 - x^2 - y^2}} \right| \leq \frac{R}{\sqrt{R^2 - x^2 - y^2}} \leq \frac{R}{\sqrt{R^2 - a^2}} \quad \text{and} \quad \left| -\frac{y}{\sqrt{R^2 - x^2 - y^2}} \right| \leq \frac{R}{\sqrt{R^2 - a^2}}.$$

The range S of X is the part of the upper hemisphere of radius R centered at the origin, that sits above the disk of radius $a > 0$ centered at the origin in the x, y plane. See Figure 7.22. A normal vector to the plane tangent to S at $X(x, y)$ is

$$X_x(x, y) \times X_y(x, y) = \left(\frac{x}{\sqrt{R^2 - x^2 - y^2}}, \frac{y}{\sqrt{R^2 - x^2 - y^2}}, 1\right)$$

and its norm is $\|X_x(x, y) \times X_y(x, y)\| = \dfrac{R}{\sqrt{R^2 - x^2 - y^2}}.$ ☐

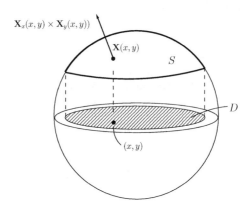

Fig. 7.22 The surface S in Example 7.29.

We call $X(x, y) = (x, y, f(x, y))$ the (x, y) parametrization of the graph of f. Here

$$X_x(x, y) = (1, 0, f_x(x, y)), \qquad X_y(x, y) = (0, 1, f_y(x, y))$$

and a normal vector at the point $(x, y, f(x, y))$ is

$$X_x(x, y) \times X_y(x, y) = (-f_x(x, y), -f_y(x, y), 1).$$

Area Now that we have a definition of a smooth surface, we define its area.

Definition 7.8. Suppose S is a smooth surface parametrized by $\mathbf{X}$ from a smoothly bounded set D in the u, v plane into $\mathbb{R}^3$. The area of S is defined as

$$\text{Area}(S) = \int_S d\sigma = \int_D \|\mathbf{X}_u(u,v) \times \mathbf{X}_v(u,v)\| \, du \, dv.$$

Take the case where S is the graph of a function f on D, and S is parametrized by $\mathbf{X}(x,y) = (x,y,f(x,y))$. Then $\|\mathbf{X}_x(x,y) \times \mathbf{X}_y(x,y)\| = \sqrt{1 + f_x^2 + f_y^2}$ and the surface area of S is

$$\text{Area}(S) = \int_D \sqrt{1 + f_x^2 + f_y^2} \, dx \, dy.$$

We show that Definition 7.8 is a reasonable definition for the area of a smooth surface by looking at some estimates for the area. Take a point $\mathbf{X}(u,v)$ on S and three nearby points $\mathbf{X}(u+\Delta u,v)$, $\mathbf{X}(u,v+\Delta v)$, $\mathbf{X}(u+\Delta u,v+\Delta v)$. See Figure 7.23.

By holding v fixed and letting u vary, $\mathbf{X}(u,v)$ parametrizes a curve in S. Since $\mathbf{X}$ is C^1, the secant vector $\mathbf{X}(u+\Delta u,v) - \mathbf{X}(u,v)$ is well approximated by the tangent vector $\mathbf{X}_u(u,v)\Delta u$. Similarly if we hold u fixed and vary v we get another curve in S with a secant vector $\mathbf{X}(u,v+\Delta v) - \mathbf{X}(u,v)$ well approximated by the tangent vector $\mathbf{X}_v(u,v)\Delta v$. By Definition 7.7, $\mathbf{X}_u(u,v)$ and $\mathbf{X}_v(u,v)$ are linearly independent. The area of the parallelogram determined by the two tangent vectors is

$$\|\mathbf{X}_u(u,v)\Delta u \times \mathbf{X}_v(u,v)\Delta v\|.$$

It is a good estimate for the area of the two triangles determined by the secant vectors in Figure 7.23.

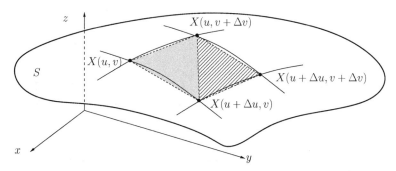

Fig. 7.23 Two secant triangles.

The sum of the areas of the secant triangles tends to the integral formula for the area of S,

$$\text{Area}(S) = \int_D \|\mathbf{X}_u \times \mathbf{X}_v\| \, du \, dv.$$

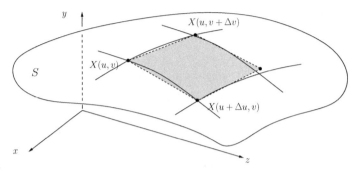

Fig. 7.24 Approximating area by parallelograms.

Example 7.30. Find the area of the sphere S of radius $R > 0$, using the parametrization

$$\mathbf{X}(u,v) = (R\cos v \sin u, R\sin v \sin u, R\cos u), \qquad (0 \le u \le \pi, \ 0 \le v \le 2\pi).$$

We get

$$\mathbf{X}_u(u,v) \times \mathbf{X}_v(u,v) = R^2 \sin u (\cos v \sin u, \sin v \sin u, \cos u)$$

and $\|\mathbf{X}_u(u,v) \times \mathbf{X}_v(u,v)\| = R^2 \sin u.$ So

$$\text{Area}(S) = \int_D \|\mathbf{X}_u \times \mathbf{X}_v\| \, du \, dv = \int_0^{2\pi} \int_0^\pi R^2 \sin u \, du \, dv$$

$$= R^2 \int_0^{2\pi} dv \int_0^\pi \sin u \, du = R^2 (2\pi) 2 = 4\pi R^2.$$

$\square$

Remark. The parametrization of the surface of a sphere in Example 7.30 is inspired by spherical coordinates, where u and v play roles similar to latitude and longitude on a map. This suggests denoting the parameters by (θ, ϕ) rather than (u,v).

Example 7.31. Find the area of the part of the upper hemisphere given by the (x,y) parametrization in Example 7.29,

$$\mathbf{X}(x,y) = \left(x, y, \sqrt{R^2 - x^2 - y^2}\right), \qquad D: x^2 + y^2 \le a^2$$

where $0 < a^2 < R^2$. In Example 7.29 we showed that

$$\|\mathbf{X}_x(x,y) \times \mathbf{X}_y(x,y)\| = \frac{R}{\sqrt{R^2 - x^2 - y^2}},$$

so

$$\text{Area}(S) = \int_S d\sigma = \int_D \frac{R}{\sqrt{R^2 - x^2 - y^2}} dx dy.$$

The integral is easier to evaluate if we change variables to polar coordinates. The region D in polar coordinates is $0 \le \theta \le 2\pi$, $0 \le r \le a$. So

$$\text{Area}(S) = \int_D \frac{R}{\sqrt{R^2 - x^2 - y^2}} dx dy = R \int_0^{2\pi} \int_0^a \frac{r \, dr d\theta}{\sqrt{R^2 - r^2}}$$

$$= 2\pi R \left[-(R^2 - r^2)^{1/2} \right]_{r=0}^{r=a} = 2\pi R (R - (R^2 - a^2)^{1/2}).$$

In Problem 7.34 we ask you to deduce from this formula a classical discovery by Archimedes. Note that as a tends to R, the area tends to $2\pi R^2$, the area of the upper hemisphere of radius R. □

Now that we have shown that the area formula works for spheres let's look at some other simple examples where we may not know the area.

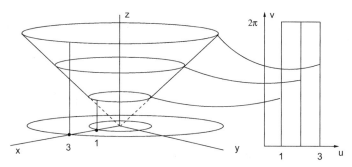

Fig. 7.25 The surface in Example 7.32 is the part of the cone between $z = 1$ and $z = 3$. The u, v rectangle is for the parametrization in Example 7.33.

Example 7.32. Let S be the part of the surface $z^2 = x^2 + y^2$ that lies between the planes $z = 1$ and $z = 3$. See Figure 7.25. Find the area of S. We need to find a parametrization of S, and the simplest approach is the (x, y) parametrization since S can be thought of as the graph of the function

$$z = f(x,y) = \sqrt{x^2 + y^2},$$

defined on the annulus D in the x, y plane where $1 \le x^2 + y^2 \le 9$. Let

$$\mathbf{X}(x,y) = (x, y, f(x,y)) = (x, y, \sqrt{x^2 + y^2})$$

on D. Then

$$\mathbf{X}_x(x,y) = (1,0,f_x(x,y)) = (1,0,x(x^2+y^2)^{-1/2})$$
$$\mathbf{X}_y(x,y) = (0,1,f_y(x,y)) = (0,1,y(x^2+y^2)^{-1/2})$$
$$\mathbf{X}_x(x,y) \times \mathbf{X}_y(x,y) = (-f_x(x,y),-f_y(x,y),1)$$
$$\|\mathbf{X}_x(x,y) \times \mathbf{X}_y(x,y)\| = \sqrt{f_x^2+f_y^2+1}$$

and we find

$$\sqrt{f_x^2+f_y^2+1} = \sqrt{\frac{x^2}{x^2+y^2}+\frac{y^2}{x^2+y^2}+1} = \sqrt{2}.$$

Therefore

$$\text{Area}(S) = \int_S d\sigma = \int_D \|\mathbf{X}_x(x,y) \times \mathbf{X}_y(x,y)\| dA$$

$$= \int_D \sqrt{2}\,dx\,dy = \sqrt{2}\,\text{Area}(D) = \sqrt{2}(\pi 3^2 - \pi 1^2) = 8\pi\sqrt{2}.$$

$\square$

Example 7.33. Another way to compute the area of S, the part of the surface $z^2 = x^2+y^2$ in Example 7.32, is to use the parametrization

$$\mathbf{X}(u,v) = (u\cos v, u\sin v, u), \qquad 1 \le u \le 3,\ 0 \le v \le 2\pi$$

suggested by polar coordinates. D is the rectangle $[1,3] \times [0,2\pi]$. When $u = k$ the points

$$\mathbf{X}(k,v) = (k\cos v, k\sin v, k)$$

are on a circle of radius k in the $z = k$ plane in $\mathbb{R}^3$ centered at $(0,0,k)$. As k increases, larger circles are mapped into higher planes. We calculate

$$\mathbf{X}_u(u,v) = (\cos v, \sin v, 1)$$
$$\mathbf{X}_v(u,v) = (-u\sin v, u\cos v, 0)$$
$$\mathbf{X}_u(u,v) \times \mathbf{X}_v(u,v) = (-u\cos v, -u\sin v, u\cos^2 v + u\sin^2 v)$$
$$\|\mathbf{X}_u(u,v) \times \mathbf{X}_v(u,v)\| = \sqrt{2}u$$

and

$$\text{Area}(S) = \int_S d\sigma = \int_D \sqrt{2}u\,du\,dv$$

$$= \int_0^{2\pi}\int_1^3 \sqrt{2}u\,du\,dv = 2\pi\sqrt{2}\frac{3^2-1^2}{2} = 8\pi\sqrt{2}.$$

$\square$

We have seen some examples where two different parametrizations of a surface are used to calculate the area for the surface. Here is another class of such examples. Take first the case where the surface S is part of a level set

$$g(x, y, z) = c$$

with $\nabla g \neq \mathbf{0}$. According to the Implicit Function Theorem, Theorem 3.13, one or more of the variables is a function of the other two, locally near any point of S. Suppose we have

$$z = f(x, y), \qquad (x, y) \text{ in } D. \tag{7.2}$$

Then $f_x = -\dfrac{g_x}{g_z}$, $f_y = -\dfrac{g_y}{g_z}$, and the area of S is

$$\int_D \sqrt{f_x^2 + f_y^2 + 1} \, dx\,dy.$$

Example 7.34. Find the area of the part of the surface $g(x, y, z) = x^2 + y^2 - z^2 = 0$ that lies over the set in the x, y plane where $1 \leq x^2 + y^2 \leq 9$. (See Examples 7.32 and 7.33.) By the Implicit Function Theorem for $z = f(x, y)$,

$$f_x = \frac{-2x}{-2z}, \qquad f_y = \frac{-2y}{-2z}.$$

So the area is

$$\int_D \sqrt{\frac{x^2}{z^2} + \frac{y^2}{z^2} + 1} \, dx\,dy = \int_D \sqrt{\frac{x^2 + y^2}{z^2} + 1} \, dx\,dy$$

$$= \int_D \sqrt{1 + 1} \, dx\,dy = \sqrt{2}\,\text{Area}(D) = 8\pi\,\sqrt{2}.$$

$\square$

Suppose that the surface S expressed in (7.2) can also be expressed as the graph of

$$y = h(x, z), \qquad (x, z) \text{ in } E.$$

Then $h_x = -\dfrac{g_x}{g_y}$, $h_z = -\dfrac{g_z}{g_y}$. We show now that

$$\int_D \sqrt{f_x^2 + f_y^2 + 1} \, dx\,dy = \int_E \sqrt{h_x^2 + h_z^2 + 1} \, dx\,dz.$$

There is a mapping $\mathbf{F}$ from E onto D,

$$\mathbf{F}(x, z) = (x, h(x, z)) = (x, y).$$

See Figure 7.26. The Jacobian of $\mathbf{F}$ is

$$JF(x,z) = \det \begin{bmatrix} 1 & 0 \\ h_x & h_z \end{bmatrix} = h_z = -\frac{g_z}{g_y}.$$

The change of variables formula, Theorem 6.12, gives

$$\int_D \sqrt{f_x^2 + f_y^2 + 1}\, dx\, dy = \int_E \sqrt{\frac{g_x^2 + g_y^2 + g_z^2}{g_z^2}} \left| \frac{-g_z}{g_y} \right| dx\, dz = \int_E \sqrt{h_x^2 + 1 + h_z^2}\, dx\, dz.$$

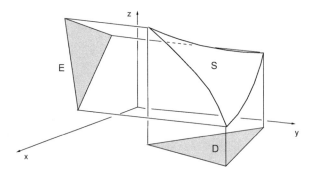

Fig. 7.26 S is both the graph of $z = f(x,y)$ and the graph of $y = h(x,z)$.

More generally suppose we have two parametrizations, one $\mathbf{X}_1(u,v)$ from D_1 onto S, and another $\mathbf{X}_2(s,t)$ from D_2 onto S, and a differentiable "change of parametrization" $\mathbf{G}$ so that

$$(s,t) = \mathbf{G}(u,v)$$
$$\mathbf{X}_2(s,t) = \mathbf{X}_2(\mathbf{G}(u,v)) = \mathbf{X}_1(u,v)$$

for all (s,t) in D_2 and (u,v) in D_1. See Figure 7.27. By the Chain Rule

$$D\mathbf{X}_2(\mathbf{G}(u,v))D\mathbf{G}(u,v) = D\mathbf{X}_1(u,v) \tag{7.3}$$

In Problem 7.45 we guide you to a proof that (7.3) implies that

$$\mathbf{X}_{2s}(u,v) \times \mathbf{X}_{2t}(u,v) \det D\mathbf{G}(u,v) = \mathbf{X}_{1u}(u,v) \times \mathbf{X}_{1v}(u,v). \tag{7.4}$$

By (7.4) and the change of variables formula,

$$\text{Area}(S) = \int_{D_2} \|\mathbf{X}_{2s}(s,t) \times \mathbf{X}_{2t}(s,t)\|\, ds\, dt$$

$$= \int_{D_1} \|\mathbf{X}_{2s}(\mathbf{G}(u,v)) \times \mathbf{X}_{2t}(\mathbf{G}(u,v))\|\, |\det D\mathbf{G}(u,v)|\, du\, dv$$

$$= \int_{D_1} \|\mathbf{X}_{1u}(u,v) \times \mathbf{X}_{1v}(u,v)\|\, du\, dv.$$

The area of S is independent of the parametrization of S.

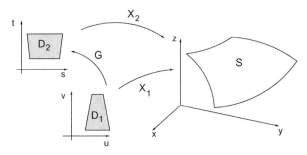

Fig. 7.27 S is parametrized by both $\mathbf{X}_1$ and $\mathbf{X}_2$.

Surface integrals. Suppose at each point (x,y,z) on a surface S we know the charge density $f(x,y,z)$ [charge/area]. What is the total charge distributed over S? Given a parametrization of S, $\mathbf{X}(u,v)$ defined on D, we can approximate the charge on the patch of S that is the image of a Δu by Δv rectangle in D by multiplying the density at a point on the patch, $f(\mathbf{X}(u,v))$, by the approximate area of the patch:

$$f(\mathbf{X}(u,v))\|\mathbf{X}_u(u,v)\Delta u \times \mathbf{X}_v(u,v)\Delta v\| = f(\mathbf{X}(u,v))\|\mathbf{X}_u(u,v) \times \mathbf{X}_v(u,v)\|\,|\Delta u \Delta v|.$$

The total charge is the sum of the charge on the patches. This motivates the definition of the surface integral of f over S.

Definition 7.9. Suppose S is a smooth surface parametrized by $\mathbf{X}$ from a smoothly bounded set D in the u, v plane to $\mathbb{R}^3$, and suppose f is a continuous function on S. The *surface integral of f over S* is defined as

$$\int_S f\,d\sigma = \int_D f(\mathbf{X}(u,v))\|\mathbf{X}_u \times \mathbf{X}_v\|\,du\,dv.$$

In particular the surface area of S is the integral of $f = 1$.

Example 7.35. Let S be the surface $x^2 + y^2 = z^2$, $1 \leq z \leq 3$ of Examples 7.32, 7.33, and 7.34. Suppose the charge density at each point (x,y,z) of S is

$$f(x,y,z) = e^{-x^2-y^2}.$$

Find the total charge on S. We parametrize S by

$$\mathbf{X}(u,v) = (u\cos v, u\sin v, u), \qquad (1 \leq u \leq 3,\ 0 \leq v \leq 2\pi).$$

The total charge is

$$\int_S f \, d\sigma = \int_D f(\mathbf{X}(u,v)) \|\mathbf{X}_u \times \mathbf{X}_v\| \, dA = \int_0^{2\pi} \int_1^3 e^{-u^2} \sqrt{2} u \, du \, dv$$

$$= 2\pi \sqrt{2} \int_1^3 e^{-u^2} u \, du = 2\pi \sqrt{2} \left[-\tfrac{1}{2} e^{-u^2} \right]_1^3 = \pi \sqrt{2} (e^{-1} - e^{-9}).$$

The *average* charge density on S, using the area we found previously, is

$$f_{\text{average}} = \frac{\int_S f \, d\sigma}{\int_S d\sigma} = \frac{\pi \sqrt{2}(e^{-1} - e^{-9})}{8\pi \sqrt{2}} = \tfrac{1}{8}(e^{-1} - e^{-9}).$$

□

Properties of surface integrals. From properties of double integrals, surface integrals have these properties: If f and g are continuous functions and c is a constant, then

- $\displaystyle \int_S cf \, d\sigma = c \int_S f \, d\sigma$

- $\displaystyle \int_S (f + g) \, d\sigma = \int_S f \, d\sigma + \int_S g \, d\sigma$

- If $m \le f \le M$ then $m \operatorname{Area}(S) \le \displaystyle \int_S f \, d\sigma \le M \operatorname{Area}(S)$.

We also define the integral over *piecewise smooth surfaces*. If S is a union of finitely many smooth surfaces $S_1, S_2, \ldots, S_m$ that intersect only pairwise on common boundaries, such as the six sides of a cube that meet along their edges, we have

$$\int_{S_1 \cup S_2 \cup \cdots \cup S_m} f \, d\sigma = \int_{S_1} f \, d\sigma + \cdots + \int_{S_m} f \, d\sigma.$$

Flux of a vector field across a surface. In Chapter 1 we defined the volumetric flow rate or *flux* of a constant velocity field $\mathbf{U}$ across the parallelogram determined by $\mathbf{V}$ and $\mathbf{W}$ in the direction of $\mathbf{V} \times \mathbf{W}$ as

$$\mathbf{U} \cdot (\mathbf{V} \times \mathbf{W}).$$

See Figure 7.28. Let $\mathbf{N}$ be the unit normal $\mathbf{N} = \dfrac{\mathbf{V} \times \mathbf{W}}{\|\mathbf{V} \times \mathbf{W}\|}$ and call the parallelogram S; then the flux of $\mathbf{U}$ across S in the direction $\mathbf{V} \times \mathbf{W}$ is

$$\mathbf{U} \cdot (\mathbf{V} \times \mathbf{W}) = \mathbf{U} \cdot \frac{\mathbf{V} \times \mathbf{W}}{\|\mathbf{V} \times \mathbf{W}\|} \|\mathbf{V} \times \mathbf{W}\| = \mathbf{U} \cdot \mathbf{N} \operatorname{Area}(S).$$

Suppose now $\mathbf{F}$ is a vector field whose domain contains a smooth surface S, parametrized by $\mathbf{X}$ from the u, v plane to $\mathbb{R}^3$. By Definition 7.7 the vectors $\mathbf{X}_u(u,v)$ and $\mathbf{X}_v(u,v)$ are linearly independent and determine a plane tangent to S at $\mathbf{X}(u,v)$. The vector

$$\mathbf{N}(\mathbf{X}(u,v)) = \frac{1}{\|\mathbf{X}_u(u,v) \times \mathbf{X}_v(u,v)\|} \mathbf{X}_u(u,v) \times \mathbf{X}_v(u,v)$$

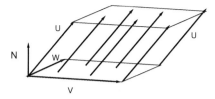

Fig. 7.28 Flux of **U** across the parallelogram determined by **V** and **W**.

is a unit normal to the tangent plane at $\mathbf{X}(u,v)$. See Figure 7.29. We call

$$\mathbf{F}(\mathbf{X}(u,v))\cdot\mathbf{N} = \mathbf{F}(\mathbf{X}(u,v))\cdot\frac{\mathbf{X}_u(u,v)\times\mathbf{X}_v(u,v)}{\|\mathbf{X}_u(u,v)\times\mathbf{X}_v(u,v)\|}$$

the *normal component* of **F** at $\mathbf{X}(u,v)$. Next we define the flux of **F** across S.

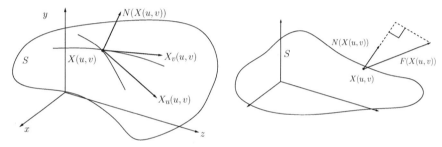

Fig. 7.29 *Left:* A unit normal vector **N** at $\mathbf{X}(u,v)$ on S. *Right:* $\mathbf{F}(\mathbf{X}(u,v))\cdot N(\mathbf{X}(u,v))$ is the normal component of **F** at $\mathbf{X}(u,v)$.

Definition 7.10. Let S be a smooth surface parametrized by **X** from a smoothly bounded set D in the u,v plane to $\mathbb{R}^3$.

Let $\mathbf{F}(x,y,z) = (f_1(x,y.z), f_2(x,y,z), f_3(x,y,z))$ be a continuous function on S. The integral of the normal component of **F**, $\mathbf{F}\cdot\mathbf{N}$, over S is called the *flux* of **F** across S in the direction of **N**. Using Definition 7.9 it is

$$\int_S \mathbf{F}\cdot\mathbf{N}\,d\sigma = \int_D \mathbf{F}(\mathbf{X}(u,v))\cdot\mathbf{N}(\mathbf{X}(u,v))\|\mathbf{X}_u(u,v)\times\mathbf{X}_v(u,v)\|\,du\,dv$$

$$= \int_D \mathbf{F}(\mathbf{X}(u,v))\cdot(\mathbf{X}_u(u,v)\times\mathbf{X}_v(u,v))\,du\,dv.$$

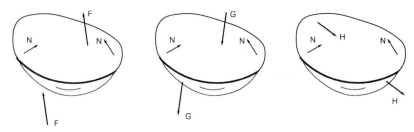

Fig. 7.30 Positive flux of **F** across S at the left, and negative flux of **G** across S at center. For the flux of **H** at the right we can't tell the sign of the flux at a glance.

Example 7.36. Let S be the surface $x^2 + y^2 = z^2$, $1 \leq z \leq 3$. Find the flux of $\mathbf{F}(x,y,z) = (-y,x,z)$ across S, $\displaystyle\int_S \mathbf{F} \cdot \mathbf{N}\, d\sigma$, where **N** points away from the z axis. We know from Example 7.33 that

$$\mathbf{X}(u,v) = (u\cos v, u\sin v, u), \qquad 1 \leq u \leq 3,\ 0 \leq v \leq 2\pi$$

is a parametrization of S and that

$$\mathbf{X}_u(u,v) = (\cos v, \sin v, 1)$$
$$\mathbf{X}_v(u,v) = (-u\sin v, u\cos v, 0)$$
$$\mathbf{X}_u(u,v) \times \mathbf{X}_v(u,v) = (-u\cos v, -u\sin v, u)$$
$$\|\mathbf{X}_u(u,v) \times \mathbf{X}_v(u,v)\| = \sqrt{2}u$$

The vector

$$\mathbf{N}(u,v) = \frac{\mathbf{X}_u(u,v) \times \mathbf{X}_v(u,v)}{\|\mathbf{X}_u(u,v) \times \mathbf{X}_v(u,v)\|} = \frac{(-u\cos v, -u\sin v, u)}{\sqrt{2}u} = \frac{1}{\sqrt{2}}(-\cos v, -\sin v, 1)$$

is normal to S. At $u = 2, v = \frac{\pi}{2}$, $\mathbf{N}(2, \frac{\pi}{2}) = \frac{1}{\sqrt{2}}(0, -1, 1)$ points toward the z axis. See Figure 7.31. To get the net flux of **F** across S in the direction away from the z axis we use the opposite unit normal

$$\mathbf{N}(u,v) = \frac{\mathbf{X}_v(u,v) \times \mathbf{X}_u(u,v)}{\|\mathbf{X}_v(u,v) \times \mathbf{X}_u(u,v)\|} = \frac{1}{\sqrt{2}}(\cos v, \sin v, -1).$$

$$\int_S \mathbf{F} \cdot \mathbf{N}\, d\sigma = \int_{v=0}^{2\pi} \int_{u=1}^{3} \mathbf{F}(\mathbf{X}(u,v)) \cdot \mathbf{N}(u,v)\|\mathbf{X}_v(u,v) \times \mathbf{X}_u(u,v)\|\, du\, dv$$

$$= \int_{v=0}^{2\pi} \int_{u=1}^{3} (-u\sin v, u\cos v, u) \cdot \frac{1}{\sqrt{2}}(\cos v, \sin v, -1)\, \sqrt{2}u\, du\, dv$$

$$= \int_{v=0}^{2\pi} \int_{u=1}^{3} (-u^2)\, du\, dv = 2\pi\left[-\tfrac{1}{3}u^3 \right]_1^3 = \tfrac{1}{3}2\pi(-26).$$

$\square$

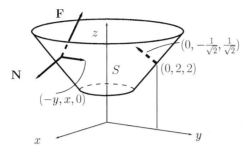

Fig. 7.31 A unit normal $\dfrac{\mathbf{X}_u(u,v) \times \mathbf{X}_v(u,v)}{\|\mathbf{X}_u(u,v) \times \mathbf{X}_v(u,v)\|}$ at the point $(0,2,2)$ in Example 7.36 points toward the "inside" of the cone, toward the z axis.

Orientation of surfaces. An important thing to notice in the definition of the flux integral $\displaystyle\int_S \mathbf{F} \cdot \mathbf{N}\, d\sigma$ is that the sign of the integrand depends on the order in which the cross product of $\mathbf{X}_u(u,v)$ and $\mathbf{X}_v(u,v)$ is taken. The resulting unit normals

$$\frac{\mathbf{X}_u(u,v) \times \mathbf{X}_v(u,v)}{\|\mathbf{X}_u(u,v) \times \mathbf{X}_v(u,v)\|}, \qquad \frac{\mathbf{X}_v(u,v) \times \mathbf{X}_u(u,v)}{\|\mathbf{X}_v(u,v) \times \mathbf{X}_u(u,v)\|}$$

can be used to distinguish the two sides of a surface. For example, a sphere S has a set of unit normals that point outward and a set that point inward. In Example 7.28 we parametrized a sphere of radius three centered at the origin by

$$\mathbf{X}(u,v) = (3\cos v \sin u, 3\sin v \sin u, 3\cos u), \quad 0 \le u \le \pi,\ 0 \le v \le 2\pi.$$

Which way do the unit normals

$$\frac{\mathbf{X}_u(u,v) \times \mathbf{X}_v(u,v)}{\|\mathbf{X}_u(u,v) \times \mathbf{X}_v(u,v)\|}$$

point? We found that $\mathbf{X}_u(u,v) \times \mathbf{X}_v(u,v) = 9\sin u(\cos v \sin u, \sin v \sin u, \cos u)$. Where $\sin u$ is positive, this is a positive multiple of $\mathbf{X}(u,v)$, a point on the sphere, so these normals point outward.

When we choose a side for the normals we say the surface is *oriented*. If a surface S is a union of n oriented surfaces $S_1, \ldots, S_n$, then S is *orientable* if we can choose unit normals so that if edges between the surfaces were smoothed, we would be able to extend the normals continuously across the smoothed edges. Let's look at another example.

Example 7.37. Let squares S_1 and S_2 be parametrized by

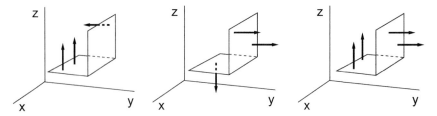

Fig. 7.32 The surfaces in Example 7.37. The left and center surfaces are oriented. The surface on the right is not oriented.

$$\mathbf{X}_1(x,y) = (x,y,4) \qquad 0 \le x \le 1,\ 2 \le y \le 3,$$
$$\mathbf{X}_2(x,z) = (x,3,z) \qquad 0 \le x \le 1,\ 4 \le z \le 5.$$

See Figure 7.32. The unit normals to S_1 are $(0,0,1)$ or $(0,0,-1)$, and unit normals to S_2 are $(0,1,0)$ or $(0,-1,0)$. For $S = S_1 \cup S_2$ to be an oriented surface we need to choose unit normals that are consistent, so

$$(0,0,-1) \text{ on } S_1,\ (0,1,0) \text{ on } S_2$$

as at the center in the figure, or

$$(0,0,1) \text{ on } S_1,\ (0,-1,0) \text{ on } S_2$$

as at the left in the figure.   □

At the right in Figure 7.32 is a surface that has two sides and is orientable, but this way of assigning unit normals does not result in an oriented surface.

Not all surfaces are orientable or have two sides. Take a strip of paper, give it a half twist and tape the two short ends together to make a Möbius band. The strip before you joined the ends was an orientable surface. It had two sides. But after you put the twist in and joined the ends any normal you try to move around the surface comes back the opposite way after one loop.

The notion of the flux of a vector field across a surface only makes sense for orientable surfaces; therefore we only work with orientable surfaces.

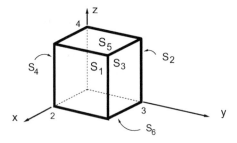

Fig. 7.33 The surface S of the box in Example 7.38.

Example 7.38. Find the flux of $\mathbf{F}(x,y,z) = (x^2, y+1, z-y)$ outward across S, the surface of the box $[0,2] \times [0,3] \times [0,4]$ shown in Figure 7.33.

$$\int_S \mathbf{F} \cdot \mathbf{N} \, d\sigma = \int_{S_1 \cup S_2 \cup S_3 \cup S_4 \cup S_5 \cup S_6} \mathbf{F} \cdot \mathbf{N} \, d\sigma$$

$$= \int_{S_1} \mathbf{F} \cdot \mathbf{N} \, d\sigma + \int_{S_2} \mathbf{F} \cdot \mathbf{N} \, d\sigma + \int_{S_3} \mathbf{F} \cdot \mathbf{N} \, d\sigma + \int_{S_4} \mathbf{F} \cdot \mathbf{N} \, d\sigma + \int_{S_5} \mathbf{F} \cdot \mathbf{N} \, d\sigma + \int_{S_6} \mathbf{F} \cdot \mathbf{N} \, d\sigma.$$

Parametrize S_1 by $\mathbf{X}_1(y,z) = (2,y,z)$. The outward unit normal vectors are $(1,0,0)$ and

$$\int_{S_1} \mathbf{F} \cdot \mathbf{N} \, d\sigma = \int_{z=0}^4 \int_{y=0}^3 (2^2, y+1, z-y) \cdot (1,0,0) \, dydz = 4(12) = 48.$$

Similarly

$$\int_{S_2} \mathbf{F} \cdot \mathbf{N} \, d\sigma = \int_{z=0}^4 \int_{y=0}^3 (0, y+1, z-y) \cdot (-1,0,0) \, dydz = 0,$$

$$\int_{S_3} \mathbf{F} \cdot \mathbf{N} \, d\sigma = \int_{z=0}^4 \int_{x=0}^2 (x^2, 3+1, z-3) \cdot (0,1,0) \, dxdz = 4(8) = 32,$$

$$\int_{S_4} \mathbf{F} \cdot \mathbf{N} \, d\sigma = \int_{z=0}^4 \int_{x=0}^2 (x^2, 1, z-0) \cdot (0,-1,0) \, dxdz = -1(8) = -8,$$

$$\int_{S_5} \mathbf{F} \cdot \mathbf{N} \, d\sigma = \int_{y=0}^3 \int_{x=0}^2 (x^2, y+1, 4-y) \cdot (0,0,1) \, dxdy = 2\left[-\tfrac{1}{2}(4-y)^2 \right]_0^3 = 15,$$

$$\int_{S_5} \mathbf{F} \cdot \mathbf{N} \, d\sigma = \int_{y=0}^3 \int_{x=0}^2 (x^2, y+1, 0-y) \cdot (0,0,-1) \, dxdy = 2\left[\tfrac{1}{2}y^2 \right]_0^3 = 9.$$

Therefore $\displaystyle\int_S \mathbf{F} \cdot \mathbf{N} \, d\sigma = 48 + 0 + 32 - 8 + 15 + 9 = 96.$ □

Alternative ways to find the flux of F across S. If S is a graph of a C^1 function $z = f(x,y)$ over D, we can also think of S as the level set of the C^1 function g from $\mathbb{R}^3$ to $\mathbb{R}$,

$$g(x,y,z) = z - f(x,y) = 0.$$

Since the gradient of g is normal at each point of S,

$$\nabla g(x,y,z) = (-f_x(x,y), -f_y(x,y), 1),$$

the two possibilities for a unit normal to S at a point $(x,y,f(x,y))$ are

$$\mathbf{N} = \pm \frac{(-f_x(x,y), -f_y(x,y), 1)}{\sqrt{f_x^2 + f_y^2 + 1}}.$$

Note that

$$\mathbf{N} \cdot (0,0,1) = \pm \frac{1}{\sqrt{f_x^2 + f_y^2 + 1}}$$

is plus or minus the cosine of the angle between the tangent plane to S at $(x, y, f(x,y))$ and the x, y plane. See Figure 7.34. That gives us a way to estimate the area of the part of S that sits above a small part of the domain D in the x, y plane. It is the area in the x, y plane times the reciprocal of the cosine of the angle between the two planes, $d\sigma \approx \sqrt{f_x^2 + f_y^2 + 1}\, dA$. So the upward flux of $\mathbf{F}$ across S is

$$\int_S \mathbf{F} \cdot \mathbf{N}\, d\sigma = \int_D \mathbf{F}(x, y, f(x,y)) \cdot (-f_x(x,y), -f_y(x,y), 1)\, dx\, dy$$

where D is the domain of f and S is the graph of f over D.

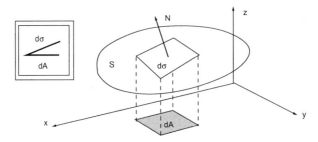

Fig. 7.34 The angle between the tangent plane and the x, y plane. $\mathbf{N}$ is drawn upward.

Example 7.39. Let the surface S be the part of the graph of $f(x,y) = x^2 + y^2$ above the rectangle $R = [0,1] \times [0,3]$, with unit normal vectors $\mathbf{N}$ having a negative z component. The flux of $\mathbf{F}(x,y,z) = (-x, 3, z)$ across S is

$$\int_S \mathbf{F} \cdot \mathbf{N}\, d\sigma = \int_D \mathbf{F}(x, y, f(x,y)) \cdot (f_x(x,y), f_y(x,y), -1)\, dx\, dy$$

$$= \int_{y=0}^3 \int_{x=0}^1 (-x, 3, x^2 + y^2) \cdot (2x, 2y, -1)\, dx\, dy = \int_{y=0}^3 \int_{x=0}^1 (-3x^2 - y^2 + 6y)\, dx\, dy = 15.$$

□

Evaluating $\displaystyle\int_S \mathbf{F} \cdot \mathbf{N}\, d\sigma$ **geometrically** Sometimes we can evaluate a flux integral without parametrizing the surface, by making use of special geometric relationships between the vector field $\mathbf{F}$ and the surface S.

Example 7.40. Let $\mathbf{F}(\mathbf{X}) = -\dfrac{\mathbf{X}}{\|\mathbf{X}\|^3}$ be the inverse square vector field. Find the flux of $\mathbf{F}$ outward across a sphere S_R of radius R centered at the origin. The direction of $\mathbf{F}(\mathbf{X})$ is in the opposite direction of $\mathbf{X}$, and $\|\mathbf{F}(\mathbf{X})\| = \|\mathbf{X}\|^{-2} = R^{-2}$.

The outward unit normal vector $\mathbf{N}(\mathbf{X}) = \dfrac{\mathbf{X}}{\|\mathbf{X}\|}$ points in the opposite direction of $\mathbf{F}(\mathbf{X})$. Therefore

$$\mathbf{F}(\mathbf{X}) \cdot \mathbf{N}(\mathbf{X}) = \|\mathbf{F}(\mathbf{X})\| \, \|\mathbf{N}(\mathbf{X})\| \cos \theta = \|\mathbf{F}(\mathbf{X})\|(1)(-1) = -R^{-2}.$$

The flux of $\mathbf{F}$ outward across S_R is

$$\int_{S_R} \mathbf{F} \cdot \mathbf{N} \, d\sigma = \int_{S_R} -\frac{1}{R^2} \, d\sigma = -\frac{1}{R^2} \operatorname{Area}(S_R) = -\frac{1}{R^2} 4\pi R^2 = -4\pi.$$

$\square$

If in Example 7.40 we had computed the flux of the vector field *into* the sphere we would have used the inward pointing unit normals and the cosine of the angle between $\mathbf{F}$ and $\mathbf{N}$ would have been 1 instead of -1. The inward flux of $\mathbf{F}$ across S is 4π.

Example 7.41. Suppose at each point of a surface S, $\mathbf{F}(\mathbf{X})$ is tangent to S. Find the flux of $\mathbf{F}$ across S. Since $\mathbf{F}(\mathbf{X})$ is tangent to S, $\mathbf{F}(\mathbf{X}) \cdot \mathbf{N}(\mathbf{X}) = 0$ at each point. So

$$\int_S \mathbf{F} \cdot \mathbf{N} \, d\sigma = 0.$$

$\square$

Example 7.42. Find the flux of $\mathbf{F}(x,y,z) = (x, y + 3x^2, z)$ outward across a sphere S of radius R centered at the origin. The outward pointing unit normal vectors are

$$\mathbf{N} = \frac{(x,y,z)}{\sqrt{x^2 + y^2 + z^2}} = \frac{1}{R}(x,y,z)$$

at every point (x,y,z) of the sphere. The flux is then

$$\int_S \mathbf{F} \cdot \mathbf{N} \, d\sigma = \int_S (x, y + 3x^2, z) \cdot \left(\frac{1}{R}(x,y,z)\right) d\sigma$$

$$= \frac{1}{R} \int_S (x^2 + y^2 + 3x^2 y + z^2) \, d\sigma = \frac{1}{R}\left(\int_S (x^2 + y^2 + z^2) \, d\sigma + 3 \int_S (x^2 y) \, d\sigma\right).$$

On S, $x^2 + y^2 + z^2 = R^2$, so the first integral is

$$\int_S (x^2 + y^2 + z^2) \, d\sigma = R^2 \int_S d\sigma = R^2 \operatorname{Area}(S) = 4\pi R^4.$$

The second integral $\displaystyle\int_S (x^2 y) \, d\sigma$ is zero due to symmetry. (See Problem 7.37.) Therefore the flux is $R^{-1}(4\pi R^4) = 4\pi R^3$.

$\square$

Problems

7.31. Set up integrals for the areas of the graphs of $z = x^2 + y^2$ and $z = x^2 - y^2$ over a smoothly bounded set D in $\mathbb{R}^2$. Show that for all D, the two graphs have the same area.

7.32. Let S_1 be the part of the plane

$$\tfrac{2}{7}x + \tfrac{3}{7}y + \tfrac{6}{7}z = 10,$$

that lies over the rectangle $0 \le x \le 1$ and $0 \le y \le 1$ in the x,y plane. Let S_2 be the part of the same plane that lies over the rectangle $0 \le y \le 1$ and $0 \le z \le 1$ in the y,z plane. Show that $\mathrm{Area}(S_1) = \tfrac{7}{6}$ and $\mathrm{Area}(S_2) = \tfrac{7}{2}$.

7.33. Suppose $a^2 + b^2 + c^2 = 1$, and let S be the part of the plane $ax + by + cz = d$ that lies over a rectangle D in the x,y plane. Show that

$$|c|\,\mathrm{Area}(S) = \mathrm{Area}(D).$$

7.34. Let R be the radius of a sphere. It is part of a classical discovery by Archimedes that the area of a spherical cap of height h, $2\pi Rh$, is the same as the area of a cylinder of the same height h and radius R.

(a) Verify that the expression $(R - (R^2 - a^2)^{1/2})$, found in Example 7.31, is the height of the spherical cap there, thus proving Archimedes' result.
(b) Without evaluating any further surface integrals, deduce another discovery by Archimedes: that the area of *every* section of height h of a sphere has the same area. See Figure 7.35.

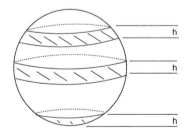

Fig. 7.35 All sections of equal height through a sphere have the same surface area, in Problem 7.34.

7.35. Let $\mathbf{X}(u,v) = (\sqrt{2}uv, u^2, v^2)$, $(1 \le u \le 2,\ 1 \le v \le 2)$.

(a) Calculate $\mathbf{X}_u$, $\mathbf{X}_v$ and $\mathbf{X}_u \times \mathbf{X}_v$.
(b) Show that $\mathbf{X}$ parametrizes a surface S in $\mathbb{R}^3$, and at each point (x,y,z) of S, $x^2 = 2yz$.
(c) Find the area of S.

(d) Calculate the integral $\int_S y \, d\sigma$.

7.36. Let S be the surface of the cylinder in $\mathbb{R}^3$ where $x^2 + y^2 = r^2$ and $0 \le z \le h$, where r and h are positive constants.

(a) Verify that a parametrization of S is

$$\mathbf{X}(u,v) = (r\cos u, r\sin u, v), \qquad 0 \le u \le 2\pi, \ 0 \le v \le h.$$

(b) Use $\mathbf{X}(u,v)$ to show that the area of S is $2\pi rh$.

(c) Evaluate $\int_S y \, d\sigma$.

(d) Evaluate $\int_S y^2 \, d\sigma$.

7.37. Let S be the unit sphere in $\mathbb{R}^3$ centered at the origin. S has a *symmetry*, that for each point (x,y,z) on S there is another point $(-x,-y,-z)$ also on S. Show that

$$\int_S x^2 y \, d\sigma = 0.$$

7.38. Let S be the unit sphere centered at the origin in $\mathbb{R}^3$. Evaluate the following items, using as little calculation as possible.

(a) $\int_S 1 \, d\sigma$

(b) $\int_S \|\mathbf{X}\|^2 \, d\sigma$

(c) Verify that $\int_S x_1^2 \, d\sigma = \int_S x_2^2 \, d\sigma = \int_S x_3^2 \, d\sigma$ using either a symmetry argument or parametrization without evaluating the integrals.

(d) Use the result of parts (b) and (c) to deduce the value of $\int_S x_1^2 \, d\sigma$.

7.39. Let S be the unit sphere centered at the origin in $\mathbb{R}^3$.

(a) Show that $x_1^4 + x_2^4 + x_3^4 + 2(x_1^2 x_2^2 + x_2^2 x_3^2 + x_3^2 x_1^2) = 1$ on S.

(b) Evaluate $\int_S x_3^4 \, d\sigma$ using spherical coordinates.

(c) Evaluate $\int_S x_1^2 x_2^2 \, d\sigma$ using items (a), (b), and a symmetry argument.

7.40. Let S be the flat parallelogram surface determined by vectors $\mathbf{V}$ and $\mathbf{W}$, and let $\mathbf{N} = \dfrac{\mathbf{V} \times \mathbf{W}}{\|\mathbf{V} \times \mathbf{W}\|}$. Verify that the flux of a constant vector field $\mathbf{F}$ across S is

$$\int_S \mathbf{F} \cdot \mathbf{N} \, d\sigma = \mathbf{F} \cdot \mathbf{N} \, \text{Area}(S).$$

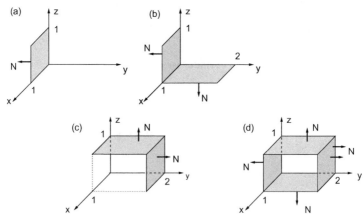

Fig. 7.36 Surfaces in Problems 7.41 and 7.42.

7.41. Find the flux of the constant vector field $\mathbf{F} = (2,3,4)$ through each of the four oriented surfaces in Figure 7.36.

7.42. Find the flux of the vector field $\mathbf{F} = (2y,3z,4x)$ through surface (a) in Figure 7.36.

7.43. Find the flux of the vector field $\mathbf{F} = (2y,3z,4x)$ across the parallelogram determined by the vectors $\mathbf{V} = (1,1,1)$ and $\mathbf{W} = (0,0,2)$ in the direction of $\mathbf{V} \times \mathbf{W}$.

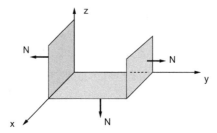

Fig. 7.37 The oriented surface in Problem 7.44.

7.44. Let S be the oriented surface shown in Figure 7.37. Determine without calculation which of the following fluxes are positive.

(a) $\displaystyle\int_S (0,1,0) \cdot \mathbf{N}\, d\sigma$

(b) $\displaystyle\int_S (0,3y,0) \cdot \mathbf{N}\, d\sigma$

(c) $\displaystyle\int_S (1,3y,0) \cdot \mathbf{N}\, d\sigma$

(d) $\displaystyle\int_S (x^3, 0, 5) \cdot \mathbf{N}\, d\sigma$

7.45. Suppose $\mathbf{A}, \mathbf{B}, \mathbf{C}, \mathbf{D}$ are vectors of $\mathbb{R}^3$ written as columns, and that

$$\underbrace{[\mathbf{A}\ \mathbf{B}]}_{3\times 2}\begin{bmatrix} a & b \\ c & d \end{bmatrix} = \underbrace{[\mathbf{C}\ \mathbf{D}]}_{3\times 2}.$$

Justify the following statements to prove that

$$(ad - bc)\mathbf{A} \times \mathbf{B} = \mathbf{C} \times \mathbf{D}. \tag{7.5}$$

Remark. We used this formula in the "change of parametrization."

(a) If we adjoin any third column $\mathbf{Y}$ to create square matrices, then

$$\underbrace{[\mathbf{A}\ \mathbf{B}\ \mathbf{Y}]}_{3\times 3}\begin{bmatrix} a & b & 0 \\ c & d & 0 \\ 0 & 0 & 1 \end{bmatrix} = \underbrace{[\mathbf{C}\ \mathbf{D}\ \mathbf{Y}]}_{3\times 3}.$$

(b) Then $\mathbf{Y} \cdot (\mathbf{A} \times \mathbf{B})(ad - bc) = \mathbf{Y} \cdot (\mathbf{C} \times \mathbf{D})$.
(c) Conclude (7.5).

7.46. As we will see in the next chapter, the flux of fluid mass across a surface S is

$$\int_S \rho \mathbf{V} \cdot \mathbf{N}\, d\sigma,$$

where ρ is the fluid density and $\mathbf{V}$ the velocity. Evaluate this integral for two cases:

(a) S is a square of area A parallel to the x, y plane with unit normal vector $\mathbf{N} = (0, 0, 1)$, ρ is constant, and $\mathbf{V} = (a, b, c)$ is also constant.
(b) S is a square of area A with a unit normal vector $\mathbf{N}$, ρ is constant, and $\mathbf{V}$ is a constant k times $\mathbf{N}$.

7.47. As we will see in the next chapter, the flux of fluid momentum across a surface S is *vector valued* (integrate componentwise):

$$\int_S (\rho \mathbf{V})\mathbf{V} \cdot \mathbf{N}\, d\sigma,$$

where ρ is the fluid density and $\mathbf{V}$ the velocity. Evaluate this integral for two cases:

(a) S is a square of area A parallel to the x, y plane with unit normal vector $\mathbf{N} = (0, 0, 1)$, ρ is constant, and $\mathbf{V} = (a, b, c)$ is also constant.
(b) S is a square of area A with a unit normal vector $\mathbf{N}$, ρ is constant, and $\mathbf{V}$ is a constant k times $\mathbf{N}$.

7.48. Let S be the part of the plane

$$3x - 2y + z = 10$$

over the square $D = [0, 1] \times [0, 1]$ in the x, y plane, and take the normal vectors $\mathbf{N}$ upward, that is with positive z component.

(a) Express the plane as the graph of a function of x and y, and find $\mathbf{N}$.
(b) Find the area of S and verify that the area of S is not less than the area of D.
(c) Let $\mathbf{F}$ be a constant vector field (a, b, c). Find the flux of $\mathbf{F}$ across S.

7.49. Let S be the sphere of radius $R > 0$ centered at the origin of $\mathbb{R}^3$. Consider the inverse square vector field $\mathbf{F}(\mathbf{X}) = \dfrac{\mathbf{X}}{\|\mathbf{X}\|^3}$. Show that the outward flux

$$\int_S \mathbf{F} \cdot \mathbf{N} \, d\sigma = 4\pi,$$

thus the flux does not depend on the radius of S. See Figure 7.38.

7.50. Let S be the sphere of radius $R > 0$ centered at the origin of $\mathbb{R}^3$. Consider a vector field of the form

$$\mathbf{F}(\mathbf{X}) = k(\|\mathbf{X}\|)\mathbf{X},$$

where k is a function only of the norm $\|\mathbf{X}\|$. Suppose the outward flux $\displaystyle\int_S \mathbf{F} \cdot \mathbf{N} \, d\sigma$ does not depend on the radius of S. Prove that $k(\|\mathbf{X}\|)$ is some multiple of $\|\mathbf{X}\|^{-3}$. See Figure 7.38.

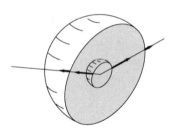

Fig. 7.38 Flux of inverse square field through sphere, in Problems 7.49 and 7.50.

7.51. Suppose surface S is parametrized by $\mathbf{X}(u, v)$, (u, v) in some smoothly bounded set D, and k is a positive number. Let T be the set of points

$$\mathbf{Y}(u, v) = k\mathbf{X}(u, v), \qquad (u, v) \text{ in } D.$$

Show that T is a smooth surface parametrized by $\mathbf{Y}$. Show that

$$\text{Area}(T) = k^2 \, \text{Area}(S).$$

Chapter 8
Divergence and Stokes' Theorems and conservation laws

Abstract Green's and Stokes' Theorems are extensions to functions of several variables of the relation between differentiation and integration.

8.1 Green's Theorem and the Divergence Theorem in $\mathbb{R}^2$

The Fundamental Theorem of single variable calculus asserts that differentiation and integration are inverse operations. More precisely, if f is a function that has continuous first derivative f' then

$$\int_a^b f'(x)\,\mathrm{d}x = f(b) - f(a). \tag{8.1}$$

The analogue of this result for C^1 functions of two variables is called the Divergence Theorem. As we will see it is obtained by applying the single variable result in each variable and integrating with respect to the other variable.

Take first the case that D is a smoothly bounded set in the plane that is both x simple and y simple. Denote by ∂D the boundary of D. Let c and d be the smallest and largest numbers y for which there is a point (x,y) contained in D. Since D is x simple, for every value y_0 between c and d, the set of points (x,y_0) in D is a single horizontal interval $[a(y_0), b(y_0)]$ and both $a(y)$ and $b(y)$ are continuous on $[c,d]$. Denote by a and b the smallest and largest numbers x for which there is a point (x,y) in D. Since D is y simple, for every value x_0 between a and b the set of points (x_0,y) in D is a single vertical interval $[c(x_0), d(x_0)]$ and both $c(x)$ and $d(x)$ are continuous on $[a,b]$. See Figure 8.1.

Figure 8.1 shows a set that is both x simple and y simple and Figure 8.2 shows two that are not. Let f be a function that has continuous first partial derivatives f_x and f_y in D. Fix y and apply the Fundamental Theorem of Calculus in one variable. We get

© Springer International Publishing AG 2017

P. D. Lax and M. S. Terrell, *Multivariable Calculus with Applications*,

Undergraduate Texts in Mathematics, https://doi.org/10.1007/978-3-319-74073-7_8

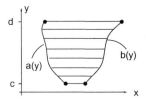
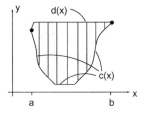

Fig. 8.1 *Left: D* is *x* simple with boundary functions $x = a(y)$ and $x = b(y)$. *Right: D* is *y* simple with boundary functions $y = c(x)$ and $y = d(x)$.

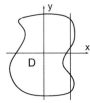
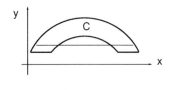

Fig. 8.2 *Left: D* is *x* simple but not *y* simple. *Right: C* is *y* simple but not *x* simple.

$$\int_{a(y)}^{b(y)} f_x(x,y)\,\mathrm{d}x = f(b(y),y) - f(a(y),y),$$

where *y* lies between *c* and *d*.

Integrating this relation with respect to *y* from *c* to *d* we get

$$\int_c^d \int_{x=a(y)}^{x=b(y)} f_x(x,y)\,\mathrm{d}x\mathrm{d}y = \int_c^d \Big(f(b(y),y) - f(a(y),y) \Big)\mathrm{d}y.$$

The iterated integral on the left is the integral of f_x over *D*, $\int_D f_x\,\mathrm{d}x\mathrm{d}y$. Next we observe that the integral on the right side is equal to the line integral

$$\int_{\partial D} f\,\mathrm{d}y$$

taken counterclockwise along the boundary of *D*, using parametrizations $\mathbf{X}_1(y) = (b(y),y)$ on the graph of $b(y)$ with *y* running from *c* to *d*, and $\mathbf{X}_2(y) = (a(y),y)$ on the graph of $a(y)$ with *y* from *d* down to *c*. Therefore the integral formula can be written as

$$\int_D f_x\,\mathrm{d}x\mathrm{d}y = \int_{\partial D} f\,\mathrm{d}y. \tag{8.2}$$

Let *g* be another continuously differentiable function on *D*. We derive an analogous formula for the integral of g_y over *D*:

$$\int_a^b \int_{c(x)}^{d(x)} g_y(x,y)\,\mathrm{d}y\mathrm{d}x = \int_a^b \Big(g(x,d(x)) - g(x,c(x)) \Big)\mathrm{d}x.$$

The left side is $\displaystyle\int_D g_y \, dx \, dy$ and the right side is equal to a line integral of g taken over ∂D in the *clockwise* direction. Taking it in the counterclockwise direction we get

$$\int_D g_y \, dx \, dy = -\int_{\partial D} g \, dx. \tag{8.3}$$

Adding the two formulas (8.2) and (8.3) we get

$$\int_D (f_x + g_y) \, dx \, dy = \int_{\partial D} f \, dy - g \, dx. \tag{8.4}$$

We restate this result in vector language. Let $\mathbf{X}(s) = (x(s), y(s))$ be the arc length parametrization of the boundary of D increasing in the counterclockwise direction. s is between 0 and the length of ∂D. Since the tangent vector $(\frac{dx}{ds}, \frac{dy}{ds})$ has length 1 at each point of ∂D, $\mathbf{N} = (\frac{dy}{ds}, -\frac{dx}{ds})$ is the outward pointing unit normal. Recall in Section 3.5 we defined $\operatorname{div}\mathbf{F} = f_x + g_y$, where $\mathbf{F} = (f, g)$. Then we can write (8.4) as

$$\int_D \operatorname{div}\mathbf{F} \, dx \, dy = \int_{\partial D} \mathbf{F} \cdot \mathbf{N} \, ds. \tag{8.5}$$

This is called the *Divergence Theorem*.

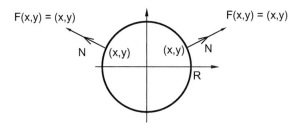

Fig. 8.3 In Example 8.1 the unit normals and $\mathbf{F}$ are parallel on the boundary of the disk.

Example 8.1. We verify the Divergence Theorem in the case $\mathbf{F}(x, y) = (x, y)$ and D is a disk of radius R centered at the origin. At each point on the boundary, the outward pointing unit normal to ∂D is parallel to $\mathbf{F}$. See Figure 8.3. Therefore at each point on ∂D, $\mathbf{F} \cdot \mathbf{N} = \|\mathbf{F}\| = R$. The line integral is

$$\int_{\partial D} \mathbf{F} \cdot \mathbf{N} \, ds = \int_{\partial D} \|\mathbf{F}\| \, ds = R \int_{\partial D} ds = R(2\pi R) = 2\pi R^2.$$

Since $\operatorname{div}\mathbf{F} = \frac{\partial}{\partial x}(x) + \frac{\partial}{\partial y}(y) = 2$, the integral of $\operatorname{div}\mathbf{F}$ over D is

$$\int_D \operatorname{div} \mathbf{F} \, dA = 2 \int_D dA = 2\pi R^2.$$

□

Example 8.2. Let $\mathbf{F}(x,y) = (x,y)$ and let D be the rectangular region where

$$-3 \leq x \leq 5 \quad \text{and} \quad -7 \leq y \leq 2.$$

Find the flux of $\mathbf{F}$ outward across ∂D. See Figure 8.4. Since $\operatorname{div} \mathbf{F} = 2$, by the Divergence Theorem

$$\int_{\partial D} \mathbf{F} \cdot \mathbf{N} \, ds = \int_D \operatorname{div} \mathbf{F} \, dA = \int_D 2 \, dA = 2 \operatorname{Area}(D) = 2(8)(9) = 144.$$

□

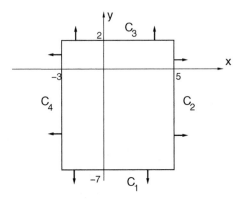

Fig. 8.4 The rectangle in Example 8.2 with outward pointing unit normals.

Example 8.3. We can also compute $\displaystyle\int_{\partial D} \mathbf{F} \cdot \mathbf{N} \, ds$ of Example 8.2 by computing four line integrals.

$$\int_{\partial D} \mathbf{F} \cdot \mathbf{N} \, ds = \int_{C_1} \mathbf{F} \cdot \mathbf{N} \, ds + \int_{C_2} \mathbf{F} \cdot \mathbf{N} \, ds + \int_{C_3} \mathbf{F} \cdot \mathbf{N} \, ds + \int_{C_4} \mathbf{F} \cdot \mathbf{N} \, ds.$$

We get

$$\int_{C_1} \mathbf{F} \cdot \mathbf{N} \, ds = \int_{-3}^{5} (x, -7) \cdot (0, -1) \, dx = \int_{-3}^{5} 7 \, dx = 56,$$

$$\int_{C_3} \mathbf{F} \cdot \mathbf{N} \, ds = \int_{-3}^{5} (x, 2) \cdot (0, 1) \, dx = 16,$$

$$\int_{C_2} \mathbf{F} \cdot \mathbf{N} \, ds = \int_{-7}^{2} (5, y) \cdot (1, 0) \, dy = \int_{-7}^{2} 5 \, dy = 45,$$

$$\int_{C_4} \mathbf{F} \cdot \mathbf{N} \, ds = \int_{-7}^{2} (-3, y) \cdot (-1, 0) \, dy = 27.$$

Therefore $\int_{\partial D} \mathbf{F} \cdot \mathbf{N} \, ds = 56 + 16 + 45 + 27 = 144$. This agrees with the calculation in Example 8.2.   □

Example 8.4. Let $\mathbf{F}(x, y) = (2x + y^2, y + \cos x)$, then $\operatorname{div} \mathbf{F} = 3$. The flux of $\mathbf{F}$ outward across the boundary of a smoothly bounded set D with area 10 that is both x simple and y simple is

$$\int_{\partial D} \mathbf{F} \cdot \mathbf{N} \, ds = \int_{D} \operatorname{div} \mathbf{F} \, dA = 3 \int_{D} dA = 3(10) = 30.$$

□

We next extend the Divergence Theorem to sets that are finite unions of smoothly bounded sets that are both x simple and y simple and have boundary arcs in common.

Let D be a smoothly bounded set in the x, y plane. Divide D into two parts D_1 and D_2 by a piecewise smooth curve C in D that connects two points of the boundary of D. See Figure 8.5.

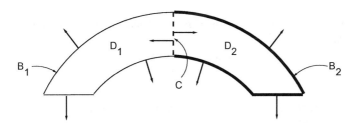

Fig. 8.5 Sets D_1 and D_2 with outward pointing normals. Curves B_1, B_2, and C are shown thin, thick, and dotted, respectively.

We show now:
If a C^1 vector field $\mathbf{F}$ satisfies

$$\int_{D_1} \operatorname{div} \mathbf{F} \, dA = \int_{\partial D_1} \mathbf{F} \cdot \mathbf{N} \, ds \quad \text{and} \quad \int_{D_2} \operatorname{div} \mathbf{F} \, dA = \int_{\partial D_2} \mathbf{F} \cdot \mathbf{N} \, ds \qquad (8.6)$$

over the two sets D_1 and D_2 then

$$\int_D \operatorname{div} \mathbf{F} \, dA = \int_{\partial D} \mathbf{F} \cdot \mathbf{N} \, ds$$

over the union $D = D_1 \cup D_2$.

To prove this, add the two equations in (8.6). The sum of the left sides is the integral of $\operatorname{div} \mathbf{F}$ over D. We claim that the sum of the right sides is the integral of $\mathbf{F} \cdot \mathbf{N}$ over the boundary of D. To see this, note in Figure 8.5 that the endpoints of the curve C divide the boundary of D into two parts B_1 and B_2. The boundary of D_1 is the union of B_1 and C, and the boundary of D_2 is the union of B_2 and C.

We note that the normal to the connecting curve C that is outward with respect to D_1 is the negative of the normal to C that is outward with respect to D_2. Therefore in the sum of the right sides of (8.6) the integrals of $\mathbf{F} \cdot \mathbf{N}$ over C as the boundary of D_1 and as boundary of D_2 cancel! The sum of the remaining integrals over the boundary of D_1 and D_2 is the integral of $\mathbf{F} \cdot \mathbf{N}$ over the boundary of D.

We can extend the Divergence Theorem to smoothly bounded sets that are the union of sets that are both x simple and y simple.

Definition 8.1. We call a smoothly bounded set D *regular* if it is the union of a finite number of smoothly bounded subsets, each of which is x simple and y simple and any two have only a boundary arc in common.

We have proved the Divergence Theorem for a vector field that is C^1 on a smoothly bounded set that is x simple and y simple. We have also shown that if a set D is a union of two subsets for which the theorem holds, then it holds for D. Using this proposition repeatedly we conclude the following theorem.

Theorem 8.1. The Divergence Theorem. *If* $\mathbf{F}$ *from* $\mathbb{R}^2$ *to* $\mathbb{R}^2$ *is* C^1 *on a regular set* D *then*

$$\int_D \operatorname{div} \mathbf{F} \, dA = \int_{\partial D} \mathbf{F} \cdot \mathbf{N} \, ds.$$

All sets of practical or theoretical importance are regular; therefore we confine our studies to regular sets.

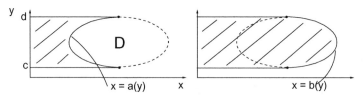

Fig. 8.6 $\text{Area}(D) = \int_c^d (b(y) - a(y)) \, dy = \int_{\partial D} x \, dy.$ See Example 8.5.

Example 8.5. For the function $\mathbf{F}(x,y) = (x,0)$, $\operatorname{div}\mathbf{F}(x,y) = 1$. Let D be a regular set in $\mathbb{R}^2$. By the Divergence Theorem

$$\operatorname{Area}(D) = \int_D 1\,dA = \int_D \operatorname{div}\mathbf{F}\,dA = \int_{\partial D} x\,dy.$$

This shows that the area of D can be computed using only measurements taken on its boundary. See Figure 8.6. □

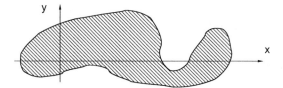

Fig. 8.7 The set D in Example 8.6.

Example 8.6. Let $\mathbf{F}(x,y) = (x,-y)$. Find the outward flux of $\mathbf{F}$ across the boundary of D shown in Figure 8.7. We compute

$$\operatorname{div}\mathbf{F} = \frac{\partial}{\partial x}(x) + \frac{\partial}{\partial y}(-y) = 1 - 1 = 0.$$

By the Divergence Theorem the outward flux of $\mathbf{F}$ across ∂D is

$$\int_{\partial D} \mathbf{F}\cdot\mathbf{N}\,ds = \int_D \operatorname{div}\mathbf{F}\,dA = \int_D 0\,dA = 0.$$

□

Next we show how to use the Divergence Theorem to find the integral of the tangential component of $\mathbf{F}$ along ∂D,

$$\int_{\partial D} \mathbf{F}\cdot\mathbf{T}\,ds.$$

Suppose $\mathbf{F}$ is C^1 on a regular set D and ∂D is traversed in a counterclockwise direction. If the outward unit normal vector at a point on ∂D is $\mathbf{N} = (n_1,n_2)$, then the unit tangent vector there is $\mathbf{T} = (t_1,t_2) = (-n_2,n_1)$. See Figure 8.8.

So for $\mathbf{F} = (f,g)$,

$$\int_{\partial D} \mathbf{F}\cdot\mathbf{T}\,ds = \int_{\partial D} (ft_1 + gt_2)\,ds = \int_{\partial D} (gn_1 - fn_2)\,ds.$$

By the Divergence Theorem and the definition of curl from Section 3.5 we have

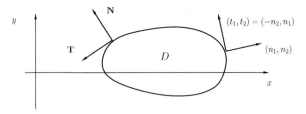

Fig. 8.8 The unit normal and tangent vectors.

$$\int_{\partial D}(gn_1+(-f)n_2)\,\mathrm{d}s=\int_D\left(\frac{\partial g}{\partial x}+\frac{\partial(-f)}{\partial y}\right)\mathrm{d}A=\int_D\left(\frac{\partial g}{\partial x}-\frac{\partial f}{\partial y}\right)\mathrm{d}A=\int_D\operatorname{curl}\mathbf{F}\,\mathrm{d}A.$$

So we have proved Green's Theorem.

Theorem 8.2. Green's Theorem *If* $\mathbf{F}=(f,g)$ *is continuously differentiable on a regular set D in* $\mathbb{R}^2$ *then*

$$\int_D\operatorname{curl}\mathbf{F}\,\mathrm{d}A=\int_{\partial D}\mathbf{F}\cdot\mathbf{T}\,\mathrm{d}s,$$

where the boundary of D is traversed in the counterclockwise direction.

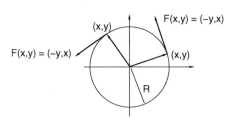

Fig. 8.9 In Example 8.7 the unit tangent and $\mathbf{F}(x,y)$ point in the same direction at each (x,y) on the circle.

Example 8.7. Find the work done in moving a particle in the counterclockwise direction along a circle of radius R centered at the origin in the presence of a force field $\mathbf{F}(x,y)=(-y,x)$. Green's Theorem can be applied as follows. The work is

$$\int_{\partial D}\mathbf{F}\cdot\mathbf{T}\,\mathrm{d}s=\int_D\left(\frac{\partial}{\partial x}(x)-\frac{\partial}{\partial y}(-y)\right)\mathrm{d}A=\int_D(1+1)\,\mathrm{d}A=2\operatorname{Area}(D)=2\pi R^2.$$

Alternatively, without using Green's Theorem note that $\mathbf{T}$ and $\mathbf{F}$ have the same direction at each point on ∂D. Hence $\mathbf{F}\cdot\mathbf{T}=\|\mathbf{F}\|=\sqrt{x^2+y^2}=R$ and

$$\int_{\partial D} \mathbf{F} \cdot \mathbf{T} \, ds = \int_{\partial D} R \, ds = R(\text{Length}(\partial D)) = R(2\pi R) = 2\pi R^2.$$

□

Example 8.8. Let $\mathbf{F}(x,y) = \left(\dfrac{-y}{x^2+y^2}, \dfrac{x}{x^2+y^2}\right)$ and let D be the disk of radius R centered at the origin. Find

$$\int_{\partial D} \mathbf{F} \cdot \mathbf{T} \, ds,$$

where we traverse ∂D in the counterclockwise direction. We cannot use Green's Theorem to evaluate this integral because the domain of $\mathbf{F}$ does not contain $(0,0)$, hence $\mathbf{F}$ is not differentiable (or even defined) on all of D. Let's compute $\displaystyle\int_{\partial D} \mathbf{F} \cdot \mathbf{T} \, ds$ and $\displaystyle\int_D (g_x - f_y) \, dA$ and compare the results.

$\mathbf{F}$ and $\mathbf{T}$ are in the same direction, and we see $\mathbf{F} \cdot \mathbf{T} = \|\mathbf{F}\| = \frac{1}{R}$. Therefore,

$$\int_{\partial D} \mathbf{F} \cdot \mathbf{T} \, ds = \int_{\partial D} \frac{1}{R} \, ds = \frac{1}{R} \text{Length}(\partial D) = \frac{1}{R} 2\pi R = 2\pi.$$

On the other hand

$$g_x = \frac{\partial}{\partial x}\left(\frac{x}{x^2+y^2}\right) = \frac{(x^2+y^2)-x(2x)}{(x^2+y^2)^2} = \frac{y^2-x^2}{(x^2+y^2)^2},$$

$$f_y = \frac{\partial}{\partial y}\left(\frac{-y}{x^2+y^2}\right) = \frac{(x^2+y^2)(-1)-(-y)(2y)}{(x^2+y^2)^2} = \frac{y^2-x^2}{(x^2+y^2)^2}.$$

Therefore, $\displaystyle\int_D (g_x - f_y) \, dA = \int_D 0 \, dA = 0$. The two integrals are not equal. □

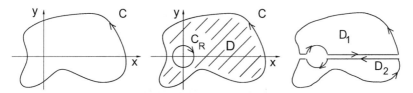

Fig. 8.10 *Left:* the loop C in Example 8.9 encircles the origin. *Center:* the boundary of D is $C \cup C_R$. *Right:* D comprises two subregions D_1 and D_2.

Example 8.9. Find $\displaystyle\int_C \mathbf{F} \cdot \mathbf{T} \, ds$, where $\mathbf{F}(x,y) = \left(\dfrac{-y}{x^2+y^2}, \dfrac{x}{x^2+y^2}\right)$ and C is the loop shown in Figure 8.10 traversed in the counterclockwise direction. $\mathbf{F}$ is not defined at $(0,0)$ but $\mathbf{F}$ is defined on the regular region D between C and C_R, a circle of radius R centered at the origin that does not intersect C. The boundary of D is $C \cup C_R$. We know from Example 8.8 that

$$\frac{\partial}{\partial x}\left(\frac{x}{x^2+y^2}\right) - \frac{\partial}{\partial y}\left(\frac{-y}{x^2+y^2}\right) = 0.$$

By Green's Theorem

$$\int_{\partial D} \mathbf{F} \cdot \mathbf{T} \, ds = \int_D 0 \, dA = 0.$$

Splitting D into two regions D_1, D_2 as in Figure 8.10 and using that C_R is oriented clockwise we get

$$0 = \int_{\partial D} \mathbf{F} \cdot \mathbf{T} \, ds = \int_{\partial D_1} \mathbf{F} \cdot \mathbf{T} \, ds + \int_{\partial D_2} \mathbf{F} \cdot \mathbf{T} \, ds$$

$$= \int_C \mathbf{F} \cdot \mathbf{T} \, ds + \int_{C_R \text{ clockwise}} \mathbf{F} \cdot \mathbf{T} \, ds$$

In Example 8.8 we saw that the integral along C_R counterclockwise is 2π, so clockwise it is -2π. Therefore $\int_C \mathbf{F} \cdot \mathbf{T} \, ds = 2\pi$.   □

Problems

8.1. Let R be the rectangular region where $|x| \le 4$ and $|y| \le 2$, U the closed unit disk centered at $\mathbf{0}$, and S the disk where $x^2 + y^2 \le 25$. We define two more sets in $\mathbb{R}^2$ as follows. Let D_1 be the closure of the set obtained by removing U from R. Let D_2 be the closure of the set obtained by removing R from S. Sketch D_1 and D_2 and show that they are regular sets.

8.2. Show that for a regular set D with boundary oriented counterclockwise,

$$\int_{\partial D} x \, dy = -\int_{\partial D} y \, dx = \text{Area}(D).$$

8.3. Use the Divergence Theorem to evaluate the outward flux

$$\int_C (x + 6y^2, y + 6x^2) \cdot \mathbf{N} \, ds$$

where

(a) C is the unit circle $x^2 + y^2 = 1$,
(b) C is the boundary of the rectangle $[a,b] \times [c,d]$.

8.4. Use the Divergence Theorem to evaluate the following integrals where D is a regular set and the unit normals $\mathbf{N}$ are outward.

(a) $\int_{\partial D} \nabla(x^2 - y^2) \cdot \mathbf{N} \, ds$

(b) $\int_{\partial D} \nabla f \cdot \mathbf{N} \, ds$, where f is a C^2 function that satisfies $f_{xx} + f_{yy} = 0$ on D.

8.5. Let D be the right half of the unit disk,

$$x^2 + y^2 \leq 1, \quad x \geq 0,$$

and let $\mathbf{F}(x,y) = (1 + x^2 y^2, 0)$. Verify the Divergence Theorem by evaluating both sides of

$$\int_{\partial D} \mathbf{F} \cdot \mathbf{N} \, ds = \int_D \operatorname{div} \mathbf{F} \, dA$$

where the unit normals $\mathbf{N}$ are outward.

8.6. Let $\mathbf{F}(x,y) = (-y,x)$, and let D be the quarter disc with outward unit normals $\mathbf{N}$,

$$x^2 + y^2 \leq 1, \quad 0 \leq x, \, 0 \leq y,$$

and denote by R the set drawn in Figure 8.11.

(a) Compute the integrals

$$\int_D \operatorname{div} \mathbf{F} \, dA, \qquad \int_{\partial D} \mathbf{F} \cdot \mathbf{N} \, ds$$

without using the Divergence Theorem.

(b) Find the flux of $\mathbf{F}$ outward across ∂R.

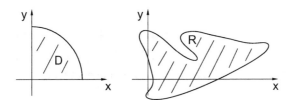

Fig. 8.11 The regions D and R in Problem 8.6.

8.7. Let D be a convex polygonal region in $\mathbb{R}^2$ with vertices $\mathbf{P}_0, \mathbf{P}_1, \ldots, \mathbf{P}_n$. Let $\mathbf{N}_i$ be the outward unit normal on the i-th edge, D_i, from $\mathbf{P}_{i-1}$ to $\mathbf{P}_i$ (D_1 is $\mathbf{P}_n$ to $\mathbf{P}_1$). Justify the following items to prove

$$\sum_{i=1}^{n} \|\mathbf{P}_i - \mathbf{P}_{i-1}\| \mathbf{N}_i = \mathbf{0}.$$

See Figure 8.12 for the case $n = 3$.

(a) Let $\mathbf{F}_1(x,y) = (1,0)$ and let the unit normals $\mathbf{N} = (n_1, n_2)$ be outward. Apply the Divergence Theorem to show that

$$\int_{\partial D} n_1 \, ds = 0,$$

(b) Let $\mathbf{F}_2(x,y) = (0,1)$ to show that

$$\int_{\partial D} n_2 \, ds = 0.$$

(c) The integral of a vector valued function is computed componentwise. Conclude that

$$\int_{\partial D} \mathbf{N} ds = \left(\int_{\partial D} n_1 \, ds, \int_{\partial D} n_2 \, ds \right) = \mathbf{0}.$$

(d) Show that $\displaystyle\int_{D_i} \mathbf{N}_i \, ds = \|\mathbf{P}_i - \mathbf{P}_{i-1}\| \mathbf{N}_i.$

(e) Show that $\displaystyle\sum_{i=1}^{n} \|\mathbf{P}_i - \mathbf{P}_{i-1}\| \mathbf{N}_i = \mathbf{0}.$

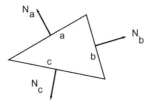

Fig. 8.12 a, b, and c are the edge lengths, and $a\mathbf{N}_a + b\mathbf{N}_b + c\mathbf{N}_c = \mathbf{0}$ in Problem 8.7.

8.8. Consider a rectangular region with edges parallel to the x and y axes.

(a) Make a sketch of such a region showing the vector field $\mathbf{F}(x,y) = (0, \cos x)$ at various points on the boundary. Explain why the net flux across the boundary is zero.

(b) Make another sketch to show that the net flux of the vector field $\mathbf{G}(x,y) = (y^2, 0)$ across the boundary is zero.

(c) Show that the net flux of the vector field $\mathbf{H}(x,y) = (y^2, \cos x)$ across the boundary of the region is zero.

8.9. Suppose C is a circle in the plane that does not go through $(0,0)$. Show that the circulation of

$$\mathbf{F}(x,y) = \left(\frac{-y}{x^2+y^2}, \frac{x}{x^2+y^2} \right)$$

along C in the counterclockwise direction is either 0 or 2π.

8.10. Suppose $\mathbf{F}$ is a C^1 vector field on an annulus D in $\mathbb{R}^2$ whose boundary consists of an inner circle C_2 of radius 1, oriented clockwise, and an outer circle C_1 of radius 3, oriented counterclockwise. If

$$\int_{C_1} \mathbf{F} \cdot \mathbf{T} ds = 11 \quad \text{and} \quad \int_{C_2} \mathbf{F} \cdot \mathbf{T} ds = -9,$$

find the average of the function curl $\mathbf{F}$ over D.

8.11. Let g from $\mathbb{R}^2$ to $\mathbb{R}$ and $\mathbf{F}$ from $\mathbb{R}^2$ to $\mathbb{R}^2$ be C^1 functions on a regular set D.

(a) Show that $\operatorname{div}(g\mathbf{F}) = g\operatorname{div}\mathbf{F} + \mathbf{F} \cdot \nabla g$.
(b) Suppose $g = 0$ on ∂D. Use the Divergence Theorem to show that

$$\int_D (\operatorname{div}\mathbf{F})g \, dA = -\int_D \mathbf{F} \cdot \nabla g \, dA.$$

(c) Suppose g is C^2, and take $\mathbf{F} = \nabla g$. If $g = 0$ at all points of the boundary of D, show that

$$\int_D (\varDelta g)g \, dA = -\int_D |\nabla g|^2 \, dA.$$

8.12. Let $g_1(x, y) = \sin x \sin y$ in the region D where $0 \le x \le \pi$ and $0 \le y \le \pi$.

(a) Verify that g_1 is zero at all points of the boundary of D, and that $\varDelta g_1 = -2g_1$.
(b) Suppose that g is some C^2 function that is zero at all points of the boundary of D, and that $\varDelta g = 2g$. Use the results of Problem 8.11 to show this is not possible unless g is identically zero.

8.13. Suppose $\mathbf{F}(x, y) = (f(x, y), g(x, y))$ is C^1 and let $\mathbf{X}(t) = (x(t), y(t))$ be a smooth curve that satisfies the differential equation

$$\mathbf{X}' = \mathbf{F}(\mathbf{X}),$$

that is,

$$x'(t) = f(x(t), y(t)), \qquad y'(t) = g(x(t), y(t)).$$

We say that $\mathbf{X}$ is a *periodic orbit* of period p if p is the smallest positive number for which

$$\mathbf{X}(t + p) = \mathbf{X}(t).$$

A periodic orbit is the boundary of a regular set. Use the Divergence Theorem to show that if $\operatorname{div}\mathbf{F} > 0$ then a curve $\mathbf{X}$ that satisfies $\mathbf{X}' = \mathbf{F}(\mathbf{X})$ cannot have a periodic orbit.

8.14. Suppose P is a regular region in the plane whose boundary is a polygon. We list the n vertices

$$(x_1, y_1), (x_2, y_2), \ldots, (x_n, y_n)$$

in order counterclockwise around the boundary. Justify the following steps to show that

$$2\operatorname{Area}(P) = (-y_1 x_2 + x_1 y_2) + (-y_2 x_3 + x_2 y_3) + \cdots + (-y_n x_1 + x_n y_1).$$

(a) $\displaystyle\int_{\partial P} -y \, dx + x \, dy = 2\operatorname{Area}(P)$
(b) If C denotes a straight segment from a point (a, b) to point (p, q) then

$$\int_C -y \, dx + x \, dy = -bp + aq.$$

8.2 The Divergence Theorem in $\mathbb{R}^3$

The Divergence Theorem can be proved for vector valued functions with three components of three variables on a smoothly bounded set D in $\mathbb{R}^3$ that is x and y and z simple. The idea of the proof is the same as for two variables. We apply the Fundamental Theorem of Calculus in one of the variables and then integrate with respect to the other two variables. Let

$$\mathbf{F}(x,y,z) = (f(x,y,z), g(x,y,z), h(x,y,z)).$$

Since D is x simple let D_{yz} be the set of all points (y,z) for which some point (x,y,z) is in D. See Figure 8.13. We assume D_{yz} is a smoothly bounded set in the y,z plane, and that

$$a(y,z) \leq x \leq b(y,z)$$

describes the interval of points (x,y,z) in D for which (y,z) is in D_{yz}.

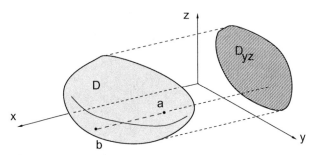

Fig. 8.13 A set D in $\mathbb{R}^3$, and D_{yz} in the y,z plane.

By the Fundamental Theorem of Calculus we have that

$$\int_{a(y,z)}^{b(y,z)} \frac{\partial f}{\partial x}(x,y,z)\,dx = f(b(y,z),y,z) - f(a(y,z),y,z).$$

Integrating this over D_{yz} we get

$$\int_D \frac{\partial f}{\partial x}\,dx\,dy\,dz = \int_{D_{yz}} \left(\int_{a(y,z)}^{b(y,z)} \frac{\partial f}{\partial x}(x,y,z)\,dx \right) dy\,dz$$

$$= \int_{D_{yz}} (f(b(y,z),y,z) - f(a(y,z),y,z))\,dy\,dz. \tag{8.7}$$

At each point on the surface $x = b(y,z)$ the cosine of the angle between the tangent plane to the surface and the y,z plane is $\mathbf{N} \cdot \mathbf{i} = n_1$, the x component of the outward pointing unit normal $\mathbf{N}$. See Figure 8.14.

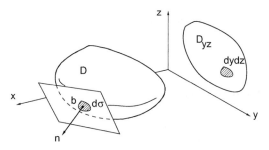

Fig. 8.14 Tangent plane and unit normal **N** where $x = b(y,z)$ and n_1 is positive; $n_1\,d\sigma = dy\,dz$.

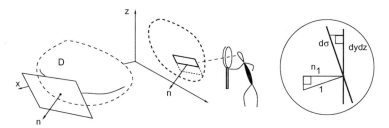

Fig. 8.15 The ratio of $dy\,dz$ to $d\sigma$ is n_1.

Therefore the relation between integrating with respect to $dy\,dz$ in the y,z plane and the corresponding surface area $d\sigma$ above it on $x = b(y,z)$ is

$$n_1\,d\sigma = dy\,dz$$

so $f(b(y,z),y,z)n_1\,d\sigma = f(b(y,z),y,z)\,dy\,dz$. Similarly the cosine of the angle between the y,z plane and the plane tangent to the $x = a(y,z)$ surface is $-n_1$, where n_1 is the x component of the outward pointing unit normal. As a result on the graph of $x = a(y,z)$ we have

$$n_1\,d\sigma = -dy\,dz$$

and $f(a(y,z),y,z)n_1\,d\sigma = -f(a(y,z),y,z)\,dy\,dz$. See Figure 8.16.

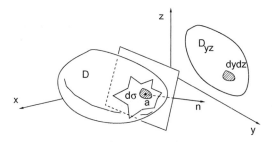

Fig. 8.16 Tangent plane and unit normal **N** where $x = a(y,z)$ and n_1 is negative; $n_1\,d\sigma = -dy\,dz$.

So summing over the graphs of $x = b(y,z)$ and $x = a(y,z)$,

$$\int_{D_{yz}} (f(b(y,z),y,z) - f(a(y,z),y,z))\,dy\,dz = \int_{\partial D} f n_1\,d\sigma.$$

Combining this with (8.7) we have

$$\int_D \frac{\partial f}{\partial x}\,dx\,dy\,dz = \int_{\partial D} f n_1\,d\sigma.$$

Now repeating the argument in the other two coordinates we have

$$\int_D \frac{\partial f}{\partial x}\,dx\,dy\,dz + \int_D \frac{\partial g}{\partial y}\,dx\,dy\,dz + \int_D \frac{\partial h}{\partial z}\,dx\,dy\,dz$$

$$= \int_{\partial D} f n_1\,d\sigma + \int_{\partial D} g n_2\,d\sigma + \int_{\partial D} h n_3\,d\sigma,$$

giving

$$\int_D \operatorname{div}\mathbf{F}\,dx\,dy\,dz = \int_{\partial D} \mathbf{F}\cdot\mathbf{N}\,d\sigma \qquad (8.8)$$

where $\mathbf{N}$ is the outward unit normal to the boundary of D. Just as in $\mathbb{R}^2$ we can extend the result (8.8) to sets D that are *regular*, i.e., a finite union of smoothly bounded sets in $\mathbb{R}^3$ that are x, y, and z simple and have only boundary points in common. This result is known as the Divergence Theorem.

Theorem 8.3. The Divergence Theorem. *Let $\mathbf{F}$ be a C^1 vector field on regular set D in $\mathbb{R}^3$ and let $\mathbf{N}$ be the unit normals to ∂D that point out of D. Then*

$$\int_{\partial D} \mathbf{F}\cdot\mathbf{N}\,d\sigma = \int_D \operatorname{div}\mathbf{F}\,dV.$$

Example 8.10. Let $\mathbf{F}(x,y,z) = (x,y,z)$ and let D be the solid rectangular box

$$2 \le x \le 4, \quad 7 \le y \le 10, \quad 1 \le z \le 5.$$

Find the flux of $\mathbf{F}$ outward across the boundary of D. By the Divergence Theorem

$$\int_{\partial D} \mathbf{F}\cdot\mathbf{N}\,d\sigma = \int_D \operatorname{div}\mathbf{F}\,dV.$$

Since $\operatorname{div}\mathbf{F} = \dfrac{\partial(x)}{\partial x} + \dfrac{\partial(y)}{\partial y} + \dfrac{\partial(z)}{\partial z} = 1 + 1 + 1 = 3$, we have

$$\int_{\partial D} \mathbf{F}\cdot\mathbf{N}\,d\sigma = \int_D 3\,dV = 3\operatorname{Vol}(D) = 3(2)(3)(4) = 72.$$

$\square$

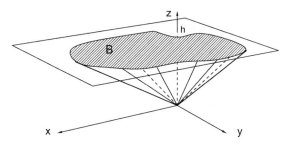

Fig. 8.17 The cone in Example 8.11.

Example 8.11. Let D be a solid cone with base B of area A and height h as shown in Figure 8.17. We use the vector field $\mathbf{F}(x,y,z) = (x,y,z)$ to show that the volume of the cone is $\frac{1}{3}hA$. Since $\operatorname{div}\mathbf{F} = 3$, the Divergence Theorem gives

$$\int_{\partial D} \mathbf{F}\cdot\mathbf{N}\,d\sigma = \int_D \operatorname{div}\mathbf{F}\,dV = 3\operatorname{Vol}(D).$$

The boundary of D consists of the base B and sides S. The unit normal to B is $\mathbf{k} = (0,0,1)$, and the normals to S are perpendicular to $\mathbf{F}$. Therefore

$$\int_{\partial D} \mathbf{F}\cdot\mathbf{N}\,d\sigma = \int_B \mathbf{F}\cdot\mathbf{k}\,d\sigma + \int_S \mathbf{F}\cdot\mathbf{N}\,d\sigma = \int_B h\,d\sigma + 0 = hA,$$

and $\operatorname{Vol}(D) = \frac{1}{3}hA$.   □

Example 8.12. Let $\mathbf{F}(x,y,z) = \dfrac{(x,y,z)}{(x^2+y^2+z^2)^{3/2}}$. The divergence of this field is zero: since

$$\frac{\partial}{\partial x}\frac{x}{(x^2+y^2+z^2)^{3/2}} = \frac{(x^2+y^2+z^2)-3x^2}{(x^2+y^2+z^2)^{5/2}},$$

$$\frac{\partial}{\partial y}\frac{y}{(x^2+y^2+z^2)^{3/2}} = \frac{(x^2+y^2+z^2)-3y^2}{(x^2+y^2+z^2)^{5/2}},$$

$$\frac{\partial}{\partial z}\frac{z}{(x^2+y^2+z^2)^{3/2}} = \frac{(x^2+y^2+z^2)-3z^2}{(x^2+y^2+z^2)^{5/2}},$$

their sum $\dfrac{\partial f_1}{\partial x} + \dfrac{\partial f_2}{\partial y} + \dfrac{\partial f_3}{\partial z} = \operatorname{div}\mathbf{F} = 0$. In the next examples we will compute the flux of $\mathbf{F}$ across various surfaces.   □

Example 8.13. Let $\mathbf{F}(x,y,z) = \dfrac{(x,y,z)}{(x^2+y^2+z^2)^{3/2}}$. Find the flux of $\mathbf{F}$ outward across a sphere S_R of radius R centered at the origin. We cannot apply the Divergence Theorem since $\mathbf{F}$ is not defined at $(0,0,0)$. So we compute the surface integral. At each point on the sphere, the direction of $\mathbf{F}$ is radial and normal to the surface, so $\mathbf{F}\cdot\mathbf{N} = \|\mathbf{F}\|$. At each point on S_R, $\|\mathbf{F}\| = \dfrac{1}{R^2}$. Therefore

$$\int_{S_R} \mathbf{F} \cdot \mathbf{N} \, d\sigma = \int_{S_R} \|\mathbf{F}\| \, d\sigma = \int_{S_R} \frac{1}{R^2} \, d\sigma = \frac{1}{R^2} 4\pi R^2 = 4\pi.$$

□

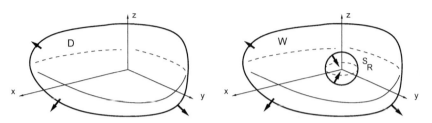

Fig. 8.18 *Left:* A regular set D with the origin interior, *Right:* A regular set W that does not contain $(0,0,0)$. See Example 8.14.

Example 8.14. Let $\mathbf{F}(x,y,z) = \dfrac{(x,y,z)}{(x^2+y^2+z^2)^{3/2}}$. Find the flux of $\mathbf{F}$ outward across the boundary of a regular set D that contains the origin in its interior. Because $\mathbf{F}$ is not defined at the origin, we cannot apply the Divergence Theorem directly to compute the flux of $\mathbf{F}$ across ∂D. Since $(0,0,0)$ is in the interior of D, there is a small sphere S_R of radius R centered at $(0,0,0)$ contained in the interior of D. See Figure 8.18. The region between S_R and ∂D is a regular set that does not contain $(0,0,0)$. Call it W. We can apply the Divergence Theorem to $\mathbf{F}$ on W and get

$$\int_{\partial W} \mathbf{F} \cdot \mathbf{N} \, d\sigma = \int_W \operatorname{div} \mathbf{F} \, dV.$$

In Example 8.12 we found $\operatorname{div} \mathbf{F} = 0$. Therefore $\int_W \operatorname{div} \mathbf{F} \, dV = 0$. By the Divergence Theorem the outward flux of $\mathbf{F}$ across

$$\partial W = (\partial D) \cup S_R$$

is also zero. Using the result from Example 8.13 we get

$$0 = \int_{\partial W} \mathbf{F} \cdot \mathbf{N} \, d\sigma = \int_{\partial D} \mathbf{F} \cdot \mathbf{N} \, d\sigma - \int_{S_R} \mathbf{F} \cdot \mathbf{N} \, d\sigma = \int_{\partial D} \mathbf{F} \cdot \mathbf{N} \, d\sigma - 4\pi.$$

Therefore $\int_{\partial D} \mathbf{F} \cdot \mathbf{N} \, d\sigma = 4\pi.$ □

Problems

8.15. Let $F(X) = X$. Use the Divergence Theorem to evaluate

$$\int_{\partial D} F \cdot N \, d\sigma$$

for each of the following sets D in $\mathbb{R}^3$.

(a) D is the unit ball, $\|X\| \leq 1$.
(b) D is a ball, $\|X - A\| \leq r$, A and $r > 0$ constant.

8.16. Let D be a regular set in $\mathbb{R}^3$, let N be the outward pointing unit normals to ∂D, and let F be a C^2 vector field. Show that

$$\int_{\partial D} (\text{curl} F) \cdot N \, d\sigma = 0.$$

8.17. Use the Divergence Theorem to evaluate the integrals over the sphere S given by $x^2 + y^2 + z^2 = 8^2$ with unit normals N pointing away from the origin.

(a) $\displaystyle\int_S (x + 2y, 3y + 4z, 5z + 6x) \cdot N \, d\sigma$

(b) $\displaystyle\int_S (1,0,0) \cdot N \, d\sigma$

(c) $\displaystyle\int_S (x,0,0) \cdot N \, d\sigma$

(d) $\displaystyle\int_S (x^2,0,0) \cdot N \, d\sigma$

(e) $\displaystyle\int_S (0,x^2,0) \cdot N \, d\sigma$

8.18. Use the Divergence Theorem to find the flux of F outward across the boundary of the rectangular box $D = [a,b] \times [c,d] \times [e,f]$.

(a) $F(x,y,z) = (p + qx + rx^2, 0, 0)$, p,q,r constants
(b) $F(x,y,z) = (0, p + qx + rx^2, 0)$, p,q,r constants
(c) $F = \nabla h$, if h is a C^2 function with $\Delta h = 0$.
(d) $F(x,y,z) = (e^x, e^y, e^z)$.

8.19. Let F be a C^1 vector field on a regular set D in $\mathbb{R}^3$. Justify the following steps to show that

$$\int_D \text{curl} F \, dV = \int_{\partial D} N \times F \, d\sigma.$$

Here N is the outward unit normal and the integral of a vector means to integrate each component of the vector.

(a) $(N \times F) \cdot C = N \cdot (F \times C)$ where $C = (c_1, c_2, c_3)$ is a constant vector.

(b) $\displaystyle\int_{\partial D} \mathbf{N}\cdot(\mathbf{F}\times\mathbf{C})\,d\sigma = \int_D \mathrm{div}\,(\mathbf{F}\times\mathbf{C})\,dV.$

(c) $\displaystyle\int_{\partial D} \mathbf{N}\cdot(\mathbf{F}\times\mathbf{C})\,d\sigma = \int_D (\mathrm{curl}\,\mathbf{F})\cdot\mathbf{C}\,dV.$

(d) $\displaystyle\int_{\partial D} (\mathbf{N}\times\mathbf{F})\cdot\mathbf{C}\,d\sigma - \int_D (\mathrm{curl}\,\mathbf{F})\cdot\mathbf{C}\,dV = 0.$

(e) $\displaystyle\left(\int_{\partial D} \mathbf{N}\times\mathbf{F}\,d\sigma - \int_D \mathrm{curl}\,\mathbf{F}\,dV\right)\cdot\mathbf{C} = 0.$

(f) $\displaystyle\int_{\partial D} \mathbf{N}\times\mathbf{F}\,d\sigma - \int_D \mathrm{curl}\,\mathbf{F}\,dV = \mathbf{0}.$

8.20. Let p be a C^1 function on a regular set D in $\mathbb{R}^3$. Justify the following steps to show that

$$\int_D \nabla p\,dV = \int_{\partial D} \mathbf{N}p\,d\sigma.$$

Here the integral of a vector means to integrate each component of the vector. Let $\mathbf{C} = (c_1, c_2, c_3)$ be a constant vector.

(a) $\displaystyle\int_{\partial D} \mathbf{N}\cdot(p\mathbf{C})\,d\sigma = \int_D \mathrm{div}\,(p\mathbf{C})\,dV = \int_D (\nabla p\cdot\mathbf{C} + p\,\mathrm{div}\,\mathbf{C})\,dV = \int_D \nabla p\cdot\mathbf{C}\,dV.$

(b) $\displaystyle 0 = \int_{\partial D} \mathbf{N}p\,d\sigma\cdot\mathbf{C} - \int_D \nabla p\,dV\cdot\mathbf{C} = \left(\int_{\partial D} \mathbf{N}p\,d\sigma - \int_D \nabla p\,dV\right)\cdot\mathbf{C}.$

(c) $\displaystyle\int_{\partial D} \mathbf{N}p\,d\sigma = \int_D \nabla p\,dV.$

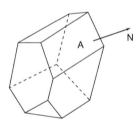

Fig. 8.19 The sum over all the faces, of face areas A times unit normal vectors $\mathbf{N}$ is the zero vector, in Problem 8.21.

8.21. Suppose a polyhedron D has faces $S_1,\ldots,S_k$. Let $\mathbf{N}_i$ be the outward unit normal vector of S_i. Justify the following items to prove that the sum of areas times unit normal vectors of the faces is zero:

$$\sum_{i=1}^{k} \mathrm{Area}\,(S_i)\mathbf{N}_i = \mathbf{0}.$$

See Figure 8.19.

(a) Use the Divergence Theorem for the constant vector field $\mathbf{F} = (1,0,0)$ to show that

$$\int_{\partial D} n_1 \, d\sigma = 0,$$

where $\mathbf{N} = (n_1, n_2, n_3)$ is the outer unit normal.

(b) Show that $\int_{\partial D} \mathbf{N} d\sigma = \mathbf{0}$, where the integral of a vector means to integrate each component of the vector.

(c) For each face of D, $\int_{S_i} \mathbf{N}_i \, d\sigma = \text{Area}(S_i)\mathbf{N}_i$.

8.22. Denote the inverse square vector field $\mathbf{F}(x,y,z) = \dfrac{(x,y,z)}{(x^2 + y^2 + z^2)^{3/2}}$ of Example 8.12 as

$$\mathbf{F}(\mathbf{X}) = \|\mathbf{X}\|^{-3}\mathbf{X},$$

and define a vector field $\mathbf{G}(\mathbf{X}) = \mathbf{F}(\mathbf{X} - \mathbf{A})$, a translation of $\mathbf{F}$ by a constant $\mathbf{A}$.

(a) What is the domain of $\mathbf{G}$?
(b) Show that $\operatorname{div}\mathbf{G} = 0$.
(c) Show that the flux

$$\int_{\partial W} \mathbf{G} \cdot \mathbf{N} d\sigma$$

of $\mathbf{G}$ outward through the boundary of a regular region W is zero or 4π according to whether $\mathbf{A}$ is outside or inside W and $\mathbf{A}$ is not on ∂W.

8.23. Let $\mathbf{G}(\mathbf{X}) = c_1\mathbf{F}(\mathbf{X} - \mathbf{A}_1) + \cdots + c_n\mathbf{F}(\mathbf{X} - \mathbf{A}_n)$ be a linear combination of inverse square vector fields, where $\mathbf{F}(\mathbf{X}) = \|\mathbf{X}\|^{-3}\mathbf{X}$.

(a) What is the domain of $\mathbf{G}$?
(b) Show that $\operatorname{div}\mathbf{G} = 0$.
(c) Show that the flux of $\mathbf{G}$ out of each regular set W is

$$\int_{\partial W} \mathbf{G} \cdot \mathbf{N} d\sigma = \sum_{\mathbf{A}_k \text{ in } W} 4\pi c_k,$$

where the sum is over indices k for which $\mathbf{A}_k$ is in the interior of W and none of the $\mathbf{A}_k$ is on the boundary.

8.24. Let $\mathbf{X} = (x_1, x_2, x_3)$ and let $\mathbf{H}(\mathbf{X}) = (\|\mathbf{X}\|^{-3}\mathbf{X})_{x_1}$, the first x_1 partial derivative of the inverse square vector field. See Figure 8.20.

(a) Show that $\mathbf{H} = \|\mathbf{X}\|^{-3}(1,0,0) - 3x_1\|\mathbf{X}\|^{-5}\mathbf{X}$.
(b) What is the domain of $\mathbf{H}$?
(c) Show that $\operatorname{div}\mathbf{H} = 0$.
(d) Use the Divergence Theorem to show that the flux of $\mathbf{H}$ out of every regular set D that does not contain the origin is zero.

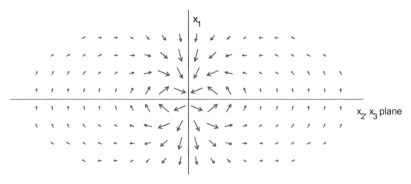

Fig. 8.20 The field **H** in Problem 8.24, drawn in any plane containing the x_1 axis.

(e) Evaluate

$$\int_{\partial B} \mathbf{H} \cdot \mathbf{N} d\sigma$$

directly when B is the region where $\|\mathbf{X}\| \leq 1$. Explain why the Divergence Theorem does not apply here.

8.25. Let u and v be C^2 on a regular set D in $\mathbb{R}^3$. Show that $\text{div}\,(v\nabla u) = v\Delta u + \nabla v \cdot \nabla u$, and use the Divergence Theorem to derive

(a) $\displaystyle\int_D (u\Delta u + |\nabla u|^2)\,dV = \int_{\partial D} u\nabla u \cdot \mathbf{N} d\sigma$

(b) $\displaystyle\int_D (v\Delta u - u\Delta v)\,dV = \int_{\partial D} (v\nabla u - u\nabla v) \cdot \mathbf{N} d\sigma.$

8.26. Let $f(x,y,z) = \sin(x)\sin(y)\sin(z)$.

(a) Show that $\Delta f = -3f$ in the cube $[0,\pi]^3$ and that $f = 0$ on the boundary of the cube.

(b) Suppose a function g has the properties $\Delta g = 3g$ in the cube and $g = 0$ on the boundary. Use the result in Problem 8.25(a) to show that g is identically 0 in the cube.

8.27. Let D be a regular set in $\mathbb{R}^3$. Suppose f is a C^2 function on D and λ a number such that

$$\Delta f = \lambda f$$

in D and $f = 0$ on ∂D. Show that if $\lambda > 0$ then f is zero in D.

8.28. Define $\mathbf{F}(\mathbf{X}) = (\|\mathbf{X}\|^2 - 2)\mathbf{X}$, where $\mathbf{X} = (x,y,z)$, and let D be the set $\|\mathbf{X}\| \leq r$ where r is a positive number. Find r so that

$$\int_D \text{div}\,\mathbf{F}\,dV = 0.$$

8.3 Stokes' Theorem

Green's Theorem says that for a C^1 vector field $\mathbf{F}(x,y) = (f_1(x,y), f_2(x,y))$ on a regular region D in the plane we have

$$\int_D \operatorname{curl} \mathbf{F} \, dA = \int_{\partial D} \mathbf{F} \cdot \mathbf{T} \, ds.$$

Suppose we have a flat surface S that lies in a plane $z = c$ in $\mathbb{R}^3$. Denote by S_{xy} the "shadow" of S in the x, y plane, that is the set of points $(x, y, 0)$ such that (x, y, c) is in S. See Figure 8.21. We see for $\mathbf{G}(x, y, z) = (g_1(x, y, z), g_2(x, y, z), g_3(x, y, z))$

$$\int_S (\operatorname{curl} \mathbf{G}) \cdot \mathbf{N} \, d\sigma = \int_S (\operatorname{curl} \mathbf{G}) \cdot (0, 0, 1) \, d\sigma$$

$$= \int_{S_{xy}} \left(\frac{\partial g_2}{\partial x}(x, y, c) - \frac{\partial g_1}{\partial y}(x, y, c) \right) d\sigma$$

$$= \int_{\partial S_{xy}} g_1(x, y, c) \, dx + g_2(x, y, c) \, dy = \int_{\partial S} \mathbf{G} \cdot \mathbf{T} \, ds.$$

Therefore

$$\int_S (\operatorname{curl} \mathbf{G}) \cdot \mathbf{N} \, d\sigma = \int_{\partial S} \mathbf{G} \cdot \mathbf{T} \, ds. \qquad (8.9)$$

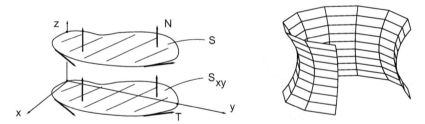

Fig. 8.21 *Left:* A flat surface S and its shadow. *Right:* A surface that is the union of polygons.

Equation 8.9 also holds for surfaces in every plane, not just in planes parallel to the x, y plane. Hence if our surface is the union of finitely many plane polygon faces S_i that meet along their edges then we can use the additivity of surface integrals and line integrals to add the formulas for the faces:

$$\int_S (\operatorname{curl} \mathbf{G}) \cdot \mathbf{N} \, d\sigma = \sum_{i=1}^{n} \int_{S_i} (\operatorname{curl} \mathbf{G}) \cdot \mathbf{N}_i \, d\sigma$$

$$= \sum_{i=1}^{n} \int_{\partial S_i} \mathbf{G} \cdot \mathbf{T}_i \, d\sigma$$

$$= \int_{\partial S} \mathbf{G} \cdot \mathbf{T} \, ds.$$

See Figure 8.21. This suggests Stokes' Theorem, that equation 8.9 holds for piece-wise smooth surfaces that are images of sets where Green's Theorem holds.

Theorem 8.4. Stokes' Theorem. *Let* **G** *be a vector field that is* C^1 *on a piece-wise smooth oriented surface* S *in* $\mathbb{R}^3$ *whose boundary* ∂S *is a piecewise smooth curve, and that the domains of the parametrizations of* S *are regular sets in* $\mathbb{R}^2$. *Then*

$$\int_S \operatorname{curl} \mathbf{G} \cdot \mathbf{N} \, d\sigma = \int_{\partial S} \mathbf{G} \cdot \mathbf{T} \, ds, \qquad (8.10)$$

where the orientation of the unit normal vector **N** *to* S *and of the unit tangent vector* **T** *to the boundary are chosen as in Figure 8.22.*

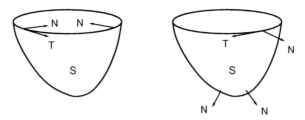

Fig. 8.22 A surface S oriented two ways, and the corresponding boundary orientations in Stokes' Theorem.

We give two proofs. The first proof approximates the surface by triangles and applies Green's Theorem to those. The second proof uses Green's Theorem on the preimages of S.

Proof. (Approximation argument) Let **X** from D in the u, v plane to $\mathbb{R}^3$ be a parametrization for one of the smooth surfaces that compose S. Since $\mathbf{X}_u(u,v)$ and $\mathbf{X}_v(u,v)$ are linearly independent, a triangular region T with vertices

$$(u_0,v_0), \quad (u_0+h,v_0), \quad (u_0,v_0+h)$$

is mapped to a "curved triangular" region $\mathbf{X}(T)$ on S with vertices

$$\mathbf{X}(u_0,v_0), \quad \mathbf{X}(u_0+h,v_0), \quad \mathbf{X}(u_0,v_0+h).$$

Denote by T' the triangle in $\mathbb{R}^3$ with those vertices. See Figure 8.23.

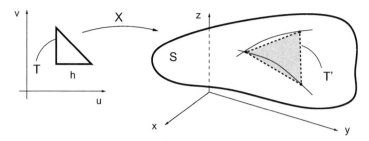

Fig. 8.23 Triangles T and T' in the proof of Stokes' Theorem.

By the Mean Value Theorem

$$\mathbf{X}(u_0 + h, v_0) - \mathbf{X}(u_0, v_0) = \mathbf{X}_u(\tilde{u}, v_0)h,$$

$$\mathbf{X}(u_0, v_0 + h) - \mathbf{X}(u_0, v_0) = \mathbf{X}_v(u_0, \tilde{v})h.$$

The norm of the cross product of these vectors is twice the area of the triangle T'. The area of T is $\frac{1}{2}h^2$. Since $\mathbf{X}_u$ and $\mathbf{X}_v$ are continuous functions, for every $\epsilon > 0$ we can choose h so small that for each point (u_1, v_1) in T

$$\int_T \|\mathbf{X}_u(u,v) \times \mathbf{X}_v(u,v) - \mathbf{X}_u(u_1,v_1) \times \mathbf{X}_v(u_1,v_1)\| \, du \, dv \leq \epsilon(\tfrac{1}{2}h^2).$$

Denote by $\mathbf{X}_1(u,v)$ the linear approximation of $\mathbf{X}(u,v)$ through T'. Since the first derivatives of $\mathbf{X}(u,v)$ are continuous, for every $\epsilon > 0$, for h small enough

$$\|\operatorname{curl} \mathbf{G}(\mathbf{X}(u,v)) - \operatorname{curl} \mathbf{G}(\mathbf{X}_1(u,v))\| < \epsilon.$$

Also since $\mathbf{X}_1(u,v)$ is the linear approximation of $\mathbf{X}(u,v)$ through T', for h small enough

$$\|\mathbf{X}(u,v) - \mathbf{X}_1(u,v)\| < \epsilon$$

for all (u,v) in the triangle T. We show that the difference between the flux of $\operatorname{curl} \mathbf{G}$ across $\mathbf{X}(T)$ and the flux of $\operatorname{curl} \mathbf{G}$ across T' is small:

$$\left| \int_{\mathbf{X}(T)} (\operatorname{curl} \mathbf{G}) \cdot \mathbf{N} \, d\sigma - \int_{T'} (\operatorname{curl} \mathbf{G}) \cdot \mathbf{N} \, d\sigma \right|$$

$$= \left| \int_T \Big((\operatorname{curl} \mathbf{G}(\mathbf{X}(u,v))) \cdot \mathbf{X}_u \times \mathbf{X}_v - (\operatorname{curl} \mathbf{G}(\mathbf{X}_1(u,v))) \cdot \mathbf{X}_{1u} \times \mathbf{X}_{1v} \Big) du \, dv \right|.$$

By the triangle inequality this is

$$\leq \left| \int_T \Big((\operatorname{curl} \mathbf{G}(\mathbf{X}(u,v)) - \operatorname{curl} \mathbf{G}(\mathbf{X}_1(u,v))) \cdot \mathbf{X}_u \times \mathbf{X}_v \Big) du\, dv \right|$$

$$+ \left| \int_T \Big((\operatorname{curl} \mathbf{G}(\mathbf{X}_1(u,v))) \cdot (\mathbf{X}_u \times \mathbf{X}_v - \mathbf{X}_{1u} \times \mathbf{X}_{1v}) \Big) du\, dv \right|$$

$$\leq \int_T \|\operatorname{curl} \mathbf{G}(\mathbf{X}(u,v)) - \operatorname{curl} \mathbf{G}(\mathbf{X}_1(u,v))\| \, \|\mathbf{X}_u \times \mathbf{X}_v\| \, du\, dv$$

$$+ \int_T \|\operatorname{curl} \mathbf{G}(\mathbf{X}_1(u,v))\| \, \|\mathbf{X}_u \times \mathbf{X}_v - \mathbf{X}_u(\tilde{u}, v_0) \times \mathbf{X}_v(u_0, \tilde{v})\| \, du\, dv$$

$$\leq k\epsilon \operatorname{Area}(T) = k\epsilon \frac{h^2}{2},$$

for some constant k that depends on the maximum of $\|\operatorname{curl} \mathbf{G}\|$. There are $\dfrac{\operatorname{Area}(D)}{\frac{1}{2} h^2}$ triangular regions in the region D. The error in using triangular regions to approximate

$$\int_{\cup_i \mathbf{X}(\mathbf{T}_i)} (\operatorname{curl} \mathbf{G}) \cdot \mathbf{N} \, d\sigma$$

is bounded by the sum of the errors in each triangle,

$$\frac{\operatorname{Area}(D)}{\frac{1}{2} h^2} \left(k\epsilon \frac{h^2}{2} \right) = \operatorname{Area}(D) k\epsilon.$$

By taking h small enough we can get the difference between the integrals less than ϵ:

$$\left| \int_{\cup_i \mathbf{X}(T_i)} (\operatorname{curl} \mathbf{G}) \cdot \mathbf{N} \, d\sigma, - \int_{\cup_i T_i'} (\operatorname{curl} \mathbf{G}) \cdot \mathbf{N} \, d\sigma \right| < \epsilon. \tag{8.11}$$

Now in each flat triangle T_i' we use Green's Theorem,

$$\int_{T_i'} (\operatorname{curl} \mathbf{G}) \cdot \mathbf{N} \, d\sigma = \int_{\partial T_i'} \mathbf{G} \cdot \mathbf{T} \, ds$$

to get

$$\int_{T_i'} (\operatorname{curl} \mathbf{G}) \cdot \mathbf{N} \, d\sigma - \int_{\partial T_i'} \mathbf{G} \cdot \mathbf{T} \, ds = 0. \tag{8.12}$$

We use an argument on the boundaries similar to the one we used for the triangular regions to show that for h small enough

$$\left| \int_{\cup_i \partial T_i'} \mathbf{G} \cdot \mathbf{T} \, ds - \int_{\cup_i \partial \mathbf{X}(T_i)} \mathbf{G} \cdot \mathbf{T} \, ds \right| < \epsilon. \tag{8.13}$$

Similarly for h small enough

$$\left| \iint_S (\operatorname{curl} \mathbf{G}) \cdot \mathbf{N} \, d\sigma - \int_{\cup_i \mathbf{X}(T_i)} (\operatorname{curl} \mathbf{G}) \cdot \mathbf{N} \, d\sigma \right| < \epsilon \qquad (8.14)$$

and

$$\left| \int_{\partial S} \mathbf{G} \cdot \mathbf{T} \, ds - \int_{\cup_i \partial \mathbf{X}(T_i)} \mathbf{G} \cdot \mathbf{T} \, ds \right| < 2\epsilon \qquad (8.15)$$

Adding relations (8.11)–(8.15) we get by triangle inequalities that

$$\left| \iint_S (\operatorname{curl} \mathbf{G}) \cdot \mathbf{N} \, d\sigma - \int_{\partial S} \mathbf{G} \cdot \mathbf{T} \, ds \right|$$

can be made as small as we like by taking h small enough. This proves (8.10) of Theorem 8.4. $\qquad \square$

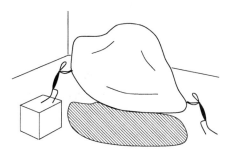

Fig. 8.24 Preparing to view Stokes' Theorem as a curved version of Green's Theorem.

The second proof uses Green's Theorem in a different way. See Figure 8.24.

Proof. We consider first the case where the surface S is the range of a single smooth parametrization $\mathbf{X}$ from the u, v plane to $\mathbb{R}^3$. So $S = \mathbf{X}(D)$, where D is a regular set in $\mathbb{R}^2$ and $\mathbf{X}(\partial D) = \partial S$. Then

$$\iint_S (\operatorname{curl} \mathbf{G}) \cdot \mathbf{N} \, d\sigma = \iint_D (\operatorname{curl} \mathbf{G})(\mathbf{X}(u,v)) \cdot \mathbf{X}_u \times \mathbf{X}_v \, du \, dv.$$

By the Chain Rule and some simplifying that we ask you to carry out in Problem 8.38 we see that

$$(\operatorname{curl} \mathbf{G})(\mathbf{X}(u,v)) \cdot \mathbf{X}_u \times \mathbf{X}_v = (\mathbf{G}(\mathbf{X}(u,v)))_u \cdot \mathbf{X}_v - (\mathbf{G}(\mathbf{X}(u,v)))_v \cdot \mathbf{X}_u.$$

We define a vector field on D in $\mathbb{R}^2$ by

$$\mathbf{F}(u,v) = (\mathbf{G}(\mathbf{X}(u,v)) \cdot \mathbf{X}_u(u,v), \mathbf{G}(\mathbf{X}(u,v)) \cdot \mathbf{X}_v(u,v));$$

then

$$\operatorname{curl}\mathbf{F}(u,v) = \frac{\partial}{\partial u}\Big(\mathbf{G}(\mathbf{X}(u,v))\cdot\mathbf{X}_v(u,v)\Big) - \frac{\partial}{\partial v}\Big(\mathbf{G}(\mathbf{X}(u,v))\cdot\mathbf{X}_u(u,v)\Big)$$

$$= \mathbf{G}(\mathbf{X}(u,v))\cdot\mathbf{X}_{vu}(u,v) + \mathbf{X}_v(u,v)\cdot\frac{\partial}{\partial u}\mathbf{G}(\mathbf{X}(u,v))$$

$$- \mathbf{G}(\mathbf{X}(u,v))\cdot\mathbf{X}_{uv}(u,v) - \mathbf{X}_u(u,v)\cdot\frac{\partial}{\partial v}\mathbf{G}(\mathbf{X}(u,v))$$

$$= (\operatorname{curl}\mathbf{G})(\mathbf{X}(u,v))\cdot\mathbf{X}_u\times\mathbf{X}_v.$$

So

$$\int_S (\operatorname{curl}\mathbf{G})\cdot\mathbf{N}\,d\sigma = \int_D (\operatorname{curl}\mathbf{G})(\mathbf{X}(u,v))\cdot\mathbf{X}_u\times\mathbf{X}_v\,du\,dv$$

$$= \int_D \operatorname{curl}\Big(\mathbf{G}(\mathbf{X}(u,v))\cdot\mathbf{X}_u(u,v),\mathbf{G}(\mathbf{X}(u,v))\cdot\mathbf{X}_v(u,v)\Big)du\,dv = \int_D \operatorname{curl}\mathbf{F}\,du\,dv.$$

By Green's Theorem

$$\int_D \operatorname{curl}\mathbf{F}\,du\,dv = \int_{\partial D}\mathbf{F}\cdot\mathbf{T}\,ds.$$

Now let $\mathbf{R}(t) = (u(t),v(t))$, $a\le t\le b$ parametrize ∂D. Then $\mathbf{X}(\mathbf{R}(t))$ parametrizes ∂S.

$$\int_{\partial D}\mathbf{F}\cdot\mathbf{T}\,ds = \int_a^b \mathbf{F}(\mathbf{R}(t))\cdot\mathbf{R}'(t)\,dt$$

$$= \int_a^b \Big(\mathbf{G}(\mathbf{X}(\mathbf{R}(t)))\cdot\mathbf{X}_u(\mathbf{R}(t)),\mathbf{G}(\mathbf{X}(\mathbf{R}(t)))\cdot\mathbf{X}_v(\mathbf{R}(t)),\Big)\cdot\mathbf{R}'(t)\,dt$$

$$= \int_a^b \Big(g_1 x_u + g_2 y_u + g_3 z_u, g_1 x_v + g_2 y_v + g_3 z_v\Big)\cdot\mathbf{R}'(t)\,dt$$

$$= \int_a^b \Big((g_1 x_u + g_2 y_u + g_3 z_u)u' + (g_1 x_v + g_2 y_v + g_3 z_v)v'\Big)dt$$

$$= \int_{\partial S} g_1\,dx + g_2\,dy + g_3\,dz = \int_{\partial S}\mathbf{G}\cdot\mathbf{T}\,ds.$$

This concludes the case where S has a single parametrization.

Suppose now that S is a union of such surfaces that meet pairwise on common edges. We assume these are oriented so that the line integrals associated to adjoining parts cancel on these common edges. See Figure 8.25. The edges that are not common to two parts constitute ∂S, so Stokes' formula for S is obtained by adding the Stokes' formulas for the parametrized parts of S.   □

In Example 3.32 we saw that if $\mathbf{F} = \nabla f$ then $\operatorname{curl}\mathbf{F} = \mathbf{0}$. We now show that under an additional condition on the domain of $\mathbf{F}$ the converse holds. Suppose that every closed loop C in the domain of $\mathbf{F}$ is the boundary ∂S of a smooth surface S in D.

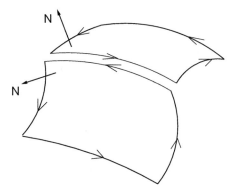

Fig. 8.25 A surface parametrized in two parts, in the proof of Stokes' Theorem. The common edge is drawn apart to show why the line integrals cancel there.

By Stokes' Theorem if $\operatorname{curl}\mathbf{F} = \mathbf{0}$ we have

$$\int_{C=\partial S} \mathbf{F} \cdot \mathbf{T}\, ds = \int_S \operatorname{curl}\mathbf{F} \cdot \mathbf{N}\, d\sigma = 0,$$

and by Theorem 7.1 there is a potential function g so that $\nabla g = \mathbf{F}$. Domains that have the property that we need are called simply connected.

Definition 8.2. We say that a set in $\mathbb{R}^n$ is *simply connected* if it is connected and if every simple closed curve in the set can be shrunk continuously within the set to a point.

We state without proof the following result: If a set is simply connected then every piecewise smooth simple closed curve in the set is the boundary of a piecewise smooth surface in the set.

So we have the following theorem.

Theorem 8.5. *Suppose a vector field* $\mathbf{F}$ *is* C^1 *on an open simply connected set* U *in* $\mathbb{R}^3$ *on which* $\operatorname{curl}\mathbf{F} = \mathbf{0}$. *Then there is a function* g *from* U *to* $\mathbb{R}$ *so that* $\nabla g = \mathbf{F}$.

Example 8.15. The set D_1 of points in $\mathbb{R}^3$ with $\|\mathbf{X}\| \neq 0$ is simply connected. The set D_2 of points in $\mathbb{R}^3$ not on the z axis is not simply connected.

If $\mathbf{F}_1$ is a C^1 vector field with $\operatorname{curl}\mathbf{F}_1 = \mathbf{0}$ on D_1, then $\mathbf{F}_1$ has a potential function.

If $\mathbf{F}_2$ is a C^1 vector field with $\operatorname{curl}\mathbf{F}_1 = \mathbf{0}$ on D_2, then we cannot conclude anything about the existence of a potential function for $\mathbf{F}_2$.   □

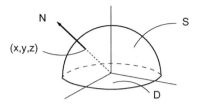

Fig. 8.26 The hemisphere in Example 8.16.

Example 8.16. Verify Stokes' Theorem for $\mathbf{F}(x,y,z) = (z^2,-2x,y^5)$ and S the upper hemisphere of radius 1 centered at the origin. See Figure 8.26. We compute $\operatorname{curl}\mathbf{F}(x,y,z) = (5y^4,2z,-2)$. At each point (x,y,z) on S, the unit normal vector is $\mathbf{N} = (x,y,z)$.

$$\int_S (\operatorname{curl}\mathbf{F}) \cdot \mathbf{N}\,d\sigma = \int_S (5y^4,2z,-2)\cdot(x,y,z)\,d\sigma$$

$$= \int_S 5y^4 x\,d\sigma + \int_S 2zy\,d\sigma - \int_S 2z\,d\sigma.$$

By symmetry the first two integrals are zero. Using the parametrization of the upper hemisphere

$$\mathbf{X}(x,y) = (x,y,\sqrt{1-x^2-y^2}), \quad 0 \le x^2+y^2 \le 1, \quad \|\mathbf{X}_x \times \mathbf{X}_y\| = \frac{1}{\sqrt{1-x^2-y^2}}$$

with domain the disk $D : 0 \le x^2 + y^2 \le 1$ we get

$$\int_S (\operatorname{curl}\mathbf{F}) \cdot \mathbf{N}\,d\sigma = -\int_S 2z\,d\sigma = -2\int_D \frac{\sqrt{1-x^2-y^2}}{\sqrt{1-x^2-y^2}}\,dx\,dy = -2\operatorname{Area}(D) = -2\pi.$$

To compute $\int_{\partial S} \mathbf{F}\cdot\mathbf{T}\,ds$ we use $\mathbf{X}(t) = (\cos t, \sin t, 0), 0 \le t \le 2\pi$ to parametrize ∂S consistent with the upward pointing unit normals on S. We get

$$\int_{\partial S} \mathbf{F}\cdot\mathbf{T}\,ds = \int_0^{2\pi} (0,-2\cos t,\sin^5 t)\cdot(-\sin t,\cos t,0)\,dt = \int_0^{2\pi} -2\cos^2 t\,dt = -2\pi.$$

$\square$

Example 8.17. Let $\mathbf{F}(x,y,z) = \left(\frac{-y}{x^2+y^2}, \frac{x}{x^2+y^2}, 0\right)$.

(a) Find the circulation of $\mathbf{F}$ around a piecewise smooth simple closed curve C_1 that does not encircle or intersect the z-axis.

(b) Find the circulation of $\mathbf{F}$ around a piecewise smooth simple closed curve C_2 that encircles the z-axis once (but does not intersect it).

See Figure 8.27. For (a) let S be a piecewise smooth surface with $\partial S = C_1$. Then since $\operatorname{curl}\mathbf{F} = \mathbf{0}$ we have by Stokes' Theorem

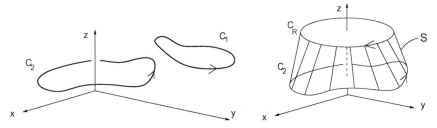

Fig. 8.27 *Left:* The curves and *Right:* the surface in Example 8.17.

$$\int_{C_1=\partial S} \mathbf{F}\cdot\mathbf{T}\,ds = \int_S \operatorname{curl}\mathbf{F}\cdot\mathbf{N}\,d\sigma = 0.$$

For (b), since $\mathbf{F}$ is not defined on the z axis, $\operatorname{curl}\mathbf{F}$ is not defined there either. Every surface with boundary C_2 would include a point on the z axis. Stokes' Theorem cannot be applied on such a surface. Consider a circle C_R of radius R centered on the z axis in a horizontal plane $z = k$ that does not intersect C_2. Let S be a piecewise smooth oriented surface whose boundary is $C_R \cup C_2$. Stokes' Theorem applied to this surface gives

$$\int_{C_2} \mathbf{F}\cdot\mathbf{T}\,ds + \int_{C_R} \mathbf{F}\cdot\mathbf{T}\,ds = \int_{\partial S=C_R\cup C_2} \mathbf{F}\cdot\mathbf{T}\,ds = \int_S \operatorname{curl}\mathbf{F}\cdot\mathbf{N}\,d\sigma = \int_S 0\,d\sigma = 0.$$

Suppose C_2 encircles the z axis "counterclockwise" as in Figure 8.27. Then C_R goes clockwise. Parametrize C_R by $\mathbf{X}(t) = (R\cos t, -R\sin t, k)$, $0 \le t \le 2\pi$. We get

$$0 = \int_{C_2} \mathbf{F}\cdot\mathbf{T}\,ds + \int_{C_R} \mathbf{F}\cdot\mathbf{T}\,ds$$

$$= \int_{C_2} \mathbf{F}\cdot\mathbf{T}\,ds + \int_0^{2\pi} \left(\frac{R\sin t}{R^2}, \frac{R\cos t}{R^2}, 0\right)\cdot(-R\sin t, -R\cos t, 0)\,dt$$

$$= \int_{C_2} \mathbf{F}\cdot\mathbf{T}\,ds + \int_0^{2\pi} (-1)\,dt = \int_{C_2} \mathbf{F}\cdot\mathbf{T}\,ds - 2\pi.$$

The circulation of $\mathbf{F}$ around C_2 is 2π. Similarly if C_2 encircles the z axis once in the other direction the circulation is -2π. □

Consider two smooth surfaces S_1 and S_2 that have the same oriented boundary

$$\partial S_1 = \partial S_2$$

as illustrated in Figure 8.28, and a vector field $\mathbf{F}$ that is C^1 on both surfaces. Since

$$\int_{\partial S_1} \mathbf{F}\cdot\mathbf{T}\,ds = \int_{\partial S_2} \mathbf{F}\cdot\mathbf{T}\,ds,$$

Stokes' Theorem implies that the fluxes of $\operatorname{curl}\mathbf{F}$ across S_1 and across S_2 are equal,

$$\int_{S_1} \text{curl}\,\mathbf{F} \cdot \mathbf{N}\,d\sigma = \int_{S_2} \text{curl}\,\mathbf{F} \cdot \mathbf{N}\,d\sigma$$

where the normal vectors $\mathbf{N}$ are consistent with the boundary orientation.

Example 8.18. Let $\mathbf{F}(x,y,z) = (z^2, -2x, y^5)$ and let S be the upper unit hemisphere as in Example 8.16. Let S_1 be the unit disk $x^2 + y^2 \le 1$ in the x, y plane, with unit normal vectors $(0, 0, 1)$. Then

$$\int_S \text{curl}\,\mathbf{F} \cdot \mathbf{N}\,d\sigma = \int_{S_1} \text{curl}\,\mathbf{F} \cdot (0, 0, 1)\,d\sigma$$

$$= -2\int_{S_1} d\sigma = -2\,\text{Area}(S_1) = -2\pi.$$

$\square$

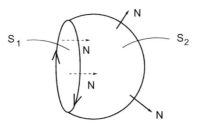

Fig. 8.28 Two surfaces S_1 and S_2 that have the same boundary.

Electromagnetism examples. Stokes' Theorem is often used in the study of electromagnetism. The electric field denoted $\mathbf{E}$ [force/charge] and magnetic field $\mathbf{B}$ [force/charge/velocity] are known to satisfy the system of Maxwell's differential equations

$$\mu_0\epsilon_0 \mathbf{E}_t = \text{curl}\,\mathbf{B} - \mu_0\mathbf{J}, \quad \mathbf{B}_t = -\text{curl}\,\mathbf{E}, \quad \text{div}\,\mathbf{E} = \frac{\rho}{\epsilon_0}, \quad \text{div}\,\mathbf{B} = 0,$$

where μ_0, ϵ_0 are constants, ρ [charge/vol] is the density of electric charge, and $\mathbf{J}$ [charge/time/area] is the current density. Several of the Problems explore relations among these to illustrate the use of vector calculus.

Problems

8.29. Use Stokes' Theorem to evaluate the line integrals

$$\int_C \mathbf{F} \cdot \mathbf{T}\,ds$$

(a) $\mathbf{F}(x,y,z) = (0,x,0)$ and C is the circle $x^2 + y^2 = 1$ in the plane $z = 0$, traversed in the counterclockwise direction as you look at the plane from the positive z axis.

(b) $\mathbf{F}(x,y,z) = (y,0,0)$ and C is the triangular path consisting of the three line segments from $\mathbf{A} = (a,0,0)$ to $\mathbf{B} = (0,b,0)$ to $\mathbf{C} = (0,0,c)$ to $\mathbf{A}$, where a,b,c are positive.

8.30. Verify Stokes' formula

$$\int_S \operatorname{curl} \mathbf{F} \cdot \mathbf{N} \, d\sigma = \int_{\partial S} \mathbf{F} \cdot \mathbf{T} \, ds$$

for the following cases by evaluating the integrals on both sides.

(a) $\mathbf{F}(x,y,z) = (-y,x,1)$, and S is the top face of the cube $[0,1]^3$, with $\mathbf{N}$ toward $+z$.

(b) $\mathbf{F}(x,y,z) = (-y,x,1)$, and S consists of the other five faces of the cube $[0,1]^3$, with $\mathbf{N}$ inward.

8.31. Verify Stokes' formula

$$\int_S \operatorname{curl} \mathbf{F} \cdot \mathbf{N} \, d\sigma = \int_{\partial S} \mathbf{F} \cdot \mathbf{T} \, ds$$

for the following cases by evaluating the integral on both sides.

(a) $\mathbf{F}(x,y,z) = (-y,x,2)$, and S is the hemisphere $x^2 + y^2 + z^2 = r^2$, $z \geq 0$, r constant, with $\mathbf{N}$ toward $+z$.

(b) $\mathbf{F}(x,y,z) = (-y,x,2)$, and S is the disk $x^2 + y^2 \leq r^2$, $z = 0$, r constant, with $\mathbf{N}$ toward $+z$.

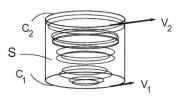

Fig. 8.29 The flow in Problem 8.32.

8.32. A C^1 vector field $\mathbf{V}$ models the fluid velocity in a rotating storm as in Figure 8.29. The surface S in the figure is a vertical cylinder open at top and bottom, where C_1 and C_2 are boundary circles, and $\mathbf{V}_1$ and $\mathbf{V}_2$ denote the values of $\mathbf{V}$ on C_1 and C_2, respectively. We suppose two properties hold:

(a) $\operatorname{curl} \mathbf{V} = (0,0,h(x,y,z))$ is parallel to the axis of the cylinder, and

(b) $\mathbf{V}_1$ and $\mathbf{V}_2$ in the figure are tangent to C_1 and C_2, respectively, and have constant norms $\|\mathbf{V}_1\| < \|\mathbf{V}_2\|$.

Use Stokes' Theorem to show that the properties are *not* consistent with each other.

8.33. Suppose S is a sphere and the vector field $\mathbf{F}$ is C^1 on S. Show that

$$\int_S \operatorname{curl} \mathbf{F} \cdot \mathbf{N} \, d\sigma = 0.$$

8.34. Let S be the hemisphere $x \le 0$ of the unit sphere centered at the origin in $\mathbb{R}^3$, oriented with normal pointing away from the origin. Evaluate

$$\int_S \operatorname{curl}(x^3 z, x^3 y, y) \cdot \mathbf{N} \, d\sigma.$$

8.35. A wire of radius R along the z axis (see Figure 8.30) carries a constant current density $\mathbf{J} = (0, 0, j)$ and there is a magnetic field of the form

$$\mathbf{B} = \begin{cases} c_1(-y, x, 0) & (r < R) \\ c_2 \frac{(-y, x, 0)}{r^p} & (r > R) \end{cases}$$

where $r = \sqrt{x^2 + y^2}$ and c_1, c_2 and p are some constants. Outside the wire $\mathbf{J} = \mathbf{0}$, and there is no dependence on the time t in this problem.

(a) Use the Maxwell equation $0 = \operatorname{curl} \mathbf{B} - \mu_0 \mathbf{J}$ inside the wire to find c_1.
(b) Find p so that $0 = \operatorname{curl} \mathbf{B} - \mu_0 \mathbf{J}$ holds outside the wire.
(c) Find c_2 so that $\mathbf{B}$ is continuous in $\mathbb{R}^3$.
(d) Let D be a disk of radius $R_1 > R$, center on the z axis, parallel to the x, y plane and with normal $\mathbf{N}$ aligned with $\mathbf{J}$. Then $\operatorname{curl} \mathbf{B}$ is not continuous on D. Show that Stokes' formula

$$\int_D \operatorname{curl} \mathbf{B} \cdot \mathbf{N} \, d\sigma = \int_{\partial D} \mathbf{B} \cdot \mathbf{T} \, ds$$

holds. Be sure to explain the meaning of the surface integral of the discontinuous function.

8.36. We say a vector field $\mathbf{F}$ is a *vector potential* of a vector field $\mathbf{G}$ if $\mathbf{G} = \operatorname{curl} \mathbf{F}$. Show that if $\mathbf{F}$ is a vector potential of $\mathbf{G}$ then the flux of $\mathbf{G}$ across a surface S depends only on the values of $\mathbf{F}$ on the boundary of S. As a result, we say if $\mathbf{G}$ has a vector potential its flux is "independent of surface."

8.37. Ampere's original law related magnetic field $\mathbf{B}$ along the boundary of each oriented surface S with the electric current $\mathbf{J}$ passing through S by

$$\int_{\partial S} \mathbf{B} \cdot \mathbf{T} \, ds = \mu_0 \int_S \mathbf{J} \cdot \mathbf{N} \, d\sigma,$$

μ_0 a constant.

(a) Use Stokes' formula to deduce

$$\operatorname{curl} \mathbf{B} = \mu_0 \mathbf{J}.$$

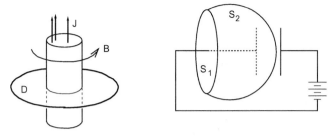

Fig. 8.30 *Left:* Current in a wire for Problem 8.35. *Right:* The displaced surface acquires displacement current in Problem 8.37.

(b) Show that part (a) contradicts the conservation of charge law

$$\rho_t + \operatorname{div} \mathbf{J} = 0,$$

where ρ is the density of electric charge that can vary with time.

Remark. Maxwell later replaced Ampere's law $\operatorname{curl} \mathbf{B} = \mu_0 \mathbf{J}$ with

$$\operatorname{curl} \mathbf{B} = \mu_0(\mathbf{J} + \epsilon_0 \mathbf{E}_t),$$

calling $\epsilon_0 \mathbf{E}_t$ the "displacement current." This change allowed the unification of light with electromagnetism. See Figure 8.30 for a sketch of a surface displaced into the region of changing magnetic field in a capacitor.

8.38. Show that if $\mathbf{G}$ is a continuously differentiable vector field in a set containing a smooth surface S, and S is parametrized by $\mathbf{X}(u,v)$ with domain D in the plane then

$$\big((\operatorname{curl} \mathbf{G})(\mathbf{X}(u,v))\big) \cdot (\mathbf{X}_u \times \mathbf{X}_v) = (\mathbf{G} \circ \mathbf{X})_u \cdot \mathbf{X}_v - (\mathbf{G} \circ \mathbf{X})_v \cdot \mathbf{X}_u.$$

We used this formula in our second proof of Stokes' Theorem.

8.39. Use Stokes' formula to deduce

$$\frac{\mathrm{d}}{\mathrm{d}t} \int_S \mathbf{B} \cdot \mathbf{N} \, \mathrm{d}\sigma = - \int_{\partial S} \mathbf{E} \cdot \mathbf{T} \, \mathrm{d}s$$

from the Maxwell law $\mathbf{B}_t = -\operatorname{curl} \mathbf{E}$. See Figure 8.31.
    Remark. This relates to generation of electric power.

8.40. Which of these sets are simply connected?

(a) The set of all points (x, y, z) in $\mathbb{R}^3$ that are not on the x axis.
(b) The set of all points in $\mathbb{R}^2$ other than $(0,0)$.
(c) The set of all points in $\mathbb{R}^3$ in a solid torus (doughnut).

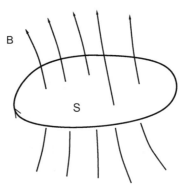

Fig. 8.31 Changing magnetic flux is related to electric field in a loop, in Problem 8.39.

8.4 Conservation laws

We illustrate an application of multivariable calculus by discussing *conservation laws*. A conservation law expresses the fact that the rate at which the total amount of some substance (mass, momentum, energy) contained in a set D changes is equal to the rate at which that substance enters the set. We denote the density of the substance as s [substance/volume], and denote the rate at which it flows [substance/area/time] as $\mathbf{F}$. Then the total amount of the substance in D at time t is

$$\int_D s\,dV.$$

The amount of substance that flows out of D per unit time is

$$\int_{\partial D} \mathbf{F} \cdot \mathbf{N}\,d\sigma$$

where the unit normal vectors $\mathbf{N}$ are outward. According to the conservation law the rate at which the amount of substance contained in D changes,

$$\frac{d}{dt}\int_D s\,dV$$

is equal to the rate at which the substance flows inward through the boundary of D. So conservation of substance is expressed by the equation

$$\frac{d}{dt}\int_D s\,dV = -\int_{\partial D} \mathbf{F} \cdot \mathbf{N}\,d\sigma. \qquad (8.16)$$

On the left we can carry out differentiation with respect to t under the integral sign, and on the right we use the Divergence Theorem to transform the integral over the boundary of D into an integral over D. We get

$$\int_D s_t \, dV = -\int_D \operatorname{div} \mathbf{F} \, dV.$$

Combining the two sides we get

$$\int_D (s_t + \operatorname{div} \mathbf{F}) \, dV = 0.$$

Since this relation holds for all regular regions D it follows that the integrand must be zero (see Problem 6.17):

$$s_t + \operatorname{div} \mathbf{F} = 0. \tag{8.17}$$

This is the differential form of a conservation law.

Conservation laws for fluid flow. We shall describe now the three basic conservation laws of mass, momentum, and energy of a fluid as instances of the general conservation law (8.17).

Fluid dynamics, the study of the flow of fluids, is an extremely interesting and important branch of science, encompassing aspects of mathematics, physics, and engineering. It is basic for understanding the flight of airplanes.

A. Conservation of mass. We denote the density [mass/volume] of a fluid by the Greek letter ρ. Flow velocity [distance/time] is denoted by $\mathbf{V} = (u, v, w)$. Therefore the rate at which material is transported is $\rho \mathbf{V}$ [mass/area/time]. Setting

$$s = \rho, \qquad \mathbf{F} = \rho \mathbf{V}$$

into the conservation law (8.17) gives the law of conservation of mass:

$$\rho_t + \operatorname{div}(\rho \mathbf{V}) = 0. \tag{8.18}$$

For an incompressible fluid the density ρ is a constant, independent of space and time; for such a fluid the equation of conservation of mass is then

$$\operatorname{div} \mathbf{V} = 0. \tag{8.19}$$

B. Conservation of momentum. As above we denote the density of the fluid by ρ and velocity by $\mathbf{V} = (u, v, w)$. The pressure is P [force/area].

We will show that

$$\rho \underbrace{(\mathbf{V}_t + \mathbf{V} \cdot \nabla \mathbf{V})}_{\text{accel.}} = -\nabla P,$$

a version of Newton's law of motion.

The density of momentum in the x direction is the x component of $\rho \mathbf{V}$, ρu. We assume it is transported into a set D by two mechanisms:

(a) It is carried by the substance flowing across ∂D into D.
(b) It is imparted by the x component of the fluid pressure acting on ∂D.

We describe now these mechanisms in detail.

a) The rate at which x momentum is carried into the domain D by the flow is

$$-\int_{\partial D} \rho u \mathbf{V} \cdot \mathbf{N} d\sigma,$$

where $\mathbf{N} = (n_1, n_2, n_3)$ is the outward unit normal.

b) The rate at which x momentum is imparted to the fluid at the boundary is the pressure force on ∂D in the x direction,

$$-\int_{\partial D} P n_1 d\sigma.$$

We assume no forces act other than pressure forces. The total rate of change of x momentum is the sum

$$-\int_{\partial D} (\rho u \mathbf{V} \cdot \mathbf{N} + P n_1) d\sigma.$$

Thus the law of conservation of x momentum in the form (8.16) is

$$\frac{d}{dt} \int_D \rho u \, dV = -\int_{\partial D} (\rho u \mathbf{V} \cdot \mathbf{N} + P n_1) d\sigma = -\int_{\partial D} (\rho u \mathbf{V} + (P, 0, 0)) \cdot \mathbf{N} d\sigma.$$

In the differential form (8.17) it is

$$(\rho u)_t + \operatorname{div} (\rho u \mathbf{V} + (P, 0, 0)) = 0.$$

We use the mass conservation equation (8.18) to express ρ_t as $-\operatorname{div} (\rho \mathbf{V})$ in the equation above; we get

$$\rho u_t - u \operatorname{div} (\rho \mathbf{V}) + \operatorname{div} (\rho u \mathbf{V}) + P_x = 0.$$

Using differentiation rules we have $\operatorname{div} (\rho u \mathbf{V}) = \nabla u \cdot (\rho \mathbf{V}) + u \operatorname{div} (\rho \mathbf{V})$. Substitute and simplifying we get

$$\rho u_t + \rho \nabla u \cdot \mathbf{V} + P_x = 0$$

or

$$\rho (u_t + u u_x + v u_y + w u_z) + P_x = 0.$$

Similar equations hold for the other two coordinates; we write all three as the vector equation

$$\rho \left(\begin{bmatrix} u_t \\ v_t \\ w_t \end{bmatrix} + u \frac{\partial}{\partial x} \begin{bmatrix} u \\ v \\ w \end{bmatrix} + v \frac{\partial}{\partial y} \begin{bmatrix} u \\ v \\ w \end{bmatrix} + w \frac{\partial}{\partial z} \begin{bmatrix} u \\ v \\ w \end{bmatrix} \right) + \begin{bmatrix} P_x \\ P_y \\ P_z \end{bmatrix} = \begin{bmatrix} 0 \\ 0 \\ 0 \end{bmatrix}$$

or

$$\rho (\mathbf{V}_t + \mathbf{V} \cdot \nabla \mathbf{V}) + \nabla P = \mathbf{0}. \qquad (8.20)$$

We ask you in Problem 8.44 to identify the vector

$$\begin{bmatrix} u_t + u u_x + v u_y + w u_z \\ v_t + u v_x + v v_y + w v_z \\ w_t + u w_x + v w_y + w w_z \end{bmatrix} = \mathbf{V}_t + \mathbf{V} \cdot \nabla \mathbf{V}$$

as the acceleration $\mathbf{X}''(t)$ of the fluid at the point (x,y,z,t). Thus equation (8.20) states that

$$(\text{density})(\text{acceleration}) + \nabla(\text{pressure}) = 0,$$

a version of Newton's law of motion.

C. Conservation of energy. The total energy in a fluid contained in a set D is the sum of its kinetic and internal energy. Internal energy density, denoted as e, is defined as internal energy per unit volume, and depends on the density of the fluid ρ and fluid pressure P. We assume energy is imparted to a set D by two mechanisms:

(a) Energy carried by the substance flowing across the boundary into D,
(b) The work done by the fluid pressure on the boundary of D.

The rate at which energy, both internal and kinetic, is flowing out of D is

$$\int_{\partial D} (e + \tfrac{1}{2}\rho \mathbf{V}\cdot\mathbf{V})\mathbf{V}\cdot\mathbf{N}\,d\sigma. \tag{8.21}$$

where $\mathbf{N}$ is the outward normal. Therefore the rate at which energy is flowing into D is the negative of (8.21).

The rate at which the fluid pressure does work on the boundary of D is

$$-\int_{\partial D} P\mathbf{V}\cdot\mathbf{N}\,d\sigma.$$

Therefore the total rate at which energy is imparted to D is

$$-\int_{\partial D} (e + \tfrac{1}{2}\rho \mathbf{V}\cdot\mathbf{V} + P)\mathbf{V}\cdot\mathbf{N}\,d\sigma;$$

the function $e + \tfrac{1}{2}\rho\mathbf{V}\cdot\mathbf{V} + P$ is called *enthalpy*.

The total energy $E(D)$ of fluid contained in D is the sum of the internal energy and kinetic energy of fluid contained in D:

$$E(D) = \int_D (e + \tfrac{1}{2}\rho\mathbf{V}\cdot\mathbf{V})\,dV.$$

Therefore the conservation law in the form (8.16) is

$$\frac{d}{dt}\int_D (e + \tfrac{1}{2}\rho\mathbf{V}\cdot\mathbf{V})\,dV = -\int_{\partial D} (e + \tfrac{1}{2}\rho\mathbf{V}\cdot\mathbf{V} + P)\mathbf{V}\cdot\mathbf{N}\,d\sigma.$$

In the differential form (8.17) it is

$$e_t + \tfrac{1}{2}\rho_t\mathbf{V}\cdot\mathbf{V} + \rho\mathbf{V}\cdot\mathbf{V}_t = -\text{div}\big((e + \tfrac{1}{2}\rho\mathbf{V}\cdot\mathbf{V} + P)\mathbf{V}\big).$$

Rearrange this equation as

$$e_t + \text{div}\,(e\mathbf{V}) = -\tfrac{1}{2}\rho_t\mathbf{V}\cdot\mathbf{V} - \rho\mathbf{V}\cdot\mathbf{V}_t - \text{div}\big((\tfrac{1}{2}\rho\mathbf{V}\cdot\mathbf{V} + P)\mathbf{V}\big).$$

We show now that the right hand side simplifies to $-P\operatorname{div}\mathbf{V}$. By the differentiation rule $\operatorname{div}(f\mathbf{W}) = f\operatorname{div}\mathbf{W} + \mathbf{W}\cdot\nabla f$ we get

$$\operatorname{div}\left((\tfrac{1}{2}\rho\mathbf{V}\cdot\mathbf{V} + P)\mathbf{V}\right) = \tfrac{1}{2}\operatorname{div}(\rho\mathbf{V})\mathbf{V}\cdot\mathbf{V} + \tfrac{1}{2}\rho\mathbf{V}\cdot\nabla(\mathbf{V}\cdot\mathbf{V}) + \operatorname{div}(P\mathbf{V}).$$

Therefore

$$e_t + \operatorname{div}(e\mathbf{V}) = -\tfrac{1}{2}\rho_t\mathbf{V}\cdot\mathbf{V} - \rho\mathbf{V}\cdot\mathbf{V}_t - \tfrac{1}{2}\operatorname{div}(\rho\mathbf{V})\mathbf{V}\cdot\mathbf{V} - \tfrac{1}{2}\rho\mathbf{V}\cdot\nabla(\mathbf{V}\cdot\mathbf{V}) - \operatorname{div}(P\mathbf{V}).$$

Using the mass conservation law $\rho_t + \operatorname{div}(\rho\mathbf{V}) = 0$ the first and third terms cancel on the right side, giving

$$e_t + \operatorname{div}(e\mathbf{V}) = -\rho\mathbf{V}\cdot\mathbf{V}_t - \tfrac{1}{2}\rho\mathbf{V}\cdot\nabla(\mathbf{V}\cdot\mathbf{V}) - \operatorname{div}(P\mathbf{V}).$$

Using again $\operatorname{div}(f\mathbf{W}) = f\operatorname{div}\mathbf{W} + \mathbf{W}\cdot\nabla f$ the right side is

$$= -\rho\mathbf{V}\cdot\mathbf{V}_t - \tfrac{1}{2}\rho\mathbf{V}\cdot\nabla(\mathbf{V}\cdot\mathbf{V}) - \mathbf{V}\cdot\nabla P - P\operatorname{div}\mathbf{V}.$$

In Problem 8.43 we ask you to verify that $\mathbf{V}\cdot\nabla(\mathbf{V}\cdot\mathbf{V}) = 2\mathbf{V}\cdot(\mathbf{V}\cdot\nabla\mathbf{V})$. Therefore the last expression becomes

$$= -\mathbf{V}\cdot(\rho\mathbf{V}_t + \rho\mathbf{V}\cdot\nabla\mathbf{V} + \nabla P) - P\operatorname{div}\mathbf{V},$$

and by the momentum conservation law (8.20) this is

$$= -P\operatorname{div}\mathbf{V}.$$

So the energy equation is

$$e_t + \operatorname{div}(e\mathbf{V}) = -P\operatorname{div}\mathbf{V}. \tag{8.22}$$

The energy equation has to be supplemented by an equation of state, that specifies the internal energy e as a function of density and pressure. The three conservation laws (8.18), (8.20), and (8.22) supplemented by an equation of state are the equations governing the flow of fluids when the only force is that due to the gradient of pressure.

Problems

8.41. The integral form of the mass conservation law is

$$\frac{\mathrm{d}}{\mathrm{d}t}\int_D \rho\,\mathrm{d}V = -\int_{\partial D}\rho\mathbf{V}\cdot\mathbf{N}\mathrm{d}\sigma.$$

Take the case where $\mathbf{V} = (u,0,0)$ is aligned with the x axis, ρ and $\mathbf{V}$ depend on x only, and the set D is a cylinder of cross section area A, with axis $a \le x \le b$. Show that the mass conservation implies

$$0 = \big(\rho(b)u(b) - \rho(a)u(a)\big)A.$$

8.42. Show for all differentiable functions ρ and $\mathbf{V}$ that

$$\rho_t + \operatorname{div}(\rho\mathbf{V}) = \rho_t + \mathbf{V}\cdot\nabla\rho + \rho\operatorname{div}\mathbf{V}.$$

8.43. Show for all differentiable functions u, v, and w that

$$u(u^2+v^2+w^2)_x + v(u^2+v^2+w^2)_y + w(u^2+v^2+w^2)_z = 2(u,v,w)\cdot\begin{bmatrix} uu_x + vu_y + wu_z \\ uv_x + vv_y + wv_z \\ uw_x + vw_y + ww_z \end{bmatrix}$$

to verify the formula $\mathbf{V}\cdot\nabla(\mathbf{V}\cdot\mathbf{V}) = 2\mathbf{V}\cdot(\mathbf{V}\cdot\nabla\mathbf{V})$ that we used in the energy equation.

8.44. Let $\mathbf{X}(t)$ be the path followed by one particle moving with the fluid. This means that the velocity agrees with that of the fluid at each point of the path:

$$\mathbf{X}'(t) = \mathbf{V}(\mathbf{X}(t),t).$$

Use the Chain Rule to show that

$$\mathbf{V}_t + \mathbf{V}\cdot\nabla\mathbf{V} = \begin{bmatrix} u_t + uu_x + vu_y + wu_z \\ v_t + uv_x + vv_y + wv_z \\ w_t + uw_x + vw_y + ww_z \end{bmatrix}$$

is the acceleration of the particle, that is,

$$\mathbf{X}''(t) = \mathbf{V}_t(\mathbf{X}(t),t) + \mathbf{V}(\mathbf{X}(t),t)\cdot\nabla\mathbf{V}(\mathbf{X}(t),t).$$

Remark: Note that $\mathbf{V}\cdot\nabla\mathbf{V} = (D\mathbf{V})\mathbf{V}$ where $D\mathbf{V}$ is the matrix derivative of u,v,w with respect to x,y,z.

8.45. Take the fluid velocity to be $\mathbf{V}(\mathbf{X}) = c\|\mathbf{X}\|^{-3}\mathbf{X}$ where c is a constant.

(a) Show that $\operatorname{div}\mathbf{V} = 0$.
(b) Show that the acceleration (See Problem 8.44) is $-2c^2\|\mathbf{X}\|^{-6}\mathbf{X}$.
(c) This is a model for flow in a conical duct. For each of the three cases indicated in Figure 8.32, indicate the direction of acceleration.

Remark: Maxwell used this flow as an analogy for a static electric field.

8.46. Use either direct calculations or the Divergence Theorem to verify that

$$\int_D \nabla P\,dV = \int_{\partial D} \mathbf{N}P\,d\sigma$$

for the functions in (a) and (b), where D is the ball of radius r given by $\|\mathbf{X}\|^2 \le r^2$.

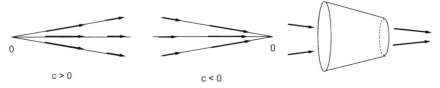

$c > 0$

$c < 0$

Fig. 8.32 Flows using **V** in Problem 8.45.

(a) $P(\mathbf{X}) = a\|\mathbf{X}\|^2$, $a > 0$ constant
(b) $P(\mathbf{X}) = \mathbf{B} \cdot \mathbf{X}$, $\mathbf{B} \neq \mathbf{0}$ constant
(c) Using the notion of pressure,

$$-\int_{\partial D} \mathbf{N} P \, d\sigma = \text{pressure force on } D,$$

which of (a) or (b) gives a nonzero force on the ball?

8.47. Let the fluid velocity be

$$\mathbf{V}(\mathbf{X}, t) = \frac{1}{1+t}\mathbf{X}.$$

(a) Describe the fluid velocity at times $t = 0$ and at $t = 1$.
(b) Show by computing the divergence div **V** that this is a compressible flow.
(c) Find a number a so that the functions

$$\rho(\mathbf{X}, t) = (1+t)^a\|\mathbf{X}\|^2, \qquad \mathbf{V}(\mathbf{X}, t) = \frac{1}{1+t}\mathbf{X}$$

satisfy the conservation of mass equation

$$\rho_t + \text{div}(\rho\mathbf{V}) = 0.$$

8.48. Take the case where the flow only depends on x and t, and the velocity is parallel to the x-axis, $\mathbf{V} = (u, 0, 0)$. Show that the mass, momentum, and energy equations become

$$\rho_t + (\rho u)_x = 0$$

$$u_t + uu_x = -\frac{P_x}{\rho}$$

$$e_t + (eu)_x = -Pu_x.$$

Remark. Problems 8.49–8.51 introduce *sound* waves. The wave equation is also discussed in Chapter 9.

8.49. For an ideal gas the pressure and energy are related to the density by

$$P = k\rho^\gamma, \qquad e = \frac{ck}{R}\rho^\gamma,$$

where k, c, R, and γ are constants. Show that for flows of an ideal gas the energy equation (8.22) is implied by the mass equation (8.18) if we set

$$\gamma = 1 + \frac{R}{c}.$$

8.50. Use the result of Problems 8.48 and 8.49 to derive the equations

$$\rho_t + (\rho u)_x = 0$$
$$u_t + uu_x = -k\gamma\rho^{\gamma-2}\rho_x$$

for flows of an ideal gas.

8.51. In the equations of Problem 8.50, consider the case where the velocity u is very small, and the density ρ is close to a constant, say

$$u = \epsilon f(x,t), \qquad \rho = \rho_0 + \epsilon g(x,t),$$

ϵ small. Show that

$$g_t + \rho_0 f_x = 0$$
$$f_t + (k\gamma\rho_0^{\gamma-2})g_x = 0$$

approximately, ignoring powers of ϵ. Deduce that g satisfies the *wave equation*

$$g_{tt} = (k\gamma\rho_0^{\gamma-1})g_{xx}.$$

8.52. If in addition to the effects we've considered, there is a gravitational acceleration $(0,-g,0)$ acting on the fluid, the conservation law in the y direction is

$$\frac{d}{dt}\int_D \rho v\, dV = -\int_{\partial D}\left(\rho v\mathbf{V} + (0,P,0)\right)\cdot\mathbf{N}\,d\sigma + \int_D -\rho g\, dV.$$

Show that the resulting differential equation becomes

$$\rho(v_t + uv_x + vv_y + wv_z + g) + P_y = 0.$$

8.5 Conservation laws and one-dimensional flows

One way of stating the Fundamental Theorem of Calculus is this:

$$\int_a^b f'(x)\,dx = f(b) - f(a), \qquad f \text{ is } C^1.$$

The higher-dimensional analogue of this result is the Divergence Theorem, which states that

$$\int_D \operatorname{div} \mathbf{F} \, dV = \int_{\partial D} \mathbf{F} \cdot \mathbf{N} \, d\sigma, \qquad \mathbf{F} \text{ is } C^1, \, \mathbf{N} \text{ is outer.} \qquad (8.23)$$

As we saw in the last section, the right side of (8.23) has an interesting interpretation when $\mathbf{F}$ is the rate of flow of some physical quantity like mass, momentum, or energy (stuff per area per time). The dot product $\mathbf{F} \cdot \mathbf{N}$ is the rate of flow in the direction $\mathbf{N}$. Therefore the integral on the right side of (8.23) is *the rate at which stuff is flowing out across the boundary of D*.

The one-dimensional situation is even simpler. There the rate of flow $f(x)$ is a scalar quantity (stuff per time), denoting the rate at which stuff is flowing *in the positive x direction*. The quantity $f(b)$ is then the rate at which stuff is flowing out of the interval $[a, b]$ at its right endpoint, and $f(a)$ is the rate at which stuff is flowing into the interval $[a, b]$ at its left endpoint. So $f(b) - f(a)$ is the *rate at which stuff is flowing out* of the interval $[a, b]$.

There is another way of calculating the rate at which stuff contained in an interval is changing. Denote by ρ the *density* (stuff per length) of the stuff under consideration (mass, momentum, energy). The total amount of stuff contained in an interval $[a, b]$ is the integral of the density

$$\int_a^b \rho(x) \, dx.$$

If density depends on time t, as well as on position x, so will the total amount of stuff contained in $[a, b]$,

$$\int_a^b \rho(x, t) \, dx.$$

The rate of change of the total amount contained in $[a, b]$ is the time derivative

$$\frac{d}{dt} \int_a^b \rho(x, t) \, dx.$$

Assume now that no chemical reactions take place, therefore stuff is not created nor destroyed. Then the only way the total amount of stuff in $[a, b]$ can change is through stuff entering or leaving across the boundary of $[a, b]$. According to the previous discussion, the rate at which stuff is leaving $[a, b]$ through its boundary is $f(b, t) - f(a, t)$. The rate at which the amount contained in the interval $[a, b]$ changes is the *negative* of the rate at which stuff leaves through the boundary,

$$\frac{d}{dt} \int_a^b \rho(x, t) \, dx = -f(b, t) + f(a, t). \qquad (8.24)$$

If we assume that ρ and f are continuously differentiable functions of t and x, we can carry out the differentiation with respect to t on the left of (8.24) under the integral sign, and rewrite the right side using the Fundamental Theorem of Calculus

$$\int_a^b \rho_t \, dx = -\int_a^b f_x \, dx. \tag{8.25}$$

We can rewrite this as

$$\int_a^b (\rho_t + f_x) \, dx = 0. \tag{8.26}$$

This relation holds for every interval $[a,b]$; from this we conclude that

$$\rho_t + f_x = 0 \tag{8.27}$$

for all values of x and t. For if there were a location x_0 and time t_0 where (8.27) is violated, say $(\rho_t + f_x)(x_0, t_0)$ is positive, then, since the partial derivatives ρ_t and f_x are assumed to be continuous, it would follow that $\rho_t + f_x$ is positive in a small enough interval $[x_0 - \epsilon, x_0 + \epsilon]$ in $[a,b]$. But then the integral

$$\int_{x_0 - \epsilon}^{x_0 + \epsilon} (\rho_t + f_x) \, dx$$

would be positive, contrary to (8.27).

As we saw in the last section, the same analysis was carried out in three dimensions to obtain the law of conservation of mass expressed as the differential equation

$$\rho_t + \operatorname{div}(\rho \mathbf{V}) = 0.$$

Deriving differential equations that the rates of flow must satisfy is only the first step in studying flows. The major task is to find solutions of these equations. This will tell us how flows behave. As an example, we present, and solve, a simplified model of the flow equations in one space dimension. The simplification is that we assume that the rate of flow [stuff/time] is a function of density alone, $f = f(\rho)$. Then equation (8.27) for the flow becomes

$$\rho_t + f(\rho)_x = 0. \tag{8.28}$$

For the sake of simplicity we also assume that f is a quadratic function of density, $f(\rho) = \frac{1}{2}\rho^2$. Then the equation for the conservation law is

$$\rho_t + \rho\rho_x = 0. \tag{8.29}$$

If the values of density ρ at time $t = 0$ are known,

$$\rho(x,0) = \rho_0(x),$$

we shall show how to use equation (8.29) to obtain the values of the density ρ for future times.

Consider functions x that satisfy the differential equation

$$\frac{dx}{dt} = \rho(x,t). \tag{8.30}$$

The graph of x is a curve on which we examine the values of ρ. Denote the starting point of a curve by x_0:

$$x(0) = x_0.$$

Next compute the derivative of $\rho(x(t),t)$ along such a curve. Using the Chain Rule we get

$$\frac{d}{dt}\rho(x(t),t) = \rho_x \frac{dx}{dt} + \rho_t. \tag{8.31}$$

According to (8.30), $\frac{dx}{dt} = \rho$. Setting this into (8.31) gives

$$\frac{d}{dt}\rho(x(t),t) = \rho_x \rho + \rho_t. \tag{8.32}$$

But according to the differential equation for the conservation law (8.29) satisfied by ρ, the right side of (8.32) is zero! This shows that $\frac{d}{dt}\rho(x(t),t) = 0$, which is the case only if the density at $x(t)$ at time t, $\rho(x(t),t)$, is independent of t.

It follows that the right side of equation (8.30) is constant and therefore the graph of x is a straight line! The speed $\frac{dx}{dt}$ with which this line propagates can be determined at $t = 0$: $\rho(x(t)x,t) = \rho(x_0,0)$, the initial value of ρ at the point x_0. Denote $\rho(x_0,0)$ as ρ_0; the solution of equation (8.30) is

$$x(t) = x_0 + \rho_0 t \tag{8.33}$$

and the solution of (8.29) satisfies

$$\rho(x_0 + \rho_0 t) = \rho_0. \tag{8.34}$$

The geometric interpretation of formula (8.33) is this: from each point x_0 of the line where $t = 0$ draw a ray propagating with speed ρ_0 in the positive t direction. Set $\rho(x,t) = \rho_0$ on this ray. Using the Inverse Function Theorem we can show that if the initial value for the density $\rho_0(x)$ is a smooth function of x, then for t sufficiently small $\rho(x,t)$, as defined here, is a smooth function of x and t, and is a solution of equation (8.29). We guide you through this argument in Problem 8.62.

The interesting question is: what happens when t is no longer sufficiently small? Suppose that for two values $x_1 < x_2$, $\rho_0(x_1)$ is greater than $\rho_0(x_2)$.

Then the rays given by equation (8.33) *intersect* at some critical positive value of t. At this point of intersection (x,t), the density $\rho(x,t)$ is defined to be both equal to $\rho_1 = \rho_0(x_1)$ and $\rho_2 = \rho_0(x_2)$, a *contradiction*. This shows that no solution of equation (8.29) with such prescribed initial values $\rho_0(x)$ exists for t greater than this critical value of t.

The following example points a way of resolving this problem. Take for the initial values ρ_0 the continuous function

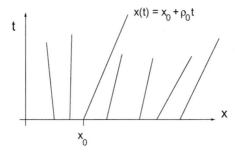

Fig. 8.33 The rays in the x,t plane on which $\rho(x,t)$ is constant, have a slope that depends on that constant. Note the time evolves upward.

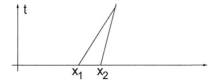

Fig. 8.34 Two rays may intersect at a certain time.

$$\rho_0(x) = \begin{cases} 1 & \text{if } x < -1, \\ -x & \text{if } -1 \le x \le 0, \\ 0 & \text{if } 0 < x. \end{cases} \tag{8.35}$$

For this choice of ρ_0 the rays are as follows.

(a) For $x_0 \le -1$, $x(t) = x_0 + t$.
(b) For $-1 < x_0 < 0$, $x(t) = x_0 - x_0 t$.
(c) For $0 \le x$, $x(t) = x_0$.

For $t < 1$ these lines look as in Figure 8.35.

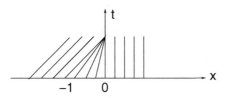

Fig. 8.35 The rays of constant $\rho(x,t)$, for the initial value in equation 8.35.

The rays don't intersect for $t < 1$, but as t tends to 1, all rays issuing from points of the interval $-1 \le x \le 0$ intersect at $x = 0$. So the value of $\rho(x,t)$ at $t = 1$ is

$$\rho(x,1) = \begin{cases} 1 & \text{for } x < 0 \\ 0 & \text{for } 0 < x, \end{cases} \tag{8.36}$$

a *discontinuous* function.

The solution ρ of $\rho_t + \rho\rho_x = 0$ for this example is indicated in Figure 8.36, for $0 \le t < 1$. It satisfies the initial values (8.35). We ask you in Problem 8.57 to verify the values of ρ indicated in the figure.

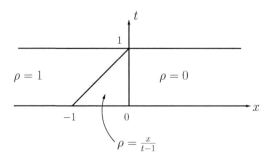

Fig. 8.36 Solution ρ of (8.29) with initial values (8.35).

We show now how to continue such a solution ρ of

$$\rho_t + f_x = 0$$

as a *discontinuous* solution. This sounds like nonsense, for a discontinuous function is not differentiable at the points of discontinuity, therefore cannot satisfy the differential equation at such points. To give meaning to the concept of a discontinuous solution we have to go back to the integral version of the conservation law, equation (8.24), from which the differential version (8.28) was derived. Whereas the differential version makes no sense for discontinuous functions, the integral version does!

Suppose ρ is a function that has a discontinuity across a smooth curve in the x, t plane. On each disk that does not intersect the curve, ρ is continuous. Describe this curve of discontinuity as $x = y(t)$ in Figure 8.37.

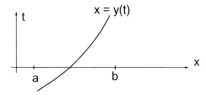

Fig. 8.37 A discontinuity of $\rho(x,t)$ can occur across a curve $x = y(t)$.

Choose the interval $[a, b]$ so that the discontinuity $y(t)$ lies between a and b during the time interval we are investigating. Since ρ is continuous on each side of the graph of $x = y(t)$ we can write the total amount of stuff in $[a, b]$ as

$$\int_a^b \rho \, dx = \int_a^{y(t)} \rho \, dx + \int_{y(t)}^b \rho \, dx.$$

We differentiate the above integral with respect to t, using the rule (see Problem 8.53)

$$\frac{d}{dt} \int_{y_1(t)}^{y_2(t)} g(x,t) \, dt = \int_{y_1(t)}^{y_2(t)} g_t(x,t) \, dt + g(y_2(t),t) y_2'(t) - g(y_1(t),t) y_1'(t)$$

for differentiating an integral whose integrand and limits both depend on t. We get

$$\frac{d}{dt} \int_a^b \rho \, dx = \frac{d}{dt} \int_a^{y(t)} \rho \, dx + \frac{d}{dt} \int_{y(t)}^b \rho \, dx$$

$$= \int_a^{y(t)} \rho_t \, dx + \rho(L) y_t + \int_{y(t)}^b \rho_t \, dx - \rho(R) y_t. \qquad (8.37)$$

Here ρ_t and y_t denote derivatives with respect to t, and $\rho(L), \rho(R)$ denote the limiting value of ρ on the left and right sides of the discontinuity.

We apply the integral conservation law (8.24) to an interval $[a,z]$ with $z < y(t)$. Since ρ is differentiable on this interval,

$$\frac{d}{dt} \int_a^z \rho \, dx = \int_a^z \rho_t \, dx = f(a) - f(z)$$

Now let z tend to $y(t)$. Then $f(z)$ tends to $f(L)$, the value of f on the left side of the discontinuity. So we get

$$\int_a^y \rho_t \, dx = f(a) - f(L)$$

Similarly

$$\int_y^b \rho_t \, dx = f(R) - f(b)$$

where $f(R)$ is the value of f on the right side of the discontinuity. Setting these relations into the right side of (8.37) yields

$$\frac{d}{dt} \int_a^b \rho \, dx = f(a) - f(L) + \rho(L) y_t + f(R) - f(b) - \rho(R) y_t. \qquad (8.38)$$

According to the integral conservation law, $\dfrac{d}{dt} \displaystyle\int_a^b \rho \, dx = f(a) - f(b)$. So we conclude from (8.38) that

$$f(R) - f(L) + (\rho(L) - \rho(R)) y_t = 0 \qquad (8.39)$$

It is convenient to denote the jump of ρ and f across the discontinuity by brackets:

$$\rho(R) - \rho(L) = [\rho], \qquad f(R) - f(L) = [f].$$

So we can rewrite (8.39) as $y_t = \dfrac{[f]}{[\rho]}$. The derivative y_t of y is the *speed* with which the discontinuity is propagating. Let's denote it by s: $s = y_t$. Then

$$s = \frac{[f]}{[\rho]} \tag{8.40}$$

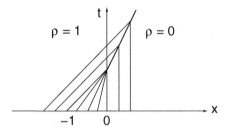

Fig. 8.38 A shock wave at speed $\frac{1}{2}$ continues the solution shown in Figures 8.35 and 8.36.

We return now to our example $\rho_t + \rho\rho_x = 0$, where $f(\rho) = \frac{1}{2}\rho^2$ and $\rho(x, 1)$ is given by the discontinuous,

$$\rho(x, 1) = \begin{cases} 1 & x < 0 \\ 0 & x > 0. \end{cases}$$

The solution $\rho(x,t)$ for $t > 1$ consists of two regions separated by a discontinuity issuing from the point $(0, 1)$; see Figure 8.38. To the left of the discontinuity $\rho(x,t) = 1$, and to the right of the discontinuity $\rho(x,t) = 0$. The discontinuity is a straight line with speed $s = \frac{[f]}{[\rho]}$. This moving discontinuity is called a *shock wave*. Here $[\rho] = 0 - 1 = -1$ and for $f = \frac{1}{2}\rho^2$, $[f] = 0 - \frac{1}{2} = -\frac{1}{2}$, so by (8.40)

$$s = \frac{[f]}{[\rho]} = \frac{-\frac{1}{2}}{-1} = \frac{1}{2}.$$

Similar calculations of discontinuous solutions of conservation laws for functions of two or three variables yield the analogue of formula (8.40), where s is the speed of propagation of the discontinuity in the direction *normal* to the discontinuity.

Problems

8.53. Apply the Fundamental Theorem of Calculus, and the Chain Rule in the form

$$\frac{d}{dt}g(a(t),b(t),t) = \nabla g \cdot (a',b',1),$$

to the function $g(r,s,t) = \int_r^s f(x,t)\,dx$ to show that

$$\frac{d}{dt}\int_{a(t)}^{b(t)} f(x,t)\,dx = f(b(t),t)b'(t) - f(a(t),t)a'(t) + \int_{a(t)}^{b(t)} f_t(x,t)\,dx,$$

for C^1 functions f.

8.54. Show that $\rho(x,t) = \dfrac{x}{t+8}$ is constant along certain rays $x = x_0 + mt$ and give a formula relating m and x_0.

8.55. Show that the function $\rho(x,t)$ of Problem 8.54 is a solution of $\rho_t + \rho\rho_x = 0$.

8.56. Verify that the conservation law

$$\frac{d}{dt}\int_a^b \rho(x,t)\,dx = -f(b,t) + f(a,t)$$

holds for the function $\rho(x,t)$ of Problem 8.54, the flux function $f = \frac{1}{2}\rho^2$, and the interval $[a,b] = [2,5]$.

8.57. For $0 \le t < 1$ let

$$\rho(x,t) = \begin{cases} 1, & x \le t-1 \\ \frac{x}{t-1}, & t-1 \le x \le 0 \\ 0, & 0 \le x \end{cases}$$

as illustrated in Figure 8.36.

(a) Show that ρ is continuous for $0 \le t < 1$.
(b) Sketch the graphs of $\rho(x,0)$ and of $\rho(x,1)$.
(c) Show that $\rho_t + \rho\rho_x = 0$ for $0 < t < 1$ except along the segments $x = t-1$ and $x = 0$.
(d) Equation (8.34) applied to this function says that

$$\rho(x_0 + (-x_0)t, t) = -x_0, \qquad (-1 < x_0 < 0).$$

Use this to derive the $\dfrac{x}{t-1}$ part of the formula for ρ.

8.58. A solution of $\rho_t + \rho\rho_x = 0$ has initial value

$$\rho(x,0) = \begin{cases} 10, & x \le 0 \\ 10 - 10x, & 0 \le x \le 1 \\ 0, & 1 \le x \end{cases}$$

At what time t does a shock wave form?

8.59. Suppose instead of the conservation law (8.24) there is some mechanism to generate stuff within the interval at rate g:

$$\frac{d}{dt} \int_a^b \rho(x,t)\,dx = -f(b,t) + f(a,t) + \int_a^b g(x,t)\,dx.$$

If this holds for all intervals $[a,b]$ and the functions are all continuously differentiable, show that the differential equation form of the conservation law becomes

$$\rho_t + f_x = g.$$

8.60. For the equation $\rho_t + (\frac{1}{2}\rho^2)_x = 0$, show that the jump speed formula $s = \frac{[f]}{[\rho]}$ gives the speed as

$$s = \tfrac{1}{2}(\rho(L) + \rho(R)),$$

the average of the left and right limits at the discontinuity.

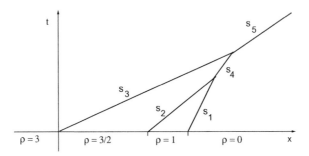

Fig. 8.39 Five shock waves for $\rho_t + \rho\rho_x = 0$ in Problem 8.61.

8.61. Use the result of Problem 8.60 to find all the shock speeds indicated in Figure 8.39. The initial value $\rho(x,0)$ is a piecewise constant function whose values are shown along the x axis.

8.62. The solution of $\rho_t + \rho\rho_x = 0$ with differentiable initial value $\rho(x,0) = \rho_0(x)$ requires that ρ be constant on rays $x = x_0 + \rho_0 t$. That means we need to find a function $\rho(x,t)$ that satisfies

$$\rho(x + \rho_0(x)t, t) = \rho_0(x).$$

Define a function $\mathbf{F}(x,t) = (x + \rho_0(x)t, t)$, so that we need $\rho \circ \mathbf{F}(x,t) = \rho_0(x)$. Justify the following steps using $\mathbf{F}$ to prove the existence of such a function ρ.

(a) Find the matrix derivative $D\mathbf{F}$.
(b) Show that $D\mathbf{F}(x,t)$ is invertible when t is small.
(c) Apply the Inverse Function Theorem to conclude that $\mathbf{F}^{-1}$ is locally defined and differentiable.

(d) The formula $\rho = \rho_0 \circ$ (the first component of $\mathbf{F}^{-1}$) defines a differentiable function that solves $\rho_t + \rho\rho_x = 0$ for small t and has $\rho(x,0) = \rho_0(x)$.

8.63. Suppose we draw two rays $x = x_0 + \rho_0 t$, one starting from $x_0 = 0$ where we assume $\rho_0 = 2$, and the second starting from $x_0 = 3$ where $\rho_0 = 1.5$. Find the critical value of t and the value of x at which these two rays intersect.

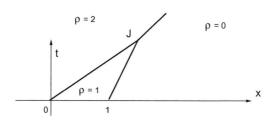

Fig. 8.40 Values $\rho(x,t)$ of the solution in Problem 8.64 are indicated in the x,t plane.

8.64. Consider the equation $\rho_t + (\frac{1}{2}\rho^2)_x = 0$ with initial data $\rho(x,0) = 2$ if $x < 0$, $\rho(x,0) = 1$ if $0 < x < 1$, and $\rho(x,0) = 0$ if $x > 1$. See Figure 8.40.

(a) Show that the speeds of the shock waves starting from $x = 0$ and $x = 1$ are $3/2$ and $1/2$.
(b) Find the point J where the shock waves meet.
(c) Find the speed of the single shock that continues the solution beyond the time at point J.
(d) If you could stand at the point $x = 2$ at time 0 and wait, would you observe one, or two, shock waves passing by?

8.65. Suppose $\rho_t + x\rho_x = 0$. Sketch curves in the (x,t) plane on which $x'(t) = x(t)$. Show that on those curves,

$$\frac{d}{dt}\rho(x(t),t) = 0,$$

so that ρ is constant on each such curve. Find $\rho(x,t)$ if $\rho(x,0) = x^2$.

Chapter 9
Partial differential equations

Abstract In this chapter we derive the laws governing the vibration of a stretched
string and a stretched membrane, and the equations governing the propagation of
heat. Like the laws of conservation of mass, momentum and energy studied in the
previous chapter, and like the electromagnetism laws, these laws are expressed as
partial differential equations. We derive some properties and some solutions of these
equations. We also state the Schrödinger equation of quantum mechanics, derive a
property of the solutions and explain the physical meaning of this property.

9.1 Vibration of a string

Imagine a string stretched along the x-axis. When we pluck the string each point of
the string vibrates in a direction perpendicular to the stretched string, call it the u
direction. See Figure 9.1. Assume the vibration is in the x, u plane. The displacement
u of each point x of the string is a function of x and t. We use Newton's law, force
equals mass times acceleration, to derive a partial differential equation satisfied by
$u(x,t)$.

Fig. 9.1 A string displacement is shown at one value of the time. The displacement and slope of
the string are assumed very small everywhere and are exaggerated here for visibility.

Let T be the magnitude of the tension force in the elastic string, that we take to
be the same at all points of the string and at all times. Consider small vibrations of
the string, where the slope of the vibrating string differs little from the direction of
the undisturbed string.

P. D. Lax and M. S. Terrell, *Multivariable Calculus with Applications*,
Undergraduate Texts in Mathematics, https://doi.org/10.1007/978-3-319-74073-7_9

Take a small piece of the string between x and $x + h$, h positive. There are two forces acting on this piece of string, the tension at each end of the piece. Let $\theta(x)$ be the angle between the tangent to the string at the position x, and the x axis,

$$\tan\theta(x) = \frac{\partial u}{\partial x}. \tag{9.1}$$

See Figure 9.2.

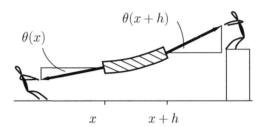

Fig. 9.2 Tension forces depend on angles $\theta(x)$, $\theta(x+h)$ at the ends of a bit of the string. The angles are drawn too large in order to make them more visible.

The forces in the (x, u) plane acting on the piece of the string at x and $x + h$ are

$$-T(\cos\theta(x), \sin\theta(x)), \qquad T(\cos\theta(x+h), \sin\theta(x+h)).$$

We consider the motion of the string only in the u direction.

The force on the small piece in the u direction at $x + h$ is $T\sin\theta(x+h)$. The force at x is $-T\sin\theta(x)$. The total force is the sum of these two forces:

$$\text{total force } = T\sin\theta(x+h) - T\sin\theta(x).$$

We use the Chain Rule and the Mean Value Theorem to express the right side as

$$\text{total force } = hT\cos(\theta)\frac{d\theta}{dx}, \tag{9.2}$$

where θ and $\dfrac{d\theta}{dx}$ are evaluated at some point between x and $x + h$. Since θ is small, we approximate $\cos\theta$ as 1, and $\tan\theta$ as θ. Since at each point the slope of the line tangent to the string is the tangent of θ we get

$$\frac{\partial u}{\partial x} = \tan(\theta(x))$$

$$= \frac{\sin\theta}{\cos\theta} \approx \frac{\theta}{1}.$$

Replacing θ by $\dfrac{\partial u}{\partial x}$ and $\cos\theta$ by 1 in formula (9.2) we get

$$\text{total force } = hT\,\frac{\partial^2 u}{\partial x^2}. \tag{9.3}$$

The mass of the piece of string of length h is hW, where W is the mass per unit length of the string. The acceleration of the piece of string in the u direction is u_{tt}. We apply the law

$$\text{total force} = \text{mass times acceleration}$$

to the motion of the piece of string. Using formula (9.3) for the total force acting on the piece of string, and hW for its mass, we get

$$hT\,\frac{\partial^2 u}{\partial x^2} = hW u_{tt}.$$

Dividing by hW gives

$$\tfrac{T}{W}\, u_{xx} = u_{tt}. \tag{9.4}$$

Since T and W are both positive, so is $\tfrac{T}{W}$. We write $c = \sqrt{\tfrac{T}{W}}$ and set it into equation (9.4); we get

$$u_{tt} - c^2 u_{xx} = 0. \tag{9.5}$$

Equation (9.5) is called the *one-dimensional wave equation*.

   It follows from equation (9.5) that c has the dimension of velocity; but velocity of what? We shall show that $\pm c$ are the velocities with which certain waves in the string propagate along the x direction.

   Example 9.1. Take $u(x,t) = \cos(x - t)$. Then u is a solution of the wave equation with $c^2 = 1$ because

$$u_{tt} - u_{xx} = -\cos(x - t) - (-\cos(x - t)) = 0.$$

This is a wave that propagates to the right at speed 1. See Figure 9.3. In Problem 9.5 we ask you to identify the speed and direction that other waves move. □

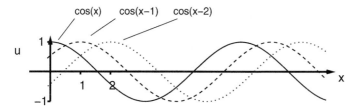

Fig. 9.3 The wave in Example 9.1 plotted at times $t = 0, 1, 2$.

Example 9.2. The function

$$u(x,t) = \cos(x - ct) + \cos(x + ct)$$

is a sum of a wave moving to the right and a wave moving to the left, both at speed c. By the addition formula for the cosine function,

$$u(x,t) = \cos x \cos(ct) + \sin x \sin(ct) + \cos x \cos(ct) - \sin x \sin(ct) = 2\cos x \cos(ct).$$

This is a wave that oscillates up and down. See Figure 9.4.   □

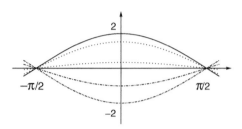

Fig. 9.4 The sum of waves in Example 9.2 is illustrated at several times.

It follows from the rules of calculus that every function of the form

$$u(x,t) = f(x - ct) + g(x + ct)$$

where f and g are twice differentiable functions of a single variable, is a solution of $u_{tt} - c^2 u_{xx} = 0$. The theorem below shows that every solution is of this form.

We introduce the following notation: D denotes the trapezoid

$$a + ct \leq x \leq b - ct, \qquad t_1 \leq t \leq t_2 \tag{9.6}$$

in the (x,t) plane. We number its sides as C_1, C_2, C_3 and C_4. See Figure 9.5.

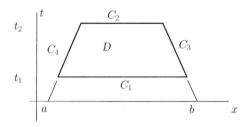

Fig. 9.5 The trapezoid defined in (9.6).

Theorem 9.1. *Every solution u of the wave equation*

$$u_{tt} - c^2 u_{xx} = 0$$

in the trapezoid D where $a + ct \le x \le b - ct$, $t_1 \le t \le t_2$ is of the form

$$u(x,t) = f(x - ct) + g(x + ct), \tag{9.7}$$

where f and g are twice differentiable functions.

We derive Theorem 9.1 from Theorem 9.2 below. First we introduce the notion of initial data. Denote by $u(x,t)$ a solution of the wave equation in the trapezoid. The pair of functions

$$u(x,t_1), \quad u_t(x.t_1), \qquad a + ct_1 \le x \le b - ct_1 \tag{9.8}$$

are called the *initial data* of u at time t_1.

Theorem 9.2. *Let u be a solution of the wave equation*

$$u_{tt} - c^2 u_{xx} = 0$$

in the trapezoid D where $a + ct \le x \le b - ct$, $t_1 \le t \le t_2$. Suppose the initial data of u at time t_1 are zero. Then $u(x,t)$ is zero in the whole trapezoid.

Before giving the proof of Theorem 9.2 we show how it implies Theorem 9.1.

Proof. (of Theorem 9.1) Let u be a solution of the wave equation in D. We show first that there is a solution v of the wave equation of the form (9.7) whose initial data are the same as the initial data of u. We construct two functions f and g that satisfy the relations

$$u(x,t_1) = f(x - ct_1) + g(x + ct_1), \qquad u_t(x,t_1) = c(g'(x + ct_1) - f'(x - ct_1)) \tag{9.9}$$

for $a + ct_1 \le x \le b - ct_1$. Differentiate the first equation with respect to x and add $\frac{1}{c}$ times the second equation to get $u_x + \frac{1}{c} u_t = 2g'$, or subtract $\frac{1}{c}$ times the second equation to get $u_x - \frac{1}{c} u_t = 2f'$. Therefore

$$g'(x + ct_1) = \tfrac{1}{2} u_x(x,t_1) + \tfrac{1}{2c} u_t(x,t_1), \qquad f'(x - ct_1) = \tfrac{1}{2} u_x(x,t_1) - \tfrac{1}{2c} u_t(x,t_1), \tag{9.10}$$

from which f and g can be determined by integration. These formulas show that, since u is twice differentiable, so are f and g.

Define the function v as

$$v(x.t) = f(x - ct) + g(x + ct). \tag{9.11}$$

Then v is a solution of the wave equation in D. It follows from (9.9) that the initial data of v and u at time t_1 are equal:

$$v(x,t_1) = u(x,t_1), \quad v_t(x,t_1) = u_t(x,t_1). \tag{9.12}$$

Next define w as the difference of u and v,

$$w(x,t) = u(x,t) - v(x,t).$$

Being the difference of two solutions, w also is a solution of the wave equation. It follows from (9.12) that the initial data of w are zero:

$$w(x,t_1) = 0, \quad w_t(x,t_1) = 0.$$

Therefore according to Theorem 9.2, $w = u - v$ is zero in the trapezoid. This shows that $u = v$ in the trapezoid; therefore u is of the form (9.7).   □

We turn now to the proof of Theorem 9.2.

Proof. Multiply the wave equation in Theorem 9.2 by u_t and integrate over the trapezoid D; we get

$$\int_D (u_t u_{tt} - c^2 u_t u_{xx}) \, dx \, dt = 0.$$

We observe that the integrand is a divergence with respect to the (x,t) variables in that order:

$$
\begin{aligned}
0 = u_t(u_{tt} - c^2 u_{xx}) &= (\tfrac{1}{2}u_t^2)_t - c^2(u_t u_x)_x + c^2 u_{tx} u_x \\
&= (-c^2 u_t u_x)_x + \tfrac{1}{2}(u_t^2 + c^2 u_x^2)_t \\
&= \mathrm{div}\,(-c^2 u_t u_x, \tfrac{1}{2}(u_t^2 + c^2 u_x^2)).
\end{aligned}
$$

Using the Divergence Theorem for $\mathbf{F} = (-c^2 u_t u_x, \tfrac{1}{2}(u_t^2 + c^2 u_x^2))$ and $\mathrm{div}\,\mathbf{F} = 0$ we get

$$
\begin{aligned}
0 = \int_D \mathrm{div}\,\mathbf{F}\, dx\, dt = {}& \int_{\partial D} (-c^2 u_t u_x, \tfrac{1}{2}(u_t^2 + c^2 u_x^2)) \cdot \mathbf{N}\, ds \\
= {}& -\int_{C_1} \tfrac{1}{2}(u_t^2 + c^2 u_x^2)\, dx + \int_{C_2} \tfrac{1}{2}(u_t^2 + c^2 u_x^2)\, dx \\
& + \int_{C_3} (-c^2 u_t u_x, \tfrac{1}{2}(u_t^2 + c^2 u_x^2)) \cdot \mathbf{N}\, ds \\
& + \int_{C_4} (-c^2 u_t u_x, \tfrac{1}{2}(u_t^2 + c^2 u_x^2)) \cdot \mathbf{N}\, ds.
\end{aligned}
$$

We have used $\mathbf{N} = (0,-1)$ on C_1 and $\mathbf{N} = (0,1)$ on C_2. Denote the integrals over C_1 and C_2 as $E(t_1)$ and $E(t_2)$. Using $\mathbf{N} = \dfrac{1}{\sqrt{1+c^2}}(1,c)$ on C_3 and $\mathbf{N} = \dfrac{1}{\sqrt{1+c^2}}(-1,c)$ on C_4 we get

$$E(t_2) - E(t_1) + \int_{C_3} \frac{1(-c^2 u_t u_x) + c\frac{1}{2}(u_t^2 + c^2 u_x^2)}{\sqrt{1+c^2}} \, ds$$

$$+ \int_{C_4} \frac{-1(-c^2 u_t u_x) + c\frac{1}{2}(u_t^2 + c^2 u_x^2)}{\sqrt{1+c^2}} \, ds = 0. \tag{9.13}$$

The integrands in both integrals in (9.13) are perfect squares, $\frac{1}{2}c(u_t \pm cu_x)^2$. Therefore both integrals are nonnegative. This proves that $E(t_2) - E(t_1)$ is nonpositive. Therefore

$$E(t_2) \le E(t_1) \tag{9.14}$$

Denote by $C(t)$ the portion of the string represented by a horizontal line segment drawn across the trapezoid at time t. The term $\int_{C(t)} \frac{1}{2} u_t^2 \, dx$ is the kinetic energy of the $C(t)$ portion of the moving string, and the term $\int_{C(t)} \frac{1}{2} c^2 u_x^2 \, dx$ is the elastic energy in the $C(t)$ portion of the stretched string. Their sum is the total energy stored in the $C(t)$ portion of the string. So inequality (9.14) says that the energy stored in the $C(t)$ portion of the string is a decreasing function of time.

In particular if $E(t_1)$ is zero, it follows that so is $E(t_2)$. Since $E(t_2)$ is the integral of $u_t^2 + c^2 u_x^2$ over C_2, it follows that if $E(t_2)$ is zero, u_x and u_t are zero in C_2.

Since the argument applies to every value of t_2, it follows that $u(x,t)$ is constant in the whole trapezoid. Since u is initially zero, that constant is zero. This completes the proof of Theorem 9.2. □

Example 9.3. We solve

$$u_{tt} = c^2 u_{xx}, \qquad u(x,0) = 0, \qquad u_t(x,0) = \cos(3x).$$

According to Theorem 9.1 we can express

$$u(x,t) = f(x - ct) + g(x + ct).$$

By equation (9.10) in the proof the data give

$$g'(x) = \frac{1}{2c} \cos(3x), \qquad f'(x) = -\frac{1}{2c} \cos(3x).$$

Integrating, we get

$$g(x) = \frac{1}{6c} \sin(3x) + c_1, \qquad f(x) = -\frac{1}{6c} \sin(3x) + c_2.$$

Since $0 = u(x,0) = f(x) + g(x) = \frac{c_1 - c_2}{2c}$ we get $c_1 = c_2$, and

$$u(x,t) = -\frac{1}{6c} \sin(3(x - ct)) + \frac{1}{6c} \sin(3(x + ct)).$$

□

Example 9.4. We imagine a sound wave bouncing off a wall at $x = 0$ where the displacement u is zero. Take $f(x) = -g(-x)$ in the wave equation solution $u(x,t) = f(x-ct) + g(x+ct)$, so that

$$u(x,t) = -g(-x+ct) + g(x+ct).$$

Then $u(0,t) = -g(ct) + g(ct) = 0$ for all t. Suppose $g(x)$ is zero on all but some positive interval, as in Figure 9.6. The resulting solution u can be viewed, for $x \geq 0$, as a wave g traveling left to a wall at $x = 0$, then reflected to the right, as an echo.   □

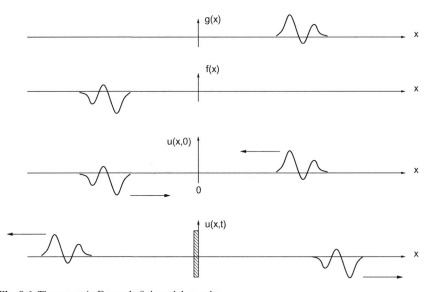

Fig. 9.6 The waves in Example 9.4 model an echo.

The string with tied ends. We turn now to studying the motion of a stretched string that is *tied* at its two ends, located at $x = 0$ and at $x = a (a > 0)$. See Figure 9.7. The displacement of the ends is zero; we express this as the boundary conditions

$$u(0,t) = 0, \quad u(a,t) = 0 \quad \text{for all } t. \tag{9.15}$$

Suppose u is of the form

$$u(x,t) = f(x-ct) + g(x+ct)$$

and satisfies the boundary conditions.

The boundary conditions state that

$$f(-ct) + g(ct) = 0, \quad f(a-ct) + g(a+ct) = 0.$$

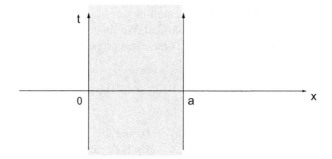

Fig. 9.7 The band $0 \leq x \leq a$, $-\infty < t < \infty$ corresponds to locations on the string at all times.

Denoting ct as x, the first relation says that $f(-x) = -g(x)$. Setting this into the second relation we get

$$-g(x - a) + g(x + a) = 0.$$

Denoting $x - a$ as y we rewrite this as

$$g(y + 2a) = g(y).$$

In words: g is a periodic function with period $2a$.

Since $f(y) = -g(-y)$, it follows that also f is periodic with period $2a$. Therefore $u(x,t) = f(x - ct) + g(x + ct)$ is a periodic function of t with period $\frac{2a}{c}$.

A function with period p also has periods $2p$, $3p$, and so forth. Thus a string that vibrates with period $\frac{2a}{c}$ also vibrates with period $\frac{2an}{c}$, n any whole number.

Example 9.5. The functions

$$\sin(x)\cos(ct), \quad \sin(2x)\cos(2ct), \quad \sin(3x)\cos(3ct),$$

and so on are vibrations of a string with ends tied at $x = 0$ and $x = \pi$. $\quad\square$

The frequency of vibration is the reciprocal of the period; so the lowest frequency is $\frac{c}{2a}$. We recall from our derivation of the wave equation that $c^2 = \frac{T}{W}$. So we get the following theorem.

Theorem 9.3. *A string of length a and weight W per unit length stretched with tension T can vibrate with the frequency $\frac{1}{2a}\sqrt{\frac{T}{W}}$ and each whole number multiple of this.*

This formula shows that the lowest frequency of vibration increases if

(a) the tension in the string is increased,
(b) the string is shortened,
(c) the string is replaced by a thinner string.

All vibrations of a stretched string are multiples of the lowest frequency; it is this property that makes string instruments musical. We show in the next section that, in contrast, a vibrating elastic sheet does not have this property.

Problems

9.1. Suppose u is a solution of $u_{tt} - c^2 u_{xx} = 0$, and that at a particular time t, the graph of u as a function of x is convex ($u_{xx} > 0$). Is the acceleration u_{tt} of the string up or down?

9.2. A violin A-string has length 330 [mm] and vibrates at 440 cycles per second. Let $A_4 = 2\pi(440)$ and

$$u(x,t) = \sin\left(\tfrac{2\pi x}{330}\right)\cos(A_4 t).$$

Find the value of $\frac{T}{W}$ so that $u_{tt} - \frac{T}{W} u_{xx} = 0$,

9.3. A violin A-string produces a vibration

$$u(x,t) = c_1 \sin\left(\tfrac{2\pi x}{330}\right)\cos(A_4 t) + c_2 \sin\left(\tfrac{2\pi(3x)}{330}\right)\cos(E_6 t)$$

where c_1 and c_2 are some constants and $u_{tt} - c^2 u_{xx} = 0$. Find $E_6 > 0$ in terms of A_4.

9.4. Show that the function $u(x,t) = A\sin(bx)\cos(bct)$ is a string vibration with ends tied at 0 and π, that is,

$$u_{tt} - c^2 u_{xx} = 0, \qquad u(0,t) = 0, \qquad u(\pi,t) = 0,$$

for certain values of the constants A, b. Show that A is arbitrary but b must be an integer.

9.5. At what speed and direction (left or right) do these waves move?

(a) $\cos(x + 3t)$
(b) $5\cos(x + 3t)$
(c) $-7\sin(t - 4x)$

9.6. Show that the following functions are solutions of $u_{tt} - c^2 u_{xx} = 0$.

(a) $\cos(x + ct)$,
(b) $u(kx, kt)$ if k is a constant and $u(x,t)$ is a solution,
(c) $au(x,t)$, if u is a solution and a a number.
(d) $u_1 + u_2$, if u_1 and u_2 are solutions.
(e) $\sin(2x - 2ct) + \cos(3x + 3ct)$

9.7. Verify that every function of the form

$$u(x,t) = f(x - ct) + g(x + ct)$$

where f and g are twice differentiable functions of a single variable, is a solution of $u_{tt} - c^2 u_{xx} = 0$.

9.8. Consider the string vibrations

$$u_1(x,t) = \sin(x - 2t)$$
$$u_2(x,t) = \sin(x + 2t)$$
$$u_3(x,t) = u_1(x,t) + u_2(x,t).$$

(a) Sketch graphs of each one as a function of x, for times $t = 0, 1, 2$.
(b) Which one moves to the left and which to the right?
(c) Show that $u_3(0,t) = 0$.
(d) Find values of a that give $u_3(a,0) = 0$, and then show that for those a, $u_3(a,t) = 0$ for all t.

9.9. Assume $2c < \pi$. Find the energy

$$E(t) = \tfrac{1}{2} \int_a^b (u_t^2 + c^2 u_x^2) \, dx$$

of the vibration $u(x,t) = \cos(x + ct)$ for:

(a) the interval $[a,b] = [0, \pi]$ at time $t = 0$,
(b) the interval $[a,b] = [ct, \pi - ct]$ at time $t = 1$.

Which is larger?

9.10. Take the case of a string with tied ends, $u(0,t) = u(a,t) = 0$.

(a) Why is u_t also zero at the ends?
(b) Show that the energy in $[0,a]$ is *conserved*, that is, $E'(t) = 0$, where

$$E(t) = \tfrac{1}{2} \int_0^a (u_t^2 + c^2 u_x^2)(x,t) \, dx.$$

9.11. Find a solution of each problem in the form

$$u(x,t) = f(x - ct) + g(x + ct).$$

(a) $u_{tt} - c^2 u_{xx} = 0$ with $u(x,0) = \sin x$ and $u_t(x,0) = 0$.
(b) $u_{tt} - u_{xx} = 0$ with $u(x,0) = 0$ and $u_t(x,0) = \cos(2x)$.
(c) $u_{tt} - 25u_{xx} = 0$ with $u(x,0) = 3\sin x + \sin(3x)$ and $u_t(x,0) = \cos(2x)$.

9.12. Suppose the wave $g(x + ct)$ in Example 9.4 is a brief sound made by an observer who waits for the echo $-g(-x + ct)$ to return, and notes the time t_1 elapsed. Show that the distance to the wall is $\tfrac{1}{2} ct_1$, and can be determined using any brief sound g (as one knows of echos).

9.13. For $x \geq 0$ let $u(x,t)$ be a solution of the wave equation as in Example 9.4. Let p be a positive number. Show that the energy remaining near the wall,

$$\frac{1}{2} \int_0^p (u_t^2 + c^2 u_x^2)\,dx,$$

is zero for large values of the time.

9.2 Vibration of a membrane

Consider an elastic membrane stretched over a frame in the x,y plane. When the membrane is displaced in the direction perpendicular to the x,y plane and then released, it will vibrate in the perpendicular direction. We derive the differential equation governing this vibration and study its solutions.

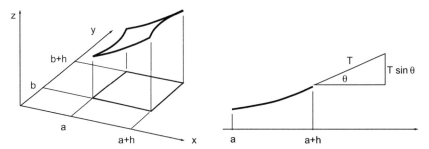

Fig. 9.8 *Left:* A small portion of the membrane. *Right:* We approximate the perpendicular component of the elastic force on the $a+h$ edge as $hT\sin\theta \approx hT\tan\theta = hTz_x(a+h,y,t)$.

Denote by $z(x,y,t)$ the displacement at time t of point (x,y) of the membrane in the perpendicular direction. See Figure 9.8. We study the motion of a small portion of the membrane

$$(x,y,z(x,y,t)), \quad a \leq x \leq a+h, \quad b \leq y \leq b+h, \quad h \text{ small}. \tag{9.16}$$

The motion of this portion is driven by the elastic forces acting on its four sides. The elastic force acting on an edge of this portion of the membrane lies in the tangent plane of the membrane and is perpendicular to the edge on which it acts; its magnitude is some constant T times the length of the edge. We shall study small vibrations, that is where the displacement $z(x,y,t)$ and its derivatives z_x and z_y are small. The vibrations are produced by forces in the direction perpendicular to the x, y plane. We analyze them similarly to the forces on a piece of string in Section 9.1. In this case the component of the force acting on the side $x = a$ is well approximated by $-hTz_x(a,y,t)$ and the force on the side $x = a+h$ is equally well approximated by $hTz_x(a+h,y,t)$.

The resultant of these forces is $hT(z_x(a+h,y,t) - z_x(a,y,t))$, and for h small is well approximated by $h^2 T z_{xx}$. Similarly the forces acting on the sides $y = b+h$ and $y = b$, in the direction perpendicular to the x, y plane, are well approximated by $h^2 T z_{yy}$. The sum of these forces is $h^2 T(z_{xx} + z_{yy})$. The mass of the portion of the membrane is $h^2 \rho$, where ρ [mass/area] is the density of the membrane. Therefore the equation governing the motion of this portion of the membrane, using Newton's law that total force = mass times acceleration, is

$$h^2 T(z_{xx} + z_{yy}) = h^2 \rho z_{tt},$$

which we rewrite as

$$z_{tt} = c^2(z_{xx} + z_{yy}) \tag{9.17}$$

where $c = \sqrt{\frac{T}{\rho}}$. This equation is called the two-dimensional *wave equation* .

We investigate now some simple motions of a membrane spanning the square

$$0 \le x \le \pi, \qquad 0 \le y \le \pi. \tag{9.18}$$

The membrane is fixed at the boundary of the square, therefore

$$z(x,y) = 0$$

on the boundary of the square. Define

$$z_1(x,y,t) = \sin(\sqrt{2}ct)\sin(x)\sin(y),$$
$$z_2(x,y,t) = \sin(\sqrt{5}ct)\sin(x)\sin(2y).$$

A simple calculation shows that both z_1 and z_2 are solutions of the wave equation (9.17), and they are zero on the boundary of the square (9.18). The solution z_1 is periodic in time, with period $\frac{2\pi}{\sqrt{2}c}$, and the solution z_2 is periodic with period $\frac{2\pi}{\sqrt{5}c}$, and all integer multiples of these periods. Since no integer multiple of $\frac{2\pi}{\sqrt{2}c}$ is equal to an integer multiple of $\frac{2\pi}{\sqrt{5}c}$. It follows that the sum

$$z_1 + z_2,$$

that is a solution of the wave equation, is not periodic in time. It can be shown that only very special solutions of the two-dimensional wave equation are periodic in time.

We express this result as follows: Vibrations of one-dimensional elastic systems are periodic in time, but the vibrations of two-dimensional systems are in general not periodic in time. This explains why all musical instruments are essentially one-dimensional vibrating systems. Violins and cellos use vibrating strings to generate sound, wind instruments like flutes and clarinets use vibrating thin columns of air to generate sound. One can point to drums as a truly two-dimensional instrument; but the sound of a drum of is muffled, without a definite pitch!

Problems

9.14. Show that the following functions are solutions of the wave equation

$$z_{tt} = c^2(z_{xx} + z_{yy}).$$

(a) $x^2 - y^2$
(b) $\cos(ct)\cos(x)$
(c) $\cos(ct)\sin(y)$
(d) $\sin(\sqrt{2}ct)\cos(x+y)$

9.15. Show that the following functions are solutions of the wave equation

$$z_{tt} = c^2(z_{xx} + z_{yy}).$$

(a) $u + v$, if u and v are solutions.
(b) kz, if k is a constant and z is a solution.
(c) $z(-y, x, t)$ if $z(x, y, t)$ is a solution, i.e., rotate $\pi/2$.
(d) $z(kx, ky, kt)$ if $z(x, y, t)$ is a solution.

9.16. Show that these functions are solutions of the wave equation

$$z_{tt} = c^2(z_{xx} + z_{yy}).$$

(a) $\cos(y + ct)$
(b) $\cos(x + y + \sqrt{2}ct)$
(c) $\cos(x - 2y - \sqrt{5}ct)$

9.17. We have said that the function

$$z_1 + z_2 = \sin(\sqrt{2}ct)\sin(x)\sin(y) + \sin(\sqrt{5}ct)\sin(x)\sin(2y)$$

is not periodic in t. Show that the function

$$\sin(1.414ct)\sin(x)\sin(y) + \sin(2.236ct)\sin(x)\sin(2y)$$

is periodic, repeating every $1000\dfrac{2\pi}{c}$ seconds.

9.18. Suppose a tension T [force/length] causes a force perpendicular to any short edge drawn in an elastic membrane, the force coplanar with the material. For example there are forces on three edges in the plane of the triangle in Figure 9.9. Denote by α and k the angle and hypotenuse length indicated. Show that the forces are as follows.

(a) The force on the right edge is $(Tk\sin\alpha, 0)$.
(b) The force on the bottom edge is $(0, -Tk\cos\alpha)$.
(c) The force on the hypotenuse is $Tk(-\sin\alpha, \cos\alpha)$.
(d) The net force on the triangle is the zero vector.

This is why we assume a constant T in the derivation of the wave equation (9.17).

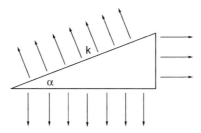

Fig. 9.9 Forces on a triangular portion of the membrane in Problem 9.18.

9.19. Let n and m be positive integers, and

$$z(x,y,t) = \cos(\sqrt{n^2 + m^2}\,t)\sin(nx)\sin(my).$$

(a) Show that z is a solution of the membrane vibration equation $z_{tt} = z_{xx} + z_{yy}$ that is zero on the boundary of the square $0 \le x \le \pi,\ 0 \le y \le \pi$.
(b) Show that the number of such solutions that have frequency $\sqrt{n^2 + m^2} \le 1000$ is roughly $\frac{1}{4}\pi(1000)^2$.

9.20. Take the case $c = 1$ so that the wave equation becomes $z_{tt} = z_{xx} + z_{yy}$, and the solutions z_1 and z_2 in the text become

$$z_1 = \sin(\sqrt{2}t)\sin(x)\sin(y), \quad z_2 = \sin(\sqrt{5}t)\sin(x)\sin(2y).$$

(a) Suppose f is a twice differentiable function of one variable. Define

$$z(x,y,t) = f(ax + by + t),$$

which is called a *traveling* wave. What is required of the constants a and b so that z is a solution of the wave equation?
(b) Verify that $\sin(u)\sin(v) = -\frac{1}{2}(\cos(u+v) - \cos(u-v))$. Use that and similar identities to express z_2 as a sum of four traveling waves.
(c) The solution $z_1 + z_2$ can be expressed as a sum of how many traveling waves?

9.21. Let $z(x,y,t)$ be a function of the form

$$z(x,y,t) = f(r)\sin(kt)$$

where $r = \sqrt{x^2 + y^2}$ and k is a constant, so that z depends only on the time and the distance to the origin. Show that z is a solution of the wave equation $z_{tt} = z_{xx} + z_{yy}$ if f satisfies

$$f''(r) + \frac{1}{r}f'(r) + k^2 f(r) = 0.$$

9.22. The wave equation for a function $u(\mathbf{X},t)$ in three space dimensions is

$$u_{tt} = c^2 \Delta u.$$

(a) Find constants k so that the function $u(\mathbf{X},t) = \cos(2x_1 + 3x_2 + 6x_3 + kt)$ is a solution of the wave equation.

(b) Find the constant vectors $\mathbf{A}$ so that C^2 functions of the form

$$u(\mathbf{X},t) = f(\mathbf{A} \cdot \mathbf{X} \pm ct)$$

are solutions of the wave equation.

Fig. 9.10 A spherically symmetric solution of the wave equation in $\mathbb{R}^3$ approaches the origin. See Problem 9.23.

9.23. Let $\rho = \sqrt{x^2 + y^2 + z^2}$ in $\mathbb{R}^3$. For waves $u(\rho,t)$ that are spherically symmetric about the origin the wave equation $u_{tt} = \Delta u$ becomes

$$u_{tt} = u_{\rho\rho} + \frac{2}{\rho} u_\rho.$$

(a) Show that C^2 functions of the form $u(\rho,t) = \dfrac{f(\rho \pm t)}{\rho}$ are solutions of the wave equation for $\rho > 0$.

(b) Suppose a solution has the form

$$u(\rho,t) = \frac{f(\rho + t)}{\rho}$$

where $f(\rho)$ is nonzero only in a small interval near $\rho = 100$, and the maximum value of $u(\rho,0)$ is 1. See Figure 9.10. Approximately what is the maximum value of $u(\rho,99)$ and where does it occur?

9.3 The conduction of heat

In this section we show that the distribution of temperature in a medium that conducts heat satisfies a partial differential equation. It is a conservation law as we have discussed in Chapter 8.

We start with the one-dimensional case, the conduction of heat in a rod. Let the rod be on the interval $[0, a]$ along the x axis. We denote by $T(x, t)$ the temperature at the position x and at time t. Assume T is a C^2 function. We assume that heat energy density is proportional to temperature, so that the heat energy in the section $[b, c]$ of the rod at time t is

$$\int_b^c p T(x, t) \, dx \tag{9.19}$$

where p is some positive constant.

Next we assume that heat energy is conducted from a hotter to a colder region, at a rate proportional to the gradient of temperature. Here "gradient" refers to the rate of change of temperature with respect to position, so in the one-dimensional case, to T_x. Therefore heat enters the section $[b, c]$ at its endpoints, at the rates

$$-r \frac{\partial T}{\partial x}(b, t) \quad \text{and} \quad r \frac{\partial T}{\partial x}(c, t) \tag{9.20}$$

where r is some positive constant. The energy conservation law states that the rate at which heat flows across the boundary of $[b, c]$ is the time derivative of the total heat energy in $[b, c]$,

$$\frac{d}{dt} \int_b^c p T(x, t) \, dx = -r \frac{\partial T}{\partial x}(b, t) + r \frac{\partial T}{\partial x}(c, t). \tag{9.21}$$

On the left side we carry out the differentiation with respect to t under the integral sign; we get

$$\int_b^c p \frac{\partial T}{\partial t} \, dx = -r \frac{\partial T}{\partial x}(b, t) + r \frac{\partial T}{\partial x}(c, t).$$

The right side is a difference of the values of the function

$$r \frac{\partial T}{\partial x}(x, t)$$

between $x = c$ and $x = b$. We express it as the integral of its derivative; we get

$$\int_b^c p \frac{\partial T}{\partial t} \, dx = \int_b^c r \frac{\partial^2 T}{\partial x^2}(x, t) \, dx.$$

We see that this is in the conservation law form of equation (8.25),

$$\int_b^c \rho_t \, dx = -\int_b^c f_x \, dx,$$

if we take $f = -rT_x$ and $\rho = pT$. We rewrite it as

$$\int_b^c \left(p\frac{\partial T}{\partial t}(x,t) - r\frac{\partial^2 T}{\partial x^2}(x,t) \right) dx = 0.$$

Since this integral is zero over all intervals $[b,c]$, it follows that the integrand is zero,

$$p\frac{\partial T}{\partial t} - r\frac{\partial^2 T}{\partial x^2} = 0.$$

We rewrite this equation as

$$T_t - hT_{xx} = 0, \tag{9.22}$$

where $h = \dfrac{r}{p}$ is a positive constant. Equation (9.22) is called the *equation of heat conduction*.

Example 9.6. Verify that the function

$$T(x,t) = e^{-ht}\sin x$$

is a solution of the heat equation (9.22). We have

$$T_t(x,t) = -he^{-ht}\sin x = -hT(x,t)$$

and since $T_{xx}(x,t) = -T(x,t)$, this is a solution. It decays toward zero as time increases, due to the exponential. See Figure 9.11.

Since $T_x = e^{-ht}\cos x$ the flow of heat energy is to the left at $x = 0$, is zero at $x = \frac{\pi}{2}$, and is to the right at $x = \pi$.   □

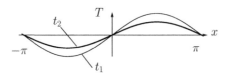

Fig. 9.11 Graphs of the heat solution $T(x,t) = e^{-ht}\sin x$ as a function of x at times $t_1 < t_2$. See Examples 9.6 and 9.7.

Example 9.7. Suppose T is a solution of the heat equation, such that at some time t the graph of T as a function of x is convex, $T_{xx} > 0$. It follows from the heat equation that $T_t > 0$. Therefore $T(x,t)$ is an increasing function of t. See the left half of Figure 9.11 where T is convex.   □

We derive now some of the basic properties of solutions of the equation of heat conduction in a rod whose endpoints are kept at some constant temperature.

Since the heat equation only involves the derivatives of T, it follows that if $T(x,t)$ is a solution of the heat equation, so is $T(x,t) - k$, where k is a constant. Choose k to be the constant temperature at the endpoints; then $T - k$ is zero at the endpoints. So it suffices to study solutions of the heat equation that are zero at the endpoints of the rod.

We show now the following property of such solutions:

Theorem 9.4. *Let T be a solution of the heat equation $T_t - hT_{xx} = 0$ for x in $[0,a]$ that is zero at the endpoints. Then the maximum of $T(x,t)$ over x in $[0,a]$ is a decreasing function of t.*

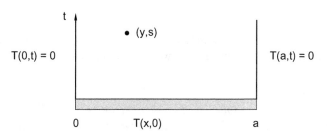

Fig. 9.12 Temperature $T(x,t)$ in a rod $0 \le x \le a$.

Proof. We show first that for $t > 0$, $T(x,t)$ is never larger than its maximum at $t = 0$. We argue indirectly; suppose for some time $s > 0$ and some y in $(0,a)$, $T(y,s)$ is larger than the maximum of $T(x,0)$. Then there are positive numbers M and ϵ such that for all x in $[0,a]$,

$$T(x,0) < M < M + \epsilon a^2 < T(y,s).$$

Let R be the rectangular set $[0,a] \times [0,s]$ and consider the function

$$u(x,t) = T(x,t) + \epsilon(x-y)^2$$

on R. Since $T(x,t)$ is zero at the endpoints and $T(x,0) < M$,

$$u(0,t) = T(0,t) + \epsilon y^2 \le \epsilon a^2$$
$$u(a,t) = T(a,t) + \epsilon(a-y)^2 \le \epsilon a^2$$
$$u(x,0) = T(x,0) + \epsilon(x-y)^2 \le M + \epsilon a^2. \qquad (9.23)$$

Since

$$u(y,s) = T(y,s) > M + \epsilon a^2$$

the maximum of u on R is at least $T(y,s)$, and according to (9.23) it must occur at some point (x,t) with $t > 0$ and x not at either endpoint of the rod. Therefore at this maximum we have $u_{xx} \le 0$ and $u_t \ge 0$. Then

$$0 \geq u_{xx} - \frac{1}{h}u_t = T_{xx} + 2\epsilon - \frac{1}{h}T_t = 2\epsilon.$$

This is a contradiction since $\epsilon > 0$.

To complete the proof suppose $0 < t_1 < t_2$ and set $v(x,t) = T(x, t_1 + t)$. Then v is a solution of the heat equation. Let $v(c,0)$ be the maximum of $v(x,0)$ on $[0,a]$. Then

$$\max_x T(x, t_2) = \max_x v(x, t_2 - t_1)$$

$$\leq v(c,0) = T(c, t_1)$$

$$\leq \max_x T(x, t_1).$$

$\square$

Since $-T(x,t)$ is a solution as well of the heat equation that is zero at $x = 0$ and $x = a$, it follows that the minimum value of $T(x,t)$ over all x is an increasing function of t. Combining the two results we deduce from Theorem 9.4 the following corollary.

Corollary 9.1. *Let T be a solution of the heat equation $T_t - hT_{xx} = 0$ for x in $[0,a]$ that is zero at the endpoints. Then the maximum with respect to x of $|T(x,t)|$ is a decreasing function of t.*

We can now prove the *uniqueness theorem*:

Theorem 9.5. *Suppose T_1 and T_2 are two solutions of the heat equation in $[0,a]$, $t \geq 0$, that are equal at $t = 0$, equal at the endpoint $x = 0$ and equal at the endpoint $x = a$. Then $T_1(x,t) = T_2(x,t)$ for all $t > 0$ and x in $[0,a]$.*

Proof. Set $T = T_1 - T_2$. Then

$$T_t - hT_{xx} = 0$$

$$T(x,0) = 0$$

$$T(0,t) = 0$$

$$T(a,t) = 0.$$

According to Corollary 9.1, $|T(x,t)|$ decreases from its initial value. But that is zero. So T is identically zero, and $T_1 = T_2$.

$\square$

Examples of the heat equation in higher dimensions The conduction of heat in a plate can be analyzed similarly to the way we analyzed the conduction of heat in a rod; we look at the flow of heat into a small portion of the plate $b \leq x \leq c$, $d \leq y \leq e$ across its boundary. We obtain an equation analogous to $T_t - hT_{xx} = 0$,

$$T_t - h\Delta T = 0, \tag{9.24}$$

where $\Delta T = T_{xx} + T_{yy}$. In Problem 9.31 we outline a derivation of this equation using the Divergence Theorem.

Example 9.8. Define $T(x,y,t) = e^{-at}\sin(bx+cy)$. We find the relation of a, b, c so that T is a solution of the equation $T_t - h\Delta T = 0$.

$$T_t - h\Delta T = -ae^{-at}\sin(bx+cy) - he^{-at}(-b^2 - c^2)\sin(bx+cy)$$

$$= (-a + (b^2 + c^2)h)T.$$

That is zero if $a = (b^2 + c^2)h$. So

$$e^{-(b^2+c^2)ht}\sin(bx+cy)$$

is a solution of the heat equation for all numbers b, c. □

Example 9.9. You have seen in Problem 4.7 that the function

$$T(\mathbf{X}, t) = (4\pi t)^{-n/2}e^{-\|\mathbf{X}\|^2/(4t)}$$

is a solution of the heat equation in n space dimensions,

$$T_t - \Delta T = 0.$$

□

Problems

9.24. Show that these functions are solutions of the heat equation $T_t = hT_{xx}$.

(a) $mx + k$ for all constants m, k
(b) $T_1 + T_2$ if T_1 and T_2 are solutions.
(c) $e^{-ht}\cos(x)$
(d) some functions $e^{-kt}\sin(mx)$; find the relation between k and m,
(e) $u(x, ht)$ where $u(x, t)$ is any solution of $u_t = u_{xx}$.

9.25. Show that these functions are solutions of the heat equation $T_t = T_{xx}$.

(a) $e^{-n^2 t}\sin(nx)$ for $n = 1, 2, 3, \ldots$,
(b) $t^p e^{-x^2/(4t)}$ for a certain exponent p; find p,
(c) $e^{-ax}\cos(ax - bt)$ for some constants a, b; find the relation between a and b.
(d) $u(kx, k^2 t)$ for any constant k if u is a solution.

9.26. The temperature at a moderate distance x below the ground is modeled as

$$T(x,t) = e^{-ax}\cos(ax - bht)$$

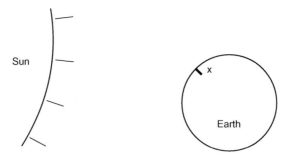

Fig. 9.13 Earth in Problem 9.26.

where the value of the constant b depends on whether we are discussing daily (sunrise, sunset) or seasonal (winter, summer) variations. See Figure 9.13.

(a) Show that $T_t = hT_{xx}$ if $b = 2a^2$.
(b) Is b larger for daily or seasonal variations?
(c) Considering the factor e^{-ax}, do daily or seasonal variations penetrate deeper into Earth?

9.27. A rod is held at zero temperature at the ends, so that

$$T_t = T_{xx}, \quad T(t,0) = T(t,\pi) = 0.$$

(a) Fill in the missing numbers so that

$$T(x,t) = (?)e^{(?)t}\sin x + (?)e^{(?)t}\sin(2x)$$

is a solution with initial value $T(x,0) = \sin x + \frac{1}{2}\sin(2x)$.
(b) Sketch the graph of T as function of x at times $t = 0, \frac{1}{2}, 1$.
(c) Which of the two terms decreases faster as t increases? Note the maximum temperature is located left of the center point of the rod initially. Does the hot spot move to the left or right as time increases?

9.28. We can calculate approximate solutions of the heat equation $T_t = T_{xx}$ by approximating the derivatives by difference quotients, where Δt and Δx are positive numbers,

$$\frac{T(x,t+\Delta t) - T(x,t)}{\Delta t} = \frac{T(x+\Delta x,t) - 2T(x,t) + T(x-\Delta x,t)}{(\Delta x)^2}.$$

(a) Solve that approximation for $T(x,t+\Delta t)$ in terms of the earlier values at time t.
(b) Take $\frac{\Delta t}{(\Delta x)^2} = \frac{1}{2}$. Show that the approximation becomes

$$T(x,t+\Delta t) = \frac{1}{2}(T(x-\Delta x,t) + T(x+\Delta x,t)).$$

(c) Initial values $T(x,0)$ are given in the table. Fill in values of T for five time steps using the approximation in (b). One of the values, $T(-\Delta x, \Delta t) = \frac{1}{2}(16+0) = 8$, is indicated as an example.

	$-6\Delta x$	$-5\Delta x$	$-4\Delta x$	$-3\Delta x$	$-2\Delta x$	$-\Delta x$	0	Δx	$2\Delta x$	$3\Delta x$	$4\Delta x$	$5\Delta x$	$6\Delta x$	$7\Delta x$
$5\Delta t$														
$4\Delta t$														
$3\Delta t$														
$2\Delta t$														
Δt						8								
0	0	0	0	0	0	0	16	16	0	0	0	0	0	0

9.29. A simple model for temperature $y(t)$ of an object at time t in an environment of temperature s is Newton's Law of Cooling,

$$y' = -k(y - s)$$

where k is a positive constant. Consider two heat equation solutions

$$T_1(x,t) = e^{-t}\sin(x), \qquad T_2(x,t) = e^{-t}\sin(x) + e^{-9t}\sin(3x)$$

for a rod located in the interval $[0,\pi]$. We imagine the environment $s = 0$, and define the average temperature of the rods to be

$$y_1(t) = \frac{1}{\pi}\int_0^\pi T_1(x,t)\,dx, \qquad y_2(t) = \frac{1}{\pi}\int_0^\pi T_2(x,t)\,dx.$$

Show that Newton's Law of Cooling holds for y_1 but not for y_2.

9.30. A steady state solution of $T_t = hT_{xx}$ is one that does not vary with time, thus $T_{xx} = 0$.

(a) Find a steady state temperature function $T(x,t)$ for $0 \le x \le a$ with $T(0,t) = 50$ and $T(10,t) = 100$.
(b) Verify that for your steady state function, the total rate that heat enters (see equation (9.20)) at left and right ends is zero,

$$-r\frac{\partial T}{\partial x}(0,t) + r\frac{\partial T}{\partial x}(a,t) = 0.$$

(c) Verify that the total heat energy (see equation (9.19)) for your steady state function,

$$\int_0^a pT(x,t)\,dx$$

is independent of the time.

9.31. In this problem you can derive the heat equation in 2 space variables

$$T_t - h\Delta T = 0.$$

The analogue of the heat conservation law (9.21) states that for all regular sets D in the plane,

$$\frac{d}{dt} \int_D pT(x,y,t)\,dA = r \int_{\partial D} \nabla T \cdot \mathbf{N}\,ds.$$

Justify the following items.

(a) The rate of change of total energy is $\int_D pT_t\,dA$.

(b) $\int_D pT_t\,dA = \int_{\partial D} \operatorname{div}(r\nabla T)\,dA$.

(c) $\int_D (pT_t - r\varDelta T)\,ds = 0$.

(d) $T_t - h\varDelta T = 0$, where $h = \frac{r}{p}$.

9.32. Suppose a solution of the two-dimensional heat equation has the form

$$T(x,y,t) = e^{-cht} f(r),$$

where r is the polar coordinate $r = \sqrt{x^2 + y^2}$. Thus T only depends on time and on the distance to the origin. Substitute into $T_t = h\varDelta T$ and use the Chain Rule to show that f satisfies

$$f''(r) + \frac{1}{r}f'(r) + cf(r) = 0.$$

9.33. Show that $T(x,y,t) = x^2 + y^2 + 4ht$ is a solution of the two-dimensional heat equation $T_t = h\varDelta T$, and describe the direction of heat conduction.

9.4 Equilibrium

In Section 9.2 we derived the differential equation governing a vibrating membrane,

$$z_{tt} = c^2(z_{xx} + z_{yy}),$$

and in Section 9.3 we derived the differential equation governing the flow of heat in a plate,

$$T_t = h(T_{xx} + T_{yy}).$$

We consider now the case of equilibrium, that is membranes in which the elastic forces are so balanced that they do not vibrate, and heat-conducting bodies in which the temperature is so balanced that it does not change.

The equations of equilibrium can be obtained from the equations of time change by simply setting the time derivatives in these equations equal to zero. So we obtain from the equation of of a vibrating membrane the equilibrium equation

$$z_{xx} + z_{yy} = 0,$$

and from the equation of heat conduction the equilibrium equation

$$T_{xx} + T_{yy} = 0.$$

We observe, with some astonishment, that except for the symbols used, these equations are *identical*. The equation

$$\Delta u = 0$$

is called the *Laplace equation* and the solutions are known as *harmonic* functions. There is no physical reason why the equilibrium of an elastic membrane and the equilibrium of heat distribution should be governed by the same equation, but they are, and so

Their mathematical theory is the same.

This is what makes mathematics a universal tool in dealing with problems of science.

We state and prove an important property of solutions of the Laplace equation.

Theorem 9.6. *Let u and v be two solutions of the Laplace equation on a connected regular set D in $\mathbb{R}^2$ that are equal on the boundary of D. Then u and v are equal in D.*

Another way of expressing this theorem is: Solutions of the Laplace equation in a regular set D in $\mathbb{R}^2$ are uniquely determined by their values on the boundary of D. We give now a mathematical proof of this proposition.

Proof. Denote by z the difference of u and v,

$$z(x,y) = u(x,y) - v(x,y).$$

Since u and v are solutions of the Laplace equation, their difference z is a solution. Since u and v are equal on the boundary of D, z is zero on the boundary. Multiply the Laplace equation $\Delta z = 0$ by z and integrate the product over D. We get

$$\int_D z\Delta z \, dx\, dy = \int_D z(z_{xx} + z_{yy}) \, dx\, dy = 0. \tag{9.25}$$

Using the product rule for div we get

$$\operatorname{div}(z\nabla z) = z\Delta z + \|\nabla z\|^2. \tag{9.26}$$

Therefore

$$0 = \int_D z\Delta z \, dx\, dy = \int_D (\operatorname{div}(z\nabla z) - \|\nabla z\|^2) \, dx\, dy.$$

Since $z = 0$ on ∂D we get from the Divergence Theorem that

$$0 = \int_{\partial D} z\nabla z \cdot \mathbf{N} \, ds - \int_D \|\nabla z\|^2 \, dx\, dy = -\int_D \|\nabla z\|^2 \, dx\, dy.$$

Therefore

$$-\int_D (z_x^2 + z_y^2)\,dx\,dy = 0. \tag{9.27}$$

One says that (9.27) has been obtained from (9.25) using "integration by parts" and the product rule for div in (9.26).

The integrand in (9.27) is a sum of squares and therefore nonnegative. Since the integral is zero, so is the integrand; therefore

$$z_x = 0, \qquad z_y = 0 \qquad \text{in } D.$$

A function z whose partial derivatives are zero in D is a constant in D; since z is zero on the boundary, that constant is zero. Since $z = u - v$, this proves that u and v are equal in D, as claimed.   □

We observe that the uniqueness result in Theorem 9.6 is intuitively clear if we interpret z as the displacement of an elastic membrane. For if the boundary of the membrane is constrained to lie in the plane $z = 0$, then the whole membrane will lie in that plane.

We note that Theorem 9.6 and its proof can be extended to functions of three, or any number, of variables.

A further basic result about the Laplace equation in D is that given any continuous function on the boundary of D, there is a solution of the Laplace equation in D with these prescribed boundary values. The result is plausible, for if we stretch a membrane over a frame on the boundary of D, the membrane will take on a shape in equilibrium. The proof of this proposition is beyond the scope of a calculus book.

Problems

9.34. Show that these functions are solutions of the Laplace equation $u_{xx} + u_{yy} = 0$.

(a) $x^2 - y^2$
(b) $x^3 - 3xy^2$
(c) $e^{-ax}\sin(by)$ for some constants a, b; find the relation between a and b.

9.35. Show that these functions are solutions of the Laplace equation $u_{xx} + u_{yy} = 0$.

(a) $u_1 + u_2$ if u_1 and u_2 are solutions
(b) $u(x\cos\theta - y\sin\theta, x\sin\theta + y\cos\theta)$ if u is a solution and θ is a constant angle, i.e., rotate the solution.
(c) the product uv of two solutions, if the gradients ∇u and ∇v are orthogonal.

9.36. Figure 9.14 shows a region bounded by two level sets of T and two curves tangent to ∇T, for a function T with $T_{xx} + T_{yy} = 0$. Justify the following items.

(a) The curves are orthogonal at the corners.
(b) The flux of ∇T toward the right is the same across the two level sets.

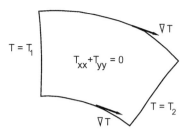

Fig. 9.14 The region in Problem 9.36.

(c) The flux of ∇T is zero across each of the other two curves.

9.37. Show that $z(x, y) = \log(x^2 + y^2)$ is a solution of the Laplace equation $\Delta z = 0$.

9.38. Recall that the Laplacian of f, Δf in any number of dimensions is $\operatorname{div} \nabla f$. Show that $\Delta(\|\mathbf{X}\|^{-1})$ is zero in three dimensions, but not in 2 or 4 dimensions.

9.39. Suppose u, v and w are C^2 functions in a regular set D in $\mathbb{R}^3$. Derive item (a) from the Divergence Theorem for the vector field $w \nabla w$, and justify the following items.

(a) If $w = 0$ on the boundary of D, then $\int_D (\Delta w) w \, dV = - \int_D \nabla w \cdot \nabla w \, dV$.

(b) If u solves Laplace's equation

$$\Delta u = 0 \text{ in } D$$
$$u = 0 \text{ on } \partial D$$

then $u(x, y, z) = 0$ at all points of D.

(c) Suppose u and v are two solutions of the Laplace equation

$$\Delta u = 0, \qquad \Delta v = 0$$

in a regular set D that are equal on the boundary of D. Show that u and v are equal in D. [Hint: apply item (b) to $u - v$.]

9.40. We derived the wave equation (9.17) from Newton's $F = ma$ law of motion, assuming that the only forces on a portion of the membrane were due to the tension. Suppose instead that the membrane is in equilibrium subject to a uniform pressure p [force/area] on the top surface. See Figure 9.15. Then the sum of edge forces on the portion $a \leq x \leq a + h$, $b \leq y \leq b + h$ are balanced by the force due to p. Justify the following statements.

(a) The upward force is $-ph^2$ when the slopes z_x, z_y are small.
(b) z satisfies the differential equation $z_{xx} + z_{yy} = \frac{p}{T}$ where T is the tension.

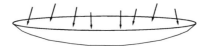

Fig. 9.15 Pressure above a membrane in Problem 9.40.

9.41. Show that the following functions are solutions of a pressurized membrane equation $z_{xx} + z_{yy} = 1$ as in Problem 9.40

(a) $\frac{1}{4}(x^2 + y^2)$
(b) $\frac{1}{6}x^2 + \frac{1}{3}y^2$
(c) $kz + (1-k)w$, if z, w are solutions and k constant,
(d) $z + w$, if z is a solution and w satisfies the Laplace equation $w_{xx} + w_{yy} = 0$.

9.42. Here $\Delta u = u_{xx} + u_{yy}$. Let $h > 0$, and denote the values of u at four points of the compass as

$$u_E = u(x+h, y)$$
$$u_S = u(x, y-h)$$
$$u_W = u(x-h, y)$$
$$u_N = u(x, y+h).$$

See Figure 9.16.

(a) Use Taylor's theorem to show that

$$\Delta u(x, y) = \frac{1}{h^2}(u_E + u_S + u_W + u_N - 4u(x, y)) + O(h^2).$$

(b) We use part (a) to approximate a solution of the equation

$$\Delta u = 1$$

in the square $0 < x < 1$, $0 < y < 1$ shown on the right side of the figure, with boundary values indicated. Use the approximation

$$\Delta u(x, y) \approx \frac{1}{h^2}(u_E + u_S + u_W + u_N - 4u(x, y))$$

to set up a system of linear equations for values u_1, u_2, u_3, u_4 at the indicated points.

9.43. Let $z(x, y) = \sin(nx)\sinh(ny)$, where n is a positive integer.

(a) Verify that $\Delta z = 0$.
(b) Verify that $z = 0$ on the boundary of the region $0 < x < \pi$, $y > 0$.
(c) Show that $z(x, y)$ is unbounded.

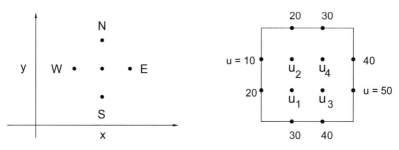

Fig. 9.16 Notation for Problem 9.42. *Left:* part (a) *Right:* part (b).

Remark: This does not happen with the heat and wave equations, and shows that we need to specify some boundary values all the way around the boundary for the Laplace equation. Or, as we may say, an initial condition cannot be specified.

9.44. We consider the vector field ∇u where u is a solution of the Laplace equation.

(a) If $\Delta u = 0$, show that ∇u has divergence equal to zero.
(b) Show that the functions $x^2 - y^2$ and $2xy$ are solutions of the Laplace equation.
(c) Show that the vector fields $\mathbf{F} = \nabla(x^2 - y^2)$ and $\mathbf{G} = \nabla(2xy)$ have divergence zero and are orthogonal to each other at each point (x, y).
(d) Show that at each point (x, y), the vectors $\mathbf{F}(x, y)$ and $\mathbf{G}(x, y)$ have the same length.
(e) Sketch the vector fields $\mathbf{F}$ and $\mathbf{G}$. These are simple models for the velocity of an incompressible fluid flow.

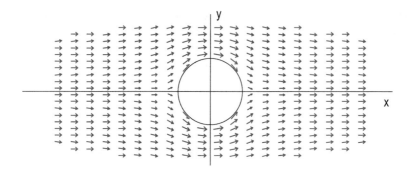

Fig. 9.17 The vector field in Problem 9.45 models incompressible fluid flow around a cylinder.

9.45. Let $u(x, y) = x + \dfrac{x}{x^2 + y^2}$.

(a) Show that u is a solution of the Laplace equation; therefore the vector field $\mathbf{F} = \nabla u$ has divergence equal to zero. See Figure 9.17.
(b) Show that $\mathbf{F}$ is tangent to the unit circle, that is, for all points on $x^2 + y^2 = 1$,

$$(x, y) \cdot \mathbf{F}(x, y) = 0.$$

(c) Show that as $x^2 + y^2$ tends to infinity, $\mathbf{F}(x,y)$ tends to $(1,0)$.

9.46. Suppose $u(x,y,z)$ is a solution of the Laplace equation in $\mathbb{R}^3$

$$u_{xx} + u_{yy} + u_{zz} = 0.$$

Define a vector field $\mathbf{F} = \nabla u$. Show that $\operatorname{div} \mathbf{F} = 0$ and $\operatorname{curl} \mathbf{F} = \mathbf{0}$.

9.5 The Schrödinger equation

The Schrödinger equation is the basic equation of quantum mechanics. It is for complex valued functions $\psi = f + \mathrm{i}g$ where f and g are real valued functions of $\mathbf{X} = (x,y,z)$ and of t, and $\mathrm{i}^2 = -1$. The function $\overline{\psi} = f - \mathrm{i}g$ is called the complex *conjugate* of ψ. Complex valued functions can be differentiated by treating i as a constant.

The Schrödinger equation is of the form

$$\mathrm{i}\psi_t = -\Delta\psi + V\psi \tag{9.28}$$

where V is a real valued function with the property that $V(\mathbf{X})$ tends to zero rapidly as $\|\mathbf{X}\|$ tends to infinity. We consider solutions ψ that tend to zero rapidly as $\|\mathbf{X}\|$ tends to infinity.

The physics interpretation of solutions of Schrödinger's equation is based on the following property of its solutions.

Theorem 9.7. *Let ψ be a solution of Schrödinger's equation whose partial derivatives tend to zero rapidly as $\|\mathbf{X}\|$ tends to infinity. Then the integral*

$$\int_{\mathbb{R}^3} |\psi(\mathbf{X},t)|^2 \, \mathrm{d}^3\mathbf{X} \tag{9.29}$$

is independent of t.

Proof. We write $\displaystyle\int_{\mathbb{R}^3} |\psi(\mathbf{X},t)|^2 \, \mathrm{d}^3\mathbf{X} = \int_{\mathbb{R}^3} \psi(\mathbf{X},t)\overline{\psi}(\mathbf{X},t) \, \mathrm{d}^3\mathbf{X}$ and differentiate with respect to t. Since ψ and its t derivative tend to zero rapidly as $\|\mathbf{X}\|$ tends to infinity, we have

$$\frac{\mathrm{d}}{\mathrm{d}t} \int_{\mathbb{R}^3} |\psi(\mathbf{X},t)|^2 \, \mathrm{d}^3\mathbf{X} = \int_{\mathbb{R}^3} \frac{\mathrm{d}}{\mathrm{d}t}\left(\psi(\mathbf{X},t)\overline{\psi}(\mathbf{X},t)\right) \mathrm{d}^3\mathbf{X}$$

$$= \int_{\mathbb{R}^3} \left(\psi_t(\mathbf{X},t)\overline{\psi}(\mathbf{X},t) + \psi(\mathbf{X},t)\overline{\psi}_t(\mathbf{X},t)\right) \mathrm{d}^3\mathbf{X}.$$

We use the Schrödinger equation to express ψ_t and $\overline{\psi}_t$. Since V is real, the integral on the right becomes

$$\int_{\mathbb{R}^3} \left(-\mathrm{i}(-\Delta\psi + V\psi)\overline{\psi} + \psi\mathrm{i}(-\Delta\overline{\psi} + V\overline{\psi}) \right) \mathrm{d}^3\mathbf{X} = \mathrm{i}\int_{\mathbb{R}^3} \left((\Delta\psi)\overline{\psi} - \psi\Delta\overline{\psi} \right) \mathrm{d}^3\mathbf{X}$$

Since ψ and its first partial derivatives with respect to x, y and z tend to zero rapidly as $\mathbf{X}$ tends to infinity, we show in Problem 9.50 that the integral on the right side is zero. It follows that

$$\frac{\mathrm{d}}{\mathrm{d}t}\int_{\mathbb{R}^3} |\psi(\mathbf{X},t)|^2\,\mathrm{d}^3\mathbf{X} = 0.$$

This proves Theorem 9.7. □

We give now the physical interpretation of Theorem 9.7. Suppose that at $t = 0$ the function ψ satisfies

$$\int_{\mathbb{R}^3} |\psi(\mathbf{X},0)|^2\,\mathrm{d}^3\mathbf{X} = 1.$$

It follows that for all t

$$\int_{\mathbb{R}^3} |\psi(\mathbf{X},t)|^2\,\mathrm{d}^3\mathbf{X} = 1.$$

Let

$$|\psi(\mathbf{X},t)|^2 = p(\mathbf{X},t).$$

For each t, $p(\mathbf{X},t)$ is a nonnegative function on $\mathbb{R}^3$ whose integral over $\mathbb{R}^3$ is 1. Such a function is a probability density function. Suppose p is integrable with respect to $\mathbf{X}$ on a set S in $\mathbb{R}^3$. Then

$$P(S,t) = \int_S p(\mathbf{X},t)\,\mathrm{d}^3\mathbf{X},$$

is the probability associated with the set S at time t. What is the physical interpretation of this probability? According to quantum mechanics, $P(S,t)$ is the probability that a particle governed by the Schrödinger equation is located at time t in the set S.

This formulation is a radical philosophical departure from the Newtonian picture; instead of having a definite position in space, at each instant of time there is only a probability of a particle's location. Many physicists had to struggle to accept such a probabilistic description, Einstein among them. He famously remarked "God does not play dice with the Universe." But the great success of quantum mechanics has led to the universal acceptance by physicists of the probabilistic interpretation of solutions of the Schrödinger equation.

Problems

9.47. Let $\phi(\mathbf{X})$ be a solution of the equation

$$E\phi = -\Delta\phi + V\phi$$

where E is a real number.

(a) Show that the function

$$\psi(\mathbf{X},t) = e^{-iEt}\phi(\mathbf{X})$$

is a solution of Schrödinger's equation (9.28).

(b) Suppose that $\int_{\mathbb{R}^3} |\phi(\mathbf{X})|^2 \, d^3\mathbf{X} = 1$. Show that the probability that the particle is located in a smoothly bounded set S is

$$P(S,t) = \int_S |\phi(\mathbf{X})|^2 \, d^3\mathbf{X},$$

independent of the time.

9.48. Define the function $\phi(\mathbf{X}) = \pi^{-1/2}z e^{-\|\mathbf{X}\|}$, where $\mathbf{X} = (x,y,z)$.

(a) Show that

$$-\phi = -\varDelta\phi + \frac{-4}{\|\mathbf{X}\|}\phi.$$

(b) Define $\psi(\mathbf{X},t) = e^{it}\phi(\mathbf{X})$ as in Problem 9.47, so that ψ is a solution of the Schrödinger equation with $V(\mathbf{X}) = \dfrac{-4}{\|\mathbf{X}\|}$. We have shown in Example 6.41 using spherical coordinates that

$$|\phi(\mathbf{X})|^2$$

is a probability density function. Set up an iterated integral for the probability that the particle described by ψ is in the set S given by $\|\mathbf{X}\| \leq 3$.

Figure 9.18 illustrates a level set of the probability density in (b).

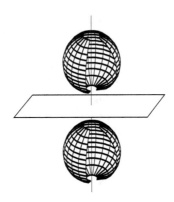

Fig. 9.18 Sketch of a level set in Problem 9.48. The plane $z = 0$ is indicated for reference.

9.49. Take $V(x,y,z) = x^2 + y^2 + z^2$ in the Schrödinger equation, and consider the corresponding equation

$$E\phi = -\varDelta\phi + (x^2 + y^2 + z^2)\phi. \tag{9.30}$$

(a) Suppose X, Y and Z are functions of x, y and z respectively, and

$$E_x X = -X'' + x^2 X$$
$$E_y Y = -Y'' + y^2 Y$$
$$E_z Z = -Z'' + z^2 Z$$

for some numbers E_x, E_y and E_z. Show that the function $\phi(x,y,z) = X(x)Y(y)Z(z)$ satisfies (9.30) for some number E expressed in terms of E_x, E_y and E_z.

(b) For each of the numbers $E_w = 1, 3$ and 5, find a function of the form

$$W(w) = (a + bw + cw^2)e^{-w^2/2}$$

that satisfies $E_w W = -W'' + w^2 W$.

(c) Using functions from (b), create functions ϕ that satisfy (9.30) with energy levels

$$E = 3,5,7,9,11,13 \text{ and } 15.$$

(d) Show that the probability density $|e^{-iEt}\phi(\mathbf{X})|^2$ associated with the $E = 3$ case is

$$|\phi|^2 = \pi^{-3/2}e^{-x^2-y^2-z^2}.$$

9.50. The integral of a complex valued function $f = a + ib$, where a and b are real valued functions, is the integral of a plus i times the integral of b.

(a) Let $\mathbf{F} = (f_1, f_2, f_3)$ have complex components f_j that are C^1 functions on $\mathbb{R}^3$, that is, the real and imaginary parts of each f_j are C^1. Then $\operatorname{div}\mathbf{F}$ and $\operatorname{curl}\mathbf{F}$ are complex. Show that the Divergence Theorem 8.3 and Stokes' Theorem 8.4 hold for complex valued functions.

(b) Let f and g be complex valued functions on $\mathbb{R}^3$ that are C^2 functions, that is, the real and imaginary parts of f and g are C^2. Then ∇f and ∇g have complex components and Δf and Δg are complex. Prove the formula that we used in the proof of Theorem 9.7: if f, g, ∇f and ∇g tend rapidly to zero as $\mathbf{X}$ tends to infinity then

$$\int_{\mathbb{R}^3} (f\Delta g - g\Delta f)\,d^3\mathbf{X} = 0.$$

Answers to selected problems

Problems of Chapter 1

Section 1.1

1.1 3.5

1.3

(a) $a(1,-1) + b(1,1) = (a+b, -a+b) = (0,0)$.
Add the equations $a+b=0$ and $-a+b=0$ to get $2b=0$, so $a=b=0$. Therefore the vectors are linearly independent.

(b) $a(1,-1) + b(1,1) = (a+b, -a+b) = (2,4)$.
Add the equations $a+b=2$ and $-a+b=4$ to get $2b=6$, so $b=3$. Then the first equation gives $a+3=2$ so $a=-1$.

(c) $a(1,-1) + b(1,1) = (a+b, -a+b) = (x,y)$. Now $2b = x+y$ so $b = \frac{1}{2}(x+y)$. Then the first equation gives $a + \frac{1}{2}(x+y) = x$, so $a = \frac{1}{2}(x-y)$.

1.5 $\ell(x,y) = ax + by$. We need $a + 2b = 3$ and $2a + 3b = 5$. Subtract twice the first from the second to eliminate a, giving $-b = -1$. Then the first equation gives $a = 1$. So $\ell(x,y) = x + y$.

1.7

(a) The line through $\mathbf{0}$ and $\mathbf{U}$ consists of all points $c\mathbf{U}$; that through $\mathbf{V}$ and $\mathbf{U}+\mathbf{V}$ is all $\mathbf{V} + d\mathbf{U}$. They don't intersect because $c\mathbf{U} = \mathbf{V} + d\mathbf{U}$ gives a nontrivial linear combination. Therefore they are parallel. Alternatively the line through $\mathbf{0}$ and $\mathbf{U}$ has slope $\dfrac{u_2}{u_1}$ (or is vertical if $u_1 = 0$), and the line through $\mathbf{V}$ and $\mathbf{U}+\mathbf{V}$ has slope $\dfrac{(u_2 + v_2) - v_2}{(u_1 + v_1) - v_1}$ (or is vertical). These are equal, so those two sides are parallel.

(b) Similar to (a).

1.9 $\mathbf{W} = -\mathbf{U} - \mathbf{V}$, so $\mathbf{U} + \mathbf{V} + \mathbf{W} = \mathbf{0}$.

© Springer International Publishing AG 2017
P. D. Lax and M. S. Terrell, *Multivariable Calculus with Applications*,
Undergraduate Texts in Mathematics, https://doi.org/10.1007/978-3-319-74073-7

1.11 $\ell(\mathbf{U}+\mathbf{V}) = (u_1+v_1)-8(u_2+v_2)$, while $\ell(\mathbf{U})+\ell(\mathbf{V}) = (u_1-8u_2)+(v_1-8v_2)$, and these are equal. $\ell(c\mathbf{U}) = cu_1-8cu_2$, while $c\ell(\mathbf{U}) = c(u_1-8u_2)$, and these are equal. So ℓ is linear.

1.13 $a+3b = 4$, $3a+b = 5$

1.15 $(4,8) = 4(1,2) = 2(2,4) = -3(1,2) + \frac{7}{2}(2,4)$ and many other combinations $\mathbf{U}$ and $\mathbf{V}$ are linearly dependent.

1.17

(a) Rotation carries $\mathbf{U}$ to $\mathbf{V}$, $\mathbf{V}$ to $\mathbf{W}$, and $\mathbf{W}$ to $\mathbf{U}$, so it carries $\mathbf{U}+\mathbf{V}+\mathbf{W}$ to itself. No vector except $\mathbf{0}$ is rotated to itself, therefore $\mathbf{U}+\mathbf{V}+\mathbf{W} = \mathbf{0}$.
(b) The sines are the y coordinates of $\mathbf{U}$, $\mathbf{V}$, and $\mathbf{W}$ where θ is the angle from the x axis to one of the vectors.
(c) The sum of n equally spaced vectors on the unit circle is the zero vector by the same rotation argument; the cosines are the x coordinates and therefore their sum is zero.

1.19 $f(.5,0) = -f(.5,0) = -100$

Section 1.2

1.21

(a) $\mathbf{U}\cdot(\mathbf{V}+\mathbf{W}) = u_1(v_1+w_1)+u_2(v_2+w_2)$ is equal to
$\quad \mathbf{U}\cdot\mathbf{V}+\mathbf{U}\cdot\mathbf{W} = (u_1v_1+u_2v_2)+(u_1w_1+u_2w_2)$.
(b) $\mathbf{U}\cdot\mathbf{V} = u_1v_1+u_2v_2 = v_1u_1+v_2u_2 = \mathbf{V}\cdot\mathbf{U}$

1.23 All but (d) because all have norm one except for the last one.

1.25 If $\ell(x,y) = ax+by$ then we are given $2a+b = 3$, $a+b = 2$. Subtract to find $a = 1$, then $b = 1$, so $\mathbf{C} = (1,1)$, $\ell(\mathbf{U}) = \mathbf{C}\cdot\mathbf{U}$.

1.27 Replace $\mathbf{V}$ by $-\mathbf{V}$ in (1.8), or expand $(u_1+v_1)^2+(u_2+v_2)^2 = u_1^2+2u_1v_1+v_1^2+u_2^2+2u_2v_2+v_2^2$. Collecting terms gives $\|\mathbf{U}+\mathbf{V}\|^2 = \|\mathbf{U}\|^2+2\mathbf{U}\cdot\mathbf{V}+\|\mathbf{V}\|^2$.

1.29

(a) $\mathbf{U}\cdot\mathbf{C} = a\mathbf{C}\cdot\mathbf{C}+b\mathbf{D}\cdot\mathbf{C} = a\mathbf{C}\cdot\mathbf{C}$, so $a = \dfrac{\mathbf{C}\cdot\mathbf{U}}{\|\mathbf{C}\|^2}$.

(b) By similar argument, or interchange the symbols, $b = \dfrac{\mathbf{D}\cdot\mathbf{U}}{\|\mathbf{D}\|^2}$.

(c) $(\frac{3}{5},\frac{4}{5})$ and $(-\frac{4}{5},\frac{3}{5})$ are orthogonal unit vectors. Therefore $a = \frac{24+36}{5} = 12$.

1.31 $\mathbf{U} = (-\frac{1}{\sqrt{2}},\frac{1}{\sqrt{2}})$, $\mathbf{V} = (2\sqrt{2},0)$.

Section 1.3

1.33 $b(\mathbf{U}+\mathbf{W},\mathbf{V}) = (u_1+w_1)v_1 = u_1v_1+w_1v_1 = b(\mathbf{U},\mathbf{V})+b(\mathbf{W},\mathbf{V})$, and $b(a\mathbf{U},\mathbf{V}) = (au_1)v_1 = a(u_1v_1) = b(\mathbf{U},\mathbf{V})$.

1.35 Because of the term qr, the variables q and r are not both in the same vector. Similarly r and p are not in the same vector.

Set $\mathbf{U} = (q, p) = (u_1, u_2)$ and $\mathbf{V} = (r, s) = (v_1, v_2)$. Then

$$f(p, q, r, s) = qr + 3rp - sp = u_1 v_1 + 3u_2 v_1 - u_2 v_2 = b(\mathbf{U}, \mathbf{V}).$$

Section 1.4

1.37

(a) $(v_1 + w_1, \ldots) = (w_1 + v_1, \ldots)$ because addition of numbers is commutative

(b) $((v_1 + u_1) + w_1, \ldots) = (v_1 + (u_1 + w_1), \ldots)$ because addition of numbers is associative

(c) $c(u_1 + v_1, \ldots) = (c(u_1 + v_1), \ldots) = (cu_1 + cv_1, \ldots) = (cu_1, \ldots) + (cv_1, \ldots) = c\mathbf{U} + c\mathbf{V}$

(d) $(c + d)(u_1, \ldots) = ((c + d)u_1, \ldots) = (cu_1 + du_1, \ldots) = (cu_1, \ldots) + (dv_1, \ldots) = c\mathbf{U} + d\mathbf{U}.$

1.39 $\mathbf{X} = c_1 \mathbf{U}_1 + c_2 \mathbf{U}_2 + c_3 \mathbf{U}_3 = (x_1, x_2, x_3) = (c_1 + c_2 + c_3, c_2 + c_3, c_3)$ gives first $c_3 = x_3$, then $c_2 = x_2 - x_3$, $c_1 = x_1 - x_2$.

1.41 If

$$c_1(1, 1, 1, 1) + c_2(0, 1, 1, 1) + c_3(0, 0, 1, 1) + c_4(0, 0, 0, 1)$$

$$= (c_1, c_1 + c_2, c_1 + c_2 + c_3, c_1 + c_2 + c_3 + c_4) = (0, 0, 0, 0)$$

then the first component gives $c_1 = 0$, the second then gives $c_2 = 0$, etc. Therefore the vectors are linearly independent.

1.43 If $a(3, 7, 6, 9, 4) + b(2, 7, 0, 1, -5) = \left(-\frac{1}{2}, -\frac{7}{2}, 3, \frac{7}{2}, 7\right)$ then the third component, $6a = 3$, shows that a must be $\frac{1}{2}$. Then the last component, $2 - 5b = 7$ gives $b = -1$. Checking the other three components shows that indeed $\frac{1}{2}\mathbf{U} - \mathbf{V} = \left(-\frac{1}{2}, -\frac{7}{2}, 3, \frac{7}{2}, 7\right)$.

1.45

(a) $\ell(c\mathbf{U}) = c_1(cu_1) + \cdots + c_n(cu_n) = cc_1 u_1 + \cdots + cc_n u_n = c\ell(\mathbf{U})$, and

(b) $\ell(\mathbf{U} + \mathbf{V}) = c_1(u_1 + v_1) + \cdots + c_n(u_n + v_n) = (c_1 u_1 + c_1 v_1) + \cdots + (c_n u_n + c_n v_n)$
$= (c_1 u_1 + \cdots + c_n u_n) + (c_1 v_1 + \cdots + c_n v_n) = \ell(\mathbf{U}) + \ell(\mathbf{V}).$

1.47 $(1, 2, 3) + (3, 2, 1) = (4, 4, 4)$ and this is $4(1, 1, 1)$. So $-4(1, 1, 1) + (1, 2, 3) + (3, 2, 1) = 0$ is a nontrivial relation among the vectors, and they are dependent.

1.49 (a), (b), and (d) are bilinear. Only (d) is symmetric, only (b) antisymmetric.

1.51 (a) and (d) only.

Section 1.5

1.53

(a) Since $\mathbf{U} \cdot (\mathbf{V} + \mathbf{W}) = u_1(v_1 + w_1) + \cdots + u_n(v_n + w_n)$
$= u_1 v_1 + \cdots + u_n v_n + u_1 w_1 + \cdots + u_n w_n = \mathbf{U} \cdot \mathbf{V} + \mathbf{U} \cdot \mathbf{W}$ the dot product is distributive and

$\mathbf{U} \cdot (c\mathbf{V}) = u_1(cv_1) + \cdots + u_n(cv_n) = c(u_1v_1 + \cdots + u_nv_n) = c\mathbf{U} \cdot \mathbf{V}$, similarly $(c\mathbf{U}) \cdot \mathbf{V} = c\mathbf{U} \cdot \mathbf{V}$, therefore b is a bilinear function.

(b) $\mathbf{U} \cdot \mathbf{V} = u_1v_1 + \cdots + u_nv_n = v_1u_1 + \cdots + v_nu_n = \mathbf{V} \cdot \mathbf{U}$ shows that the dot product is commutative and b is symmetric.

1.55 Three dot products are set equal to zero:

$$w_1 + 2w_2 - 2w_5 = 0$$
$$-2w_1 + w_2 + 2w_3 = 0$$
$$-2w_2 + w_3 + 2w_5 = 0$$

Since w_4 is not in the system $\mathbf{W} = (0, 0, 0, 1, 0)$ satisfies the equations.

1.57 (b) and (c) only

1.59 Add the identities $(u_k - v_k)^2 = u_k^2 - 2u_kv_k + v_k^2$ for $k = 1, \ldots, n$ and recognize $\sum_{k=1}^{n} u_k^2 = \|\mathbf{U}\|^2$, $\sum_{k=1}^{n} u_kv_k = \mathbf{U} \cdot \mathbf{V}$ and $\sum_{k=1}^{n} v_k^2 = \|\mathbf{V}\|^2$.

1.61

(a) $(c, c, c, \cdots, c)$
(b) $nc^2 = 1$ gives $c = n^{-1/2}$.
(c) $c = n^{-1/2}$ tends to zero as n tends to infinity.

1.63 Equations for a vector $\mathbf{V}$ orthogonal to $\mathbf{W}_1$ and $\mathbf{W}_2$ are:

$$v_1 + v_2 + v_3 = 0, \qquad v_2 + v_3 + v_4 = 0.$$

With any choice of v_2 and v_3 these equations give $v_1 = v_4 = -v_2 - v_3$. Therefore

$$\mathbf{V} = (-v_2 - v_3, v_2, v_3, -v_2 - v_3).$$

Take for example v_2 and v_3 as $1, 0$ and $0, 1$. This gives independent vectors $\mathbf{V} = (-1, 1, 0, -1)$ and $(-1, 0, 1, -1)$.

1.65

(a) $\mathbf{C} = (1 - h, 1, 2), \ \mathbf{D} = (1 + h, 1, 2)$.
(b) $2h$
(c) The icosahedron is regular, so all edges have the same length, $2h$.
(d) $\|\mathbf{D} - \mathbf{A}\|^2 = (2h)^2$, so $(1 + h - 2)^2 + (1 - (1 - h))^2 + (2 - 1)^2 = (h - 1)^2 + h^2 + 1 = 4h^2$, or $h^2 + h - 1 = 0$. Then $h = \frac{1}{2}(1 \pm \sqrt{1 + 4})$, and it must be the plus sign since h is positive.

1.67

(a) By the triangle inequality $|a| = |a - b + b| \le |a - b| + |b|$
(b) Subtract $|b|$ from both sides in (a). That gives $|a| - |b| \le |a - b|$.
(c) Switch the numbers a and b in part (b) to get $|b| - |a| \le |a - b|$. Combining that with part (b) gives $\big||a| - |b|\big| \le |a - b|$.

(d) Mimic parts (a)–(c): $\|X\| = \|X - Y + Y\| \le \|X - Y\| + \|Y\|$, subtract and switch X and Y to get $\big|\|X\| - \|Y\|\big| \le \|X - Y\|$

Section 1.6

1.69 Denote $U = (u_1, u_2)$ etc. Then

(a) $\det\begin{bmatrix} u_1 + w_1 & v_1 \\ u_2 + w_2 & v_2 \end{bmatrix} = (u_1 + w_1)v_2 - (u_2 + w_2)v_1 = (u_1 v_2 - u_2 v_1) + (w_1 v_2 - w_2 v_1)$
$= \det[U\ V] + \det[W\ V]$, and similarly for $\det[U\ V + W]$.

(b) $\det\begin{bmatrix} cu_1 & v_1 \\ cu_2 & v_2 \end{bmatrix} = (cu_1)v_2 - (cu_2)v_1 = c\det[U\ V]$ and similarly $\det[U\ cV] = c\det[U\ V]$.

1.71 (a) 1 (b) -1 (c) 1 (d) -6 (e) -6

1.73

(a) In a permutation $p_1 \cdots p_{n+1}$, move the number $n + 1$ past k smaller numbers to its right; this can be done using k transpositions. Then the first n numbers remaining to the left of $n + 1$ are some permutation of $123\cdots n$. Complete the argument inductively since none of the numbers $123\cdots n$ will need to be moved to the right of $n + 1$.

(b) In 1237456 there are three cases where a larger number is to the left of a smaller one: 74, 75, 76. So $s(1237456) = -1$.

(c) In 1273456 there are four cases where a larger number is to the left of a smaller one: 73, 74, 75, 76. So $s(1273456) = 1$.

1.75 The signature of permutation $p = p_1 p_2 \cdots p_n$ is the number $s(p)$ that gives equality in

$$\prod_{i<j}(x_{p_i} - x_{p_j}) = s(p)\prod_{i<j}(x_i - x_j).$$

The composite of p and $q = q_1 q_2 \cdots q_n$ can be expressed as

$$pq = p_{q_1} p_{q_2} \cdots p_{q_n}.$$

Denote $x_{p_{q_i}} = y_{q_i}$, that is, $x_{p_k} = y_k$ for any k. Then

$$\prod_{i<j}(x_{p_{q_i}} - x_{p_{q_j}}) = \prod_{i<j}(y_{q_i} - y_{q_j}) = s(q)\prod_{i<j}(y_i - y_j) = s(q)\prod_{i<j}(x_{p_i} - x_{p_j}) = s(q)s(p)\prod_{i<j}(x_i - x_j).$$

This shows that $s(pq) = s(q)s(p)$.

1.77 $\det(E_1, E_3, E_2) = \det\begin{bmatrix} 1 & 0 & 0 \\ 0 & 0 & 1 \\ 0 & 1 & 0 \end{bmatrix} = -1$ and the permutation 132 is one transposition from 123, so $s(132) = -1$.

Section 1.7

1.79

(a) $\frac{1}{2}(\text{base})(\text{height}) = \frac{1}{2}\|U\|(\|V\|\sin\theta)$

(b) $\sqrt{1 - \left(\frac{\mathbf{U \cdot V}}{\|\mathbf{U}\|\|\mathbf{V}\|}\right)^2} = \sqrt{1 - \cos^2\theta} = \sin\theta$

(c) $\frac{1}{2}\|\mathbf{U}\|(\|\mathbf{V}\| \sin\theta) = \frac{1}{2}\|\mathbf{U}\|\|\mathbf{V}\| \sqrt{1 - \left(\frac{\mathbf{U \cdot V}}{\|\mathbf{U}\|\|\mathbf{V}\|}\right)^2} = \frac{1}{2}\sqrt{\|\mathbf{U}\|^2\|\mathbf{V}\|^2 - (\mathbf{U \cdot V})^2}$

(d) $\|\mathbf{U}\|^2\|\mathbf{V}\|^2 - (\mathbf{U \cdot V})^2 = (u_1^2 + u_2^2)(v_1^2 + v_2^2) - (u_1v_1 + u_2v_2)^2$
$= u_1^2v_2^2 + u_2^2v_1^2 - 2u_1v_1u_2v_2 = (u_1v_2 - u_2v_1)^2$

1.81 Area is twice the area of the triangle $((0,0)(1,3)(2,1))$,
so it is $2(\frac{1}{2})|(1)(1) - (3)(2)| = 5$.

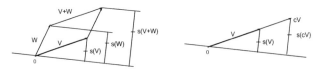

Fig. 9.19 Figure for Problem 1.83.

1.83 See Figure 9.19.
The figure shows $s(\mathbf{V} + \mathbf{W}) = s(\mathbf{V}) + s(\mathbf{W})$ for vectors on the same side of the line
at left, and $s(c\mathbf{V}) = cs(\mathbf{V})$ for $c > 0$ at the right.

Section 1.8

1.85 (a) $\begin{bmatrix} 3 \\ -1 \end{bmatrix}$ (b) 0 (c) -10 (d) b_{ij}

1.87 The function M is defined by $M(\mathbf{V}) = v_1\mathbf{M}_1 + \cdots + v_n\mathbf{M}_n$ where the $\mathbf{M}_k$ are
vectors. M is linear because:

$$M(c\mathbf{V}) = (cv_1)\mathbf{M}_1 + \cdots + (cv_n)\mathbf{M}_n = c(v_1\mathbf{M}_1 + \cdots + v_n\mathbf{M}_n) = cM(\mathbf{V}),$$

and

$$M(\mathbf{V} + \mathbf{W}) = (v_1 + w_1)\mathbf{M}_1 + \cdots + (v_n + w_n)\mathbf{M}_n$$
$$= v_1\mathbf{M}_1 + \cdots + v_n\mathbf{M}_n + w_1\mathbf{M}_1 + \cdots + w_n\mathbf{M}_n = M(\mathbf{V}) + M(\mathbf{W}).$$

1.89 (a) $\mathbf{X} \cdot \left(\begin{bmatrix} 1 & 0 \\ 0 & 2 \end{bmatrix}\mathbf{Y}\right)$ (b) $\mathbf{X} \cdot \left(\begin{bmatrix} 0 & 1 \\ -1 & 0 \end{bmatrix}\mathbf{Y}\right)$ $\mathbf{X} \cdot \left(\begin{bmatrix} 1 & 3 \\ 1 & -1 \end{bmatrix}\mathbf{Y}\right)$

1.91 $\mathbf{N}(\mathbf{MX}) = \mathbf{N}\sum_j x_j\mathbf{ME}_j = \sum_j x_j\mathbf{N}(\mathbf{ME}_j)$ and
$(\mathbf{NM})\mathbf{X} = \sum_j x_j(\mathbf{NM})\mathbf{E}_j$. So we must show that $\mathbf{N}(\mathbf{ME}_j) = (\mathbf{NM})\mathbf{E}_j$:
$(\mathbf{NM})\mathbf{E}_j$ is the j-th column of $\mathbf{NM}$ that has i-th coordinate $\sum_h n_{ih}m_{hj}$.
$\mathbf{N}(\mathbf{ME}_j)$ is $\mathbf{N}$ times the j-th column of $\mathbf{M}$, that has the same i-th coordinate because
the i-th row of $\mathbf{N}$ is $(n_{i1}, n_{i2}, \ldots)$.
Therefore $\mathbf{N}(\mathbf{MX}) = (\mathbf{NM})\mathbf{X}$.

1.93

(a) By Property (v) of the determinant, since $\det A = 0$ the columns of A are linearly dependent. By Theorem 1.20 there is a vector W not representable as AV. This proves (a).

(b) For any vector U, BU is some vector and for any such vector, by part (a), $A(BU) \neq W$. This proves (b).

By Theorem 1.20 the columns of AB are linearly dependent. By Property (v) then $\det(AB) = 0$.

Section 1.9

1.95

(a) $x = 0$

(b) Take $N = (0, 1, 1) \times (-3, 0, 0) = (0, -3, 3)$. The equation is $-3y + 3z = 0$.

(c) Since the planes are parallel we can take the same normal $N = (1, -3, 5)$. The equation is $(1, -3, 5) \cdot (X - (1, 1, 1)) = 0$ or $x - 3y + 5z = 3$.

1.97 All of them.

1.99 All of them.

1.101

(a) $X = (0, s, s)$

(b) $X = (-3t, 0, 0)$

(c) $X = (-3t, s, s)$.

1.103 Using equation (1.37),

$$\det[U \; V \; W] = u_1(v_2 w_3 - v_3 w_2) - u_2(v_1 w_3 - v_3 w_1) + u_3(v_1 w_2 - v_2 w_1).$$

Since $\det[U \; V \; W] = U \cdot V \times W$ we get

$$V \times W = (v_2 w_3 - v_3 w_2, -(v_1 w_3 - v_3 w_1), v_1 w_2 - v_2 w_1) = (v_2 w_3 - v_3 w_2, v_3 w_1 - v_1 w_3, v_1 w_2 - v_2 w_1).$$

1.105

(a) $(1, 0, 0) \times (0, 1, 0) = i \times j = k$

(b) $j \times (i + k) = -i \times j + j \times k = -k + i$

(c) $(2i + 3k) \times (ai + bj + ck) = 2bi \times j + 2ci \times k + 3ak \times i + 3bk \times j$
$= 2bk - 2cj + 3aj - 3bi = -3bi + (3a - 2c)j + 2bk$

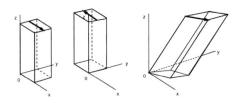

Fig. 9.20 Figure for Problem 1.107.

1.107 See Figure 9.20.

(a) $\mathbf{U} \cdot (\mathbf{V} \times \mathbf{W}) = \det \begin{bmatrix} 2 & 0 & 0 \\ 0 & 2 & 0 \\ 0 & 0 & 7 \end{bmatrix} = 28.$

(b) $\mathbf{U} \cdot (\mathbf{V} \times \mathbf{W}) = \det \begin{bmatrix} -2 & 0 & 0 \\ 0 & 2 & 0 \\ 0 & 0 & 7 \end{bmatrix} = -28.$

(c) $\mathbf{U} \cdot (\mathbf{V} \times \mathbf{W}) = \det \begin{bmatrix} 2 & 1 & 7 \\ 1 & 2 & 7 \\ 0 & 0 & 7 \end{bmatrix} = 21.$

Problems of Chapter 2

Section 2.1

2.1

(a) $(1,2) \cdot \mathbf{X}$

(b) $(1,2,0) \cdot \mathbf{X}$

(c) $\begin{bmatrix} 1 & 0 \\ 1 & 2 \end{bmatrix} \mathbf{X}$

(d) $\begin{bmatrix} 1 & 0 & 0 \\ 1 & 2 & 0 \end{bmatrix} \mathbf{X}$

(e) $\begin{bmatrix} 0 & -1 & 0 & 1 \\ -1 & 0 & 1 & 0 \\ 5 & 1 & 0 & 0 \\ 1 & 1 & 1 & 1 \end{bmatrix} \mathbf{X}$

(f) $\begin{bmatrix} 1 & 0 \\ 5 & 0 \\ 0 & -1 \\ -2 & 0 \\ 0 & 1 \end{bmatrix} \mathbf{X}$

(g) $\begin{bmatrix} 1 & 0 & 0 & 0 & 0 \\ 5 & 0 & 0 & 0 & 0 \\ 0 & -1 & 0 & 0 & 0 \\ -2 & 0 & 0 & 0 & 0 \\ 0 & 1 & 0 & 0 & 0 \end{bmatrix} \mathbf{X}$

2.3 The level set $f = 0$ is the origin together with all the points $\mathbf{X}$ with $\|\mathbf{X}\| \geq 1$. The level sets $f = c$ with $0 < c < 1$ are (a) the two points $\pm\sqrt{c}$, (b) the sphere of radius $\sqrt{c}$, and (c) the sphere in $\mathbb{R}^5$ centered at the origin of radius $\sqrt{c}$. All other level sets are empty.

2.5 $\ell(x,y,z) = 2x + 3y - z$ defines a function $\ell : \mathbb{R}^3 \to \mathbb{R}$ for which $\ell(x,y,z) = 0$ gives the plane $z = 2y + 3y$. Any nonzero multiple of $2x + 3y - z$ would also work.

2.7

(a) Subtract $\|\mathbf{A}\|^2 + 2\mathbf{A} \cdot \mathbf{U}$ from each side of $\|\mathbf{A} + \mathbf{U}\|^2 = \|\mathbf{A}\|^2 + 2\mathbf{A} \cdot \mathbf{U} + \|\mathbf{U}\|^2$, giving

$$\|\mathbf{A} + \mathbf{U}\|^2 - (\|\mathbf{A}\|^2 + 2\mathbf{A} \cdot \mathbf{U}) = \|\mathbf{U}\|^2.$$

Since the first term $\|\mathbf{A} + \mathbf{U}\|^2$ is equal to $f(\mathbf{X})$ and the second is $g(\mathbf{X})$, this proves what we want.

(b) We are assuming $\|\mathbf{X} - \mathbf{A}\| < 10^{-2}$. That is, $\|\mathbf{U}\| < 10^{-2}$. Part (a) gives $f(\mathbf{X}) - g(\mathbf{X}) = \|\mathbf{U}\|^2$, so

$$\left| f(\mathbf{X}) - g(\mathbf{X}) \right| = \|\mathbf{U}\|^2 < 10^{-4}.$$

2.9 Take $x^2 - y^2 = c(x^2 + y^2)$ when $y \neq 0$ and $c \neq 1$. Divide by y^2 to get $\left(\frac{x}{y}\right)^2 - 1 = c\left(\left(\frac{x}{y}\right)^2 + 1\right)$, from which $\left(\frac{x}{y}\right)^2 = \frac{1+c}{1-c}$. This gives two lines (minus origin) $x = \pm\frac{1+c}{1-c}y$. Note that $\frac{1+c}{1-c}$ can be *any* nonzero.

2.11

(a) Level sets are the origin, unit circle, and circle of radius $\sqrt{2}$.
(b) Level sets are the origin, unit circle, and circle of radius 2.
(c) Level sets are empty, unit circle, and circle of radius $\frac{1}{\sqrt{2}}$.

2.13

(a) Level sets are spheres of radius $\frac{1}{\sqrt{2}}$, 1, $\sqrt{2}$ centered at the origin.
(b) Level sets are spheres of radius $\frac{1}{2}$, 1, 2.
(c) Level sets are spheres of radius $\sqrt{2}$, 1, $\frac{1}{\sqrt{2}}$.
(d) $\mathbf{X} = a\mathbf{U} + \mathbf{V}$, $f(\mathbf{X}) = a\mathbf{U} \cdot \mathbf{U} + \mathbf{U} \cdot \mathbf{V} = a = c$. So $\mathbf{X} = c\mathbf{U} + \mathbf{V}$, $\mathbf{V}$ any vector orthogonal to $\mathbf{U}$. This is a plane through $c\mathbf{U}$ and orthogonal to $\mathbf{U}$.

2.15 According to Theorem 2.1, $\mathbf{L}(\mathbf{X}) = \begin{bmatrix} c_{11}x_1 + \cdots + c_{1n}x_n \\ c_{21}x_1 + \cdots + c_{2n}x_n \\ c_{m1}x_1 + \cdots + c_{mn}x_n \end{bmatrix}$. Since this is equal to

$$x_1 \begin{bmatrix} c_{11} \\ c_{21} \\ \vdots \\ c_{m1} \end{bmatrix} + \cdots + x_n \begin{bmatrix} c_{1n} \\ c_{2n} \\ \vdots \\ c_{mn} \end{bmatrix}$$

the numbers multiplying x_1 are the first column, x_2 the second, and so on. So $\mathbf{L}(\mathbf{X}) = x_1\mathbf{V}_1 + \cdots + x_n\mathbf{V}_n$.

2.17 We have $f(\mathbf{X}) = 1$ and $g(\mathbf{X}) = \|\mathbf{X}\|^{-2}$ for all $\mathbf{X} \neq \mathbf{0}$. So the level set $f(\mathbf{X}) = 1$ is $\mathbb{R}^3 - \{\mathbf{0}\}$, $g(\mathbf{X}) = 1$ is the unit sphere in $\mathbb{R}^3$ where $\|\mathbf{X}\| = 1$, $g(\mathbf{X}) = 2$ is the sphere of radius $\frac{1}{\sqrt{2}}$, and $g(\mathbf{X}) = 4$ is the sphere of radius $\frac{1}{2}$. There are no points where $f = 2$ because $f = 1$ everywhere it is defined.

2.19

(a) $\mathbf{F}(t,\theta) = (t\cos\theta, t\sin\theta)$.
(b) The rectangle maps to a segment of the unit disk

2.21 $\|L(x,y,z)\| = x^2 + z^2 + y^2 = 1$, so L maps the sphere into itself. The right half-plane goes into the unit disk. The unit disk goes into the left half-plane.

2.23

(a) This uses two statements of Taylor's Theorem

$$f(u) = f(0) + f'(\theta_1)u, \qquad f(u) = f(0) + f'(0)u + \tfrac{1}{2}f''(\theta_2)u^2$$

using the first, and then second derivative remainders.

(b) Apply part (a) with $a = \|\mathbf{A}\|$ and $u = 2\mathbf{A} \cdot \mathbf{U} + \|\mathbf{U}\|^2$.

(c) Reading from the first derivative Taylor formula we see $\mathbf{L}_1(\mathbf{U}) = \|\mathbf{A}\|^{-3}\mathbf{U}$,

$$\text{``large''} = \left(-\tfrac{3}{2}(\|\mathbf{A}\|^2 + \theta_1)^{-5/2}(2\mathbf{A} + \mathbf{U}) \cdot \mathbf{U}\right)(\mathbf{A} + \mathbf{U})$$

with norm less than some multiple of $\|\mathbf{U}\|$, and $\mathbf{L}_2(\mathbf{U}) = \frac{\mathbf{U}}{\|\mathbf{A}\|^3} - 3\frac{\mathbf{A} \cdot \mathbf{U}}{\|\mathbf{A}\|^5}\mathbf{A}$,

$$\text{``small''} = -\tfrac{3}{2}\|\mathbf{A}\|^{-5}\|\mathbf{U}\|^2 + \tfrac{15}{8}(\|\mathbf{A}\|^2 + \theta_2)^{-7/2}(2\mathbf{A} \cdot \mathbf{U} + \|\mathbf{U}\|^2)^2(\mathbf{A} + \mathbf{U}),$$

with norm less than some multiple of $\|\mathbf{U}\|^2$.

Section 2.2

2.25

(a) Combine continuity of g and of $\mathbf{F}$: given $\epsilon > 0$ there is δ so that
if $\|\mathbf{F}(\mathbf{B}) - \mathbf{F}(\mathbf{A})\| < \delta$ then $|g(\mathbf{F}(\mathbf{B})) - g(\mathbf{F}(\mathbf{A}))| < \epsilon$
and there is η so that if $\|\mathbf{B} - \mathbf{A}\| < \eta$ then $|\mathbf{F}(\mathbf{B}) - \mathbf{F}(\mathbf{A})| < \delta$. Therefore if
$\|\mathbf{B} - \mathbf{A}\| < \eta$ then $|g(\mathbf{F}(\mathbf{B})) - g(\mathbf{F}(\mathbf{A}))| < \epsilon$.

(b) Linear functions are continuous.

(c) Combine (a) and (b).

(d) Take $g(x, y) = xy$ in (a). To show that g is continuous at (a, b), show that if
$\|(x, y) - (a, b)\| < \delta$ then

$$|xy - ab| = |xy - xb + xb - ab| < (|a| + \delta)\delta + \delta b \le M\delta.$$

2.27

(a) $\epsilon/2$

(b) 3δ

(c) We know that $|x - a| < \|(x, y) - (a, b)\|$ and $|y - b| < \|(x, y) - (a, b)\|$. If
$\|(x, y) - (a, b)\| < m$ then

$$|\cos(2x)\cos(3y) - \cos(2a)\cos(3b)| = |(\cos(2x) - \cos(2a))\cos(3y) + \cos(2a)(\cos(3y) - \cos(3b))|$$

$$\le |\cos(2x) - \cos(2a)||\cos(3y)| + |\cos(2a)||\cos(3y) - \cos(3b)|$$

$$< |\cos(2x) - \cos(2a)| + |\cos(3y) - \cos(3b)| < 2m + 3m.$$

Take $m = \epsilon/5$.

2.29

(a) True. Observe that 0 is between the values $f(1, 0)$ and $f(0, \tfrac{1}{4})$. Consider the
continuous function of one variable $g(t) = f(\mathbf{X}(t))$ where $\mathbf{X}(t)$ parametrizes the
line segment from $(1, 0)$ to $(0, \tfrac{1}{4})$. By the Intermediate Value Theorem for continuous functions of one variable, there is a t_1 where $g(t_1) = 0$.

(b) True.

(c) False, that is, there are such functions f for which this does not hold.

(d) True; this is part of being continuous at $(0, \tfrac{1}{4})$.

2.31 These are all compositions and products and sums of continuous functions. Therefore they are continuous.

2.33 The graph is part of a plane in $\mathbb{R}^3$. The maximum value of f occurs, by inspection, at the boundary circle $x_1^2 + x_2^2 = 2$ where x_2 is as large as possible. This is the point $(x_1, x_2) = (0, \sqrt{2})$, so the maximum value is $\sqrt{2}$.

2.35 The inequality $x^2 + y^2 < 1$ confines (x, y) to the open unit disk, and places no restriction on z. Around any point of the unit disk you can center a small open disk contained in the unit disk. For each point (x, y, z) then there is a small open ball of that same small radius centered at (x, y, z) and contained in the cylinder. Therefore the cylinder is open.

2.37

(a) Boundary is the three edges $x = 0$, $y = 0$, and $x + y = 1$.
(b) The point is .0001 from the left edge, and .0001 *vertically* thus $.0001/\sqrt{2} >$ .00005 distance from the hypoteneuse. Take $r = .00001$. If $\mathbf{Q}$ is any point within distance r of $(.0001, .9998)$ then $\mathbf{Q}$ is certainly inside a square

$$.0001 - r \le x \le .0001 + r, \qquad .9998 - r \le y \le .9998 + r,$$

that is,
$$.00009 \le x \le .00011, \qquad .99979 \le y \le .99981.$$

The top right point of that square is $(.00011, .99981)$, still inside T because

$$.00011 + .99981 = .999982 < 1.$$

Since this little square is contained in T, a disk of radius r centered at $(.0001, .9998)$ is also contained in T.

2.39

(a) c can be taken to be the matrix norm, $\sqrt{1^2 + 5^2 + 5^2 + 1^2} = \sqrt{52}$, or any larger number.
(b) $d = c$, using linearity of $\mathbf{F}$.
(c) Yes, by part (b): if $\|\mathbf{X} - \mathbf{Y}\| < \frac{\epsilon}{\sqrt{52}}$ then $\|\mathbf{F}(\mathbf{X}) - \mathbf{F}(\mathbf{Y})\| < \epsilon$.

2.41

(a) If $\mathbf{C} = \mathbf{0}$ the function is constant so is uniformly continuous. Otherwise:

$$|f(\mathbf{X}) - f(\mathbf{Y})| = |\mathbf{C} \cdot (\mathbf{X} - \mathbf{Y})| \le \|\mathbf{C}\| \|\mathbf{X} - \mathbf{Y}\|.$$

Let $\epsilon > 0$. For any $\mathbf{X}$ and $\mathbf{Y}$ with $\|\mathbf{X} - \mathbf{Y}\| < \dfrac{\epsilon}{\|\mathbf{C}\|}$, then $|f(\mathbf{X}) - f(\mathbf{Y})| < \epsilon$.
(b) One way: this is a polynomial in the variables $(x_1, \ldots, x_n, y_1, \ldots, y_n)$, therefore continuous. Or we can prove directly that g is continuous at each point $(\mathbf{U}, \mathbf{V})$:

$$g((\mathbf{X}, \mathbf{Y})) - g(\mathbf{U}, \mathbf{V}) = \mathbf{X} \cdot \mathbf{Y} - \mathbf{U} \cdot \mathbf{V} = \mathbf{X} \cdot (\mathbf{Y} - \mathbf{V}) + (\mathbf{X} - \mathbf{U}) \cdot \mathbf{V}$$

has absolute value

$$\leq \|\mathbf{X}\|\|\mathbf{Y}-\mathbf{V}\| + \|\mathbf{X}-\mathbf{U}\|\|\mathbf{V}\|$$

If $\|(\mathbf{X},\mathbf{Y})-(\mathbf{U},\mathbf{V})\| < \delta$, then each of $\|\mathbf{X}-\mathbf{U}\|$ and $\|\mathbf{Y}-\mathbf{V}\|$ is less than δ, and our estimate becomes

$$|g((\mathbf{X},\mathbf{Y}))-g(\mathbf{U},\mathbf{V}))| \leq \|\mathbf{X}\|\delta + \delta\|\mathbf{V}\|. \leq (\|\mathbf{U}\|+\delta)\delta + \delta\|\mathbf{V}\|.$$

This is sufficient to show that g is continuous at $(\mathbf{U},\mathbf{V})$, but since the estimate depends on $(\mathbf{U},\mathbf{V})$ we don't get uniform continuity.

2.43 Since $\|\mathbf{X}\| \geq 2$ you have $\dfrac{1}{\|\mathbf{X}\|} \leq \frac{1}{2}$. Similarly for Y. Therefore

$$\left| \frac{1}{\|\mathbf{X}\|} - \frac{1}{\|\mathbf{Y}\|} \right| = \left| \frac{\|\mathbf{Y}\|-\|\mathbf{X}\|}{\|\mathbf{X}\|\|\mathbf{Y}\|} \right| \leq \frac{1}{2}\frac{1}{2}|\|\mathbf{X}\|-\|\mathbf{Y}\|| \leq \frac{1}{4}\|\mathbf{X}-\mathbf{Y}\|$$

This shows uniform continuity.

Section 2.3

2.45 θ runs from 0 to π, and $r > 0$, so your sketch ought to show a "rectangle" of height π that extends infinitely far to the right.

2.47 (a) $0 \leq r < 1$, any θ.
(b) $0 < r$ and $0 < \theta < \frac{\pi}{2}$.

2.49 (a) $1 = rr^{-1} = r(a+b\cos\theta) = ar+bx = a\sqrt{x^2+y^2}+bx$.
(b) Subtract bx and square: $1-bx = a\sqrt{x^2+y^2}$, $1-2bx+b^2x^2 = a^2(x^2+y^2)$
So $1 = (a^2-b^2)x^2+2bx+a^2y^2$. Complete the square to get

$$1 = (a^2-b^2)\left(x+\frac{b}{a^2-b^2}\right)^2 - \frac{b^2}{a^2-b^2} + a^2y^2.$$

Add $\dfrac{b^2}{a^2-b^2}$ to get $\frac{a^2}{a^2-b^2} = (a^2-b^2)\left(x+\frac{b}{a^2-b^2}\right)^2 + a^2y^2$. Multiply by $\dfrac{a^2-b^2}{a^2}$ to get

$$1 = \frac{\left(x+\frac{b}{a^2-b^2}\right)^2}{\left(\frac{a}{a^2-b^2}\right)^2} + \frac{y^2}{(\frac{1}{\sqrt{a^2-b^2}})^2}.$$

That is an equation for an ellipse.

2.51 (a) (iii) (b) (i) (c) (ii) (d) (iv)

2.53

(a) 1, spheres
(b) $s_2(1,\phi,\theta) = 0$.

(c) zero.

(d) $3\cos^2\phi - 1$ is zero where $\cos\phi = \pm\sqrt{1/3}$, a cone.

(e) negative because the $\cos^2\phi$ is less than $1/3$ there.

2.55

(a) In the expression $(xu - vy, yu + xv)$, each of the two components is a continuous function of (x, y, u, v). So multiplication zw is a continuous function of z and w. In particular, the function z^2 is continuous, and then $z^3 = z^2 z$ is a product of continuous functions so it is continuous, and so forth for all the powers of z. Also multiplying z^k by p_k is continuous. And the sum of continuous functions is continuous as we already know. Therefore polynomials are continuous.

(b) Squaring doubles the angle θ and squares the polar r:

$$(r(\cos\theta + i\sin\theta))^2 = r^2(\cos(2\theta) + i\sin(2\theta)).$$

Therefore to find a square root of a complex number, take half the angle, and the square root of the absolute value. Similarly to find a cube root, divide the angle by three and take the cube root of the absolute value, etc.

(c) $|x + iy| = \sqrt{x^2 + y^2}$ is the same as the norm in $\mathbb{R}^2$, that we know is continuous. When $z = r(\cos\theta + i\sin\theta)$ you have $|z| = r$, so $|zw| = |z||w|$ follows from

$$r_1(\cos\theta_1 + i\sin\theta_1)r_2(\cos\theta_2 + i\sin\theta_2) = r_1 r_2(\cos(\theta_1 + \theta_2) + i\sin(\theta_1 + \theta_2)).$$

(d) In the first part you can take $P = |p_{n-1}| + \cdots + |p_0|$.

(e) We are assuming $|p(z)|$ is positive, and we know it is continuous and tends to infinity as $|z|$ tends to infinity. Therefore f is continuous on all of $\mathbb{R}^2$ and tends to zero as $|z|$ tends to infinity.

(f) By the product rule

$$\frac{d}{da}\left(q(a) + q'(a)(z - a) + q''(a)\frac{(z-a)^2}{2!} + \cdots + q^{(n)}(a)\frac{(z-a)^n}{n!}\right)$$

$$= q'(a) + q''(a)(z - a) + q'(a)(-1) + q'''(a)\frac{(z-a)^2}{2!} + q''(a)(-(z-a))$$

$$+ \cdots + q^{(n+1)}(a)\frac{(z-a)^n}{n!} + q^{(n)}(a)\frac{-(z-a)^{n-1}}{(n-1)!}.$$

Since q has degree n, $q^{(n+1)}(a) = 0$, and all other terms cancel in pairs, so the derivative is zero. The contradiction follows from part (d), using a rather than z as the variable. Finally taking $a = z$ evaluates the right hand side as $q(a)$ since all other terms become 0.

(g) The contradiction arises because the terms indicated as $\cdots$ all involve higher than the k-th power of ϵ. The sum of these terms is smaller than $m\epsilon^k$ when ϵ is sufficiently small, giving $|p(z + \epsilon h)| < m$, contradicting part (e).

2.57 The Cartesian coordinates of the points are

$$(\sin\phi_1\cos\theta_1, \sin\phi_1\sin\theta_1, \cos\phi_1), \qquad (\sin\phi_2\cos\theta_2, \sin\phi_2\sin\theta_2, \cos\phi_2),$$

so the dot product is

$$\sin\phi_1\sin\phi_2(\cos\theta_1\cos\theta_2 + \sin\theta_1\sin\theta_2) + \cos\phi_1\cos\phi_2$$

$$= \sin\phi_1\sin\phi_2\cos(\theta_2 - \theta_1) + \cos\phi_1\cos\phi_2$$

$$= \sin\phi_1\sin\phi_2\cos(\theta_2-\theta_1)+\cos(\phi_2-\phi_1)-\sin\phi_1\sin\phi_2 = \cos(\phi_2-\phi_1)-\sin\phi_2\sin\phi_1(1-\cos(\theta_2-\theta_1)).$$

Problems of Chapter 3

Section 3.1

3.1 (a) $6x$, (b) 4. These are the x and y partial derivatives of $3x^2 + 4y$,

3.3

(a) $f_x(x,y) = -2xe^{-x^2-y^2}$, $f_y(2,0) = 0$,
(b) $xe^y + e^x$
(c) $-y\sin(xy) + \cos(xy) - xy\sin(xy)$

3.5

(x,y)	$(.1,.2)$	$(.01,.02)$
$(1+x+3y)^2$	2.89	1.1449
$1+x+3y$	1.7	1.07
$1+2x+6y$	2.4	1.14

3.7 Level sets of m are closer, and ∇m is longer.

3.9

(a) Any linear function is given by some matrix.
(b) A vector tends to zero if and only if each component tends to zero.
(c) With these substitutions the fraction becomes the definition of partial derivative. The fraction tends to f_{i,x_j} by definition. And the j-th component of row $\mathbf{C}_i$ is the (i,j) entry of the matrix $\mathbf{C}$.

3.11 $\nabla f(x,y) = (-\sin(x+y), -\sin(x+y))$ so $f_x - f_y = -\sin(x+y) - (-\sin(x+y)) = 0$. $\nabla g(x,y) = (2\cos(2x-y), -\cos(2x-y))$ so $g_x + 2g_y = 2\cos(2x-y) + 2(-\cos(2x-y)) = 0$.

3.13 $\nabla g(x,y) = (ae^{ax+by}, be^{ax+by}) = e^{ax+by}(a,b)$

Section 3.2

3.15

(a) $z = f(0,0) + f_x(0,0)x + f_y(0,0)y = x$
(b) $z = f(1,2) + f_x(1,2)(x-1) + f_y(1,2)(y-2) = 5 + (x-1) + 4(y-2)$

3.17
(a) If $a^2 + b^2 = 1$ then

$$f(a,b) = e^{-(a^2+b^2)} = e^{-1}, \qquad g(a,b) = e^{-1}$$

(b) $\nabla f(x,y) = e^{-(x^2+y^2)}(-2x, -2y)$ and $\nabla g(x,y) = -e^{-1}(x^2+y^2)^{-2}(2x, 2y)$

(c) At point (a,b) where $a^2 + b^2 = 1$ you have by part (b)

$$\nabla f(a,b) = e^{-1}(-2a,-2b), \qquad \nabla g(a,b) = -e^{-1}(2a,2b)$$

and $f(a,b) = g(a,b)$ by part (a). Since the values and gradients of f and g agree, the graphs are tangent.

Section 3.3

3.19

(a) $f_{xx} = 2$, $f_{xy} = 0$, $f_{yy} = -2$
(b) $f_{xx} = 2$, $f_{xy} = 0$, $f_{yy} = 2$
(c) $f_{xx} = 2$, $f_{xy} = 2$, $f_{yy} = 2$
(d) $f_{xx} = e^{-x}\cos y$, $f_{xy} = -e^{-x}\sin y$, $f_{yy} = -e^{-x}\cos y$
(e) $f_{xx} = -a^2 e^{-ay}\sin(ax)$, $f_{xy} = -a^2 e^{-ay}\cos(ax)$, $f_{yy} = a^2 e^{-ay}\sin(ax)$

3.21 (a) (b) and (c) only

3.23 $\nabla f(x,y) = (3+y, 2+x)$, $\nabla f(1,1) = (4,3)$. The directional derivatives are

$$\nabla f \cdot \mathbf{V} = 4,\ 5,\ 3,\ 0,\ -4.$$

The second one, the largest, is equal to $\|\nabla f(1,1)\|$ because $\mathbf{V}$ is in the direction of $\nabla f(1,1)$.

3.25

(a) If $u(x,t) = f(x-3t)$ then according to the Chain Rule,
$u_x = f'(x-3t)$, $u_t = f'(x-3t)(-3)$, so $u_t + 3u_x = 0$.
(b) $D_\mathbf{V} u = \mathbf{V} \cdot \nabla u = \frac{1}{\sqrt{10}}(3,1) \cdot (u_x, u_t) = \frac{1}{\sqrt{10}}(3u_x + u_t) = 0$.
(c) The slope of a line (x horizontal, t vertical) parallel to V is $1/3$, as is the line that goes through $(x-3t,0)$ and (x,t).
(d) Since $u(x,t)$ is constant on the line through $(x-3t,0)$, it follows that $u(x,t) = u(x-3t,0)$. This shows that every solution is a function of $x-3t$.

3.27

(a) $(x^2+y^2)^3 = r^6$, and $((x^2+y^2)^3)_x = 6x(x^2+y^2)^2$, similarly for y. Then

$$x\left(6x(x^2+y^2)^2\right) + y\left(6y(x^2+y^2)\right) = (6x^2+6y^2)(x^2+y^2)^2 = 6r^2 r^4 = 6r^5 r = r\frac{d}{dr}(r^6).$$

(b) $r = \sqrt{x^2+y^2}$, so if $f(x,y) = g(r)$ then $f_x = g'(r)r_x = g'(r)\frac{x}{r}$, and similarly for f_y.
Then
$$xf_x + yf_y = g'(r)\left(x\frac{x}{r} + y\frac{y}{r}\right) = g'(r)r.$$

(c) $f_{xx} = \left(g'(r)\frac{x}{r}\right) = g''(\frac{x}{r})^2 + g'\frac{r-\frac{xx}{r}}{r^2}$, $f_{yy} = g''(\frac{y}{r})^2 + g'\frac{r-\frac{yy}{r}}{r^2}$, so $f_{xx}+f_{yy} = g'' + r^{-1}g'$.

3.29 Differentiate the change of variables $x = r\cos\theta$, $y = r\sin\theta$ to get

$$x_r = \cos\theta, \quad y_r = \sin\theta, \quad x_\theta = -r\sin\theta, \quad y_\theta = r\cos\theta.$$

The Chain Rule gives for any differentiable function v of x and y

$$v_r = v_x\cos\theta + v_y\sin\theta$$
$$v_\theta = v_x(-r\sin\theta) + v_y r\cos\theta$$

Multiply the first by r and use the change of variables again to get

$$rv_r = xv_x + yv_y$$
$$v_\theta = -yv_x + xv_y$$

The formula for v_r applied to u gives $u_r = u_x\cos\theta + u_y\sin\theta$, and the r partial of this is

$$u_{rr} = u_{rx}\cos\theta + u_{ry}\sin\theta.$$

The formula for v_r applied to u_x and u_y gives

$$u_{rr} = (u_{xx}\cos\theta + u_{xy}\sin\theta)\cos\theta + (u_{yx}\cos\theta + u_{yy}\sin\theta)\sin\theta.$$

The same process gives

$$u_{\theta\theta} = -r\cos\theta\, u_x - yu_{x\theta} - r\sin\theta\, u_y + xu_{y\theta}$$

$$= -xu_x - yu_y - y(-yu_{xx} + xu_{yx}) + x(-yu_{yx} + xu_{yy}).$$

This gives $u_{\theta\theta} = -ru_r + r^2(u_{xx} + u_{yy})$. Divide by r^2 and add u_{rr} to get

$$\frac{1}{r^2}u_{\theta\theta} + u_{rr} = -\frac{1}{r}u_r + \Delta u,$$

or $\Delta u = u_{rr} + \frac{1}{r}u_r + \frac{1}{r^2}u_{\theta\theta}$.

3.31

(a) $u_x = 2x$, $v_y = 2x = u_x$, $u_y = -2y$, $v_x = 2y = -u_y$
(b) $u_{xx} + u_{yy} = v_{yx} + (-v_{xy}) = 0$
(c) $w(x,y) = u^2 - v^2 = (x^2 - y^2)^2 - (2xy)^2 = x^4 - 6x^2y^2 + y^4$,
    so $w_{xx} + w_{yy} = 12x^2 - 12y^2 - 12x^2 + 12y^2 = 0$.
(d) $w(x,y) = r(p(x,y), q(x,y))$ gives $w_x = r_x p_x + r_y q_x$ and $w_y = r_x p_y + r_y q_y$. Then

$$w_{xx} = (r_{xx}p_x + r_{xy}q_x)p_x + r_x P_{xx} + (r_{yx}p_x + r_{yy}q_x)q_x + r_y q_{xx}$$

and

$$w_{yy} = (r_{xx}p_y + r_{xy}q_y)p_y + r_x P_{yy} + (r_{yx}p_y + r_{yy}q_y)q_y + r_y q_{yy}.$$

Therefore

$$w_{xx} + w_{yy} = (p_x^2 + p_y^2)r_{xx} + (2q_xp_x + 2p_yq_y)r_{xy} + (q_x^2 + q_y^2)r_{yy} + r_x(p_{xx} + p_{yy}) + r_y(q_{xx} + q_{yy}).$$

Since $p_x = q_y$ and $p_y = -q_x$ the last expression is

$$w_{xx} + w_{yy} = (p_x^2 + p_y^2)(r_{xx} + r_{yy}) + r_x(p_{xx} + p_{yy}) + r_y(q_{xx} + q_{yy}).$$

But $p_{xx} + p_{yy}$, $q_{xx} + q_{yy}$ and $r_{xx} + r_{yy}$ are all zero, so $w_{xx} + w_{yy} = 0$.

3.33

(a) $(x + iy)^2 = x^2 - y^2 + 2ixy$ so $u = x^2 - y^2$ and $v = 2xy$.

(b) $(z + h)^2 = z^2 + 2zh + h^2$. The term $2zh$ is a complex multiple of h, and h^2 is small in the sense that $\dfrac{h^2}{h} = h$ tends to zero when h does. Therefore $(z^2)' = 2z$.

(c) If h is real then $z + h = (x + h) + iy$ and by Taylor's Theorem applied to u and v you have

$$f(z + h) = u(x + h, y) + iv(x + h, y) = u(x, y) + u_x h + i(v(x, y) + v_x h) + s$$

where s is small. That says $f(z + h) = f(z) + (u_x + iv_x)h + s$, so $f' = u_x + iv_x$.

(d) Take $h = ik$ pure imaginary. Then

$$f(z + h) = u(x, y + k) + iv(x, y + k) = u(x, y) + ku_y + i(v(x, y) + v_y k) + s$$

$$= f(z) + (u_y + iv_y)k + s = f(z) + (u_y + iv_y)(-ih) + s.$$

Therefore $f' = -iu_y + v_y$.

(e) $f' = u_x + iv_x = -iu_y + v_y$ gives $u_x = v_y$ and $u_y = -v_x$.

(f) If C^2 then $u_{xx} = (v_y)_x = v_{xy} = (-u_y)_y$ so $u_{xx} + u_{yy} = 0$,
similarly $\Delta v = 0$. For z^2 you have $\Delta(x^2 - y^2) = 2 - 2 = 0$, $\Delta(2xy) = 0 + 0$,
and for $z^3 = x^3 + 3x^2iy - 3xy^2 - iy^3$ you have $\Delta(x^3 - 3xy^2) = 6x + (-6x) = 0$ and
$\Delta(3x^2y - y^3) = 6y - 6y = 0$.

Section 3.4

3.35

(a) $DF = \begin{bmatrix} e^x \cos y & -e^x \sin y \\ e^x \sin y & e^x \cos y \end{bmatrix}$

(b) $\det DF(x, y) = e^{2x}$, $\det DF(a, b) = e^{2a}$ is not zero for any a.

(c) $F(x, y + 2\pi) = F(x, y)$ because of the sine and cosine.

3.37

(a) $(-1)^4 + 2(-1)(1) + 1 = 0$ and $1 = 1$

(b) $DF = \begin{bmatrix} 4x^3 + 2y & 2x \\ 0 & 1 \end{bmatrix}$

(c) $\det DF(-1, 1) = \det \begin{bmatrix} -2 & -2 \\ 0 & 1 \end{bmatrix} = -2 \neq 0$

(d) $DF^{-1}(0, 1) = DF(-1, 1)^{-1} = \begin{bmatrix} -\frac{1}{2} & -1 \\ 0 & 1 \end{bmatrix}$

$$\mathbf{F}^{-1}(u,v) \approx \mathbf{F}^{-1}(0,1) + \begin{bmatrix} -\frac{1}{2} & -1 \\ 0 & 1 \end{bmatrix} \begin{bmatrix} u-0 \\ v-1 \end{bmatrix} \text{ gives}$$

$$\mathbf{F}^{-1}(.2, 1.01) \approx (-1,1) + \begin{bmatrix} -\frac{1}{2} & -1 \\ 0 & 1 \end{bmatrix} \begin{bmatrix} .2 \\ .01 \end{bmatrix} = (-1.11, \ 1.01).$$

3.39

(a) No, since $f_z(1,2,-3) = 0$.

(b) Yes, since $f_y(1,0,3) \neq 0$.

(c) Yes, since $f_y(1,2,3) \neq 0$.

(d) No, because the y values are different:

$$0 = g_{\text{part(b)}}(1,3) \neq 2 = g_{\text{part(c)}}(1,3).$$

(e) No, because $f_x(1,0,3) = 0$.

3.41

(a) $3(0) + (4)^2 + 4(-4) = 0$, $4(0)^3 + 4 + (-4) = 0$.

(b) $\begin{bmatrix} f_{1y_1} & f_{1y_2} \\ f_{2y_1} & f_{2y_2} \end{bmatrix} = \begin{bmatrix} 3 & 2y_2 \\ 12y_1^2 & 1 \end{bmatrix}$ and at $(y_1, y_2) = (0,4)$ this is $\begin{bmatrix} 3 & 8 \\ 0 & 1 \end{bmatrix}$ that is invertible. So there is a function **G**.

(c) From $f_1(x, g_1(x), g_2(x)) = 0$, $f_2(x, g_1(x), g_2(x)) = 0$ we get

$$\begin{bmatrix} f_{1x} \\ f_{2x} \end{bmatrix} + \begin{bmatrix} f_{1y_1} & f_{1y_2} \\ f_{2y_1} & f_{2y_2} \end{bmatrix} \begin{bmatrix} g_1' \\ g_2' \end{bmatrix} = \begin{bmatrix} 0 \\ 0 \end{bmatrix}.$$

At $x = -4$ this gives

$$\begin{bmatrix} 4 \\ 1 \end{bmatrix} + \begin{bmatrix} 3 & 8 \\ 0 & 1 \end{bmatrix} \begin{bmatrix} g_1'(-4) \\ g_2'(-4) \end{bmatrix} = \begin{bmatrix} 0 \\ 0 \end{bmatrix}$$

so

$$\begin{bmatrix} g_1'(-4) \\ g_2'(-4) \end{bmatrix} = -\begin{bmatrix} 3 & 8 \\ 0 & 1 \end{bmatrix}^{-1} \begin{bmatrix} 4 \\ 1 \end{bmatrix} = -\begin{bmatrix} \frac{1}{3} & -\frac{8}{3} \\ 0 & 1 \end{bmatrix} \begin{bmatrix} 4 \\ 1 \end{bmatrix} = \begin{bmatrix} \frac{4}{3} \\ -1 \end{bmatrix}.$$

3.43

(a) $\begin{bmatrix} 2y_1 & 0 \\ y_2 & y_1 \end{bmatrix}$

(b) The determinant is not zero when $y_1 \neq 0$.

(c) $y = \pm\sqrt{1-x^2}$, $v = -\frac{ux}{y} = \mp\frac{ux}{\sqrt{1-x^2}}$

(d) $x = \pm\sqrt{1-y^2}$, $u = -\frac{vy}{x} = \mp\frac{vy}{\sqrt{1-y^2}}$

Section 3.5

3.45

(a) $\text{div}\,(3\mathbf{F} + 4\mathbf{G})(1,2,-3) = 3\,\text{div}\,\mathbf{F}(1,2,-3) + 4\,\text{div}\,\mathbf{G}(1,2,-3)$
   $= 3(6) + 4(1^2 - 2(-3)) = 46$

(b) $\text{div}\,\text{curl}\,\mathbf{F}(x,y,z) = \text{div}\,(5y + 7z, 3x, 0) = \frac{\partial}{\partial x}(5y + 7z) + \frac{\partial}{\partial y}(3x) + \frac{\partial}{\partial z}(0) = 0$

(c) $\mathrm{curl}\,(3\mathbf{F}+4\mathbf{G})(1,2,-3) = 3\mathrm{curl}\,\mathbf{F}(1,2,-3)+4\mathrm{curl}\,\mathbf{G}(1,2,-3)$
$= 3(5(2)+7(-3),3(1),0)+4(5,7,9) = (-13,37,36)$

3.47 The vectors $\mathbf{H}(1,0) = (0,1)$, $\mathbf{H}(0,1) = (-1,0)$, $\mathbf{H}(-1,0) = (0,-1)$, and $\mathbf{H}(0,-1) = (1,0)$ suggest counterclockwise rotation. But:

$$\mathrm{curl}\,\mathbf{H} = \left(\frac{x}{(x^2+y^2)^p}\right)_x - \left(\frac{-y}{(x^2+y^2)^p}\right)_y$$

$$= \frac{1}{(x^2+y^2)^p} - 2px\frac{x}{(x^2+y^2)^{p+1}} + \frac{1}{(x^2+y^2)^p} - 2py\frac{y}{(x^2+y^2)^{p+1}}$$

$$= \frac{(x^2+y^2)-2px^2+(x^2+y^2)-2py^2}{(x^2+y^2)^{p+1}} = \frac{(2-2p)}{(x^2+y^2)^p}$$

The sign depends on p: negative when $p = 1.05$ or any $p > 1$, positive when $p = .95$ or any $p < 1$, and the curl is zero when $p = 1$.

3.49

(a) $n_x = (p_{1y}-v)_x = p_{1yx}-v_x = p_{1xy}-v_x = u_y-v_x = 0$
(b) $p_{2x} = p_{1x} = u$, $p_{2y} = p_{1y}-c_y = v+n-c_y = v$
(c) $m_x = (p_{2z}-w)_x = p_{2xz}-w_x = u_z-w_x = 0$ and $m_y = (p_{2z}-w)_y = p_{2yz}-w_y = v_z-w_y = 0$.
(d) $p_{3z} = p_{2z} - \frac{df}{dz} = w+m-m = w$

3.51

(a) The Chain Rule gives $\mathbf{X}'' = (D\mathbf{V})\mathbf{X}'+\mathbf{V}_t = \mathbf{V}_t+(D\mathbf{V})\mathbf{V}$.
(b) Since $\mathbf{X}(t) = \mathbf{C}_1+t^{-1}\mathbf{C}_2$, $\mathbf{V}(\mathbf{X}(t),t)) = \mathbf{X}'(t) = -t^{-2}\mathbf{C}_2$ and $\mathbf{X}'' = 2t^{-3}\mathbf{C}_2)$.
Since $\mathbf{V}(\mathbf{X}) = t^{-1}(\mathbf{C}_1-\mathbf{X})$, you have $\mathbf{V}_t = -t^{-2}(\mathbf{C}_1-\mathbf{X})$ and $D\mathbf{V} = -t^{-1}I$. Then

$$\mathbf{V}_t(\mathbf{X}(t),t))+D\mathbf{V}(\mathbf{X}(t),t))\mathbf{V}(\mathbf{X}(t),t)) = -t^{-2}(\mathbf{C}_1-\mathbf{X})-t^{-1}\mathbf{V}(\mathbf{X}(t),t))$$

$$= -t^{-2}(\mathbf{C}_1-(\mathbf{C}_1+t^{-1}\mathbf{C}_2))-t^{-1}(-t^{-2}\mathbf{C}_2) = -2t^{-3}\mathbf{C}_2$$

and this is equal to $\mathbf{X}''(\mathbf{X}(t),t)$.

3.53 The first term is, by the Mean Value Theorem,

$$-(f(x_0+\Delta x,y_m,z_m)-f(x_0,y_m,z_m))(1,0,0)\Delta y\Delta z = (-f_x(\bar{x},y_m,z_m),0,0)\Delta x\Delta y\Delta z$$

for some point $\bar{z}$ between z_0 and $z_0+\Delta z$. Similarly for the other two terms.

3.55 For each $k = 1,2,3$, the product rule gives $(uv)_{x_kx_k} = u_{x_kx_k}v+2u_{x_k}v_{x_k}+uv_{x_kx_k}$. Sum over k.

3.57 Begin with the right side. First calculate, writing terms in cyclic 123123 order,

$$(\mathrm{div}\,\mathbf{G})\mathbf{F}+\mathbf{G}\cdot\nabla\mathbf{F}$$

$$= ((g_{1x_1} + g_{2x_2} + g_{3x_3})f_1 + (g_1 f_{1x_1} + g_2 f_{1x_2} + g_3 f_{1x_3}),$$

$$(g_{2x_2} + g_{3x_3} + g_{1x_1})f_2 + (g_2 f_{2x_2} + g_3 f_{2x_3} + g_1 f_{2x_1}),$$

$$(g_{3x_3} + g_{1x_1} + g_{2x_2})f_3 + (g_3 f_{3x_3} + g_1 f_{3x_1} + g_2 f_{3x_2}))$$

Interchange $\mathbf{F}$ and $\mathbf{G}$ and subtract to get

$$(\operatorname{div}\mathbf{G})\mathbf{F} + \mathbf{G}\cdot\nabla\mathbf{F} - \big((\operatorname{div}\mathbf{F})\mathbf{G} + \mathbf{F}\cdot\nabla\mathbf{G}\big)$$

$$= (g_{2x_2}f_1 - f_{2x_2}g_1 + g_2 f_{1x_2} - f_2 g_{1x_2} + g_{3x_3}f_1 - f_{3x_3}g_1 + g_3 f_{1x_3} - f_3 g_{1x_3},$$

$$g_{3x_3}f_2 - f_{3x_3}g_2 + g_3 f_{2x_3} - f_3 g_{2x_3} + g_{1x_1}f_2 - f_{1x_1}g_2 + g_1 f_{2x_1} - f_1 g_{2x_1},$$

$$g_{3x_3}f_2 - f_{3x_3}g_2 + g_3 f_{2x_3} - f_3 g_{2x_3} + g_{1x_1}f_2 - f_{1x_1}g_2 + g_1 f_{2x_1} - f_1 g_{2x_1})$$

Group terms using the product rule:

$$= ((f_1 g_2 - f_2 g_1)_{x_2} - (f_3 g_1 - f_1 g_3)_{x_3},$$

$$(f_2 g_3 - f_3 g_2)_{x_3} - (f_1 g_2 - f_2 g_1)_{x_1},$$

$$(f_3 g_1 - f_1 g_3)_{x_1} - (f_2 g_3 - f_3 g_2)_{x_2}) = \operatorname{curl}(\mathbf{F}\times\mathbf{G}).$$

3.59

(a) Identity (g):

$$\operatorname{curl}\operatorname{curl}\mathbf{F}$$

$$= ((f_{2x_1} - f_{1x_2})_{x_2} - (f_{1x_3} - f_{3x_1})_{x_3},$$

$$(f_{3x_2} - f_{1x_3})_{x_3} - (f_{2x_1} - f_{1x_2})_{x_1},$$

$$(f_{1x_3} - f_{1x_1})_{x_1} - (f_{3x_2} - f_{2x_3})_{x_2})$$

$$= (f_{2x_1x_2} - f_{1x_2x_2} - f_{1x_3x_3} + f_{3x_1x_3},$$

$$f_{3x_2x_3} - f_{2x_3x_3} - f_{2x_1x_1} + f_{1x_2x_1},$$

$$f_{1x_3x_1} - f_{3x_1x_1} - f_{3x_2x_2} + f_{2x_3x_2})$$

$$+ (f_{1x_1x_1} - f_{1x_1x_1}, f_{2x_2x_2} - f_{2x_2x_2}, f_{3x_3x_3} - f_{3x_3x_3})$$

$$= \nabla(\operatorname{div}\mathbf{F}) - \varDelta\mathbf{F}$$

(b) Identity (h):

$$\varDelta(fg) = (fg)_{x_1 x_1} + (fg)_{x_2 x_2} + (fg)_{x_3 x_3}$$

$$= f_{x_1 x_1}g + 2 f_{x_1}g_{x_1} + f g_{x_1 x_1} + \cdots = g\varDelta f + 2\nabla f\cdot\nabla g + f\varDelta g.$$

(c) Identity (i): By Identity (d) and $\operatorname{curl}\nabla g = 0$ we have $\operatorname{curl}(f\nabla g) = \nabla f\times\nabla g$. Since $\nabla f\times\nabla g$ is a curl and $\operatorname{div}\operatorname{curl}\mathbf{H} = 0$ we get $\operatorname{div}(\nabla f\times\nabla g) = 0$.

Problems of Chapter 4

Section 4.1

4.1

(a) $f_x = -x(x^2 + y^2 + z^2)^{-3/2}$, $f_{xx} = -(x^2 + y^2 + z^2)^{-3/2} + 3x^2(x^2 + y^2 + z^2)^{-5/2}$ and by symmetry $f_{zz} = -(x^2 + y^2 + z^2)^{-3/2} + 3z^2(x^2 + y^2 + z^2)^{-5/2}$

(b) 0

(c) 0

(d) $h_{x_j} = a_j$ is constant, so second derivatives $h_{x_j x_k}$ are 0

4.3 $f_x(x,y) = \frac{3}{2}(x^2 + y^2)^{1/2}(2x) = 3x(x^2 + y^2)^{1/2}$, and when $(x,y) \neq (0,0)$

$$f_{xy}(x,y) = \frac{3}{2}x(x^2 + y^2)^{-1/2}(2y) = 3xy(x^2 + y^2)^{-1/2}.$$

4.5

(a) $g_x = ae^{ax+by+cz+dw} = ag$

(b) $a^2 d^2 e^{ax+by+cz+dw}$, $b^2 c^2 e^{ax+by+cz+dw}$.

(c) $a^2 d^2 + b^2 c^2 - 2abcd = (ad - bc)^2 = 0$ so $ad = bc$.

(d) $a^2 d^2 + b^2 c^2 - 2 = 0$.

4.7 $u_t = -\frac{n}{2}t^{-1-n/2}e^{-\|\mathbf{X}\|^2/4t} + \frac{\|\mathbf{X}\|^2}{4t^2}t^{-n/2}e^{-\|\mathbf{X}\|^2/4t} = -\frac{n}{2}t^{-1}u + \frac{\|\mathbf{X}\|^2}{4t^2}u.$

Similarly $u_{x_j} = -\frac{x_j}{2t}u$, $u_{x_j x_j} = -\frac{1}{2t}u + (-\frac{x_j}{2t})^2 u$, so

$$u_{x_1 x_1} + \cdots + u_{x_n x_n} = n(-\frac{1}{2t})u + \frac{x_1^2 + \cdots + x_n^2}{4t^2}u = u_t.$$

4.9 Differentiate $r^2 = x_1^2 + \cdots + x_n^2$ to get $r_{x_k} = \frac{x_k}{r}$. Therefore $(r^p)_{x_k} = pr^{p-2}x_k$. Then by the product rule $(r^p)_{x_k x_k} = p(p-2)r^{p-4}x_k x_k + pr^{p-2}$. Sum over k to get

$$\Delta(r^p) = p(p-2)r^{p-4}r^2 + npr^{p-2} = p(p-2+n)r^{p-2}.$$

This is zero if $p = 2 - n$.

4.11

(a) $((x^2 + y^2 + z^2)^{1/2})_x = x(x^2 + y^2 + z^2)^{-1/2}$ and

$$((x^2 + y^2 + z^2)^{1/2})_{xx} = (x^2 + y^2 + z^2)^{-1/2} - x^2(x^2 + y^2 + z^2)^{-3/2}.$$

So by symmetry $\Delta((x^2 + y^2 + z^2)^{1/2}) = (3 - 1)(x^2 + y^2 + z^2)^{-1/2}$.

(b) $u_x = v_r r_x = v_r \frac{x}{r}$ and $u_{xx} = v_{rr} \frac{x^2}{r^2} + r^{-1} v_r - v_r \frac{x^2}{r^3}$; similarly for u_{yy} and u_{zz}. Then

$$\Delta u = v_{rr} \frac{x^2 + y^2 + z^2}{r^2} + 3r^{-1} v_r - v_r \frac{x^2 + y^2 + z^2}{r^3} = v_{rr} + 2r^{-1} v_r.$$

(c) Using parts (a) and (b),

$$\Delta \Delta r = \Delta(2r^{-1}) = 2((-1)(-2)r^{-3} + 2r^{-1}(-1)r^{-2}) = 2(2-2)r^{-3} = 0.$$

Section 4.2

4.13

(a) $\nabla f(x,y) = (-2 + 2x, 4 - 8y + 3y^2) = (-2 + 2x, (y-2)(3y-2))$ is zero at $(1,2)$ and at $(1, \frac{2}{3})$.

(b) $\mathcal{H}(x,y) = \begin{bmatrix} 2 & 0 \\ 0 & -8+6y \end{bmatrix}$, $\mathcal{H}(1, \frac{2}{3}) = \begin{bmatrix} 2 & 0 \\ 0 & -4 \end{bmatrix}$, $\mathcal{H}(1,2) = \begin{bmatrix} 2 & 0 \\ 0 & 4 \end{bmatrix}$.

(c) $\mathcal{H}(1,2)$ is positive definite because $\mathbf{U} \cdot \mathcal{H}(1,2)\mathbf{U} = 2u^2 + 4v^2$ is positive except at the origin. $\mathcal{H}(1, \frac{2}{3})$ is indefinite because $2u^2 - 4v^2$ has both positive and negative values. Therefore f has a local minimum at $(1,2)$ and a saddle at $(1, \frac{2}{3})$.

4.15

(a) $3x_1^2 + 4x_1 x_2 + x_2^2 = \mathbf{X} \cdot \left(\begin{bmatrix} 3 & 2 \\ 2 & 1 \end{bmatrix} \mathbf{X} \right)$

(b) $-x_1^2 + 5x_1 x_2 + 3x_2^2 = \mathbf{X} \cdot \left(\begin{bmatrix} -1 & \frac{5}{2} \\ \frac{5}{2} & 3 \end{bmatrix} \mathbf{X} \right)$

4.17 $\nabla f = (-3x^2 + 2x + y, x + 6y)$. The first derivatives at $(0,0)$ are zero, and

$$\begin{bmatrix} f_{xx}(0,0) & f_{xy}(0,0) \\ f_{yx}(0,0) & f_{yy}(0,0) \end{bmatrix} = \begin{bmatrix} 2 & 1 \\ 1 & 6 \end{bmatrix}$$

is positive definite. Therefore $f(0,0)$ is a local minimum.

4.19 Complete the square: $x^2 + qxy + y^2 = (x + \frac{1}{2}qy)^2 + (1 - \frac{q^2}{4})y^2$. That is positive definite if $|q| < 2$.

4.21 At $(0,0)$, $\begin{bmatrix} f_{xx} & f_{xy} \\ f_{xy} & f_{yy} \end{bmatrix} = \begin{bmatrix} 2 & 0 \\ 0 & 4 \end{bmatrix}$, $\begin{bmatrix} g_{xx} & g_{xy} \\ g_{xy} & g_{yy} \end{bmatrix} = \begin{bmatrix} 0 & 0 \\ 0 & 4 \end{bmatrix}$. The corresponding quadratic forms are

$$S_f(u,v) = 2u^2 + 4v^2$$

which is positive definite, and $S_g(u,v) = 4v^2$ which is not because the values $S_g(u,0)$ are not positive.

4.23 The closest point is $(-\frac{1}{2}, 1, \frac{1}{2})$. This is also a normal vector to the plane because the equation for the plane can be written as

$$-\frac{1}{2}x + y + \frac{1}{2}z = \frac{3}{2}.$$

A sketch makes clear that the distance to the origin is minimum when that distance is measured along a line normal to the plane.

Section 4.3

4.25

(a) $\nabla f(x,y,z) = (\frac{e^y}{1+x}, \log(1+x)e^y, \cos z)$ is nowhere the zero vector, so there are no local extrema.

(b) $f(0,0,0) = 0$, $\nabla f(0,0,0) = (1,0,1)$, and

$$\mathcal{H}f(x,y,z) = \begin{bmatrix} -\frac{e^y}{(1+x)^2} & \frac{e^y}{1+x} & 0 \\ \frac{e^y}{1+x} & \log(1+x)e^y & 0 \\ 0 & 0 & -\sin z \end{bmatrix}, \quad \mathcal{H}f(0,0,0) = \begin{bmatrix} -1 & 1 & 0 \\ 1 & 0 & 0 \\ 0 & 0 & 0 \end{bmatrix},$$

so

$$p_2(\mathbf{H}) = f(0,0,0) + \nabla f(0,0,0) \cdot \mathbf{H} + \tfrac{1}{2}\mathbf{H}^T \mathcal{H}f(0,0,0)\mathbf{H} = h_1 + h_3 + \tfrac{1}{2}(-h_1^2 + 2h_1h_2).$$

4.27

(a) $\nabla f(x,y,z) = (2x+y, x+2z+y, 2y+3z^2)$. This is $(0,0,0)$ when

$$y = -2x, \; z = -\tfrac{1}{2}x+x, \; 2(-2x)+3(\tfrac{1}{4}x^2) = 0,$$

giving $(x,y,z) = (0,0,0)$ and $(x,y,z) = \mathbf{A} = (\frac{16}{3}, -\frac{32}{3}, \frac{8}{3})$.

(b)

$$\mathcal{H}f(x,y,z) = \begin{bmatrix} 2 & 1 & 0 \\ 1 & 1 & 2 \\ 0 & 2 & 6z \end{bmatrix}, \quad \mathcal{H}f(0,0,0) = \begin{bmatrix} 2 & 1 & 0 \\ 1 & 1 & 2 \\ 0 & 2 & 0 \end{bmatrix}, \quad \mathcal{H}f(\mathbf{A}) = \begin{bmatrix} 2 & 1 & 0 \\ 1 & 1 & 2 \\ 0 & 2 & 16 \end{bmatrix}.$$

For $\mathcal{H}f(\mathbf{A})$ you have determinants $2 > 0$, $1 > 0$, $-2(4)+(16) = 8 > 0$ so $f(\mathbf{A})$ is a local minimum.

(c) For $\mathcal{H}f(\mathbf{0})$ the determinants are $2 > 0$, $1 > 0$, and $-8 < 0$ so the determinant test for positive definiteness doesn't help. But $f(0,0,z) = z^3$ has the same sign as z so $f(0,0,0)$ is a saddle.

4.29

(a) $f(\mathbf{X}) = (x_1^2 + \cdots)^{-1/2}$ so $f_{x_j}(\mathbf{X}) = -\tfrac{1}{2}(x_1^2 + \cdots)^{-3/2}2x_j = -(x_1^2 + \cdots)^{-3/2}x_j$ and

$$f_{x_j x_k}(\mathbf{X}) = 3(x_1^2 + \cdots)^{-5/2}x_k x_j - (x_1^2 + \cdots)^{-3/2}\frac{\partial x_j}{\partial x_k} = 3(x_1^2 + \cdots)^{-5/2}x_k x_j - (x_1^2 + \cdots)^{-3/2}\delta_{jk}$$

where δ_{jk} is 1 if $j = k$ else 0.

(b) $f(\mathbf{A}) = \|\mathbf{A}\|^{-1}$ and from part (a), $f_{x_j}(\mathbf{A}) = \|\mathbf{A}\|^{-3}a_j$ and

$$f_{x_j x_k}(\mathbf{A}) = 3\|\mathbf{A}\|^{-5}a_k a_j - \|\mathbf{A}\|^{-3}\delta_{jk} = \|\mathbf{A}\|^{-5}(3a_k a_j - \|\mathbf{A}\|^2\delta_{jk}).$$

So the Taylor approximation is

$$\|\mathbf{A}+\mathbf{H}\|^{-1} \approx p_2(\mathbf{A}+\mathbf{H}) = \|\mathbf{A}\|^{-1} + \|\mathbf{A}\|^{-3} \sum_{j=1}^{3} a_j h_j + \tfrac{1}{2}\|\mathbf{A}\|^{-5} \sum_{j,k=1}^{3} ((3a_k a_j - \|\mathbf{A}\|^2 \delta_{jk})) h_j h_k$$

$$= \|\mathbf{A}\|^{-1} + \|\mathbf{A}\|^{-3}\mathbf{A}\cdot\mathbf{H} + \tfrac{1}{2}\|\mathbf{A}\|^{-5}(3(\mathbf{A}\cdot\mathbf{H})^2 - \|\mathbf{A}\|^2\|\mathbf{H}\|^2).$$

4.31 $p_1(\mathbf{X}) = x_1 + x_2 + x_3 + x_4 = p_2(\mathbf{X}) = p_3(\mathbf{X})$ and $p_4(\mathbf{X}) = x_1 + x_2 + x_3 + x_4 + x_1 x_2 x_3 x_4 = p_5(\mathbf{X})$.

Section 4.4

4.33 $\nabla(3x^2 + 2xy + 3y^2) = \lambda\nabla(x^2 + y^2)$ gives $(6x + 2y, 2x + 6y) = \lambda(2x, 2y)$. So we have to solve

$$(6-2\lambda)x + 2y = 0, \qquad 2x + (6-2\lambda)y = 0, \qquad x^2 + y^2 = 1.$$

We get $y = (-3+\lambda)x$, $x + (3-\lambda)(-3+\lambda)x = 0$, so $x = 0$ or $1 - (\lambda-3)^2 = 0$. So $\lambda = 2$ or 4. Using 4, $y = x$ and $Q(\frac{1}{\sqrt{2}}, \frac{1}{\sqrt{2}}) = (3+2+3)\frac{1}{2} = 4 = Q(\frac{-1}{\sqrt{2}}, \frac{-1}{\sqrt{2}})$. That is the maximum because using $\lambda = 2$ you find $y = -x$ and $Q(\frac{1}{\sqrt{2}}, -\frac{1}{\sqrt{2}}) = (3-2+3)\frac{1}{2} = 2 = Q(\frac{-1}{\sqrt{2}}, \frac{1}{\sqrt{2}})$.

4.35 We solve $\nabla(\|\mathbf{X}-\mathbf{A}\|^2) = \lambda\nabla(\mathbf{C}\cdot\mathbf{X})$ with $\mathbf{C}\cdot\mathbf{X} = 0$, or

$$2(\mathbf{X}-\mathbf{A}) = \lambda\mathbf{C}, \qquad \mathbf{C}\cdot\mathbf{X} = 0.$$

This gives

$$\mathbf{X} = \mathbf{A} + \tfrac{1}{2}\lambda\mathbf{C}, \qquad \mathbf{C}\cdot\mathbf{X} = \mathbf{C}\cdot\mathbf{A} + \tfrac{1}{2}\lambda\|\mathbf{C}\|^2 = 0.$$

Therefore

$$\lambda = -\frac{2\mathbf{C}\cdot\mathbf{A}}{\|\mathbf{C}\|^2}, \qquad \mathbf{X} = \mathbf{A} - \frac{\mathbf{C}\cdot\mathbf{A}}{\|\mathbf{C}\|^2}\mathbf{C},$$

and the minimum distance is $\dfrac{|\mathbf{C}\cdot\mathbf{A}|}{\|\mathbf{C}\|}$.

4.37 We maximize $f(x,y) = [x\ y]\begin{bmatrix} 1 & 2 \\ 2 & -2 \end{bmatrix}\begin{bmatrix} x \\ y \end{bmatrix} = x^2 + 4xy - 2y^2$ subject to $g(x,y) = x^2 + y^2 = 1$. We solve $\nabla f = \lambda\nabla g$ or

$$(2x+4y, 4x-4y) = \lambda(2x, 2y).$$

Note this says that $2\mathbf{A}\mathbf{X} = \lambda 2\mathbf{X}$. Dividing by 2, these give

$$(1-\lambda)x + 2y = 0$$

$$2x - (2+\lambda)y = 0.$$

Either $(x,y) = (0,0)$ or the determinant

$$(1-\lambda)(-2-\lambda) - 4 = 0.$$

We get $\lambda = 2$ or -3. Therefore 2 is the maximum, and it occurs where

$$-x + 2y = 0.$$

This gives $x = 2y$ with $x^2 + y^2 = 5y^2 = 1$, or

$$\mathbf{A}\begin{bmatrix} \frac{2}{\sqrt{5}} \\ \frac{1}{\sqrt{5}} \end{bmatrix} = 2\begin{bmatrix} \frac{2}{\sqrt{5}} \\ \frac{1}{\sqrt{5}} \end{bmatrix}.$$

4.39 Let $f(x,y,z) = (x+1)^2 + y^2 + z^2$ and $g(x,y,z) = x^3 + y^2 + z^2 = 1$. Find solutions to $\nabla f = \lambda \nabla g$ and $x^3 + y^2 + z^2 = 1$.

$$2(x+1) = \lambda(3x^2), \quad 2y = \lambda(2y), \quad 2z = \lambda(2z).$$

Case 1: If y or z is not 0 then $\lambda = 1$ and $x = \frac{2 \pm \sqrt{4-4(3)(-2)}}{6} = \frac{1 \pm \sqrt{7}}{3}$.

Since $y^2 + z^2 = 1 - (\frac{1+\sqrt{7}}{3})^3$ is negative, the plus sign is not possible. Therefore only $x = \frac{1-\sqrt{7}}{3}$ is possible. Then

$$y^2 + z^2 = 1 - (\tfrac{1-\sqrt{7}}{3})^3 = 1 + (\tfrac{\sqrt{7}-1}{3})^3,$$

$$f(\tfrac{1-\sqrt{7}}{3}, y, z) = (1 + \tfrac{1-\sqrt{7}}{3})^2 + y^2 + z^2 = (1 + \tfrac{1-\sqrt{7}}{3})^2 + 1 + (\tfrac{\sqrt{7}-1}{3})^3 < (\tfrac{2}{3})^2 + 1 + (\tfrac{2}{3})^3 < 3.$$

Case 2: If $y = 0$ and $z = 0$ then $x^3 = 1$ and $x = 1$ and $f(1,0,0) = (1+1)^2 = 4$.

Therefore $f(x,y,z) = (x+1)^2 + y^2 + z^2$ subject to $x^3 + y^2 + z^2 = 1$ has a minimum on the circle of points

$$x = \tfrac{1-\sqrt{7}}{3}, \qquad y^2 + z^2 = 1 + (\tfrac{\sqrt{7}-1}{3})^3.$$

Problems of Chapter 5

Section 5.1

5.1

(a) velocity $\mathbf{X}'(t) = \mathbf{0}$, acceleration $\mathbf{X}''(t) = \mathbf{0}$, speed $\|\mathbf{X}'(t)\| = 0$
(b) $(1,1,1)$, $\mathbf{0}$, $\sqrt{3}$
(c) $(-1,-1,1)$, $\mathbf{0}$, $\sqrt{3}$
(d) $(1,2,3)$, $\mathbf{0}$, $\sqrt{14}$
(e) $(1,2t,3t^2)$, $(0,2,6t)$, $\sqrt{1+4t^2+9t^4}$

5.3

(a) $\mathbf{V}(t) = \mathbf{X}'(t) = r\omega(-\sin(\omega t), \cos(\omega t))$ so $\mathbf{V}(t) \cdot \mathbf{X}(t) = 0$
(b) $\|\mathbf{V}(t)\| = r\omega \underbrace{\|(-\sin(\omega t), \cos(\omega t))\|}_{1}$

(c) $\mathbf{X}''(t) = -r\omega^2(\cos(\omega t), \sin(\omega t))$ is a negative multiple of $\mathbf{X}(t)$, so is toward the origin.

(d) $\|\mathbf{X}''(t)\| = r\omega^2 \underbrace{\|(\cos(\omega t), \sin(\omega t))\|}_{1}$

5.5 Differentiate $\mathbf{X} \cdot \mathbf{X} =$ constant to find $\mathbf{X}' \cdot \mathbf{X} = 0$. Since $\mathbf{X}(t)$ is on the sphere, this means $\mathbf{X}'(t)$ is tangent.

5.7 $\mathbf{X}(t) = \mathbf{A} + \mathbf{B}t + \frac{1}{2m}\mathbf{F}t^2$ with constant $\mathbf{F}$ gives $\mathbf{X}'(t) = \mathbf{B} + \frac{1}{m}\mathbf{F}t$ and $\mathbf{X}''(t) = \frac{1}{m}\mathbf{F}$ so $\mathbf{F} = m\mathbf{X}''$. $\mathbf{X}(0) = \mathbf{A}$, $\mathbf{X}'(0) = \mathbf{B}$.

5.9

(a) $\mathbf{X}(t)$ is a multiple of $\mathbf{A}$, so the motion is along the line through the origin containing $\mathbf{A}$ (or remains at $\mathbf{0}$ if $\mathbf{A} = \mathbf{0}$). Many planes contain the line.

(b) Same as (a).

(c) $\mathbf{X}(t)$ is a linear combination of $\mathbf{A}$ and $\mathbf{B}$ so lies in the plane through the origin containing $\mathbf{A}$ and $\mathbf{B}$. If it happens that $\mathbf{A}$ and $\mathbf{B}$ are linearly dependent, then the motion is along a line (or remains at $\mathbf{0}$) and there are many planes containing it.

5.11 $\mathbf{X}(0) = \mathbf{C}$, $\mathbf{X}'(t) = e^{-kt}\mathbf{D}$, $\mathbf{X}'(0) = \mathbf{D}$, $\mathbf{X}''(t) = -ke^{-kt}\mathbf{D} = -k\mathbf{X}'(t)$. As t increases from 0 to infinity, $\mathbf{X}(t)$ goes from $\mathbf{C}$ to $\mathbf{C} + \frac{1}{k}\mathbf{D}$, so the particle is displaced $\frac{1}{k}\mathbf{X}'(0)$.

5.13

(a) $\mathbf{V} \times (0,0,b) = (v_1\mathbf{i} + v_2\mathbf{j} + v_3\mathbf{k}) \times (b\mathbf{k}) = -bv_1\mathbf{j} + bv_2\mathbf{i} = (bv_2, -bv_1, 0)$

(b) If $\mathbf{X}(t) = (a\sin(\omega t), a\cos(\omega t), bt)$ then

$$\mathbf{V}(t) = \mathbf{X}'(t) = (a\omega\cos(\omega t), -a\omega\sin(\omega t), b), \quad \mathbf{V}'(t) = \mathbf{X}''(t) = (-a\omega^2\sin(\omega t), -a\omega^2\cos(\omega t), 0)$$

and

$$(bv_2(t), -bv_1(t), 0) = (-ba\omega\sin(\omega t), -ba\omega\cos(\omega t), 0)$$

so $\mathbf{V}' = \mathbf{V} \times \mathbf{B}$ as required.

(c) The force and acceleration are horizontal and point toward the x_3 axis.

Section 5.2

5.15

(a) From $(x(t), y(t)) = (a\cos\omega t, a\sin\omega t)$ we get

$$x^2 + y^2 = a^2\cos^2\omega t + a^2\sin^2\omega t = a^2$$

so $r = a$.

(b) From $(x(t), y(t)) = (a\cos\omega t, a\sin\omega t)$ we get

$$(x', y') = (-a\omega\sin\omega t, a\omega\cos\omega t), \qquad (x'', y'') = (-a\omega^2\cos\omega t, -a\omega^2\sin\omega t).$$

Observe that $(x'', y'') = -\omega^2(x, y)$. Using that with part (a) we get

$$x'' + \frac{x}{(x^2 + y^2)^{3/2}} = (-\omega^2 + a^{-3})x, \qquad y'' + \frac{y}{(x^2 + y^2)^{3/2}} = (-\omega^2 + a^{-3})y,$$

These are zero if $\omega^2 a^3 = 1$.

(c) The constant A is defined as

$$A = xy' - yx' = (a\cos\omega t)(a\omega\cos\omega t) - (a\sin\omega t)(-a\omega\sin\omega t) = a^2\omega(\cos^2\omega t + \sin^2\omega t) = a^2\omega.$$

5.17

(a) The error is small compared to h by definition of derivative.

(b) See Problem 1.80.

(c) This is direct from equation (5.11).

(d) The rate of change of area is $\frac{1}{2}A$ area/time, and T is the time required for one loop, so the area must be $\frac{1}{2}AT$.

5.19 We are assuming

$$f'' + \frac{f}{(f^2 + g^2)^{3/2}} = 0, \qquad g'' + \frac{g}{(f^2 + g^2)^{3/2}} = 0.$$

Replacing f by $-f$ has the same effect as multiplying the first equation by -1, so f can be replaced by $-f$. The original value of $xy' - yx' = fg' - gf'$.

The new value is $xy' - yx' = (-f)g' - g(-f') = -(fg' - gf')$, that is positive if the original value is negative.

5.21

(a) $-pr$ is positive so $q - y$ must be positive.

(b) On the y axis, $r = |y|$, so $-p|y| = q - y$ has a positive solution $y = \dfrac{q}{1-p}$ and a negative solution $y = \dfrac{q}{1+p}$. But since the orbit is left-right symmetric (due to the x^2), it can only be an ellipse.

(c) One axis has length $\dfrac{q}{1-p} - \dfrac{q}{1+p} = 2\dfrac{pq}{1-p^2}$.

The other must be at the middle where $y = \dfrac{1}{2}\left(\dfrac{q}{1-p} + \dfrac{q}{1+p}\right) = \dfrac{q}{1-p^2}$ and there

$$-pr = q - y = q - \frac{q}{1-p^2} = \frac{-p^2 q}{1-p^2}.$$

So at the middle

$$x = \pm\sqrt{r^2 - y^2} = \pm\sqrt{\frac{p^2 q^2 - q^2}{(1-p^2)^2}} = \pm\frac{q}{\sqrt{p^2-1}}.$$

Therefore the semiminor axis is $\dfrac{q}{\sqrt{p^2-1}}$, the semimajor $\dfrac{-pq}{p^2-1}$.

(d) Using (5.18) and (5.22) we get $A = \sqrt{\dfrac{q}{-p}}$, therefore $\dfrac{q}{\sqrt{p^2-1}} = \dfrac{\sqrt{q}}{\sqrt{p^2-1}}A\sqrt{-p}$

$= A(\text{semimajor axis})^{1/2}$.

(e) Therefore $\frac{1}{2}AT = \pi(\text{semiminor})(\text{semimajor}) = \pi A(\text{semimajor axis})^{3/2}$.

5.23

(a) Since $\nabla\|\mathbf{X}\| = \|\mathbf{X}\|^{-1}\mathbf{X}$, the Chain Rule gives $\nabla(\frac{1}{4}\|\mathbf{X}\|^{-4}) = -\|\mathbf{X}\|^{-5}\nabla\|\mathbf{X}\| = -\|\mathbf{X}\|^{-6}\mathbf{X}$.

(b) The energy $\frac{1}{2}\|\mathbf{X}'\|^2 - \frac{1}{4}\|\mathbf{X}\|^{-4}$ is constant; since $\mathbf{X}' = (-a\sin\theta, a\cos\theta)\theta'$ it is

$$\tfrac{1}{2}a^2(\theta')^2 - \tfrac{1}{4}a^{-4}((1+\cos\theta)^2 + \sin^2\theta)^{-2} = \tfrac{1}{2}a^2(\theta')^2 - \tfrac{1}{4}a^{-4}(2+2\cos\theta)^{-2}.$$

(c) Using (b), $\dfrac{1}{2}a^2k^2 - \dfrac{1}{16a^4} = 0$ so $k^2 = \dfrac{1}{8a^6}$, or $k = \dfrac{1}{a^3\sqrt{8}}$.

(d) Your sketch will show that $f(\theta) = \theta + \sin\theta$ is an increasing function of θ on $[0,\pi]$, so it has an inverse. Then solve $\theta + \sin\theta = kt$ by setting $\theta(t) = f^{-1}(kt)$.

(e) This motion follows a semicircle into the origin at the time t when $\theta(t) = \pi$ because at that time $\mathbf{X}(t) = (a + a\cos\pi, a\sin\pi) = (0,0)$. Since $\theta(t)$ is defined by $\theta + \sin\theta = kt$ and $\theta(t) = \pi$ we get $\pi + \sin\pi = kt$, or $t = \pi/k$.

5.25

$$\mathbf{Y}''(t) = ab^2\mathbf{X}''(bt) = -ab^2\frac{\mathbf{X}}{\|\mathbf{X}\|^3} = -ab^2\frac{a^{-1}\mathbf{Y}}{\|a^{-1}\mathbf{Y}\|^3} = -a^3b^2\frac{\mathbf{Y}}{\|\mathbf{Y}\|^3}.$$

That is $-\dfrac{\mathbf{Y}}{\|\mathbf{Y}\|^3}$ if $a^3b^2 = 1$.

Problems of Chapter 6

Section 6.1

6.1

(a) 14

(b) Yes

6.3

(a) 36, $\frac{36}{4} = 9$;

(b) 64, $\frac{64}{4} = 16$;

(c) 39, $\frac{39}{4} = 9.75$;

(d) $39 > 64 - 36$.

6.5 The h-cubes in $D \cup E$ that intersect the common boundary have total volume less than some multiple sh, where s only depends on the smooth common surface. Since this tends to 0 with h the volumes add.

6.7

(a) The plate measures 15 by 21. Using the upper and lower bounds, the mass is between $2(15)(21) = 630$ and $7(15)(21) = 2205$.

(b) D_1 measures 15 by 9, and D_2 15 by 12. So bounds give

$$2(15)(9) = 270 \le \text{mass}(D_1) \le 4(15)(9) = 540, \quad 720 \le \text{mass}(D_2) \le 1260.$$

Additive property gives $990 \le \text{mass}(D) \le 1800$.

(c) $2(9)(12) + 2(3)(9) + 4(5)(12) + 6(10)(12) = 1230 \le \text{mass}(D)$
$$\le 4(9)(12) + 2(3)(9) + 6(5)(12) + 7(10)(12) = 1686$$

6.9

(a) $\int_C \delta \, dV = \delta \, \text{Vol}(C) = 200(.05) = 10$, and the sum is 30, so $\int_D \rho \, dV = 20$.

(b) $\rho_{\min}(.05) \le \int_D \rho \, dV = 20 \le \rho_{\max}(.05)$. Therefore $\rho_{\min} \le \frac{20}{.05} = 400$ and $\rho_{\max} \ge \frac{20}{.05} = 400$.

Section 6.2

6.11

(a) $\int_D 1 \, dA = \text{Area}(D)$. From single variable calculus we know this area is (use integration by parts)

$$\int_1^2 \log x \, dx = [x \log x]_1^2 - \int_1^2 x \frac{1}{x} \, dx = -1 + 2 \log 2$$

(b) $\int_D x \, dA$, D the rectangle $0 \le x \le 3$, $-1 \le y \le 1$. The graph of $z = x$ is a plane and $z \ge 0$ on D, so $\int_D x \, dA$ is the volume under the graph of $z = x$ over D. That region R is a wedge.

$$\int_D x \, dA = \text{Vol}(R) = \tfrac{1}{2}(3)(3)(2) = 9.$$

(c) half the volume inside the unit ball, $\tfrac{1}{2}\tfrac{4}{3}\pi$

(d) same as part (c)

6.13 (b), (c), and (d) only

6.15 Let D be the unit disk centered at the origin.

$$\int_D (y^3 + 3xy + 2) \, dA = \int_D y^3 \, dA + 3 \int_D xy \, dA + 2 \int_D dA.$$

By symmetry $\int_D y^3\, dA = 0$ and $\int_D xy\, dA = 0$ and $\int_D dA = \text{Area}(D)$, so $\int_D (y^3 + 3xy + 2)\, dA = 2\pi.$

6.17

(a) Suppose $f(a,b) = p > 0$. Since f is continuous (take $\epsilon = \frac{1}{2}p$) there is an $r > 0$ so that if
$$(x-a)^2 + (y-b)^2 < r^2$$
then
$$|f(x,y) - f(a,b)| < \frac{1}{2}p.$$
Therefore $f(x,y) > \frac{1}{2}p$ for all (x,y) in D.
(b) Part (b) is the lower bound property.
(c) Parts (a) and (b) give $\int_{\text{disk}} f\, dA > 0$ which contradicts $\int_{\text{any}} f\, dA = 0$. So f cannot be positive at any point of $\mathbb{R}^2$.
(d) Apply (c) to the function $-f$. By part (c), $-f$ cannot be positive at any point of $\mathbb{R}^2$. Therefore f cannot be negative at any point.

6.19 The limits on the sum are $i^2 + j^2 \le 10h^{-2}$. When $j = 0$ this gives $i^2 h^2 \le 10$ or $-\sqrt{10} \le ih \le \sqrt{10}$. Similarly when $i = 0$, $-\sqrt{10} \le jh \le \sqrt{10}$. So it is the second integral.

6.21

(a) The sine function only takes values from -1 to 1, so these are bounds on the integrand.
(b) $J = .423 + \int_C \sin\left(\frac{1}{(1-x^2)(1-y^2)}\right) dA$, where C consists of all points of the square that are within .001 of the right edge or top edge. Since $\text{Area}(C) < .002$ the last integral is between $-.002$ and $.002$, and we get
$$.421 < J < .425.$$

6.23

(a) (i) $g \le f$ is given.
  (ii) Use Theorem 6.8 part (c): since $0 \le f - g$, $0 \le \int (f-g)\, dA$
  (iii) Use Theorem 6.8 parts (a) and (b): $\int (f-g)\, dA = \int f\, dA - \int g\, dA$
  (iv) Add $\int g\, dA$ to both sides of the inequality.
(b) (i) Because $C \subset D$.
  (ii) Create a Riemann sum for $\int_D f\, dA$ using the same points as for the Riemann sum for $\int_C f\, dA$, then any additional terms in the D sum are nonnegative because $f \ge 0$.
  (iii) The sums for D tend to $\int_D f\, dA$, and they are all larger than the C sum.

(iv) By part (c), $\int_D f\,dA$ is an upper bound for the C sums, so the limit $\int_C f\,dA \le$ $\int_D f\,dA.$

Section 6.3

6.25 $\int_{-1}^{1}\left(\int_0^{\sqrt{1-x^2}} y\,dy\right)dx = \int_{-1}^{1}\frac{1}{2}(1-x^2)\,dx = \frac{1}{2}(2-\frac{2}{3}) = \frac{2}{3}$

6.27

(a) In D, $x \le 0 \le \sqrt{y}$, so $x \le \sqrt{y}+x \le \sqrt{y}$. Therefore

$$\int_D x\,dA \le \int_D (\sqrt{y}+x)\,dA \le \int_D \sqrt{y}\,dA.$$

(b) In D, $-1 \le x \le y-1$ and $0 \le y \le 1$.

$$\int_D x\,dA = \int_0^1\left[\int_{-1}^{y-1} x\,dx\right]dy = \int_0^1 \frac{1}{2}((y-1)^2-1)\,dy = \frac{1}{2}\left[\frac{1}{3}(y-1)^3-y\right]_0^1 = -\frac{1}{2}+\frac{1}{6} = -\frac{1}{3}.$$

$$\int_D \sqrt{y}\,dA = \int_0^1\left[\int_{-1}^{y-1} \sqrt{y}\,dx\right]dy = \int_0^1 \sqrt{y}(y-1-(-1))\,dy = \left[\frac{2}{5}y^{5/2}\right]_0^1 = \frac{2}{5}.$$

$$\int_D (\sqrt{y}+x)\,dA = \frac{2}{5}-\frac{1}{3}.$$

6.29

$$\int_R \sin y\,dA = \int_0^1\left[\int_{x=y^2}^1 \sin y\,dx\right]dy = \int_0^1 (1-y^2)\sin y\,dy$$

$$= \left[(1-y^2)(-\cos y)\right]_0^1 - \int_0^1 2y\cos y\,dy = 1-2\left(\left[y\sin y\right]_0^1 - \int_0^1 \sin y\,dy\right)$$

$$= 1-2\sin(1)-2(\cos(1)-1) = 3-2\sin(1)-2\cos(1).$$

6.31 $1,3,2,5,1$:

$$\int_{[0,2]\times[-1,(1)]} (3xy^2+5x^4y^3)\,dA = (3)\int_0^{(2)} x\,dx\int_{-1}^1 y^2\,dy + (5)\int_0^2 x^4\,dx\int_{-1}^{(1)} y^3\,dy.$$

Section 6.4

6.33

(a) Corresponding parts of the triangles are proportional, so

$$\frac{kh^2+hc}{\ell+ha} = \frac{hc}{ha}.$$

Solve to get $\ell = \frac{h^2ka}{c}$.

(b) Similarly $\dfrac{m}{hb} = \dfrac{h^2k + hc}{hc}$, giving $m = hb + \dfrac{b}{c}kh^2$.

(c) The change in area is

$$\tfrac{1}{2}m(\ell + ha) - \tfrac{1}{2}(hahb) = \tfrac{1}{2}(hb + \tfrac{b}{c}kh^2)(\dfrac{h^2ka}{c} + ha) - \tfrac{1}{2}(hahb) = \tfrac{1}{2}k\dfrac{ab}{c}h^3 + k^2\dfrac{ab}{c}h^4.$$

If h is small this is less than a multiple of h^3.

6.35 $\det \begin{bmatrix} 2u & -2v \\ 2v & 2u \end{bmatrix} = 4(u^2 + v^2)$, that is 4 at $(1,0)$. The vertices of the triangle map to $(1,0)$, $(1.21,0)$, and $(.99,.2)$, so the area is about four times as large.

6.37

$$\int_D e^{-\|X\|}\,dA = \int_0^{2\pi}\int_a^b e^{-r}r\,dr\,d\theta = \int_0^{2\pi}d\theta\int_a^b e^{-r}r\,dr.$$

Using integration by parts,

$$\int_a^b e^{-r}r\,dr = \Big[(-1)re^{-r}\Big]_a^b + \int_a^b e^{-r}\,dr = \Big[-re^{-r} - e^{-r}\Big]_a^b = (-1-b)e^{-b} + (1+a)e^{-a}.$$

So

$$\int_D e^{-\|X\|}\,dA = 2\pi((-1-b)e^{-b} + (1+a)e^{-a}).$$

Fig. 9.21 Initials in Problem 6.39.

6.39 See Figure 9.21. (a) stretches horizontally by a factor of 5, and 3 vertically (b) interchanges u, v then goes by (a) (c) right edge $(u = 1)$ goes to the bottom $(y = 0)$, and left edge to the top

6.41

(a) $(5)(3) + \dfrac{5^2}{2}\dfrac{3^2}{2} =_? 15 + (15)^2\dfrac{1}{2}\dfrac{1}{2}$ yes.
(b) off by a minus sign
(c) $(5)(3) + \dfrac{5^2}{2}\dfrac{3^2}{2} =_? 15 + 15^2(\tfrac{1}{2})(1 - \tfrac{1}{2})$ yes.

Section 6.5

6.43 $D_n = [0,n] \times [0,1]$ is an increasing sequence of bounded rectangles whose union is D.

(a) $\displaystyle\int_{D_n} e^{-x}\,dx\,dy = \int_0^1 (-e^{-n} + 1)\,dy \to 1$ so $\displaystyle\int_D e^{-x}\,dx\,dy = 1$.

(b) $\int_{D_n} e^{-x\sqrt{y}}\,dx\,dy = \int_0^1 \dfrac{-e^{-n\sqrt{y}}+1}{\sqrt{y}}\,dy$. Set $y = t^2$, then this integral is equal to

$$\int_0^1 (-e^{-nt}+1)2\,dt = \dfrac{e^{-n}-1}{n}2+2 \to 2$$

so $\int_D e^{-x\sqrt{y}}\,dx\,dy = 2$.

(c) In each subrectangle where $\frac{k-1}{n} \le y \le \frac{k}{n}$ you have $e^{-xy} \ge e^{-xk/n}$. Then

$$\int_0^n \int_{\frac{k-1}{n}}^{\frac{k}{n}} e^{-xy}\,dy\,dx \ge \int_0^n \dfrac{1}{n}e^{-xk/n}\,dx = \tfrac{1}{k}(1-e^{-k}) > \tfrac{1}{2k}.$$

Adding these, $\int_{D_n} e^{-xy}\,dA > \tfrac{1}{2}(1+\tfrac{1}{2}+\tfrac{1}{3}+\cdots+\tfrac{1}{n})$, which diverges. So $\int_D e^{-xy}\,dx\,dy$
does not exist.

6.45 The third and fourth only.

6.47 Set $x = \frac{y}{\sqrt{4t}}$ in $\int_{-\infty}^{\infty} e^{-x^2}\,dx = \sqrt{\pi}$ to get $\sqrt{4\pi t}$.

6.49 First proof: If $f(\mathbf{X}) = p > 0$, then by the definition of continuity (take $\epsilon = \frac{p}{2}$)
there is a small disk centered at $\mathbf{X}$ where $f > \frac{p}{2}$, so $f = f_+$ in that disk. Therefore f_+
is continuous in that disk, and in particular at $\mathbf{X}$. If $f(\mathbf{X}) < 0$ there is a disk where
$f < 0$, so in that disk $f_+ = 0$. Therefore f_+ is continuous in that disk, and in particular
at $\mathbf{X}$. If $f(\mathbf{X}) = 0$ let $\epsilon > 0$, then there is a small disk centered at $\mathbf{X}$ where $|f| < \epsilon$.
Since $f_+(\mathbf{Y})$ is equal to either 0 or $f(\mathbf{Y})$ at every point $\mathbf{Y}$ of that disk, we also have
$|f_+(\mathbf{Y}) - f_+(\mathbf{X})| = |f_+(\mathbf{Y})| < \epsilon$ there. Therefore f_+ is continuous at $\mathbf{X}$.
  Second proof: $f_+ = \tfrac{1}{2}(f+|f|)$ is a sum of continuous functions, so it is continuous.

6.51

(a) The set is the part of the rectangle $[0,1] \times [0,2]$ below the line $y = 2 - x$.
(b)

$$\int_{x+y\le 2} p(x,y)\,dA = 0 + \int_0^1 \int_0^{2-x} \dfrac{2x+2-y}{4}\,dy\,dx$$

$$= \int_0^1 \left[\dfrac{(2x+2)y - \tfrac{1}{2}y^2}{4}\right]_{y=0}^{2-x} dx = \int_0^1 \dfrac{(2x+2)(2-x) - \tfrac{1}{2}(2-x)^2}{4}\,dx = \tfrac{19}{24}.$$

6.53 The probability that (x,y) is not in D is 1 minus the probability that (x,y)
is in D.

6.55

(a) $\int_0^{2\pi} \int_0^n \left(\dfrac{r}{1+r^4}\right)^2 r\,dr\,d\theta = 2\pi \int_0^n \dfrac{r^3}{(1+r^4)^2}\,dr = 2\pi\left[-\tfrac{1}{4}(1+r^4)^{-1}\right]_0^n$ converges as $n \to \infty$.

(b) e^{-r} goes to zero much faster than $\dfrac{r}{1+r^4}$ so it is also integrable.

(c) Since $|x| \le \sqrt{x^2+y^2} = r$, this function is integrable by comparison with part (a) also.

(d)

$$\int_0^{2\pi} \int_0^n (ye^{-r})^2\, r\,dr\,d\theta = 2\pi \int_0^n y^2 e^{-2r} r\,dr \le 2\pi \int_0^n r^2 e^{-2r} r\,dr$$

$$= 2\pi \int_0^n e^{-2r} r^3\,dr \le 2\pi \int_0^n e^{-r}\max(e^{-r}r^3)\,dr \le (\text{const.}) \int_0^n e^{-r}\,dr$$

converges as $n \to \infty$.

Section 6.6

6.57

(a) a^5 by definition.

(b) $\displaystyle\int_0^a x_1^2\,dx_1 (a^4) = \tfrac13 a^7$ by iterated integral.

(c) $\tfrac13 a^7 - \tfrac13 a^7 + 7a^3 \int_0^a x_5\,dx_5 \int_0^a x_3\,dx_3 = 7a^3(\tfrac12 a^2)^2 = \tfrac74 a^7$ by part (b) and iterated integral.

6.59

(a)

$$\int_{[1,2]\times[3,5]\times[-1,10]} xz^2\,dV = \int_1^2 x\,dx \int_3^5 dy \int_{-1}^{10} z^2\,dz = \frac32(2)\frac{1000+1}{3} = 1001.$$

(b)

$$\int_D xz^2\,dV = \int_{x=1}^{x=2} \int_{y=3}^{y=5} \int_{z=x+y}^{z=10} xz^2\,dz\,dy\,dx = \int_{x=1}^{x=2} \int_{y=3}^{y=5} \tfrac13 x(1000-(x+y)^3)\,dy\,dx$$

$$= \int_{x=1}^{x=2} \tfrac13 x\left[1000y - \tfrac14(x+y)^4\right]_{y=3}^{y=5}\,dx = \int_{x=1}^{x=2} \tfrac13 x\left(2000 - \tfrac14((x+5)^4 - (x+3)^4)\right)\,dx$$

$$= \tfrac13 \int_1^2 \left(2000x - \tfrac{x}{4}(4x^3(5-3) + 6x^2(5^2-3^2) + 4x(5^3-3^3) + 5^4 - 3^4\right)\,dx \approx 821.6$$

6.61 Using spherical coordinates

$$\int_{\|X\|\le R} e^{-\sqrt{x^2+y^2+z^2}}\,dV = 4\pi \int_0^R e^{-\rho}\rho^2\,d\rho = 4\pi\left[(-\rho^2 - 2\rho - 2)e^{-\rho}\right]_0^R = 4\pi((-R^2 - 2R - 2)e^{-R} + 2)$$

6.63

(a) $\displaystyle\int_0^n e^{-2\rho}\,d\rho = -\frac12(e^{-2n} - 1)$ tends to $\tfrac12$ as n tends to infinity.

(b) Integrate by parts.

(c) Let n tend to infinity in part (b).

(d) Using part (c) and (a) we get $i_1 = \frac{1}{2}i_0 = \frac{1}{4}$. Using part (c) again $i_2 = \frac{2}{3}i_1 = \frac{1}{4}$. Then $i_3 = \frac{3}{2}i_2 = \frac{3}{8}$. So $i_4 = \frac{4}{3}i_3 = \frac{3}{4}$.

6.65

$$\int_{\mathbb{R}^3} f\, dV = \lim_{n\to\infty} \int_0^{2\pi} \int_0^\pi \int_p^n e^{-\rho}\rho^2 \sin\phi\, d\rho d\phi d\theta = \lim_{n\to\infty} 4\pi\left[-e^{-\rho}(\rho^2+2\rho+2)\right]_p^n = 4\pi e^{-p}(p^2+2p+2).$$

6.67

(a) $\int_{[0,1]^n} x_1^2\, d^n\mathbf{X} = \int_{[0,1]^{n-1}} \left(\int_0^1 x_1^2\, dx_1\right) dx_2\cdots dx_n = \int_{[0,1]^{n-1}} \frac{1}{3}\, dx_2\cdots dx_n = \frac{1}{3}$

(b) $\frac{n}{3}$

(c) $\frac{n}{3}$

(d) $\dfrac{n\int_{[0,2]^n} x_1^2\, d^n\mathbf{X}}{\mathrm{Vol}\,([0,2]^n)} = \dfrac{n\frac{8}{3}2^{n-1}}{2^n} = \frac{4}{3}n.$

Problems of Chapter 7

Section 7.1

7.1 Write $\mathbf{X}(t) = \mathbf{A} + t(\mathbf{B} - \mathbf{A})$, $\mathbf{Y}(u) = \mathbf{B} + \frac{1}{2}u(\mathbf{A} - \mathbf{B})$.
Then $\|\mathbf{X}'(t)\| = \|\mathbf{B} - \mathbf{A}\|$, $\|\mathbf{Y}'(u)\| = \frac{1}{2}\|\mathbf{A} - \mathbf{B}\| = \frac{1}{2}\|\mathbf{X}'(t)\|$.

(a) $\int_C ds = \int_0^1 \|\mathbf{X}'(t)\|\, dt = \|\mathbf{B} - \mathbf{A}\|$, and $\int_C ds = \int_0^2 \|\mathbf{Y}'(u)\|\, du = 2(\frac{1}{2})\|\mathbf{B} - \mathbf{A}\|$.

(b) $\int_C y\, ds = \int_0^1 (a_2 + t(b_2 - a_2))\|\mathbf{X}'(t)\|\, dt = (a_2 + \frac{1}{2}(b_2 - a_2))\|\mathbf{B} - \mathbf{A}\|$

$= \frac{1}{2}(a_2 + b_2)\|\mathbf{B} - \mathbf{A}\|$ and $\int_C y\, ds = \int_0^2 (b_2 + \frac{1}{2}u(a_2 - b_2))\|\mathbf{Y}'(u)\|\, du = (2b_2$

$+ \frac{2^2-0^2}{4}(a_2 - b_2))\frac{1}{2}\|\mathbf{A} - \mathbf{B}\| = \frac{1}{2}(a_2 + b_2)\|\mathbf{B} - \mathbf{A}\|$.

(c) Both parametrizations agree about the integral of y and about the length of C, so both give average $= \frac{1}{2}(a_2 + b_2)$.

7.3 $\mathbf{X}(t) = \mathbf{X}(ks) = \mathbf{Y}(s) = \mathbf{A} + ks(\mathbf{B} - \mathbf{A})$.

$$\mathrm{Length}\,(C) = \int_0^1 \|\mathbf{X}'(t)\|\, dt = \|\mathbf{B} - \mathbf{A}\|.$$

and $\mathbf{Y}'(s) = k(\mathbf{B} - \mathbf{A})$ is a unit vector if $k = \|\mathbf{B} - \mathbf{A}\|^{-1}$. Then since $0 \le t \le 1$, s runs from 0 to $1/k = \|\mathbf{B} - \mathbf{A}\|$. So $\mathbf{Y}(\mathrm{Length}\,(C)) = \mathbf{B}$.

7.5 Parametrize the segment C from $\mathbf{A}$ to $\mathbf{B}$ as $\mathbf{X}(t) = \mathbf{A} + t(\mathbf{B} - \mathbf{A})$, $0 \le t \le 1$. Then the average of x_i is

$$\frac{\int_C x_i \, ds}{\text{Length}(C)} = \frac{1}{\|\mathbf{B}-\mathbf{A}\|} \int_0^1 (a_i + t(b_i - a_i)) \|\mathbf{B}-\mathbf{A}\| \, dt = \left[a_i t + \tfrac{1}{2} t^2 (b_i - a_i) \right]_0^1 = \tfrac{1}{2}(b_i + a_i).$$

7.7 $\displaystyle \int_C y^2 \, dx + x \, dy = \int_{(0,0) \to (1,0)} y^2 \, dx + x \, dy + \int_{(1,0) \to (1,1)} y^2 \, dx + x \, dy$

$\displaystyle + \int_{(1,1) \to (0,0)} y^2 \, dx + x \, dy$

Parametrizations for the 3 segments are $(x,y) = (t,0)$, $(1,t)$, $(1-t, 1-t)$ with $0 \le t \le 1$ in each case. We get the integral

$$= \int_0^1 (0 \, dt + 0) + \int_0^1 ((t^2)(0) + 1 \, dt) + \int_0^1 ((1-t)^2 + (1-t))(-dt) = 0 + 1 - \tfrac{1}{3} - \tfrac{1}{2} = \tfrac{1}{6}.$$

7.9 $\mathbf{G} \cdot \mathbf{N} = (f_2, -f_1) \cdot (t_2, -t_1) = f_2 t_2 + f_1 t_1 = \mathbf{F} \cdot \mathbf{T}.$

7.11 $\mathbf{X}'(t) = (1, f'(t))$ so $\|\mathbf{X}'(t)\| = \sqrt{1 + (f'(t))^2}$. For the segment $y = 3x, 0 \le x \le 1$,

$$\int_0^1 \sqrt{1 + 3^2} \, dt = \sqrt{1 + 3^2}$$

agrees with the length that is given by the Pythagorean theorem.

7.13

(a) A rotation by $\pi/2$ changes x^2 to y^2 and takes C_1 to itself without stretching, so

$$\int_{C_1} x^2 \, ds = \int_{C_1} y^2 \, ds.$$

Therefore $\tfrac{1}{2} \displaystyle\int_{C_1} (x^2 + y^2) \, ds = \int_{C_1} x^2 \, ds.$

(b) Same reason.

(c) Holds because $(x^2 + y^2)^{10} = 1$.

7.15

(a) $\displaystyle \int p \, dx = \int (y,0) \cdot \mathbf{T} \, ds$. The integrals have opposite signs for physical reasons and because $\mathbf{F} = (y,0)$ is in the direction of $\mathbf{T}$ on K roughly, and opposite that of $\mathbf{T}$ on E, that is, $\mathbf{F} \cdot \mathbf{T} > 0$ on K and $\mathbf{F} \cdot \mathbf{T} < 0$ on E.

(b) The coordinates in the figure are x, p instead of x, y, so the work done is the same as $\displaystyle\int_{K \cup E} y \, dx$. Problem 7.14 shows that this is the area between the graphs.

7.17

(a) This is because for small $\Delta\theta$ the segments are small, so can be used in a good Riemann sum approximation.

(b) $\mathbf{U} = \frac{(-y,x)}{r} \frac{1}{r} = \mathbf{W} \frac{1}{r}.$

(c) The factors $\mathbf{T}\Delta s$ are approximations for arc length segments, and the dot products give the projection in the $\mathbf{W}$ direction. Dividing by r gives the change in θ by definition of radian measure.

(d) This combines parts (a) and (c).

7.19

(a) $\int_0^3 x\,dx = \frac{1}{2}3^2$, yes equal

(b) $\int_0^3 x\,d(\frac{4}{3}x) = \frac{4}{3}\frac{1}{2}3^2$, not equal to $(3)(4)$.

(c) $\int_0^3 dx + 5d(\frac{4}{3}x) = (1 + 5(\frac{4}{3}))(3)$, yes, equal to $3 + (5)(4)$.

Section 7.2

7.21

(a) An antiderivative for $g_x = \dfrac{x}{x^2 + y^2}$ is $\frac{1}{2}\log(x^2 + y^2) + h(y)$. But checking g_y we see that we can take $h = 0$. So a potential is $g(x,y) = \frac{1}{2}\log(x^2 + y^2)$, defined on $\mathbb{R}^2 - \mathbf{0}$.

(b) Due to algebraic simplification the rules for $\nabla \tan^{-1}(\frac{y}{x})$ and $\mathbf{F}(x,y)$ are the same where their domains agree. But the domain of $\tan^{-1}(\frac{y}{x})$ is $\mathbb{R}^2$ minus the y axis, not $\mathbb{R}^2 - \mathbf{0}$. So

$$\nabla \tan^{-1}\left(\tfrac{y}{x}\right) \neq \left(\frac{-y}{x^2 + y^2}, \frac{x}{x^2 + y^2}\right).$$

7.23

(a) Not.

(b) $\left[xy - 3z^2 \cos y\right]_{(0,0,0)}^{(a,b,c)} = ab - 3c^2 \cos b,$

(c) $\left[x + y\right]_{(0,0,0)}^{(a,b,c)} = a + b$

7.25

(a) The integral is equal to $\displaystyle\int_C \nabla(\|\mathbf{X}\|^{-1}) \cdot \mathbf{T}\,ds = \left[\|\mathbf{X}\|^{-1}\right]_{(1,1,2)}^{(2,2,1)} = \frac{1}{3} - \frac{1}{\sqrt{6}}.$

(b) The integral has not been defined since the integrand is not defined at $(0,0,0)$.

7.27 The Chain Rule together with $\|\mathbf{X}\|^{-3}\mathbf{X} = \nabla(-\|\mathbf{X}\|^{-1})$ gives

$$\|\mathbf{X} - \mathbf{P}\|^{-3}(\mathbf{X} - \mathbf{P}) = \nabla(-\|\mathbf{X} - \mathbf{P}\|^{-1})$$

for any constant $\mathbf{P}$. Then

$$c\|\mathbf{X} - \mathbf{P}\|^{-3}(\mathbf{X} - \mathbf{P}) = \nabla(-c\|\mathbf{X} - \mathbf{P}\|^{-1})$$

for any constant c. Take k different such expressions and add them to get

$$\sum_{j=1}^{k} c_j\|\mathbf{X} - \mathbf{P}_j\|^{-3}(\mathbf{X} - \mathbf{P}_j) = \nabla\left(-\sum_{j=1}^{k} c_j\|\mathbf{X} - \mathbf{P}_j\|^{-1}\right).$$

7.29

(a) By the Chain Rule, $-\mathbf{F}(\mathbf{X}) = \nabla(\|\mathbf{X}\|^{-1})$ and $\mathbf{F}(x+h,y,z) = \nabla(-\|(x+h,y,z)\|^{-1})$,
    so $\mathbf{F}(x+h,y,z) - \mathbf{F}(\mathbf{X}) = \nabla(-\|(x+h,y,z)\|^{-1} + \|\mathbf{X}\|^{-1})$.

(b) The limit is $\frac{\partial}{\partial x}\mathbf{F}$ which is conservative because of the interchange of mixed
    partial derivatives: $\frac{\partial}{\partial x}\nabla g = \nabla\frac{\partial}{\partial x}g$.

(c) $\frac{\partial}{\partial x}\mathbf{G}$ is conservative because of the interchange of mixed partial derivatives:
    $\frac{\partial}{\partial x}\nabla g = \nabla\frac{\partial}{\partial x}g$. Similarly for $\frac{\partial}{\partial y}\mathbf{G}$ and $\frac{\partial}{\partial z}\mathbf{G}$.

Section 7.3

7.31 Let $f(x,y) = x^2 + y^2$ and $g(x,y) = x^2 - y^2$. Then $\nabla f(x,y) = (2x,2y)$ and $\nabla g(x,y) = (2x,-2y)$. The areas of the graphs are

$$\int_D \sqrt{1 + f_x^2 + f_y^2}\,dA = \int_D \sqrt{1 + 4x^2 + 4y^2}\,dA$$

and

$$\int_D \sqrt{1 + g_x^2 + g_y^2}\,dA = \int_D \sqrt{1 + 4x^2 + 4y^2}\,dA.$$

These are equal.

7.33 S is the graph over D of $z = \frac{1}{c}(d - ax - by)$. So $z_x = -\frac{a}{c}$, $z_y = -\frac{b}{c}$ and

$$\text{Area}(S) = \int_D \sqrt{(\tfrac{a^2}{c^2} + \tfrac{b^2}{c^2} + 1}\,dA = \int_D \sqrt{\tfrac{1-c^2}{c^2} + 1}\,dA = \tfrac{1}{|c|}\text{Area}(D).$$

7.35

(a) $\mathbf{X}_u(u,v) = (\sqrt{2}v, 2u, 0)$, $\mathbf{X}_v(u,v) = (\sqrt{2}u, 0, 2v)$, $\mathbf{X}_u \times \mathbf{X}_v = (4uv, -2\sqrt{2}v^2, -2\sqrt{2}u^2)$

(b) $\mathbf{X}_u$ and $\mathbf{X}_v$ are bounded because $1 \le u \le 2$, $1 \le v \le 2$, and are linearly
    independent by inspection since u and v are not zero. $\mathbf{X}$ is one to one because if

$$(\sqrt{2}uv, u^2, v^2) = (\sqrt{2}u_1 v_1, u_1^2, v_1^2)$$

with u, u_1, v, v_1 positive then

$$u_1 = \sqrt{u_1^2} = \sqrt{u^2} = u, \qquad v_1 = \sqrt{v_1^2} = \sqrt{v^2} = v.$$

So the range S of $\mathbf{X}$ is a smooth surface. At any point of S, $(x,y,z) = (\sqrt{2}uv, u^2, v^2)$,
you have $x^2 - 2yz = 2u^2v^2 - 2u^2v^2 = 0$.

(c) The area of S is

$$\int_S d\sigma = \int_1^2 \int_1^2 \|\mathbf{X}_u \times \mathbf{X}_v\|\,du\,dv = \int_1^2 \int_1^2 \sqrt{(4uv)^2 + (-2\sqrt{2}v^2)^2 + (-2\sqrt{2}u^2)^2}\,du\,dv$$

$$= \int_1^2 \int_1^2 \sqrt{16u^2v^2 + 8v^4 + 8u^4}\,du\,dv = \int_1^2 \int_1^2 \sqrt{8}(u^2 + v^2)\,du\,dv$$

$$= \sqrt{8} \int_1^2 \left[\tfrac{1}{3} u^3 + v^2 u \right]_{u=1}^2 dv = \sqrt{8} \int_1^2 (\tfrac{7}{3} + v^2) dv = \sqrt{8}(\tfrac{7}{3} + \tfrac{7}{3}) = \sqrt{8} \tfrac{14}{3}.$$

(d)

$$\int_S y \, d\sigma = \int_1^2 \int_1^2 u^2 \|X_u \times X_v\| \, du \, dv = \int_1^2 \int_1^2 u^2 \sqrt{8}(u^2 + v^2) \, du \, dv$$

$$= \sqrt{8} \int_1^2 \left[\tfrac{1}{5} u^5 + \tfrac{1}{3} v^2 u^3 \right]_{u=1}^2 dv = \sqrt{8} \int_1^2 (\tfrac{31}{5} + \tfrac{7}{3} v^2) dv = \sqrt{8}(\tfrac{31}{5} + \tfrac{72}{3^2}).$$

7.37 Define $G(x, y, z) = (-x, -y, -z)$. Using any parametrization $X_1(u, v) = (x_1(u, v), y_1(u, v), z_1(u, v))$ we get

$$\int_S x^2 y \, d\sigma = \int \int x_1^2(u, v) y_1(u, v) \| \cdots \| \, du \, dv$$

and using $X_2(u, v) = G \circ X_1 = (-x_1(u, v), -y_1(u, v), -z_1(u, v))$ we get

$$\int_S x^2 y \, d\sigma = \int x_1^2(u, v)(-y_1(u, v)) \| \cdots \| \, du \, dv.$$

Since these are equal the integral is zero.

7.39

(a) $x_1^2 + x_2^2 + x_3^2 = 1$. Square this to get

$$1 = (x_1^2 + x_2^2 + x_3^2)^2 = x_1^4 + x_2^4 + x_3^4 + 2(x_1^2 x_2^2 + x_2^2 x_3^2 + x_3^2 x_1^2).$$

(b) $x_3 = z = \cos\phi$, so

$$\int x_3^4 \, d\sigma = \int_0^{2\pi} \int_0^\pi (\cos\phi)^4 \sin\phi \, d\phi d\theta = \tfrac{1}{5}(-\cos^5(\pi) + \cos^5(0)) 2\pi = \tfrac{4}{5}\pi$$

(c) Integral of expression (a) using symmetry and Area$(S) = 4\pi$ gives

$$3(\tfrac{4}{5}\pi) + 2(3) \int_{S^2} x_1^2 x_2^2 \, d\sigma = 4\pi,$$

so $\int_{S^2} x_1^2 x_2^2 \, d\sigma = \tfrac{1}{6}(4 - \tfrac{12}{5})\pi = \tfrac{4}{15}\pi.$

7.41 (a) $F \cdot N \text{Area}(S) = (2, 3, 4) \cdot (0, -1, 0)(1) = -3$

(b) $-3 + (2, 3, 4) \cdot (0, 0, -1)(2) = -3 - 8 = -11$

(c) $+11$

(d) $-11 + 11 = 0$

7.43 Use parametrization $X(u, v) = uV + vW = (u, u, u + 2v)$, $0 < u < 1$, $0 < v < 1$. Then $X_u = V$, $X_v = W$, $X_u \times X_v = V \times W = (2, -2, 0)$, and the flux is

$$\int F \cdot (X_u \times X_v) \, du \, dv = \int_0^1 \int_0^1 (2(u)(2) + 3(u + 2v)(-2)) \, du \, dv$$

$$= \int_0^1 \int_0^1 (-2u - 12v) \, du \, dv = -1 - 6 = -7.$$

7.45

(a) Multiplying we get

$$[A\ B\ Y] \begin{bmatrix} a & b & 0 \\ c & d & 0 \\ 0 & 0 & 1 \end{bmatrix} = \left[aA + cB \quad bA + dB \quad \begin{bmatrix} y_1 \\ y_2 \\ y_3 \end{bmatrix} \right].$$

This is equal to $[C\ D\ Y]$ by assumption.

(b) Take the determinant of each side and use the multiplication rule for determinants. Use the definition of cross product $\det[P\ Q\ R] = R \cdot (P \times Q)$.

(c) Use the fact that if $Y \cdot X = Y \cdot Z$ for all Y, then $X = Z$. This can be seen by taking $Y = (1,0,0), (0,1,0)$ and $(0,0,1)$.

7.47

(a) $\int_S (\rho V) V \cdot N \, d\sigma = \int_S (\rho(a,b,c))(a,b,c) \cdot (0,0,1) \, d\sigma = \rho(a,b,c)c \int_S d\sigma = \rho c A(a,b,c).$

(b) $\int_S (\rho V) V \cdot N \, d\sigma = \int_S (\rho k N) k N \cdot N \, d\sigma = \rho k^2 N \int_S d\sigma = \rho k^2 A N.$

7.49 Denote by a the radius of S, then $N = X/a$, and on S, $\|X\| = a$ so

$$\int_S F \cdot N \, d\sigma = \int_S \frac{X}{a^3} \cdot \frac{X}{a} \, d\sigma = \int_S \frac{a^2}{a^4} \, d\sigma = a^{-2}(4\pi a^2) = 4\pi.$$

7.51 $Y(u,v)$ is a constant times the parametrization $X(u,v)$ so Y is differentiable on the same set D, is one to one, and $Y_u = kX_u$ and $Y_v = kX_v$ are linearly independent. Therefore Y is a parametrization and T is a smooth surface. Also

$$Y_u(u,v) \times Y_v(u,v) = (kX_u) \times (kX_v(u,v)) = k^2 X_u \times X_v(u,v)$$

so

$$\text{Area}(T) = \int_D \|Y_u \times Y_v\| \, du \, dv = \int_D k^2 \|X_u \times X_v\| \, du \, dv = k^2 \, \text{Area}(S).$$

Problems of Chapter 8

Section 8.1

8.1 One way is shown in Figure 9.22. $D_1 = \overline{R - U}$ is a union of 6 parts that are x simple and y simple. $D_2 = \overline{S - R}$ is a union of 4.

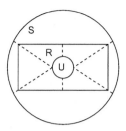

Fig. 9.22 Figure for Problem 8.1.

8.3 The divergence of the field is 2, so flux across C is

(a) $2\,\text{Area(disk)} = 2\pi$,

(b) $2\,\text{Area(rectangle)} = 2(b-a)(d-c)$.

8.5 $\text{div}\,\mathbf{F} = 2xy^2$

$$\int_D \text{div}\,\mathbf{F}\,dA = \int_0^1 \int_{-\sqrt{1-x^2}}^{\sqrt{1-x^2}} 2xy^2\,dydx = \int_0^1 \frac{4x}{3}(1-x^2)^{3/2}\,dx = \left[\tfrac{4}{3}(1-x^2)^{5/2}\tfrac{2}{5}\tfrac{-1}{2}\right]_0^1 = \tfrac{4}{15}.$$

Parametrize the vertical part of the boundary as $(x,y) = (0, 1-t)$, $0 \le t \le 2$ and the curved part as

$$(x,y) = (\cos t, \sin t), \qquad -\tfrac{\pi}{2} \le t \le \tfrac{\pi}{2}.$$

Then

$$\int_{\partial D} \mathbf{F}\cdot\mathbf{N}\,ds = \int_{\partial D} (1 + x^2y^2)\,dy = \int_0^2 (-dt) + \int_{-\pi/2}^{\pi/2} (1 + \cos^2 t \sin^2 t)\cos t\,dt$$

$$= -2 + \int_{-\pi/2}^{\pi/2} (1 + (1 - \sin^2 t)\sin^2 t)\cos t\,dt = -2 + \left[\sin t + \tfrac{1}{3}\sin^3 t - \tfrac{1}{5}\sin^5 t\right]_{-\pi/2}^{\pi/2}$$

$$= -2 + 2(1 + \tfrac{1}{3} - \tfrac{1}{5}) = \tfrac{4}{15}.$$

8.7

(a) $\displaystyle\int_{\partial D} (1,0)\cdot\mathbf{N}\,ds = \int_D \text{div}\,(1,0)\,dA = \int 0\,dA = 0$

(b) $\displaystyle\int_{\partial D} (0,1)\cdot\mathbf{N}\,ds = \int_D 0\,dA = 0$

(c) So $\displaystyle\int_{\partial D} \mathbf{N}\,ds = \left(\int_{\partial D} n_1\,ds, \int_{\partial D} n_2\,ds\right) = 0$

(d) $\mathbf{N}_i$ is constant on D_i so $\displaystyle\int_{D_i} \mathbf{N}_i\,ds = \mathbf{N}_i\,\text{Length}\,(D_i) = \|\mathbf{P}_i - \mathbf{P}_{i-1}\|\mathbf{N}_i$

(e) $\int_{\partial D} \mathbf{N}\,ds = 0 = \sum \int_{D_i} \mathbf{N}_i\,ds = \sum \|\mathbf{P}_i - \mathbf{P}_{i-1}\|\mathbf{N}_i$.

8.9 If the loop does not surround $(0,0)$ then the integral is zero by Green's Theorem since $\text{curl}\,\mathbf{F} = 0$ throughout the region enclosed. If the loop does surround $(0,0)$ then the integral is 2π by Example 8.9.

8.11

(a) Adding the product rule $(gf_j)_{x_j} = g_{x_j}f_j + gf_{j,x_j}$ for $j = 1,2,3$ you get

$$\text{div}\,(g\mathbf{F}) = g\,\text{div}\,\mathbf{F} + \mathbf{F}\cdot\nabla g.$$

(b) Apply the Divergence Theorem to $g\mathbf{F}$. The integral over the boundary is zero since $g = 0$ there, so the result follows from part (a).

(c) This follows from part (b) because when $\mathbf{F} = \nabla g$, $\text{div}\,\mathbf{F} = \text{div}\,\nabla g = \Delta g$.

8.13 Suppose there were a periodic orbit. Denote by D the region with the orbit as its boundary. Then

$$0 < \int_D \text{div}\,\mathbf{F}\,dA = \int_{\partial D} \mathbf{F}\cdot\mathbf{N}\,ds.$$

Use the periodic solution to parametrize the boundary. That could be either clockwise or counterclockwise, but in either case Green's Theorem gives

$$0 \neq \int_0^P f(x(t), y(t))\,dy - g(x(t), y(t))\,dx = \int_0^P x'(t)y'(t)\,dt - y'(t)x'(t)\,dt = 0,$$

a contradiction.

Section 8.2

8.15 Since $\text{div}\,\mathbf{X} = 3$, the integrals are 3 times the volume of D. (a) 4π, (b) $4\pi r^3$.

8.17

(a) $\displaystyle\int_{\text{ball}} (1 + 3 + 5)\,dV = 9(\tfrac{4}{3}\pi 8^2)$

(b) 0

(c) $\displaystyle\int_{\text{ball}} (1)\,dV = \tfrac{4}{3}\pi 8^2$

(d) $\displaystyle\int_{\text{ball}} (2x)\,dV = 0$ by symmetry.

(e) 0

8.19

(a) This is a property of the dot and cross product,

$$(N \times F) \cdot C = \det[N\ F\ C] = N \cdot (F \times C).$$

(b) Use the Divergence Theorem.

(c) Use

$$\text{div}(F \times C) = \text{div}(f_2 c_3 - f_3 c_2, f_3 c_1 - f_1 c_3, f_1 c_2 - f_2 c_1)$$

$$= (f_{2,1} - f_{1,2})c_3 + (-f_{3,1} + f_{1,3})c_2 + (f_{3,2} - f_{2,3})c_1 = (\text{curl}\,F) \cdot C$$

and part (a).

(d) Use parts (a) and (c).

(e) Factor out the constant C.

(f) Since C was any constant vector that means the vector in parentheses in part (e) must be the zero vector.

8.21

(a) By the Divergence Theorem, $\displaystyle\int_{\partial D}(1,0,0)\cdot N\,d\sigma = \int_D 0\,dV$, or $\displaystyle\int_{\partial D} n_1\,d\sigma = 0$.

(b) Using $(0,1,0)$ and $(0,0,1)$ similarly each component of $\displaystyle\int_{\partial D} N\,d\sigma$ is zero.

(c) A face S of D is a flat surface with a constant normal vector. So

$$\int_S N\,d\sigma = N\int_S d\sigma = \text{Area}(S)N.$$

8.23

(a) all points of $\mathbb{R}^3$ except the A_k

(b) $\text{div}\,F = 0$ so by the Chain Rule $\text{div}\,G(X) = c_1\,\text{div}\,F(X - A_1) + \cdots = c_1(0) + \cdots = 0$

(c) $\displaystyle\int_D F(X)\cdot N\,d\sigma = 4\pi$ if 0 is in the interior of D and 0 is 0 is in the exterior, so

$$\int_{\partial W} G\cdot N\,d\sigma = \sum_k c_k \underbrace{\int_{\partial W} F(X - A_k)\cdot N\,d\sigma}_{0\ \text{or}\ 4\pi} = \sum_{A_k\ \text{in}\ W} 4\pi c_k.$$

8.25 For C^2 functions

$$\text{div}(v\nabla u) = (vu_x)_x + (vu_y)_y + (vu_z)_z = v_x u_x + v_y u_y + v_z u_z + vu_{xx} + vu_{yy} + vu_{zz} = \nabla v\cdot\nabla u + v\Delta u$$

so the Divergence Theorem in a regular set D

$$\int_D \text{div}(v\nabla u)\,dV = \int_{\partial D} v\nabla u\cdot N\,d\sigma$$

gives

$$\int_D \left(\nabla v \cdot \nabla u + v \Delta u \right) dV = \int_{\partial D} v \nabla u \cdot \mathbf{N} d\sigma.$$

Taking $v = u$ gives part (a). Since $\nabla u \cdot \nabla v = \nabla v \cdot \nabla u$, we can interchange u and v and subtract to get part (b),

$$\int_D (v \Delta u - u \Delta v) \, dV = \int_{\partial D} (v \nabla u - u \nabla v) \cdot \mathbf{N} d\sigma.$$

8.27 The identity

$$\int_D (f \Delta f + |\nabla f|^2) \, dV = \int_{\partial D} f \nabla f \cdot \mathbf{N} d\sigma$$

gives

$$\int_D (\lambda f^2 + |\nabla f|^2) \, dV = 0.$$

If $\lambda > 0$ then that integrand is nonnegative, so it must be zero in D. Therefore f is zero in D.

Section 8.3

8.29

(a) $\displaystyle\int_C \mathbf{F} \cdot \mathbf{T} ds = \int_{\text{disk}} \text{curl} \, \mathbf{F} \cdot \mathbf{N} d\sigma = \int_{\text{disk}} (0,0,1) \cdot (0,0,1) d\sigma = \pi$

(b) Since C is oriented from $\mathbf{A}$ to $\mathbf{B}$ to $\mathbf{C}$ to $\mathbf{A}$ we need $\mathbf{N}$ to point away from $\mathbf{0}$.

$$\int_C \mathbf{F} \cdot \mathbf{T} ds = \int_{\substack{\text{triangular surface}}} \text{curl} \, \mathbf{F} \cdot \mathbf{N} d\sigma = \int_{\substack{\text{triangular surface}}} (0,0,-1) \cdot \frac{(\mathbf{B}-\mathbf{A}) \times (\mathbf{C}-\mathbf{A})}{\|(\mathbf{B}-\mathbf{A}) \times (\mathbf{C}-\mathbf{A})\|} d\sigma$$

$$= \int_{\substack{\text{triangular surface}}} (0,0,-1) \cdot \frac{(bc,ac,ab)}{\|(\mathbf{B}-\mathbf{A}) \times (\mathbf{C}-\mathbf{A})\|} d\sigma$$

$$= -\frac{ab}{\|(\mathbf{B}-\mathbf{A}) \times (\mathbf{C}-\mathbf{A})\|} \text{Area(triangle)} = -\frac{ab}{\|(\mathbf{B}-\mathbf{A}) \times (\mathbf{C}-\mathbf{A})\|} \tfrac{1}{2}\|(\mathbf{B}-\mathbf{A}) \times (\mathbf{C}-\mathbf{A})\| = -\tfrac{1}{2}ab$$

8.31

(a) $\text{curl} \, \mathbf{F} = (0,0,2)$. On the hemisphere $z = g(x,y) = \sqrt{r^2 - x^2 - y^2}$ and $\mathbf{N} = (x,y,z)/r$.

$$d\sigma = \sqrt{1 + g_x^2 + g_y^2} \, dx \, dy = \sqrt{1 + \frac{x^2}{g^2} + \frac{y^2}{g^2}} \, dx \, dy = \frac{r}{g(x,y)} \, dx \, dy.$$

So

$$\int_S \text{curl} \, \mathbf{F} \cdot \mathbf{N} d\sigma = \int_S (0,0,2) \cdot \frac{(x,y,g(x,y))}{r} \, d\sigma = \int_D 2 \frac{g(x,y)}{r} \frac{r}{g(x,y)} \, dx \, dy$$

where D is the disk of radius r. This is equal to $2 \, \text{Area}(D) = 2\pi r^2$. Since $z = 0$ on ∂S, the line integral is, using Green's Theorem and $\partial S = \partial D$,

$$\int_{\partial S} \mathbf{F} \cdot \mathbf{T} \, ds = \int_{\partial S} -y \, dx + x \, dy + 2 \, dz = \int_{\partial S} -y \, dx + x \, dy = \int_D 2 \, dA = 2\pi r^2$$

(b) The line integral is the same as in part (a) and the surface integral is

$$\int_D (0,0,2) \cdot (0,0,1) \, ds = 2 \, \text{Area}(D) = 2(\pi r^2).$$

8.33 Write the sphere as the union of hemispheres H and K. Then H and K have the same but oppositely oriented boundary circle, so by Stokes' Theorem applied to each of H and K,

$$\int_S \text{curl} \, \mathbf{F} \cdot \mathbf{N} \, d\sigma = \int_H \text{curl} \, \mathbf{F} \cdot \mathbf{N} \, d\sigma + \int_K \text{curl} \, \mathbf{F} \cdot \mathbf{N} \, d\sigma = \int_{\partial H} \mathbf{F} \cdot \mathbf{T} \, ds + \int_{\partial K} \mathbf{F} \cdot \mathbf{T} \, ds = 0.$$

8.35

(a) The Maxwell equation gives $0 = (0,0,2)c_1 - \mu_0 \mathbf{J}$, so $c_1 = \frac{1}{2}\mu_0 j$.

(b) We know that $\text{curl} \left(\frac{(-y,x,0)}{r^2} \right) = 0$ and $\mathbf{J} = \mathbf{0}$ so $p = 2$.

(c) We need for continuity on $r = R$ that $\frac{1}{2}\mu_0 j = \frac{c_2}{R^2}$, so $c_2 = \frac{1}{2}\mu_0 j R^2$.

(d) The surface integral side of Stokes' formula is the sum over the continuous parts

$$\int_D \text{curl} \, \mathbf{B} \cdot \mathbf{N} \, d\sigma = \int_{r \leq R} \mu_0 \mathbf{J} \cdot \mathbf{N} \, d\sigma + \int_{r \geq R} \mathbf{0} \cdot \mathbf{N} \, d\sigma = \mu_0 j \pi R^2 + 0.$$

The line integral side of Stokes' formula is, since $\mathbf{T} = \frac{(-y,x,0)}{R_1}$,

$$\int_{\partial D} \mathbf{B} \cdot \mathbf{T} \, ds = \int_{\partial D} c_2 \frac{\mathbf{T}}{r} \cdot \mathbf{T} \, ds = \frac{1}{2}\mu_0 j R^2 \int_{\partial D} \frac{1}{R_1} \, ds$$

$$= \frac{1}{2}\mu_0 j R^2 \frac{2\pi R_1}{R_1} = \mu_0 j \pi R^2.$$

These are equal, so Stokes' formula holds in this discontinuous case.

8.37

(a) Stokes' Theorem applied to Ampère's original law gives

$$\int_S \text{curl} \, \mathbf{B} \cdot \mathbf{N} \, d\sigma = \mu_0 \int_S \mathbf{J} \cdot \mathbf{N} \, d\sigma$$

for all S, so $\text{curl} \, \mathbf{B} = \mu_0 \mathbf{J}$.

(b) $\rho_t = -\text{div} \, \mathbf{J} = -(\mu_0)^{-1} \text{div} \, \text{curl} \, \mathbf{B} = 0$ does not allow any time dependence of the charge density.

8.39 $\frac{d}{dt} \int_S \mathbf{B} \cdot \mathbf{N} \, d\sigma = \int_S \mathbf{B}_t \cdot \mathbf{N} \, d\sigma$. By Maxwell's equation this is equal to $\int_S (-\text{curl} \, \mathbf{E}) \cdot \mathbf{N} \, d\sigma$. By Stokes' Theorem this is equal to $- \int_{\partial S} \mathbf{E} \cdot \mathbf{T} \, ds$.

Section 8.4

8.41 The left hand side is zero because the time derivative is zero. The integral over the boundary of the cylinder becomes $-\rho(b)u(b)A(1,0,0)$ at the right end plus $-\rho(a)u(a)A(-1,0,0)$ at the left, so the conservation law is

$$0 = \left(-\rho(b)u(b) + \rho(a)u(a)\right)(A,0,0).$$

8.43

$$\mathbf{V} \cdot \nabla(\mathbf{V} \cdot \mathbf{V}) = u(u^2 + v^2 + w^2)_x + v(u^2 + v^2 + w^2)_y + w(u^2 + v^2 + w^2)_z$$

$$= 2u(uu_x + vv_x + ww_x) + 2v(uu_y + vv_y + ww_y) + 2w(uu_z + vv_z + ww_z)$$

$$= 2u(uu_x + vu_y + wu_z) + 2v(uv_x + vv_y + wv_z) + 2w(uw_x + vw_y + ww_z) = 2\mathbf{V} \cdot (\mathbf{V} \cdot \nabla \mathbf{V}).$$

8.45 (c) The acceleration is to the left in the left figure, and to the right in the others.

8.47

(a) $\mathbf{V}(\mathbf{X},0) = \mathbf{X}$, $\mathbf{V}(\mathbf{X},1) = \frac{1}{2}\mathbf{X}$, both radial outward, slower after one second.
(b) $\operatorname{div} \mathbf{V} = \frac{1}{1+t}\operatorname{div}(x,y,z) = \frac{3}{1+t}$ is not zero, so the flow is compressible.
(c) $\rho_t + \rho \operatorname{div} \mathbf{V} + \mathbf{V} \cdot \nabla \rho = a(1+t)^{a-1}\|\mathbf{X}\|^2 + (1+t)^a\|\mathbf{X}\|^2 \frac{3}{1+t} + \frac{1}{1+t}\mathbf{X} \cdot \left((1+t)^a 2\mathbf{X}\right)$
  $= (1+t)^{a-1}\|\mathbf{X}\|^2(a+3+2)$. So take $a = -5$.

8.49

$$e_t + \operatorname{div}(e\mathbf{V}) + P\operatorname{div} \mathbf{V} = \frac{ck}{R}\gamma\rho^{\gamma-1}\rho_t + \frac{ck}{R}\operatorname{div}(\rho^\gamma \mathbf{V}) + k\rho^\gamma \operatorname{div} \mathbf{V}$$

$$= \frac{ck}{R}\gamma\rho^{\gamma-1}\rho_t + \frac{ck}{R}\operatorname{div}(\rho^{\gamma-1}(\rho\mathbf{V})) + k\rho^\gamma \operatorname{div} \mathbf{V}$$

$$= \frac{ck}{R}\gamma\rho^{\gamma-1}\rho_t + \frac{ck}{R}\left(\rho^{\gamma-1}\operatorname{div}(\rho\mathbf{V}) + \rho\mathbf{V} \cdot \nabla(\rho^{\gamma-1})\right) + k\rho^\gamma \operatorname{div} \mathbf{V}$$

$$= \frac{ck}{R}\gamma\rho^{\gamma-1}\rho_t + \frac{ck}{R}\left(\rho^{\gamma-1}\operatorname{div}(\rho\mathbf{V}) + (\gamma-1)\rho^{\gamma-1}\mathbf{V} \cdot \nabla\rho\right) + k\rho^\gamma \operatorname{div} \mathbf{V}$$

Using the mass equation $\rho_t = -\operatorname{div}(\rho\mathbf{V})$ that becomes

$$= \frac{ck}{R}(-\gamma+1)\rho^{\gamma-1}\operatorname{div}(\rho\mathbf{V}) + \frac{ck}{R}(\gamma-1)\rho^{\gamma-1}\mathbf{V} \cdot \nabla\rho + k\rho^\gamma \operatorname{div} \mathbf{V} = \frac{ck}{R}(-\gamma+1)\rho^\gamma \operatorname{div} \mathbf{V} + k\rho^\gamma \operatorname{div} \mathbf{V}.$$

This is zero if $\gamma = 1 + \frac{R}{c}$.

8.51 $\rho_t + (\rho u)_x = \epsilon g_t + (\rho_0 \epsilon f + \epsilon^2 fg)_x = \epsilon(g_t + \rho_0 f_x)$ if we ignore ϵ^2.

$$u_t + uu_x + k\gamma\rho^{\gamma-2}\rho_x = \epsilon f_t + (\epsilon f)(\epsilon f_x) + k\gamma(\rho_0 + \epsilon g)^{\gamma-2}(\epsilon g_x) = \epsilon(f_t + k\gamma\rho_0^{\gamma-2}g_x)$$

if we ignore ϵ^2, because the Mean Value Theorem gives $(a+\epsilon g)^b = a^b + b(a+\theta g)^{b-1}\epsilon$ for some θ between 0 and ϵ. Then

$$g_{tt} = -\rho_0 f_{xt} = \rho_0 (k\gamma \rho_0^{\gamma-2}) g_{xx}.$$

Section 8.5

8.53 Using

$$\frac{d}{dr} \int_r^s f(x)\,dx = -f(r)$$

we get $\nabla g = \left(-f(r,t), f(s,t), \int_r^s f_t\,dx \right)$. Then the Chain Rule gives the result as $\nabla g \cdot (a', b', 1)$.

8.55 $\rho_t + \rho\rho_x = -\dfrac{x}{(t+8)^2} + \dfrac{x}{t+8}\dfrac{1}{t+8} = 0.$

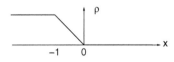

Fig. 9.23 The graph of $\rho(x,0)$ for Problem 8.57.

8.57

(a) The three parts of the formula for ρ are continuous in their respective sets, and agree on the common boundaries of those sets, so ρ is continuous.

(b) The graph of $\rho(x,0)$ is shown in Figure 9.23. Your graph of $\rho(x,1)$ ought to show the constant 1 for $x < 0$, the constant 0 for $x > 0$, and a point of discontinuity at $x = 0$.

(c) In the regions where ρ is constant, the partial derivatives are zero, so $\rho_t + \rho\rho_x = 0$. In the region where $\rho = \dfrac{x}{t-1}$, we have

$$\rho_t + \rho\rho_x = -\frac{x}{(t-1)^2} + \frac{x}{t-1}\frac{1}{t-1} = 0.$$

(d) In

$$\rho(x_0 + (-x_0)t, t) = -x_0$$

set $x = x_0 + (-x_0)t$. That gives $x_0 = \dfrac{x}{1-t}$, so

$$\rho(x,t) = -\frac{x}{1-t}.$$

8.59 The Fundamental Theorem gives

$$\int_a^b (\rho_t + f_x - g)\,dx = 0$$

and it holds for all $[a,b]$, with continuous integrand, only if that integrand is zero.

8.61 $s_1 = \frac{1}{2}(1+0) = \frac{1}{2}$, $s_2 = \frac{1}{2}(\frac{3}{2}+1) = \frac{5}{4}$, $s_3 = \frac{1}{2}(3+\frac{3}{2}) = \frac{9}{4}$, $s_4 = \frac{1}{2}(0+\frac{3}{2}) = \frac{3}{4}$, $s_5 = \frac{1}{2}(0+3) = \frac{3}{2}$,

8.63 $x = 2t$ and $x = 3+1.5t$ intersect, have the same x, t values, when $x = 2t = 3+1.5t$. That gives $t = 6$, then $x = 12$.

8.65 The curves satisfying $x' = x$ are $(x,t) = (ce^t, t)$, one for each number c. These rise from $(c,0)$ on the horizontal x axis and separate as t increases vertically. ρ is constant on each such curve when

$$\frac{d}{dt}\rho(x(t),t) = \rho_x x' + \rho_t = \rho_t + x\rho_x = 0.$$

If $\rho(x,0) = x^2$ we get $\rho(ce^t, t) = \rho(c,0) = c^2 = ((ce^t)e^{-t})^2$. Therefore $\rho(x,t) = (xe^{-t})^2 = x^2 e^{-2t}$.

Problems of Chapter 9

Section 9.1

9.1 Since $u_{xx} > 0$ the wave equation implies that also $u_{tt} > 0$, so the acceleration is upward. This agrees exactly with our derivation of the equation based on tension forces, because these forces on the two ends of any bit of the string have an upward resultant when the shape is convex.

9.3 $u_{tt} = -A_4^2\left(c_1 \sin\left(\frac{2\pi x}{330}\right)\cos(A_4 t)\right) - E_6^2\left(c_2 \sin\left(\frac{2\pi(3x)}{330}\right)\cos(E_6 t)\right)$ and
$u_{xx} = -\left(\frac{2\pi}{330}\right)^2\left(c_1 \sin\left(\frac{2\pi x}{330}\right)\cos(A_4 t)\right) - \left(\frac{2\pi(3)}{330}\right)^2\left(c_2 \sin\left(\frac{2\pi(3x)}{330}\right)\cos(E_6 t)\right)$. Then
$u_{tt} = c^2 u_{xx}$ gives

$$c^2\left(\frac{2\pi}{330}\right)^2 = A_4^2, \qquad c^2\left(\frac{2\pi(3)}{330}\right)^2 = E_6^2.$$

Divide these to get $\dfrac{E_6^2}{A_4^2} = 3^2$, so $E_6 = 3A_4$.

9.5

(a) speed 3 to the left
(b) speed 3 to the left
(c) speed 1/4 to the right

9.7 By the Chain Rule, $u_t = -cf'(x-ct) + cg'(x+ct)$, $u_x = f'(x-ct) + g'(x+ct)$,

$$u_{tt} = c^2 f''(x-ct) + c^2 g''(x+ct), \qquad u_{xx} = f''(x-ct) + g''(x+ct).$$

Therefore $u_{tt} = c^2 u_{xx}$.

9.9 $u_t = -c\sin(x+ct)$, $u_x = -\sin(x+ct)$, so $u_t^2 + c^2 u_x^2 = 2c^2 \sin^2(x+ct)$.

(a) $E(0) = c^2 \displaystyle\int_0^\pi \sin^2 x \, dx$

(b) $E(1) = c^2 \displaystyle\int_c^{\pi-c} \sin^2(x+c) \, dx$. $E(0)$ is larger because

$$E(1) = c^2 \int_0^{\pi-2c} \sin^2 \theta \, d\theta < c^2 \int_0^\pi \sin^2 \theta \, d\theta = E(0).$$

9.11

(a) $u(x,0) = \sin x = f(x) + g(x)$ and $u_t(x,0) = 0 = -cf'(x) + cg'(x)$. Then $f' = g'$ so $f = g + C$, then $\sin x = 2g(x) + C$. So

$$g(x) = \tfrac{1}{2}(\sin x - C), \quad f(x) = \tfrac{1}{2}(\sin x + C), \quad u(x,t) = \tfrac{1}{2}\sin(x - ct) + \tfrac{1}{2}\sin(x + ct).$$

(b) Here $c = 1$. $u(x,0) = 0 = f(x) + g(x)$ and $u_t(x,0) = \cos(2x) = -f'(x) + g'(x)$. Then $f = -g$ so $\cos(2x) = 2g'(x)$. Thus $g(x) = \tfrac{1}{4}\sin(2x) + C$. So

$$g(x) = \tfrac{1}{4}\sin(2x) + C, \quad f(x) = -\tfrac{1}{4}\sin(2x) - C, \quad u(x,t) = -\tfrac{1}{4}\sin(2(x-t)) + \tfrac{1}{4}\sin(2(x+t)).$$

(c) Similarly

$$u(x,t) = \tfrac{3}{2}\left(\sin(x - 5t) + \sin(x + 5t)\right) + \tfrac{1}{2}\left(\sin(3(x - 5t)) + \sin(3(x + 5t))\right)$$

$$+ \tfrac{1}{25}\left(-\sin(2(x - 5t)) + \sin(2(x + 5t))\right).$$

9.13 $u(x,t) = -g(-x + ct) + g(x + ct)$ and its derivatives are nonzero only when $-x + ct$ or $x + ct$ is in the short interval where $g \neq 0$. As t tends to infinity, with x confined to $[0, p]$, neither of those numbers is in that interval. Therefore the integral is zero.

Section 9.2

9.15

(a) $u_{tt} + v_{tt} = c^2(u_{xx} + v_{xx} + u_{yy} + v_{yy})$; add the wave equations for u and v
(b) Let $u = kz$. Then $u_{tt} = kz_{tt} = kc^2 \Delta z = c^2 \Delta(kz) = c^2 \Delta u$.
(c) Set $w(x,y,t) = z(-y, x, t)$. Then

$$w_{tt}(x,y,t) = z_{tt}(-y,x,t) \qquad w_x(x,y,t) = z_y(-y.x.t) \qquad w_{xx}(x,y,t) = z_{yy}(-y.x.t)$$

$$w_y(x,y,t) = -z_x(-y.x.t) \qquad w_{yy}(x,y,t) = (-1)^2 z_{xx}(-y.x.t)$$

so $w_{tt} = z_{tt}(-y,x,t) = c^2(z_{xx} + z_{yy})(-y,x,t) = c^2(w_{yy} + w_{xx})$
(d) Set $w(x,y,t) = z(kx, ky, kt)$, then $w_{tt}(x,y,t) = k^2 z_{tt}(kx, ky, kt)$, $w_{xx} = k^2 z_{xx}$, $w_{yy} = k^2 z_{yy}$, so $w_{tt} - c^2(w_{xx} + w_{yy}) = k^2(z_{tt} - c^2(z_{xx} + z_{yy})) = 0$.

9.17 $\sin(1.414c(t + 1000\dfrac{2\pi}{c})) = \sin(1.414ct + 1414(2\pi)) = \sin(1.414ct)$, and similarly for the 2.236 term. The 1000 insures that we've added a whole multiple of 2π.

9.19

(a) $z_{tt} = -(n^2 + m^2)z$, $z_{xx} = -n^2z$, $z_{yy} = -m^2z$, so $z_{tt} = z_{xx} + z_{yy}$. Since n and m are positive integers, from $\sin(n\pi) = \sin(m\pi) = \sin(0) = 0$ you have the boundary value 0.

(b) There is one such solution for each integer pair (n,m) inside the first quadrant of the disk of radius 1000. Since each (n,m) sits at the corner of a 1×1 square, the number of them is roughly equal to the area of the quarter disk.

9.21 $z_x = f'r_x \sin(kt)$, and $r_x = \frac{1}{2}(x^2 + y^2)^{-1/2}(2x) = x/r$.

$z_{xx} = (f'' \frac{x^2}{r^2} + f'(r)\frac{r - x(x/r)}{r^2})\sin(kt)$. Similarly $z_{yy} = (f''\frac{y^2}{r^2} + f'(r)\frac{r - y(y/r)}{r^2})\sin(kt)$. Adding,

$$z_{xx} + z_{yy} = (f''(r) + \frac{1}{r}f'(r))\sin(kt).$$

Since $z_{tt} = -k^2 f(r)\sin(kt)$ we see that $f'' + \frac{1}{r}f' + k^2 f = 0$ implies $z_{tt} = z_{xx} + z_{yy}$.

9.23

(a) $u_\rho = -\rho^{-2}f(\rho \pm t) + \rho^{-1}f'(\rho + t)$ and
$u_{\rho\rho} = 2\rho^{-3}f(\rho \pm t) - 2\rho^{-2}f'(\rho + t) + \rho^{-1}f''(\rho + t)$. So

$$u_{\rho\rho} + \frac{2}{\rho}u_\rho = \rho^{-1}f''(\rho + t).$$

But $u_{tt} = \rho^{-1}f''$, so u is a solution of the wave equation.

(b) $u(\rho, 0) = \rho^{-1}f(\rho)$ has maximum of 1 near $\rho = 100$. Therefore the maximum of f is approximately 100 near $\rho = 100$. Then

$$u(\rho, 99) = \rho^{-1}f(\rho + 99)$$

has a maximum of approximately 100 near $\rho = 1$. That is, the maximum of the wave was 1 on a spherical shell of radius 100 and becomes after 99 seconds a maximum of 100 on a shell of radius 1.

Section 9.3

9.25

(a) $(e^{-n^2 t}\sin(nx))_t - (e^{-n^2 t}\sin(nx))_{xx} = (-n^2 + n^2)e^{-n^2 t}\sin(nx) = 0$

(b) $(t^p e^{-x^2/(4t)})_t = pt^{p-1}e^{-x^2/(4t)} + t^p \frac{x^2}{4t^2}e^{-x^2/(4t)}$ and $(t^p e^{-x^2/(4t)})_x = -\frac{x}{2t}t^p e^{-x^2/(4t)}$,
$(t^p e^{-x^2/(4t)})_{xx} = -\frac{1}{2t}t^p e^{-x^2/(4t)} + (\frac{x}{2t})^2 t^p e^{-x^2/(4t)}$. Therefore $p = -\frac{1}{2}$.

(c) $(e^{-ax}\cos(ax - bt))_t = be^{-ax}\sin(ax - bt)$,
$(e^{-ax}\cos(ax - bt))_x = -ae^{-ax}\cos(ax - bt) - ae^{-ax}\sin(ax - bt)$,

$(e^{-ax}\cos(ax - bt))_{xx} = a^2 e^{-ax}\cos(ax - bt) + 2a^2 e^{-ax}\sin(ax - bt) - a^2 e^{-ax}\cos(ax - bt)$.

Therefore $b = 2a^2$

(d) With $T(x,t) = u(kx, k^2 t)$ you get $T_t = k^2 u_t(kx, k^2 t)$, $T_{xx} = k^2 u_{xx}(kx, k^2 t)$,
so $T_t - T_{xx} = k^2(u_t - u_{xx}) = 0$.

9.27

(a) $1, -1, \frac{1}{2}, -4$: $T(x,t) = e^{-t}\sin x + \frac{1}{2}e^{-4t}\sin(2x)$

(b) Your sketch ought to indicate the progression described in part (c).

(c) The second term decreases much faster because of the e^{-4t} factor. So the first term is more important as t increases, and the hot spot moves toward the center, to the right.

9.29 $y_1(t) = \frac{1}{\pi}\int_0^{\pi} e^{-t}\sin(x)\,dx = \frac{2}{\pi}e^{-t}$ satisfies $y_1' = -\frac{2}{\pi}e^{-t} = -(y_1 - 0)$, so Newton's Law of Cooling holds for y_1.

$y_2(t) = \frac{1}{\pi}\int_0^{\pi} (e^{-t}\sin(x) + e^{-9t}\sin(3x))\,dx = \frac{2}{\pi}e^{-t} + \frac{2}{3\pi}e^{-9t}$. We observe that Newton's Law of Cooling

$$y_2'(t) = -\frac{2}{\pi}e^{-t} - \frac{6}{\pi}e^{-9t} = ? -k(\frac{2}{\pi}e^{-t} + \frac{2}{3\pi}e^{-9t})$$

does not hold for y_2 for any choice of k.

9.31

(a) Take the t derivative under the integral sign.

(b) Use the Divergence Theorem in $\mathbb{R}^2$.

(c) Properties and definition of Laplacian

(d) Since D is arbitrary, the integrand must be zero.

9.33 $(x^2 + y^2 + 4ht)_{xx} = 2 = (x^2 + y^2 + 4ht)_{yy}$, and $(x^2 + y^2 + 4ht)_t = 4h$. So T is a solution. The heat flux is $-r\nabla T = -r(2x, 2y)$ so is toward the origin, that happens to be the coldest place.

Section 9.4

9.35

(a) $\Delta(u_1 + u_2) = u_{1xx} + u_{2xx} + u_{1yy} + u_{2yy} = \Delta u_1 + \Delta u_2 = 0 + 0 = 0$.

(b) Write $v(x,y) = u(x\cos\theta - y\sin\theta, x\sin\theta + y\cos\theta)$. Then

$$v_x = u_x\cos\theta + u_y\sin\theta, \quad v_y = -u_x\sin\theta + u_y\cos\theta,$$

$$v_{xx} = (u_{xx}\cos\theta + u_{xy}\sin\theta)\cos\theta + (u_{yx}\cos\theta + u_{yy}\sin\theta)\sin\theta,$$

$$v_{yy} = -(-u_{xx}\sin\theta + u_{xy}\cos\theta)\sin\theta + (-u_{yx}\sin\theta + u_{yy}\cos\theta)\cos\theta$$

So $v_{xx} + v_{yy} = u_{xx}(\cos^2\theta + \sin^2\theta) + u_{xy}(0) + u_{yy}(\cos^2\theta + \sin^2\theta) = u_{xx} + u_{yy} = 0$.

(c) $\Delta(uv) = (\Delta u)v + 2\nabla u \cdot \nabla v + u(\Delta v) = 0 + 0 + 0 = 0$.

9.37 $z(x,y) = \log(x^2 + y^2)$, so

$$z_x = \frac{2x}{x^2 + y^2}, \quad z_y = \frac{2y}{x^2 + y^2}, \quad z_{xx} = \frac{2}{x^2 + y^2} - \frac{2x(2x)}{(x^2 + y^2)^2}, \quad z_{yy} = \frac{2}{x^2 + y^2} - \frac{2y(2y)}{(x^2 + y^2)^2},$$

so

$$\Delta z = \frac{4}{x^2 + y^2} - \frac{2(2x^2 + 2y^2)}{(x^2 + y^2)^2} = 0.$$

9.39

(a) $\operatorname{div}(w\nabla w) = w\Delta w + \nabla w \cdot \nabla w$, so by the Divergence Theorem

$$0 = \int_{\partial D} w\nabla w \cdot \mathbf{N}\, d\sigma = \int_D (w\Delta w + \nabla w \cdot \nabla w)\, dV$$

(b) Use part (a) with $w = u$. That gives $0 = -\int_D \|\nabla u\|^2\, dV$. So $\|\nabla u\| = 0$. That makes u constant in D, but since u is zero on the boundary it must be zero everywhere.

(c) $u - v$ solves the Laplace equation and since it is zero on the boundary it is zero everywhere by part (b).

9.41 (a) $\frac{1}{4}(2+2) = 1$ (b) $\frac{1}{6}(2) + \frac{1}{3}(2) = 1$ (c) $k(z_{xx}+z_{yy}) + (1-k)(w_{xx}+w_{yy}) = k+(1-k) = 1$ (d) $z_{xx}+w_{xx}+z_{yy}+w_{yy} = z_{xx}+z_{yy}+w_{xx}+w_{yy} = 1+0 = 1$

9.43

(a) $z_{xx} = -n^2 z$, $z_{yy} = n^2 z$, so $\Delta z = 0$.

(b) $z(0,y) = \sin(0)\sinh(ny) = 0$, $z(\pi,y) = \sin(\pi)\sinh(ny) = 0$, $z(x,0) = \sin(nx)\sinh(0) = 0$.

(c) $z(\frac{\pi}{2n},y) = \sinh(ny) = \frac{1}{2}(e^{ny}-e^{-ny})$ is nearly $\frac{1}{2}e^{ny}$ that is very large when y is large positive.

9.45

(a) $\mathbf{F} = \nabla u = \nabla\left(x+\frac{x}{x^2+y^2}\right) = \left(1+\frac{y^2-x^2}{(x^2+y^2)^2}, \frac{-2xy}{(x^2+y^2)^2}\right) = (1,0)+r^{-4}(y^2-x^2,-2xy)$
where $r^2 = x^2+y^2$. So

$$\Delta u = \operatorname{div}\mathbf{F} = 0+r^{-4}\operatorname{div}(y^2-x^2,-2xy)-4r^{-5}\nabla(r)\cdot(y^2-x^2,-2xy)$$

$$= r^{-4}(-2x-2x)-4r^{-5}\frac{(x,y)}{r}\cdot(y^2-x^2,-2xy) = r^{-4}(-4x)-4r^{-6}(-xy^2-x^3)$$

$$= 4xr^{-4}(-1+r^{-2}(y^2+x^2)) = 0$$

(b) $(x,y)\cdot\mathbf{F}(x,y) = (x,y)\cdot((1,0)+r^{-4}(y^2-x^2,-2xy))$
$= x+r^{-4}(-xy^2-x^3) = x+r^{-4}(-xr^2)$. This is zero when $r = 1$.

(c) Since $|x| \le r$ and $|y| \le r$, $\|(y^2-x^2,-2xy)\|$ is less than a multiple of r^2. Therefore $\|\mathbf{F}-(1,0)\| = \|r^{-4}(y^2-x^2,-2xy)\|$ tends to zero as r tends to infinity.

Section 9.5

9.47

(a) $\psi_t = -iEe^{-iEt}\phi$ and $\Delta\psi = e^{-iEt}\Delta\phi$. Therefore

$$i\psi_t = Ee^{-iEt}\phi = e^{-iEt}(-\Delta\phi+V\phi) = -\Delta\psi+Ve^{-iEt}\phi = -\Delta\psi+V\psi.$$

(b) $|\psi(\mathbf{X},t)|^2 = \overline{\psi}\psi = e^{iEt}\overline{\phi(\mathbf{X})}e^{-iEt}\phi(\mathbf{X}) = |\phi(\mathbf{X})|^2$ does not depend on t. Therefore the probability $P(S,t)$ that is an integral of that, does not depend on t.

9.49

(a) $\phi_x = X'YZ$, $-\phi_{xx} = -X''YZ = (E_x - x^2)XYZ$ etc., so

$$-\Delta\phi + (x^2 + y^2 + z^2)\phi = -(X''YZ + XY''Z + XYZ'') + (x^2 + y^2 + z^2)XYZ = (E_x + E_y + E_z)\phi.$$

Therefore $E = E_x + E_y + E_z$.

(b) $W(w) = (a + bw + cw^2)e^{-w^2/2}$ gives

$$W' = (b + 2cw - w(a + bw + cw^2))e^{-w^2/2} = (b + (2c - a)w - bw^2 - cw^3)e^{-w^2/2}$$

so

$$-W'' + w^2 W$$

$$= \left(-\left((2c - a) - 2bw - 3cw^2 - w(b + (2c - a)w - bw^2 - cw^3)\right) + w^2(a + bw + cw^2)\right)e^{-w^2/2}$$

$$= (-2c + a + 3bw + 5cw^2)e^{-w^2/2}$$

For this to be $E_w W$ requires

$$-2c + a + 3bw + 5cw^2 = E_w(a + bw + cw^2).$$

Looking at the w^2 terms we find that $E_w = 5$ or $c = 0$. In case $E_w = 5$ we get $b = 0$, a arbitrary, and $c = -2a$.
In case $c = 0$ we need

$$a = E_w a, \qquad 3b = E_w b.$$

So $E_w = 3$ or $b = 0$. In case $E_w = 3$ we find $a = 0$ and b arbitrary.
In case $b = 0$ we find $E_w = 1$ and a arbitrary.
We have found solutions

$$W = ae^{-w^2/2}, \; E_w = 1; \qquad W = bwe^{-w^2/2}, \; E_w = 3; \qquad W = a(1 - 2w^2)e^{-w^2/2}, \; E_w = 5.$$

(c) Taking for $X(x)$, $Y(y)$, $Z(z)$ the various functions found in (b) as listed in the table by their E values, we get solutions ϕ with $E = E_x + E_y + E_z$.

X	1	1	1	1	1	1	3	5	3	3
Y	1	3	3	5	5	5	5	5	3	5
Z	1	1	3	1	3	5	5	5	3	3
E	3	5	7	7	9	11	13	15	9	13

(d) The $E = 3$ case is (some multiple of) $\phi(x, y, z) = e^{-x^2/2}e^{-y^2/2}e^{-z^2/2}$ so $|\phi|^2 = e^{-(x^2 + y^2 + z^2)}$. We know that to normalize this requires a factor of $\pi^{-3/2}$.

Index